THE ENERGY OF PHYSICS

PART II

ELECTRICITY AND MAGNETISM

BY Christopher J. Fischer
UNIVERSITY OF KANSAS

Bassim Hamadeh, CEO and Publisher
Kassie Graves, Director of Acquisitions
Jamie Giganti, Senior Managing Editor
Jess Estrella, Senior Graphic Designer
Mark Combes, Senior Field Acquisitions Editor
Sean Adams, Project Editor
Luiz Ferreira, Senior Licensing Specialist
Allie Kiekhofer, Associate Editor

Printed in the United States of America

ISBN: 978-1-63189-856-3 (pbk) / 978-1-63189-857-0 (br)

To my mother, who taught me to tie my shoes,
And to my father, who taught me to use an RPN calculator.
Thank you for these and all other life lessons you taught.

CONTENTS

FOREWORD

This book is intended for the second semester of a two semester calculus-based physics curriculum. As such, I assume that students using this book are comfortable with simple differentiation and integration presented in the first textbook in this series and are now ready for more advanced applications of integration; these will be reviewed in Chapter 1.

This textbook is divided into three sections. In the first section (Chapter 2 through Chapter 4) the concepts of potential and field are used to understand the gravitational, electric, and magnetic interactions. The second section (Chapter 5 through Chapter 7) contains a discussion of how the electric and magnetic interactions influence each other, and the electric and magnetic properties of material. The focus of the final section (Chapter 8 through Chapter 12) is on applications of the electric and magnetic interactions to electric circuits and optics.

This textbook has four appendices. Appendix A contains information about SI units and physical constants, and a review of basic and applied mathematics, including trigonometry. Vectors, coordinate systems, and elementary vector calculus are reviewed in Appendix B. The derivations of several equations and alternative solutions to example problems are collected together in Appendix C. Finally, a review of complex exponentials and their application to electric circuits is given in Appendix D. Please consult these appendices continually as you progress through the textbook.

I would like to thank all of the people who contributed to this book and/or supported me as I wrote it. Thanks to Professor Antonik for the suggestion of using the Bohr model of the hydrogen atom as a continuing example problem throughout the book. I would also like to thank my department chair, Professor Hume Feldman, for fostering a creative and supportive environment in the department.

Finally, I would like to thank my family for their support, especially my wife and daughters. I love all of you!

CHAPTER 1

Introduction

1-1 Introduction

In this textbook, we will learn the laws of electricity and magnetism, and the applications of these laws to electrical circuits and optics. It is worth emphasizing, however, that studying these laws has a much wider benefit than simply understanding how electronic devices function or how a camera forms an image. Indeed, the laws of electricity and magnetism also explain fundamental interactions between the molecules and atoms all around us. For example, it is differences in the magnitude and range of these interactions that distinguish solids from liquids. Furthermore, as with our study of classical mechanics, a major focus of this textbook is the continued practice and development of our applied mathematics skills. This will be useful not only in this and other physics classes but also for general critical thinking and problem solving strategies.

1-2 Energy and Forces

When learning the new material in this textbook, we will frequently refer back to and build upon our knowledge of classical mechanics, especially the concepts of energy and force. In classical mechanics, we classified a force as either conservative or non-conservative based upon whether it was associated with a potential energy. We further defined conservative forces as the gradient of their associated potential energies using Equation 1-1.

$$\vec{F} = -\nabla U \tag{1-1}$$

In Equation 1-1, the symbol "∇" denotes the gradient operator (see Section B-8). According to Equation 1-1 conservative forces are always directed/oriented so as to minimize the potential energy of the object upon which they act. This is consistent with the principle of minimum potential energy.

> *Systems will always arrange themselves in order to minimize their total potential energy.*

In this course, however, we will take this discussion of energy and force one step further to understand how long-range forces (such as the gravitational force, the electric force, and the magnetic force) work. Specifically, how does one object "sense" the presence of another object when the two objects are not in contact with each other? Over the next three chapters, we will introduce the concepts of *potential* and *field* to answer this question. While these concepts are to some extent simple abstractions of energy and force, they also represent something more profound and fundamental. Furthermore, although our motivation for introducing potential and field is to help us understand long-range interactions, the contact forces we studied in classical mechanics (such as the normal force or friction forces) are ultimately explained in terms of electric and magnetic interactions and thus are also best understood in terms of potential and field.

Finally, it is important to note that we are not replacing the conservation laws that formed the basis of most of our analysis in classical mechanics; we will take full advantage of energy and momentum conservation in discussions and problem solving in this textbook. As we shall see, however, the fact that the electric and magnetic interactions influence each other requires us to be careful when describing our systems and determining whether or not they are isolated.

1-3 Applied Integration

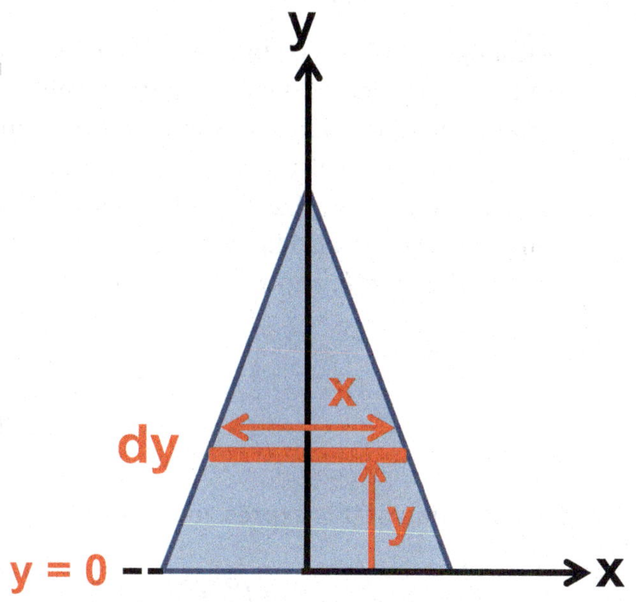

FIGURE 1.1: Determining the area of a triangle.

The divide between understanding the principles of physics *qualitatively* and solving an associated problem *quantitatively* is often bridged by calculus. In classical mechanics, the acceleration of a system is determined by differentiating the equation for the energy of that system. Similarly, the macroscopic thermodynamic parameters describing a system (temperature, pressure, *etc.*) are determined through differentiation of the equation for the entropy of the system. In this textbook, we turn our attention to applications of *integration*. It is therefore worth reviewing a few examples of applied integration.

Let's begin by calculating the area of a triangle; the height and maximum width (*i.e.*, base) of the triangle will be denoted by the variables h and b, respectively. As shown in Figure 1.1, we can determine the area of the triangle by modeling the triangle as a collection of infinitesimally small rectangles.

In Figure 1.1, we have defined a reference frame for this system with an x-axis parallel to the base of the triangle and a y-axis perpendicular to the x-axis; the origin of the y-axis is at the base of the triangle. As shown in Figure 1.1, we will use the variable x to denote the width of a rectangle, the variable y to denote its position above the base, and the variable dy to denote its height. The infinitesimal area of each rectangle, denoted by dA, is thus

$$dA = xdy$$

The area of the triangle is the summation of the infinitesimal areas of each of the rectangles constituting the triangle. In the limit that these rectangles become truly infinitesimally small (*i.e.*, in the limit that dy approaches zero), this summation becomes an integral.

$$A = \int dA \quad \rightarrow \quad A = \int x\,dy$$

Although all of the rectangles in our integration have the same height (dy), they have different widths (x) depending upon their distance from the bottom of the triangle. In other words, the variable x in our integration is a function of the variable y. Thus, the next step in our calculation is to relate the variables x and y. We know from the shape of the triangle (*i.e.*, that the sides of the triangle are straight lines) that x will be a linear function of y. We also know from the definition of the reference frame in Figure 1.1 that $x = 0$ when $y = h$, and $x = b$ when $y = 0$. Therefore,

$$x = \left(-\frac{b}{h}\right)y + b$$

Substitution of this expression gives us

$$A = \int_0^h \left(\left(-\frac{b}{h}\right)y + b\right)dy \quad \rightarrow \quad A = \left(\left(-\frac{b}{h}\right)\frac{y^2}{2} + by\right)\Bigg|_0^h$$

$$A = \left(-\frac{b}{h}\right)\frac{h^2}{2} + bh \quad \rightarrow \quad A = -\frac{bh}{2} + bh$$

$$A = \frac{bh}{2}$$

The area of the triangle is one-half the product of the base and height of triangle, as expected.

Now let's calculate the area of a disk with a radius R. As shown in Figure 1.2, we can model the disk as a collection of infinitesimally small thin rings of radius r and width dr.

The area of the disk is the sum of the infinitesimal areas of each of these rings. Each of these infinitesimal areas is the product of the circumference and width of the associated ring.

$$dA = 2\pi r\,dr$$

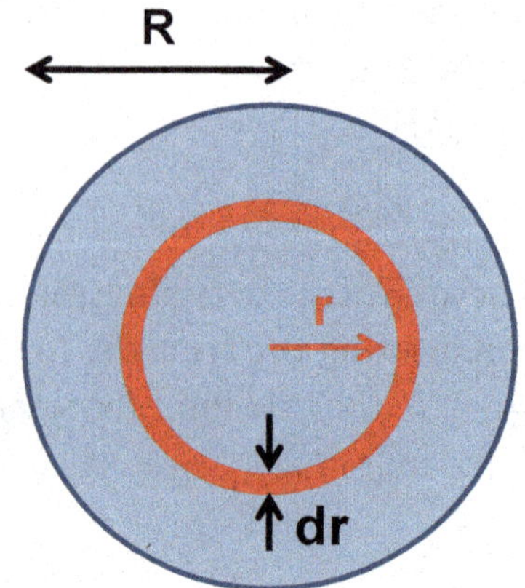

FIGURE 1.2: Determining the area of a disk.

In the limit that these infinitesimal areas become truly infinitesimally small (*i.e.*, in the limit that dr approaches zero), their summation becomes an integral.

$$A = \int dA \quad \rightarrow \quad A = \int 2\pi r\, dr$$

Integration over the entire radius of the disk then yields.

$$A = 2\pi \int_0^R r\, dr \quad \rightarrow \quad A = 2\pi \left(\frac{r^2}{2} \right) \Bigg|_0^R \quad \rightarrow \quad A = 2\pi \left(\frac{R^2}{2} - 0 \right)$$

$$A = \pi R^2$$

The area of the disk is the product of π and the square of the radius of the disk, as expected.

Example 1-1:

Problem: Where is the center of mass of a uniformly dense rigid rod of length L and mass M?

Solution: In this calculation, we can model the rigid rod as consisting of many infinitesimal "bits" of mass; each of these bits of mass is so small that it can be treated as a point particle. We recall from classical mechanics that the center of mass for a system of point particles can be determined using the following equation:

$$\vec{r}_{CM} = \frac{\sum_{i=1}^{n} m_i \vec{r}_i}{\sum_{i=1}^{n} m_i}$$

In this equation, $\vec{r}_i$ is the position of the ith point particle, m_i is the mass of the ith point particle, and $\vec{r}_{CM}$ is the position of the center of mass for the system. In the limit that these bits of mass are truly infinitesimally small, this summation becomes an integral.

$$\vec{r}_{CM} = \frac{\int \vec{r}\, dm}{\int dm}$$

The denominator in this equation is simply the total mass of the rod.

$$\int dm = M \quad \rightarrow \quad \vec{r}_{CM} = \frac{\int \vec{r}\, dm}{M}$$

In Figure 1.3, we have defined a reference frame for this system with a 1-dimenstional x-axis parallel to the rod and an origin at one end of the rod; the variable x denotes the distance of an infinitesimal bit of mass from the origin. The equation for the center of mass of the rod in this reference frame is

$$x_{CM} = \frac{\int x\, dm}{M}$$

We can then express the infinitesimal bit of mass in this equation in terms of the linear mass density of the rod, denoted by λ, and an infinitesimal distance along the rod (or infinitesimal length of the rod), denoted by dx.

$$dm = \lambda dx$$

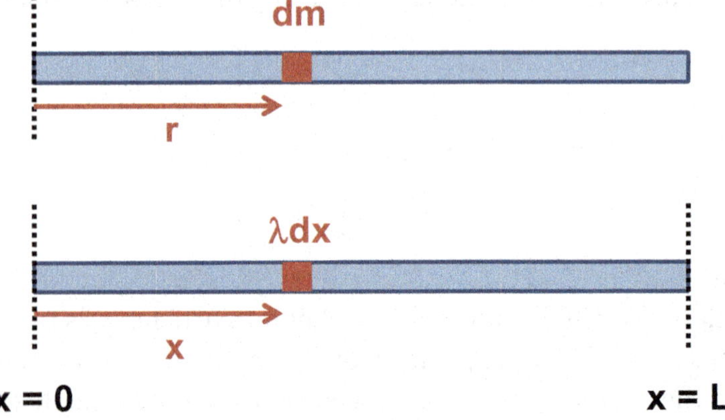

FIGURE 1.3: Determining the center of mass of a rigid rod.

The units of λ are kg/m. With these definitions, the integral with respect to mass can be written as an integral with respect to the length of the rod.

$$\int x\,dm = \int_0^L x\lambda\,dx \quad \rightarrow \quad x_{CM} = \frac{\int_0^L x\lambda\,dx}{M}$$

Since the rod is uniformly dense, λ is a constant.

$$x_{CM} = \left(\frac{\lambda}{M}\right)\int_0^L x\,dx \quad \rightarrow \quad x_{CM} = \left(\frac{\lambda}{M}\right)\left(\frac{x^2}{2}\right)\Bigg|_0^L \quad \rightarrow \quad x_{CM} = \frac{\lambda L^2}{2M}$$

The final step in our solution is to express our answer in terms of the mass of the rod. This requires us to determine the relationship between λ and M, which we obtain by integration.

$$dm = \lambda dx \quad \rightarrow \quad \int dm = \int \lambda\,dx \quad \rightarrow \quad \int_0^M dm = \lambda\int_0^L dx \quad \rightarrow \quad M = \lambda L \quad \rightarrow \quad \lambda = \frac{M}{L}$$

Substitution then gives us

$$x_{CM} = \left(\frac{M}{L}\right)\frac{L^2}{2M} \quad \rightarrow \quad x_{CM} = \frac{L}{2}$$

As expected, the center of mass is at the geometric center of the rod.

Finally, let's determine the moment of inertia of a uniformly dense rigid rod rotated around one end; the variables M and L will denote the mass and length of the rod, respectively. In this calculation, we can model the rigid rod as consisting of many infinitesimal "bits" of mass; each of these bits of mass is so small that it can be treated as a point particle. The infinitesimal moment of inertia from each of these infinitesimal bits of mass is therefore

$$dI = r^2 dm$$

In this equation, the variable r denotes the distance between the bit of mass and the axis of rotation. The moment of inertia for the rod is the sum of each of these infinitesimal moments of inertia. In the limit that these bits of mass are truly infinitesimally small, this summation becomes an integral.

$$I = \int dI \quad \rightarrow \quad I = \int r^2 \, dm$$

As discussed in Example 1-1, we can express the infinitesimal amount of mass in this equation in terms of the linear mass density of the rod, denoted by λ, and an infinitesimal distance along the rod (or infinitesimal length of the rod), denoted by dx.

$$dm = \lambda dx$$

Let's again use the reference frame shown in Figure 1.3, which has an x-axis parallel to the rod and an origin at one end of the rod; the variable x denotes the distance of an infinitesimal bit of mass from the origin. With these definitions, the integral with respect to mass can be written as an integral with respect to the length of the rod.

$$I = \int_0^L x^2 \lambda \, dx$$

Since the rod is uniformly dense, λ is a constant.

$$I = \lambda \int_0^L x^2 \, dx \quad \rightarrow \quad I = \lambda \left(\frac{x^3}{3} \right) \Bigg|_0^L \quad \rightarrow \quad I = \frac{\lambda L^3}{3}$$

The final step in our solution is to express our answer in terms of the mass of the rod. As shown in Example 1-1, the relationship between λ and M is found through integration.

$$dm = \lambda dx \quad \rightarrow \quad \int dm = \int \lambda \, dx \quad \rightarrow \quad \int_0^M dm = \lambda \int_0^L dx \quad \rightarrow \quad M = \lambda L \quad \rightarrow \quad \lambda = \frac{M}{L}$$

Substitution then gives us

$$I = \frac{L^3}{3}\left(\frac{M}{L}\right) \quad \rightarrow \quad I = \frac{1}{3}ML^2$$

As expected, this is identical to the expression we learned in classical mechanics.

Example 1-2

Problem: The area mass density[1] of a thin, solid disk decreases linearly with distance from the normal axis through the center of the disk according to the following equation:

$$\sigma = \sigma_0 - \gamma r$$

The variable r in this equation is the radial distance from the normal axis through the center of the disk, the variable σ_0 is the area mass density when $r = 0$, and the variable γ is the rate of change of the area mass density with respect to radial distance from the normal axis through the center of the disk. What is the moment of inertia for rotations of this disk around the normal axis through the center of the disk? The total mass and radius of the disk are M and R, respectively.

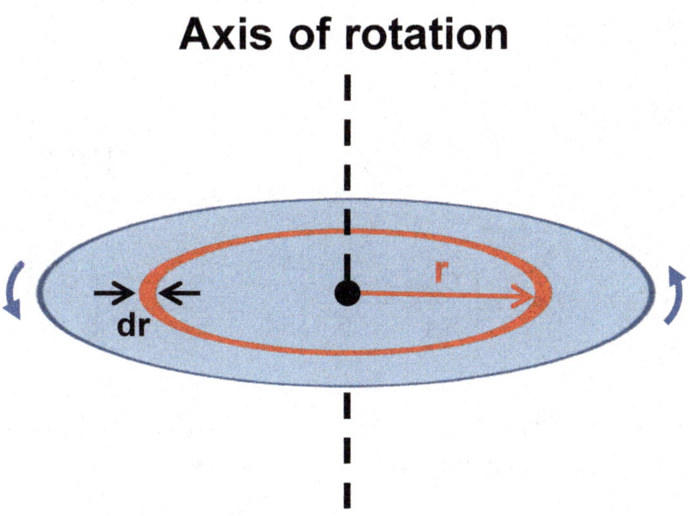

FIGURE 1.4: Determining the moment of inertia of a spinning disk.

Solution: We can model the disk as a collection of infinitesimally small, thin rings of radius r and width dr as shown in Figure 1.4.

1 The units of area mass density are kg/m^2.

The infinitesimal area of each of those rings is

$$dA = 2\pi r\, dr$$

The infinitesimal mass of each of those rings is the product of their area and the area mass density.

$$dm = \sigma dA \quad \rightarrow \quad dm = (\sigma_0 - \gamma r)(2\pi r\, dr) \quad \rightarrow \quad dm = 2\pi\left(\sigma_0 r - \gamma r^2\right) dr$$

$$dm = 2\pi\left(\sigma_0 r - \gamma r^2\right) dr$$

The moment of inertia for the disk is the sum of the infinitesimal moments of inertia from each of these infinitesimal bits of mass. In this calculation the variable r also denotes the distance from each bit of mass to the axis of rotation.

$$I = \int dI \quad \rightarrow \quad I = \int r^2\, dm \quad \rightarrow \quad I = \int r^2 2\pi\left(\sigma_0 r - \gamma r^2\right) dr$$

$$I = 2\pi \int_0^R \left(\sigma_0 r^3 - \gamma r^4\right) dr$$

$$I = 2\pi\left(\frac{\sigma_0 r^4}{4} - \frac{\gamma r^5}{5}\right)\Bigg|_0^R \quad \rightarrow \quad I = 2\pi\left(\frac{\sigma_0 R^4}{4} - \frac{\gamma R^5}{5}\right) \quad \rightarrow \quad I = \frac{\pi R^4}{10}\left(5\sigma_0 - 4\gamma R\right)$$

The total mass of the disk is determined by integrating the infinitesimal bits of mass that constitute the disk.

$$M = \int dm \quad \rightarrow \quad M = \int_0^R 2\pi\left(\sigma_0 r - \gamma r^2\right) dr \quad \rightarrow \quad M = 2\pi\left(\frac{\sigma_0 r^2}{2} - \frac{\gamma r^3}{3}\right)\Bigg|_0^R$$

$$M = 2\pi\left(\frac{\sigma_0 R^2}{2} - \frac{\gamma R^3}{3}\right) \quad \rightarrow \quad M = \frac{\pi R^2}{3}\left(3\sigma_0 - 2\gamma R\right) \quad \rightarrow \quad \sigma_0 = \frac{M}{\pi R^2} + \frac{2\gamma R}{3}$$

Substitution then gives us

$$I = \frac{\pi R^4}{10}\left(5\left(\frac{M}{\pi R^2} + \frac{2\gamma R}{3}\right) - 4\gamma R\right) \quad \rightarrow \quad I = \frac{\pi R^4}{10}\left(\frac{5M}{\pi R^2} + \frac{10\gamma R - 12\gamma R}{3}\right)$$

$$I = \frac{1}{2}MR^2 - \gamma\left(\frac{\pi R^5}{15}\right)$$

Note that when the area mass density of the disk is constant (*i.e.*, when $\gamma = 0$), we have

$$\gamma = 0 \quad \rightarrow \quad I = \frac{1}{2}MR^2$$

As expected.

1-4 Units

In this textbook, we will use the International System of Units (also known as the metric system) just as we did in classical mechanics. Recall that this system of units is typically referred to by the acronym "SI" derived from its French name, Système International d'unités. A list of the seven basic SI units and several additional derived units is shown in Section A-1. As shown in Section A-1, our list of derived units has expanded considerably from what we used in classical mechanics. Also, please note that many of these derived units can be expressed in many ways. While these many varied expressions and relationships are necessary for the correct unit conversion in the problems we that will encounter, we should not be overwhelmed by them. Indeed, the beauty of the metric system is that our final answer will always have the correct units as long as we always work in only SI units during our calculations.

1-5 Looking Ahead

The concepts of potential and field complement the energy-based approach to mechanics learned in classical mechanics, and together with all the conservation laws of classical mechanics provide the fundamental way to think about physics. Not only do they have immediate application to circuits, optics, and advanced topics in physics (such as quantum mechanics and relativity), but they are also an important part of the mathematical abstraction that is inherent in more advanced science and engineering courses.

Summary

- **Isolated system:** a system that cannot exchange energy nor interact via forces with the outside environment. The total energy of an isolated system is therefore constant.

- **Energy conservation:** energy can neither be created nor destroyed, but can be converted from one type to another or transferred between objects. This is often referred to as the 1st law of thermodynamics.

- **The Principle of Minimum Potential Energy:** all systems will arrange themselves in order to minimize their total potential energy. This follows from the 2nd law of thermodynamics.

- **Conservative forces are associated with potential energy functions.**

$$\vec{F} = -\nabla U$$

The work done by a conservative force acting on an object is equal to the negative of the change in the associated potential energy of the object.

$$W_{conservative} = -\Delta U$$

Thus, the work done by a conservative force acting on an object is independent of the path of the object's displacement.

Problems

1. The linear mass density of a rigid rod increases linearly with respect to the distance from the one end of the rod (denoted as $x = 0$) according to the following equation:

$$\lambda = \lambda_0 + \gamma x$$

In this equation, λ_0 is the linear mass density at one end of the rod and γ is the rate of change of the linear mass density of the rod with respect to position from this end of the rod. The total length and mass of the rod are L and M, respectively. Where is the center of mass of the rod?

2. The base and height of a uniformly dense triangle are denoted by b and h, respectively. How far above the base is the center of mass of this triangle?

3. What is the moment of inertia of a uniformly dense, thin rod of length L and total mass M rotated around its center of mass?

4. A uniformly dense, thin rigid rod with a length L and total mass M forms a cantilever, as shown in the figure. Show that the net gravitational torque acting on the rod can be modeled as through the gravitational force acting on the rod acted on the only center of mass of the rod. The equation for torque is

$$\vec{\tau} = \vec{r} \times \vec{F}$$

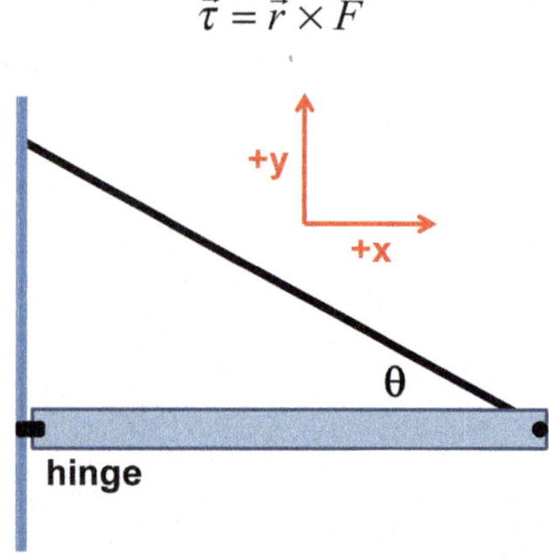

Assume that the force of gravity acts perpendicular to the length of the rod at all locations along the rod.

5. The linear mass density of a rigid rod decreases linearly with respect to the distance from the center (denoted as $x = 0$) of the rod according to the following equation:

$$\lambda = \begin{cases} \lambda_0 - \gamma x & x \geq 0 \\ \lambda_0 + \gamma x & x < 0 \end{cases}$$

In this equation, λ_0 is the linear mass density at the center of the rod and γ is the rate of change of the linear mass density of the rod with respect to position from the center of the rod. What is the moment of inertia for rotation of this rod around its center of mass? The total mass and length of the rod are M and L, respectively.

6. The linear mass density of a rigid rod decreases linearly with respect to the distance from the center (denoted as $x = 0$) of the rod according to the following equation:

$$\lambda = \begin{cases} \lambda_0 - \gamma x & x \geq 0 \\ \lambda_0 + \gamma x & x < 0 \end{cases}$$

In this equation, λ_0 is the linear mass density at the center of the rod and γ is the rate of change of the linear mass density of the rod with respect to position from the center of the rod. What is the moment of inertia for rotation of this rod around one end of the rod? The total mass and length of the rod are M and L, respectively.

CHAPTER 2

Gravitational Interaction

2-1 Introduction

A block released from rest above the ground will fall down to ground. We can explain why this occurs using our knowledge of energy and forces. For example, we could argue that the block will fall down to the ground since by doing so the gravitational potential energy of the block will be minimized[1]. Similarly, we could argue that the block falls down to the Earth because the force of gravity exerted by the Earth on the block "pulls" the block down. While these explanations are valid, they nevertheless leave unresolved an important question: how do the block and the Earth *sense* the presence of one another when they are not in contact with each other? In other words, how is the long-distance interaction between the

1 Recall that it is the gravitational potential energy of the system consisting of the Earth and the block together that is minimized. Indeed, the Earth is moving "up" toward the block as the block falls down toward the Earth.

Earth and the block, expressed either in terms of the gravitational potential energy or the gravitational force between these objects, propagated or communicated between these objects? In physics, the ability of an object to exert a long-range influence on another object is described using the concepts of *potential* and *field*.

2-2 Gravitational Potential

The gravitational interaction between objects gives rise to a potential energy that depends upon the masses of the objects and the distances between them. The equation for the gravitational potential energy of two *point*[2] particles, A and B, with masses m_A and m_B, separated by a distance r, is given in Equation 2-1.

$$U_g = -G\frac{m_A m_B}{r} \tag{2-1}$$

In Equation 2-1, the variable G is the gravitational constant of the universe.

$$G = 6.67\times10^{-11}\,\frac{\mathrm{J\,m}}{\mathrm{kg}^2}$$

In order to explain the origin of this gravitational potential energy (*i.e.*, in order to explain how the two objects *sense* each other's presence), we say that every bit of mass alters the space around it. This alteration (or distortion) of space can then be detected by other bits of mass which, in turn, gives rise to the gravitational interaction between two objects (Equation 2-1). We refer to this alteration as the ***gravitational potential*** surrounding the object.

> *Gravitational potential:* an intrinsic property of mass that describes the alteration of space associated with mass. The detection of the gravitational potential surrounding one object by another object is the origin of the gravitational interaction (potential energy and force) between the two objects. Gravitational potential is a scalar quantity.

---

2 Recall that in the point particle approximation we treat objects as though they were points without any physical size. In this approximation, the location of the point corresponding to an object is at the center of mass of the object.

The gravitational potential surrounding a point particle, denoted by V_g, is defined in Equation 2-2[3].

$$V_g = -G\frac{m}{r} \tag{2-2}$$

In Equation 2-2, the variable *m* is the mass of the point particle and *r* is the distance from the point particle to the position where the gravitational potential is measured. The units of gravitational potential are energy divided by mass (J/kg). Therefore, we can think of the gravitational potential as the gravitational potential energy per unit mass, as we will discuss further in Section 2-4.

Example 2-1

Problem: What is the gravitational potential of a 2 kg point particle at a distance of 5 m from the particle?

Solution: According to Equation 2-2 the gravitational potential is

$$V_g = -\left(6.67\times10^{-11}\,\frac{\text{J m}}{\text{kg}^2}\right)\left(\frac{2\,\text{kg}}{5\,\text{m}}\right)$$

$$V_g = -2.67\times10^{-11}\,\frac{\text{J}}{\text{kg}}$$

We see from Equation 2-2 that the gravitational potential surrounding a point particle is spherically symmetric around the point particle. In other words, all locations the same distance from the point particle, in all directions, have the exact same gravitational potential. We refer to a collection of locations with the same gravitational potential as an ***equipotential surface*** for the gravitational potential.

Equipotential surface: a region in space where every location has the same potential. An equipotential surface for the gravitational potential refers to a

3 We will discuss the origin of this Equation in Section 2-5.

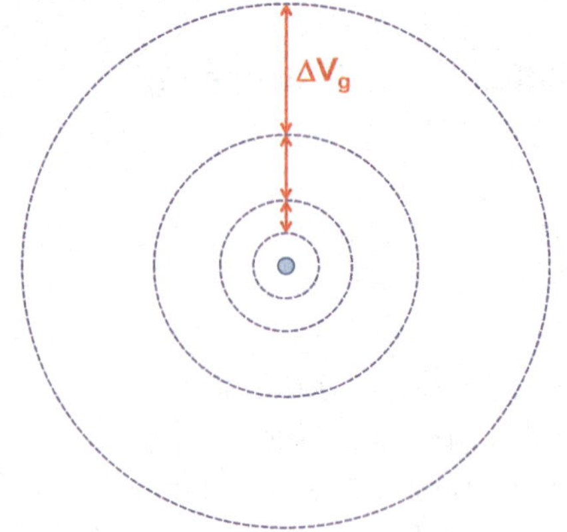

FIGURE 2.1: Equipotential surfaces (dashed lines) for the gravitational potential surrounding a point particle. The difference in gravitational potential between each set of adjacent equipotential surfaces has the same magnitude, ΔV_g, in this figure.

region in space where every location has the same gravitational potential.

A few equipotential surfaces for a point particle are shown (in cross-section) in Figure 2.1. Although the difference in the gravitational potential between each set of adjacent equipotential surfaces has the same magnitude in Figure 2.1, the separation distance between adjacent equipotential surfaces increases with increasing distance from the point particle due to the "$1/r$" dependence of Equation 2-2.

The gravitational potential of a system of point particles is the summation of the gravitational potential from each of the point particles.

Example 2-2

Problem: A system consists of two identical point particles, A and B, separated by a distance d as shown in Figure 2.2. What is the gravitational potential of this system at the position P?

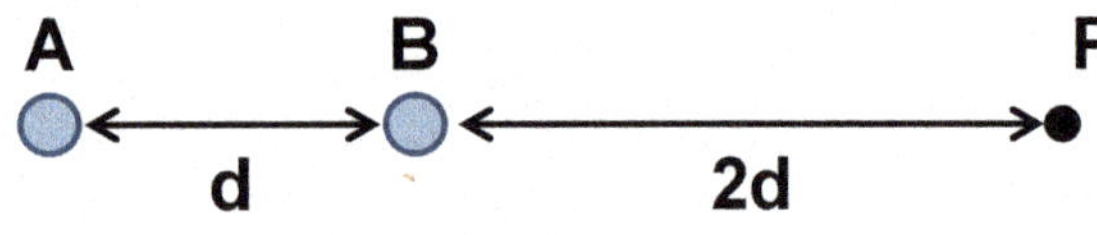

FIGURE 2.2: A system of two point particles.

Solution: The net gravitational potential of this system at the position P is the sum of the gravitational potential of each of the point particles at that location.

$$\left(V_g\right)_{net,P} = \left(V_g\right)_{A,P} + \left(V_g\right)_{B,P}$$

The variables $\left(V_g\right)_{A,P}$, $\left(V_g\right)_{B,P}$, and $\left(V_g\right)_{net,P}$ in this equation denote the gravitational potential of particle A at position P, the gravitational potential of particle B and position P, and the net gravitational potential at position P. Let's define the mass of each point particle to be m. Then, according to Equation 2-2, we have

$$\left(V_g\right)_{net,P} = \left(-G\frac{m}{d+2d}\right) + \left(-G\frac{m}{2d}\right) \rightarrow \left(V_g\right)_{net,P} = -\frac{Gm}{d}\left(\frac{1}{3}+\frac{1}{2}\right)$$

$$\left(V_g\right)_{net,P} = -\frac{5Gm}{6d}$$

Example 2-3

Problem: A system consists of two point particles, A and B, separated by a distance of 10 m as shown in Figure 2.3. The mass of particle A is 2 kg and the mass particle B is 3 kg. What is the gravitational potential of this system at the position P?

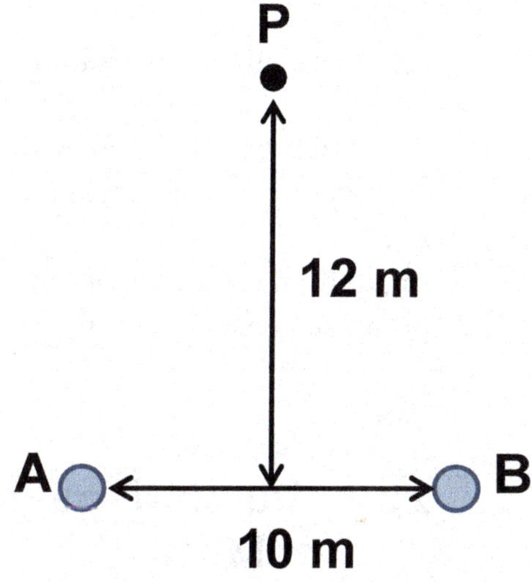

FIGURE 2.3: The system of two point particles in Example 2-3.

Solution: The net gravitational potential of this system at the position P is the sum of the gravitational potential of each of the point particles at that location.

$$\left(V_g\right)_{net,P} = \left(V_g\right)_{A,P} + \left(V_g\right)_{B,P}$$

Let's define the masses of the two point particles to be m_A and m_B, and the distances between the point particles and the position P to be r_A and r_B. Then, according to Equation 2-2, we have

$$\left(V_g\right)_{net,P} = \left(-G\frac{m_A}{r_A}\right) + \left(-G\frac{m_B}{r_B}\right) \rightarrow \left(V_g\right)_{sys} = -G\left(\frac{m_A}{r_A}+\frac{m_B}{r_B}\right)$$

The distances r_A and r_B can be found using trigonometry, as shown in Figure 2.4.

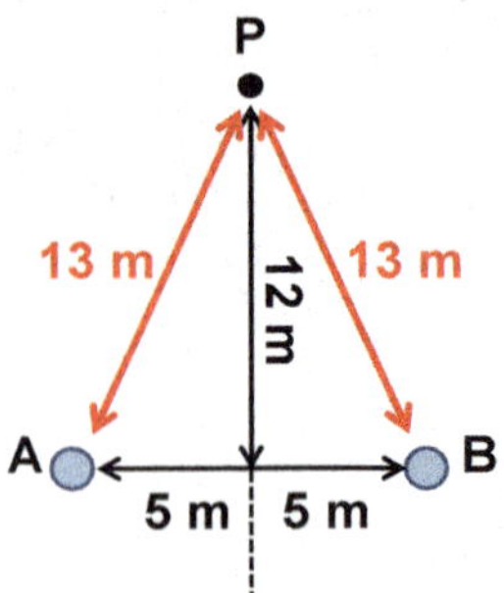

FIGURE 2.4: The distances between the point particles and the observation point for the system in Example 2-3.

Substitution then gives us

$$\left(V_g\right)_{net,P} = -\left(6.67\times10^{-11}\frac{\text{J m}}{\text{kg}^2}\right)\left(\frac{2\,\text{kg}}{13\,\text{m}}+\frac{3\,\text{kg}}{13\,\text{m}}\right)$$

$$\left(V_g\right)_{net,P} = -2.57\times10^{-11}\frac{\text{J}}{\text{kg}}$$

More formally, we refer to the gravitational potential as a ***scalar field***.

> *Scalar field:* a mathematical function or operation that assigns a scalar value to every point in a region of space.

In addition to the gravitational potential, another example of a scalar field is the temperature at each point in a room.

2-3 Gravitational Potential for Continuous Distributions of Mass

As shown in Section 1-3, the moment of inertia of a rod or a disk can be determining by summing the moments of inertia of the infinitesimal bits of mass, represented by rectangles or rings, respectively, which constitute these objects. Similarly, the gravitational potential for a system consisting of a continuous distribution of mass, such as a book or a table, is determined from the gravitational potential of each infinitesimal bit of mass that constitutes the system. *Each infinitesimal bit of mass can be modeled as a point particle whose gravitational potential is calculated using Equation 2-2.* Specifically, if the infinitesimal mass of one of these "bits" is dm, then the gravitational potential associated with this bit is

$$dV_g = -G\frac{dm}{r}$$

In this equation, the variable r is the distance between dm (*i.e.*, the location of the infinitesimal bit of mass) and the location where the gravitational potential is measured. The total gravitational potential from the distribution of mass is determined by summing the gravitational potential of each infinitesimal bit of mass; since these bits of mass are so small, this summation is an integral.

$$V_g = \int dV_g \rightarrow V_g = \int -G\frac{dm}{r} \rightarrow V_g = -G\int \frac{dm}{r} \quad \textbf{(2-3)}$$

The integral in Equation 2-3 is typically simplified by relating dm to an infinitesimal measure of distance, area, or volume using the appropriate mass density (Table 2-1)[4].

TABLE 2-1. Variable substitutions for integrals involving differential mass

Type of integral	Density used	Relation to dm	Units
linear integral	λ (linear mass density)	$dm = \lambda \cdot dx$	kg/m
area integral	σ (area mass density)	$dm = \sigma \cdot dA$	kg/m^2
volume integral	ρ (volume mass density)	$dm = \rho \cdot dV$	kg/m^3

FIGURE 2.5: Determining the gravitational potential from a rigid rod.

Consider, for example, the system shown in Figure 2.5 consisting of a thin and uniformly dense rod of length L and mass M. We can determine the gravitational potential of the rod at a distance d from one end (*i.e.*, at the position P in Figure 2.5) using Equation 2-3.

Since the rod is uniformly dense, the linear mass density of the rod (λ) is constant throughout the rod[5].

$$\lambda = \frac{M}{L}$$

4 Recall that this is also how calculations of the moment of inertia and the center of mass are performed (see Section 1-3).

5 See Example 1-1.

This allows us to express the infinitesimal amount of mass in Equation 2-3 in terms of the linear mass density of the rod and an infinitesimal distance along the rod, *dx*, as shown in Figure 2.6[6].

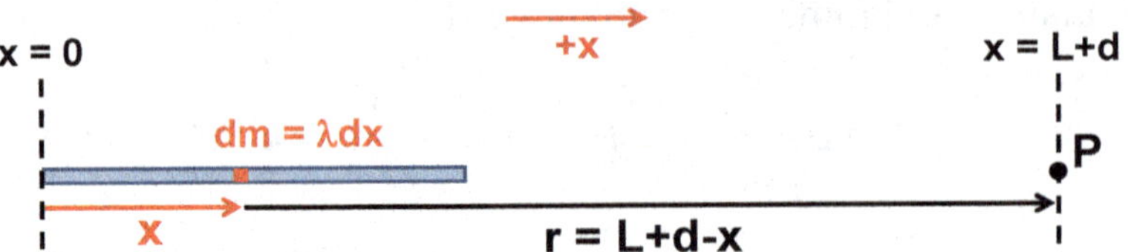

FIGURE 2.6: Determining the gravitational potential from an infinitesimal bit of mass constituting the rod in Figure 2.5.

In Figure 2.6 we have defined an *x*-axis to be parallel to the length of the rod, with an origin at one end of the rod and a positive direction pointing from the rod to the position P. With these definitions, the integral with respect to mass in Equation 2-3 can be written as an integral with respect to the length of the rod.

$$V_g = -G\int\frac{dm}{r} \rightarrow V_g = -G\int_0^L \frac{\lambda dx}{L+d-x} \rightarrow V_g = -G\lambda\int_0^L \frac{dx}{L+d-x}$$

Performing the integration gives us

$$V_g = G\lambda\left(\ln\left(L+d-x\right)\Big|_0^L\right) \rightarrow V_g = G\lambda\ln\left(\frac{d}{L+d}\right)$$

We then arrive at the final answer by substituting for the linear mass density.

$$V_g = \frac{GM}{L}\ln\left(\frac{d}{L+d}\right)$$

6 See Example 1-1.

Example 2-4

Problem: A system consists of a thin and uniformly dense rod of length L and mass M, as shown in Figure 2.7.

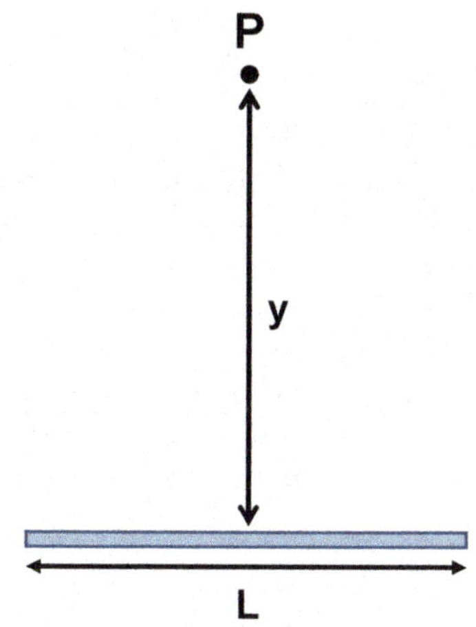

FIGURE 2.7: The rigid rod in Example 2-4.

What is the gravitational potential of this system at the position P that lies on the perpendicular bisector of the rod?

Solution: As shown previously, the integral with respect to mass in Equation 2-3 can be changed into an integral with respect to the length of the rod using the linear mass density of the rod (λ). Let's define a reference frame for our system (Figure 2.8) with an x-axis parallel to the rod, an origin at the center of the rod, and a positive direction pointing to the right in Figure 2.7. Similarly, we will define a y-axis to be perpendicular to the x-axis, with an origin also at the center of the rod and a positive direction pointing from the rod to the position P. The variable x in this reference frame will denote the position of an infinitesimal bit of mass as shown in Figure 2.8.

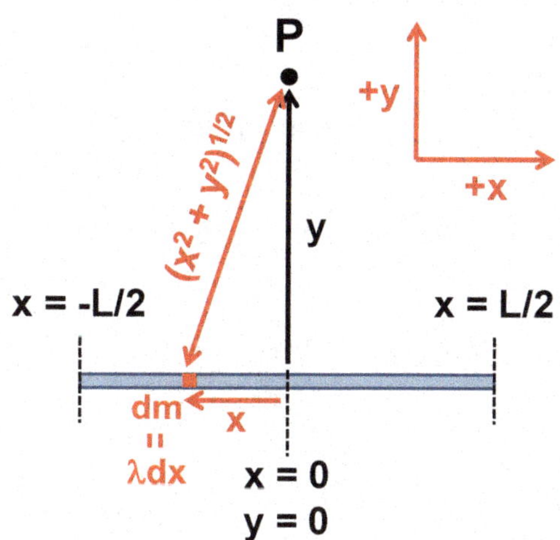

FIGURE 2.8: Determining the gravitational potential for the system in Example 2-4.

The distance between each infinitesimal bit of mass and the position P (the variable r in Equation 2-3) can be related to the variables x and y using trigonometry as shown in Figure 2.8.

$$V_g = -G\int \frac{dm}{r} \rightarrow V_g = -G\int \frac{\lambda dx}{\left(x^2+y^2\right)^{\frac{1}{2}}} \rightarrow V_g = -G\lambda\int \frac{dx}{\left(x^2+y^2\right)^{\frac{1}{2}}}$$

$$V_g = -G\lambda \int_{-\frac{L}{2}}^{\frac{L}{2}} \frac{dx}{\left(x^2 + y^2\right)^{\frac{1}{2}}}$$

Performing the integration gives us

$$V_g = -G\lambda \left(\ln\left(x + \left(x^2 + y^2\right)^{\frac{1}{2}} \right)\Bigg|_{-\frac{L}{2}}^{\frac{L}{2}} \right) \rightarrow V_g = -G\lambda \ln\left(\frac{\frac{L}{2} + \left(\frac{L^2}{4} + y^2\right)^{\frac{1}{2}}}{-\frac{L}{2} + \left(\frac{L^2}{4} + y^2\right)^{\frac{1}{2}}} \right)$$

$$V_g = -\frac{GM}{L} \ln\left(\frac{\left(L^2 + 4y^2\right)^{\frac{1}{2}} + L}{\left(L^2 + 4y^2\right)^{\frac{1}{2}} - L} \right)$$

We can simplify this expression a bit further[7] to give us.

$$V_g = -\frac{2GM}{L} \ln\left(\frac{2y}{\left(L^2 + 4y^2\right)^{\frac{1}{2}} - L} \right)$$

A similar approach is used when the distribution of mass is not uniformly dense. For example, let's calculate the gravitational potential for the system in Example 2-4 when the linear mass density of the rod decreases linearly with respect to the distance from the center ($x = 0$) of the rod in Figure 2.8.

$$\lambda = \begin{cases} \lambda_0 - \gamma x & x \geq 0 \\ \lambda_0 + \gamma x & x < 0 \end{cases}$$

7 See Section C-1.

In this equation, λ_0 is the linear mass density at the center of the rod, and γ is the rate of change of the linear mass density of the rod with respect to position from the center of the rod. Substitution of this linear mass density yields

$$V_g = -G\left(\int_{-\frac{L}{2}}^{0} \frac{(\lambda_0 + \gamma x)\,dx}{(x^2 + y^2)^{\frac{1}{2}}} + \int_{0}^{\frac{L}{2}} \frac{(\lambda_0 - \gamma x)\,dx}{(x^2 + y^2)^{\frac{1}{2}}} \right)$$

After completing the integration and simplifying the result, we are left with

$$V_g = -G\left(2\gamma y - \gamma(L^2 + 4y^2)^{\frac{1}{2}} + 2\lambda_0 \ln\left(\frac{2y}{(L^2 + 4y^2)^{\frac{1}{2}} - L} \right) \right)$$

This expression does not depend explicitly on the mass of the rod. In order to rewrite our result in terms of this mass, let's determine an expression for λ_0 in terms of γ and the total mass of the rod (M) using integration[8].

$$\int dm = \int \lambda\,dx \quad \rightarrow \quad \int_0^M dm = \int_{-\frac{L}{2}}^{0} (\lambda_0 + \gamma x)\,dx + \int_{0}^{\frac{L}{2}} (\lambda_0 - \gamma x)\,dx$$

$$M = \lambda_0 L - \frac{\gamma L^2}{4} \quad \rightarrow \quad \lambda_0 = \frac{\gamma L^2 + 4M}{4L}$$

Substitution gives us

$$V_g = -G\left(2\gamma y - \gamma(L^2 + 4y^2)^{\frac{1}{2}} + \left(\frac{\gamma L^2 + 4M}{2L} \right) \ln\left(\frac{2y}{(L^2 + 4y^2)^{\frac{1}{2}} - L} \right) \right)$$

8 See Section 1-3.

$$V_g = -G\left(\frac{2M}{L}\ln\left(\frac{2y}{\left(L^2+4y^2\right)^{\frac{1}{2}}-L}\right)+\gamma\left(2y-\left(L^2+4y^2\right)^{\frac{1}{2}}+\left(\frac{L^2}{2L}\right)\ln\left(\frac{2y}{\left(L^2+4y^2\right)^{\frac{1}{2}}-L}\right)\right)\right)$$

As expected, this expression is identical to the result of Example 2-4 when $\gamma = 0$.

2-4 The Relationship Between Gravitational Potential and Gravitational Potential Energy

The gravitational interaction between two point particles can be considered to arise from each particle sensing (or reacting to) the gravitational potential of the other particle. The gravitational potential energy of this interaction between point particles is simply the product of the mass of one particle and the gravitational potential of the other particle at that location.

$$U_g = mV_g \tag{2-4}$$

It is important to emphasize that Equation 2-4 is valid only when the gravitational potential is constant over the entire size of the object, as in the case of the interaction between two point particles. Otherwise, the gravitational potential energy would involve integration over the distributions of mass of the objects and variations in the gravitational potentials of the objects.

It follows from Equation 2-4 that the gravitational potential energy between two point particles, A and B, with masses m_A and m_B, respectively, separated by a distance r can be thought to arise from the interaction of object A with the gravitational potential of object B or from the interaction of object B with the gravitational potential of object A.

$$U_g = m_A V_{g,B} = m_B V_{g,A} = -G\frac{m_A m_B}{r}$$

This result is, as expected, exactly the same as Equation 2-1. Furthermore, as stated before, we see from Equation 2-4 that the gravitational potential can be considered to be the gravitational potential energy per unit mass, consistent with its units (J/kg).

Example 2-5

Problem: What is the gravitational potential energy of a system shown in Figure 2.9 consisting of a 2 kg point particle and a uniformly dense 4 kg rod? The 2 kg point particle is located along the perpendicular bisector of the rod.

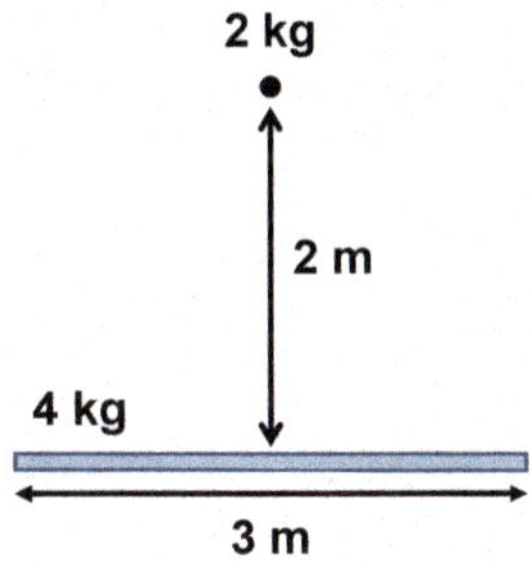

FIGURE 2.9: The system in Example 2-5.

Solution: The gravitational potential of the rod along its perpendicular bisector was calculated in Example 2-4 to be

$$\left(V_g\right)_{rod} = -\frac{2GM_{rod}}{L_{rod}} \ln\left(\frac{2y}{\left(L_{rod}^2 + 4y^2\right)^{\frac{1}{2}} - L_{rod}}\right)$$

Substitution gives us

$$\left(V_g\right)_{rod} = -\frac{2\left(6.67\times10^{-11}\,\frac{\mathrm{J\,m}}{\mathrm{kg}^2}\right)(4\,\mathrm{kg})}{3\,\mathrm{m}} \ln\left(\frac{2(2\,\mathrm{m})}{\left((3\,\mathrm{m})^2 + 4(2\,\mathrm{m})^2\right)^{\frac{1}{2}} - 3\,\mathrm{m}}\right)$$

$$\left(V_g\right)_{rod} = -1.23\times10^{-10}\,\frac{\mathrm{J}}{\mathrm{kg}}$$

Thus, from Equation 2-4, we have that the gravitational potential energy of this system is

$$U_g = m_{particle}\left(V_g\right)_{rod} \quad\rightarrow\quad U_g = (2\,\mathrm{kg})\left(-1.23\times10^{-10}\,\frac{\mathrm{J}}{\mathrm{kg}}\right)$$

$$U_g = -2.46\times10^{-10}\,\mathrm{J}$$

As shown in Section C-2, the same answer is obtained using the mass of the rod and the potential of the point particle.

Now let's imagine that the point particle in Example 2-5 moves from being 2 m away from the rod to being 1 m away from the rod, as shown in Figure 2.10, while the rod remains fixed in place. What is the change in the gravitational potential energy of the system?

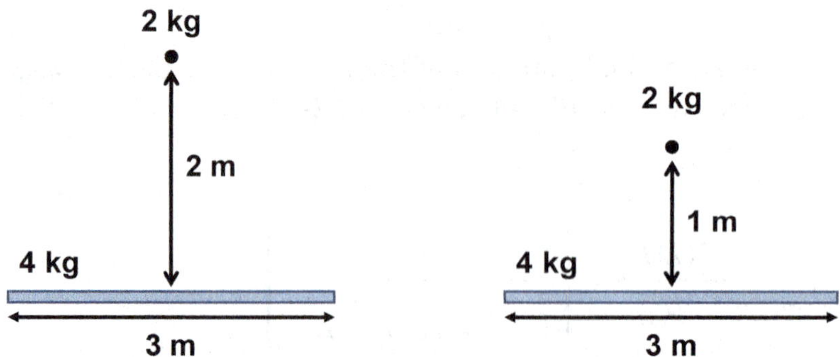

FIGURE 2.10: Determining the change in gravitational potential energy associated with the reconfiguration of the objects in a system.

It follows from Equation 2-4 that the change in the gravitational potential energy of this system is equal to the product of the mass of the point particle and the change in the gravitational potential experienced by that point particle.

$$\Delta U_g = m\Delta V_g \tag{2-5}$$

The change in the gravitational potential experienced by this point particle is the difference between the final and initial values[9] of the gravitational potential experienced by this point particle.

$$\Delta\left(V_g\right)_{rod} = \left(V_g\right)_{rod,f} - \left(V_g\right)_{rod,i}$$

$$\left(V_g\right)_{rod,i} = -1.23 \times 10^{-10}\,\frac{\text{J}}{\text{kg}}$$

9 We will use the subscripts *f* and *i* to denote final and initial values of variables.

$$\left(V_g\right)_{rod,f} = -\frac{2\left(6.67\times10^{-11}\frac{\text{J m}}{\text{kg}^2}\right)(4\text{kg})}{3\text{m}}\ln\left(\frac{2(1\text{m})}{\left((3\text{m})^2+4(1\text{m})^2\right)^{\frac{1}{2}}-3\text{m}}\right)$$

$$\left(V_g\right)_{rod,f} = -2.13\times10^{-10}\frac{\text{J}}{\text{kg}}$$

Hence,

$$\Delta\left(V_g\right)_{rod} = -9\times10^{-11}\frac{\text{J}}{\text{kg}}$$

Therefore, according to Equation 2-5, the change in the gravitational potential energy for the system for this movement of the point particle is

$$\Delta U_g = (2\text{kg})\left(-9\times10^{-11}\frac{\text{J}}{\text{kg}}\right) \rightarrow \Delta U_g = -1.8\times10^{-10}\text{J}$$

Of course, we would have obtained the same result had we considered the point particle to remain fixed and the rigid rod to be moving.

2-5 Gravitational Field

An expression for the gravitational force is found by substituting Equation 2-1 into Equation 1-1; because the gravitational potential is spherically symmetric we will use Equation B-7 for the gradient.

$$\vec{F}_g = -\nabla U_g \rightarrow \vec{F}_g = \left(-\frac{\partial U_g}{\partial r}\right)\hat{r} + \left(-\frac{1}{r}\frac{\partial U_g}{\partial \theta}\right)\hat{\theta} + \left(-\frac{1}{r\sin\theta}\frac{\partial U_g}{\partial \varphi}\right)\hat{\varphi}$$

$$\vec{F}_g = \left(-\frac{\partial}{\partial r}\left(-G\frac{m_A m_B}{r}\right)\right)\hat{r} + \left(-\frac{1}{r}\frac{\partial}{\partial \theta}\left(-G\frac{m_A m_B}{r}\right)\right)\hat{\theta} + \left(-\frac{1}{r\sin\theta}\frac{\partial}{\partial \varphi}\left(-G\frac{m_A m_B}{r}\right)\right)\hat{\varphi}$$

$$\vec{F}_g = \left(-G\frac{m_A m_B}{r^2}\right)\hat{r} \tag{2-6}$$

The unit vector $\hat{r}$ in Equation 2-6 points along an axis extending through the two point particles and for each point particle away from the other point particle (see Figure 2.11). Therefore, the negative sign in Equation 2-6 indicates that the gravitational force will always pull objects toward each another.

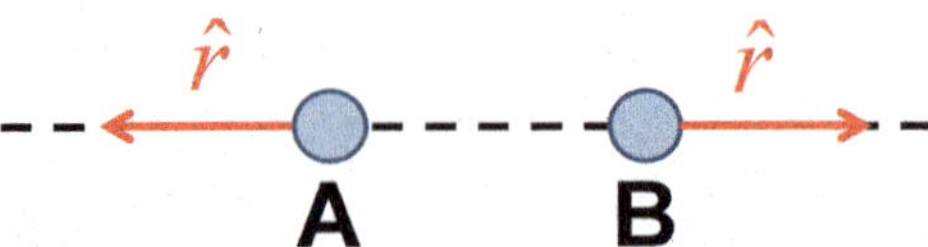

FIGURE 2.11: Definition for the unit vector in Equation 2-6.

We see from Equation 2-4 that the gravitational force acting on a point particle can also be expressed in terms of the gravitational potential experienced by the point particle.

$$\vec{F}_g = -\nabla U_g \quad \rightarrow \quad \vec{F}_g = -\nabla\left(mV_g\right) \quad \rightarrow \quad \vec{F}_g = -m\nabla V_g \tag{2-7}$$

Specifically, the gravitational force is proportional to the gradient of the gravitational potential. The negative sign in Equation 2-7 indicates that the direction of the gravitational force acting on the point particle will always cause the point particle to move toward lower gravitational potential. This is consistent with the principle of minimum potential energy[10] which states that systems will always arrange themselves in order to minimize their total potential energy[11].

Example 2-6

Problem: What is the gravitational force experienced by the 2 kg point particle in Figure 2.9?

Solution: The gravitational potential of the rod was determined to be

$$\left(V_g\right)_{rod} = -\frac{2GM_{rod}}{L_{rod}}\ln\left(\frac{2y}{\left(L_{rod}^2 + 4y^2\right)^{\frac{1}{2}} - L_{rod}}\right)$$

10 This is a restatement of the 2nd law of thermodynamics.

11 Recall from Equation 2-4 that gravitational potential is directly proportional to gravitational potential energy.

From Equation 2-7, we have

$$\vec{F}_g = -m_{particle} \nabla \left(V_g\right)_{rod}$$

The gravitational potential depends upon only the distance y separating the point particle from the rod; all of the other variables in the equation, such as the length of the rod, are constants. Since the gravitational potential is expressed in terms of rectangular coordinates we will use Equation B-5 to determine the gravitational force acting on the point particle.

$$\vec{F}_g = -m_{particle} \left[\left(\frac{\partial}{\partial x} \left(V_g\right)_{rod} \right) \hat{x} + \left(\frac{\partial}{\partial y} \left(V_g\right)_{rod} \right) \hat{y} + \left(\frac{\partial}{\partial z} \left(V_g\right)_{rod} \right) \hat{z} \right]$$

$$\frac{\partial}{\partial x} \left(V_g\right)_{rod} = \frac{\partial}{\partial z} \left(V_g\right)_{rod} = 0$$

$$\vec{F}_g = \left[-m_{particle} \frac{\partial}{\partial y} \left(-\frac{2GM_{rod}}{L_{rod}} \ln \left(\frac{2y}{\left(L_{rod}^2 + 4y^2\right)^{\frac{1}{2}} - L_{rod}} \right) \right) \right] \hat{y}$$

$$\vec{F}_g = \left[-\left(\frac{2Gm_{particle} M_{rod}}{y\left(L_{rod}^2 + 4y^2\right)^{\frac{1}{2}}} \right) \right] \hat{y}$$

Substitution gives us

$$\vec{F}_g = \left[-\left(\frac{2\left(6.67 \times 10^{-11} \frac{\mathrm{J\,m}}{\mathrm{kg}^2}\right)(2\,\mathrm{kg})(4\,\mathrm{kg})}{(2\,\mathrm{m})\left((3\,\mathrm{m})^2 + 4(2\,\mathrm{m})^2\right)^{\frac{1}{2}}} \right) \right] \hat{y}$$

$$\vec{F}_g = -\left(1.07 \times 10^{-10}\,\mathrm{N}\right) \hat{y}$$

The negative sign in our solution indicates that the direction of the gravitational force is directed toward the rod[12].

The gradient of the gravitational potential in Equation 2-7 is referred to as the ***gravitational field***.

Gravitational field: an intrinsic property of mass that describes the alteration of space associated with mass. The gravitational field is a vector quantity obtained from the gradient of the gravitational potential.

The gravitational field, denoted by $\vec{g}$, can be expressed as

$$\vec{g} = -\nabla V_g \tag{2-8}$$

For example, the gravitational field surrounding a point particle can be found through the substitution of Equation 2-2 into Equation 2-8 using Equation B-7.

$$\vec{g} = \left(-\frac{\partial}{\partial r}\left(-G\frac{m}{r}\right)\right)\hat{r} + \left(-\frac{1}{r}\frac{\partial}{\partial \theta}\left(-G\frac{m}{r}\right)\right)\hat{\theta} + \left(-\frac{1}{r\sin\theta}\frac{\partial}{\partial \varphi}\left(-G\frac{m}{r}\right)\right)\hat{\varphi}$$

$$\vec{g} = \left(-G\frac{m}{r^2}\right)\hat{r} \tag{2-9}$$

As shown in Equation 2-9, the magnitude of the gravitational field surrounding a point particle is inversely proportional to the square of the distance away from the particle; the units of the magnitude of the gravitational field are typically expressed as N/kg. Thus, the magnitude of the gravitational field is measure of the gravitational force per unit mass. Furthermore, since the unit vector $\hat{r}$ in Equation 2-9 points away from the point particle, the direction of the gravitational field is always towards the particle. This makes sense since the gravitational force will always cause two objects to move toward each other.

The gravitational field surrounding a point particle is shown in Figure 2.12. In this representation, the direction of the gravitational field at several positions is indicated

---

12 See the definition of the reference frame for the system in Figure 2.8.

by the direction of an arrow and the magnitude of the gravitational field by the length of the arrow. As shown in Figure 2.12, and as follows from Equation 2-8, the direction of the gravitational field at any position is always perpendicular to the equipotential surface for the gravitational potential associated with that position and always points towards lower gravitational potential.

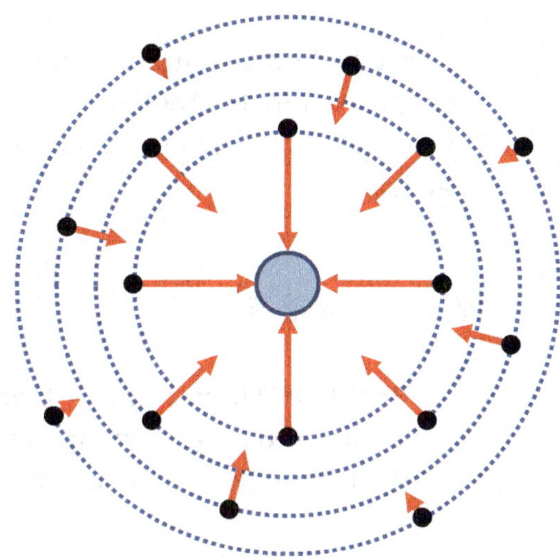

FIGURE 2.12: The gravitational field at various locations (black circles) around a point particle. All locations on the same dashed line are at the same gravitational potential.

In a certain sense, objects are sensitive to the gravitational fields around them since these fields indicate both the magnitude and the direction of the object's response to the gravitational potential at its location. In other words, they indicate how the object will move so as to minimize its gravitational potential energy. Indeed, we see from Equation 2-7 and Equation 2-8 that the force of gravity acting on a point particle is equal to the product of the particle's mass and the gravitational field experienced by the particle.

$$\vec{F}_g = m\vec{g} \tag{2-10}$$

It follows from Equation 2-10 that the gravitational field at any location is the acceleration due to gravity that a point particle would experience at that location. If the only force acting on the point particle was the force of gravity, then from Newton's 2nd law we have

$$\vec{F}_g = m\vec{a} \quad \rightarrow \quad m\vec{g} = m\vec{a} \quad \rightarrow \quad \vec{g} = \vec{a}$$

This result is also consistent with the units of the magnitude of the gravitational field.

$$1\frac{\text{N}}{\text{kg}} = 1\frac{\frac{\text{kgm}}{\text{s}^2}}{\text{kg}} = 1\frac{\text{m}}{\text{s}^2}$$

Thus, in addition to being a measure of the gravitational force per unit mass at a given location, the gravitational field is also a measure of the acceleration due to gravity at that location.

Just as the gravitational potential of a system of point particles is the summation of the gravitational potential from each of the point particles, the gravitational field of a system of point particles is the summation of the gravitational field of each of the point particles.

Example 2-7

Problem: Three point particles, A, B, and C, with masses $m_A = 3$ kg, $m_B = 9$ kg, and $m_C = 6$ kg, are arranged as shown in Figure 2.13. What is the gravitational field at the position P due to these particles?

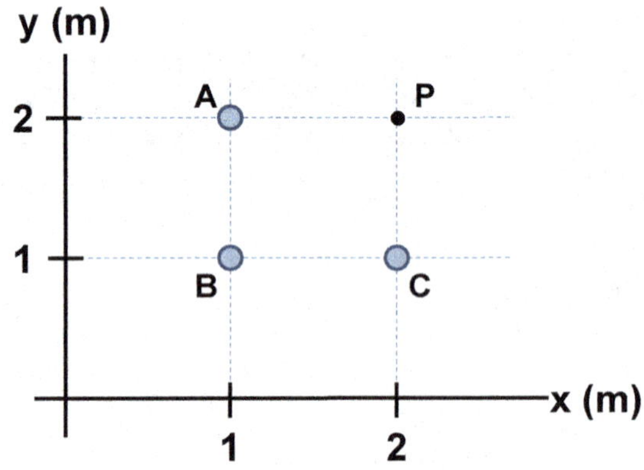

FIGURE 2.13: The system of three point particles in Example 2-7.

Solution: The gravitational field at the position P is the sum of the gravitational fields from each of the point particles at that location. Furthermore, from Equation 2-9, we know that the gravitational field for each point particle is directed toward the point particle. Thus, for the reference frame shown in Figure 2.13 we have

$$\vec{g}_{A,P} = \left(6.67\times10^{-11}\frac{\text{J m}}{\text{kg}^2}\right)\left(\frac{3\text{kg}}{(1\text{m})^2}\right)(-\hat{x})$$

The variable $\vec{g}_{A,P}$ in this equation is the gravitational field at P caused by particle A. Simplifying this equation gives us

$$\vec{g}_{A,P} = \left(-2.0\times10^{-10}\frac{\text{N}}{\text{kg}}\right)\hat{x}$$

Similarly,

$$\vec{g}_{B,P} = \left(6.67\times10^{-11}\frac{\text{J m}}{\text{kg}^2}\right)\left(\frac{9\text{kg}}{(1\text{m})^2+(1\text{m})^2}\right)\left(-(\cos45°)\hat{x}-(\sin45°)\hat{y}\right)$$

$$\vec{g}_{B,P} = \left(-2.12 \times 10^{-10} \frac{\text{N}}{\text{kg}}\right)\hat{x} + \left(-2.12 \times 10^{-10} \frac{\text{N}}{\text{kg}}\right)\hat{y}$$

$$\vec{g}_{C,P} = \left(6.67 \times 10^{-11} \frac{\text{J m}}{\text{kg}^2}\right)\left(\frac{6\,\text{kg}}{(1\,\text{m})^2}\right)(-\hat{y})$$

$$\vec{g}_{C,P} = \left(-4.0 \times 10^{-10} \frac{\text{N}}{\text{kg}}\right)\hat{y}$$

The net gravitational field at point P is thus

$$\vec{g}_{net,P} = \vec{g}_{A,P} + \vec{g}_{B,P} + \vec{g}_{C,P}$$

$$\vec{g}_{net,P} = \left(-2.0 \times 10^{-10} \frac{\text{N}}{\text{kg}}\right)\hat{x} + \left(-2.12 \times 10^{-10} \frac{\text{N}}{\text{kg}}\right)\hat{x} + \left(-2.12 \times 10^{-10} \frac{\text{N}}{\text{kg}}\right)\hat{y} + \left(-4.0 \times 10^{-10} \frac{\text{N}}{\text{kg}}\right)\hat{y}$$

$$\vec{g}_{net,P} = \left(-4.12 \times 10^{-10} \frac{\text{N}}{\text{kg}}\right)\hat{x} + \left(-6.12 \times 10^{-10} \frac{\text{N}}{\text{kg}}\right)\hat{y}$$

The net gravitational field and the individual gravitational fields from the point particles are shown in Figure 2.14.

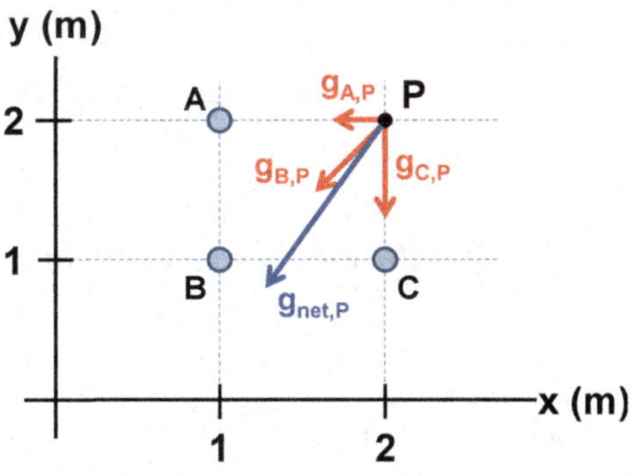

FIGURE 2.14: The net gravitational field at the position P is the summation of the gravitational fields at the position P from each of the point particles in the system.

Example 2-8

Problem: Three point particles, A, B, and C, with masses $m_A = 3$ kg, $m_B = 9$ kg, and $m_C = 6$ kg, are arranged as shown in Figure 2.13. What is the gravitational field experienced by particle A? What is the gravitational force acting on particle A? What is the gravitational potential experienced by particle A?

Solution: The gravitational field experienced by particle A is the summation of the gravitational field from particle B and the gravitational field from particle C.

$$\vec{g}_{net,A} = \vec{g}_{B,A} + \vec{g}_{C,A}$$

$$\vec{g}_{B,A} = \left(6.67 \times 10^{-11}\,\frac{\text{J m}}{\text{kg}^2}\right)\left(\frac{9\,\text{kg}}{(1\text{m})^2}\right)(-\hat{y}) \rightarrow \vec{g}_{B,A} = \left(-6.0 \times 10^{-10}\,\frac{\text{N}}{\text{kg}}\right)\hat{y}$$

$$\vec{g}_{C,A} = \left(6.67 \times 10^{-11}\,\frac{\text{J m}}{\text{kg}^2}\right)\left(\frac{6\,\text{kg}}{(1\text{m})^2 + (1\text{m})^2}\right)\left((\cos 45°)\hat{x} - (\sin 45°)\hat{y}\right)$$

$$\vec{g}_{C,A} = \left(1.4 \times 10^{-10}\,\frac{\text{N}}{\text{kg}}\right)\hat{x} + \left(-1.4 \times 10^{-10}\,\frac{\text{N}}{\text{kg}}\right)\hat{y}$$

Thus,

$$\vec{g}_{net,A} = \left(-6.0 \times 10^{-10}\,\frac{\text{N}}{\text{kg}}\right)\hat{y} + \left(1.4 \times 10^{-10}\,\frac{\text{N}}{\text{kg}}\right)\hat{x} + \left(-1.4 \times 10^{-10}\,\frac{\text{N}}{\text{kg}}\right)\hat{y}$$

$$\vec{g}_{net,A} = \left(1.4 \times 10^{-10}\,\frac{\text{N}}{\text{kg}}\right)\hat{x} + \left(-7.4 \times 10^{-10}\,\frac{\text{N}}{\text{kg}}\right)\hat{y}$$

The gravitational force experienced by particle A is found using Equation 2-10.

$$\left(\vec{F}_g\right)_{net,A} = m_A \vec{g}_{net,A} \rightarrow \left(\vec{F}_g\right)_{net,A} = (3\text{kg})\left(\left(1.4 \times 10^{-10}\,\frac{\text{N}}{\text{kg}}\right)\hat{x} + \left(-7.4 \times 10^{-10}\,\frac{\text{N}}{\text{kg}}\right)\hat{y}\right)$$

$$\left(\vec{F}_g\right)_{net,A} = \left(4.2 \times 10^{-10}\,\text{N}\right)\hat{x} + \left(-2.2 \times 10^{-9}\,\text{N}\right)\hat{y}$$

The gravitational potential experienced by particle A is the sum of the gravitational potential from particle B and the gravitational potential from particle C.

$$\left(V_g\right)_{net,A}=\left(V_g\right)_{B,A}+\left(V_g\right)_{C,A} \rightarrow \left(V_g\right)_{net,A}=\left(-G\frac{m_B}{r_B}\right)+\left(-G\frac{m_C}{r_C}\right)$$

$$\left(V_g\right)_{net,A}=-G\left(\frac{m_B}{r_B}+\frac{m_C}{r_C}\right)$$

$$\left(V_g\right)_{net,A}=-\left(6.67\times10^{-11}\frac{\mathrm{J\,m}}{\mathrm{kg}^2}\right)\left(\frac{9\,\mathrm{kg}}{1\,\mathrm{m}}+\frac{6\,\mathrm{kg}}{\left((1\,\mathrm{m})^2+(1\,\mathrm{m})^2\right)^{\frac{1}{2}}}\right)$$

$$\left(V_g\right)_{net,A}=-8.83\times10^{-10}\,\frac{\mathrm{J}}{\mathrm{kg}}$$

Since the gravitational field is determined from the gradient of the gravitational potential, many gravitational potentials can be associated with the same gravitational field. For example, the gravitational potential surrounding a point particle could be written as Equation 2-2 plus any constant scalar, denoted as *C*.

$$V_g'=V_g+C \rightarrow \vec{g}=-\nabla\left(V_g+C\right) \rightarrow \vec{g}=-\nabla V_g-\nabla C$$

Since the gradient of a constant is zero, the gravitational field associated with this definition of the gravitational potential is the same as Equation 2-9.

$$\nabla C=0 \rightarrow \vec{g}=-\nabla V_g \rightarrow \vec{g}=-\frac{\partial}{\partial r}\left(-G\frac{m}{r}\right)\hat{r} \rightarrow \vec{g}=\left(-G\frac{m}{r^2}\right)\hat{r}$$

This makes sense to us since we remember from classical mechanics that it is only a change in energy, and not the absolute value of the energy, that is important for kinematics. For example, the acceleration of a system was found through differentiation of the equation for the energy of that system. Furthermore, since changes in energy are independent of the zero-point of the energy, the associated forces responsible for the acceleration of the system (through Newton's 2nd law) must also be independent of the zero-point of the energy.

Therefore, it is most accurate to say that our definition of the gravitation field in Equation 2-9 represents the fundamental and measurable quantity (through its associated force in Equation 2-10), and the gravitational potential defined in Equation 2-2 represents only one possible gravitational potential. By writing the gravitational potential using Equation 2-2, we have specifically chosen to define the zero-point for the gravitational potential to occur infinitely far away from the point particle. Only with this additional specification can we write down a *unique* equation for the gravitational potential[13].

Indeed, we can also express the relationship between gravitational field and gravitational potential in Equation 2-8 as Equation 2-11[14].

$$\Delta V_g = -\int \vec{g} \bullet d\vec{s} \tag{2-11}$$

The differential vector $d\vec{s}$ in Equation 2-11 is a small displacement through the gravitational field. Writing the relationship between field and potential this way best conveys the concept that the gravitational field or a change in gravitational potential are the physically relevant quantities for describing the gravitational interaction.

We can use Equation 2-11 to define the gravitational potential surrounding a point particle. Specifically, we define the gravitational potential surrounding a point particle to be zero when measured infinitely far away from the point particle.

$$-\int_{\infty}^{r} \vec{g} \bullet d\vec{s} = V_g(r) - V_g(\infty) \quad \rightarrow \quad -\int_{\infty}^{r} \vec{g} \bullet d\vec{s} = V_g(r) - 0$$

$$V_g(r) = -\int_{\infty}^{r} \vec{g} \bullet d\vec{s}$$

13 We will return to this discussion again when we learn about the fields and potentials associated with the electric and magnetic interactions.

14 See Section C-3 for the derivation of Equation 2-11.

Substitution of Equation 2-9 into this expression gives us[15]

$$V_g(r) = -\int_{\infty}^{r}\left[\left(-G\frac{m}{r^2}\right)\hat{r}\right]\bullet\left[(dr)\hat{r} + (rd\theta)\hat{\theta} + (r\sin\theta d\varphi)\hat{\varphi}\right] \rightarrow V_g(r) = (Gm)\int_{\infty}^{r}\left(\frac{dr}{r^2}\right)$$

$$V_g(r) = (Gm)\left(-\frac{1}{r}\Bigg|_{\infty}^{r}\right) \rightarrow V_g(r) = -G\frac{m}{r}$$

As anticipated, this result is identical to Equation 2-2.

Example 2-9

Problem: What is the average magnitude of the gravitational field between the two equipotential surfaces of the gravitational potential in Figure 2.15?

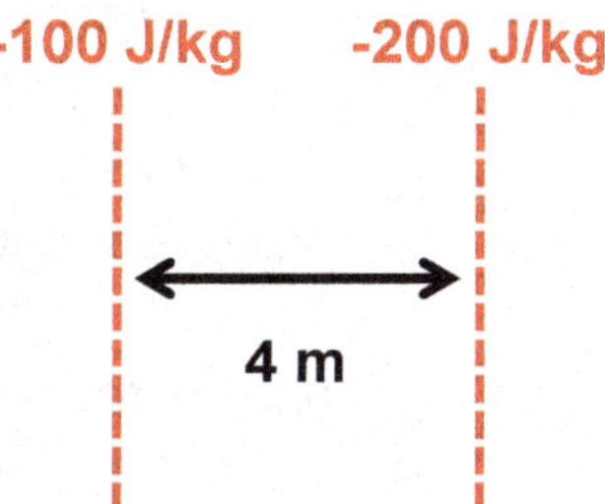

FIGURE 2.15: The two equipotential surfaces in Example 2-9.

Solution: Although the magnitude of the gravitational field at all positions in-between the two equipotential surfaces of the gravitational potential cannot be determined from the information given in the problem, we can nevertheless determine an average magnitude for the gravitational field across the entire separation distance between the two surfaces using Equation 2-11.

$$\Delta V_g = -\int \vec{g}\bullet d\vec{s} \rightarrow \Delta V_g = -\left(\vec{g}_{avg}\right)\bullet\left(\int d\vec{s}\right) \rightarrow \Delta V_g = -\left(\vec{g}_{avg}\right)\bullet\left(\Delta\vec{s}\right)$$

Let's define the origin of the s-axis in this reference frame to be at the -100 J/kg equipotential surface. The positive direction for the s-axis will point toward the -200 J/kg equipotential surface. Let's further assume that the direction of the gravitational field is the same as the direction of the s-axis. Then,

$$\left(\vec{g}_{avg}\right)\bullet\left(\Delta\vec{s}\right) = g_{avg}\Delta s \rightarrow g_{avg} = -\frac{\Delta V_g}{\Delta s}$$

15 Since Equation 2-9 is spherically symmetric we will use spherical coordinates to define the differential displacement (Section B-5).

Substitution gives us

$$g_{avg} = -\frac{-200\frac{\text{J}}{\text{kg}} - \left(-100\frac{\text{J}}{\text{kg}}\right)}{4\,\text{m} - 0\,\text{m}} \quad \rightarrow \quad g_{avg} = 25\frac{\text{N}}{\text{kg}}$$

The positive direction in our result indicates that the direction of the gravitational field is indeed the same as the direction of the s-axis. Therefore, the gravitational field points from high gravitational potential to low gravitational potential, as expected[16].

Finally, we note that the gravitational field is called a ***vector field***.

Vector field: a mathematical function or operation that assigns a vector value to every point in a region of space.

In addition to the gravitational field, another example of a vector field is the wind velocity in different towns and cities in a state.

2-6 The Gravitational Field for Continuous Distributions of Mass

The gravitational field for a system consisting of a continuous distribution of mass is determined from the gravitational field of each infinitesimal bit of mass that constitutes the system. *Each infinitesimal bit of mass can be modeled as a point particle whose gravitational field is calculated using Equation 2-9.* Specifically, if the infinitesimal mass of one of these "bits" is dm, then the gravitational field associated with this bit is

$$d\vec{g} = \left(-G\frac{dm}{r^2}\right)\hat{r}$$

16 Notice that −200 J/kg is a *lower* gravitational potential than −100 J/kg.

In this equation the variable r is the distance between dm and the location where the gravitational field is measured. The total gravitational field from the distribution of mass is determined by summing the gravitational field of each infinitesimal bit of mass; since these bits of mass are so small, this summation is an integral.

$$\vec{g} = \int d\vec{g} \quad \rightarrow \quad \vec{g} = -G\left(\int \frac{dm}{r^2}\hat{r}\right) \tag{2-12}$$

Let's revisit the system shown in Figure 2.5 consisting of a thin, rigid, and uniformly dense rod of length L and mass M. We can now use Equation 2-12 to determine the gravitational field at the position P. As shown in Figure 2.16, the gravitational field is the sum of the gravitational field from each "bit" of mass that constitutes the rod.

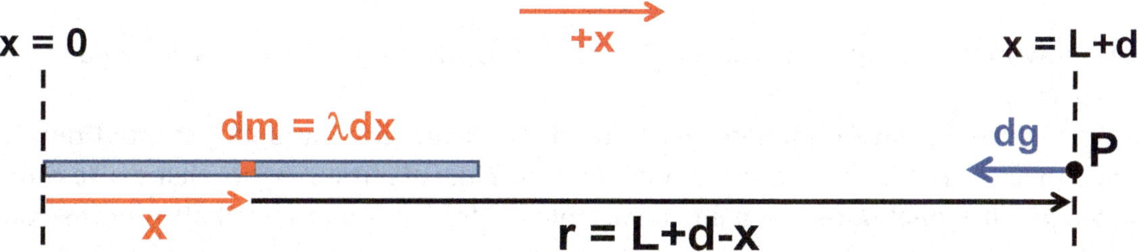

FIGURE 2.16: The net gravitational field at the position P is the sum of the gravitational field at point *P* from each bit of mass constituting the rod.

Let's define a reference frame for this system with an x-axis parallel to the length of the rod, with an origin at one end of the rod and a positive direction pointing from the rod to the position P. With these definitions, the infinitesimal gravitational field at the position P associated with an infinitesimal bit of mass dm located at position x is

$$d\vec{g} = \left(G\frac{dm}{r^2}\right)(-\hat{x}) \quad \rightarrow \quad d\vec{g} = -\left(G\frac{\lambda dx}{(L+d-x)^2}\right)\hat{x}$$

The net gravitational field at the position P is

$$\vec{g} = \int d\vec{g} \quad \rightarrow \quad \vec{g} = \left(\int_0^L\left(-G\frac{\lambda dx}{(L+d-x)^2}\right)\right)\hat{x} \quad \rightarrow \quad \vec{g} = \left(-G\lambda\int_0^L\frac{dx}{(L+d-x)^2}\right)\hat{x}$$

$$\vec{g} = \left(-G\lambda\left(\frac{1}{L+d-x}\right)\Bigg|_0^L\right)\hat{x} \;\rightarrow\; \vec{g} = \left(-G\lambda\left(\frac{L}{d(L+d)}\right)\right)\hat{x}$$

$$\lambda = \frac{M}{L} \;\rightarrow\; \vec{g} = \left(-\frac{GM}{d(L+d)}\right)\hat{x}$$

It is worth noting that this result gives a reasonable prediction in the limit of large values of *d*.

$$d >> L \;\rightarrow\; \vec{g} \approx \left(-\frac{GM}{d^2}\right)\hat{x}$$

In other words, when viewed from large distances the rod appears to be a point particle, as expected.

We also could have obtained this solution from the gravitational potential determined in Section 2-3. In order to apply Equation 2-8, let's first rewrite that expression for the gravitational potential using a variable "*x*" for the separation between the rod and the observation point rather than the fixed distance "*d*". This allows us to express the gravitational potential for an arbitrary separation distance as

$$V_g = G\lambda \ln\left(\frac{x}{L+x}\right)$$

Since this gravitational potential is expressed in terms of rectangular coordinates we will use Equation B-5 to determine the associated gravitational field using Equation 2-8.

$$\vec{g} = \left(-\frac{\partial V_g}{\partial x}\right)\hat{x} + \left(-\frac{\partial V_g}{\partial y}\right)\hat{y} + \left(-\frac{\partial V_g}{\partial z}\right)\hat{z}$$

$$\frac{\partial V_g}{\partial y} = \frac{\partial V_g}{\partial z} = 0 \;\rightarrow\; \vec{g} = \left(-G\lambda\left(\frac{L+x}{x}\right)\left(-\frac{x}{(L+x)^2} + \frac{1}{L+x}\right)\right)\hat{x}$$

$$\vec{g}=\left(-G\lambda\left(\frac{L}{x(L+x)}\right)\right)\hat{x} \rightarrow \vec{g}=\left(-G\frac{M}{L}\left(\frac{L}{x(L+x)}\right)\right)\hat{x} \rightarrow \vec{g}=\left(-\frac{GM}{x(L+x)}\right)\hat{x}$$

This is the equation for the gravitational field of the rod as a function of the separation distance from the rod. The gravitational field when $x = d$ is

$$x=d \rightarrow \vec{g}=\left(-\frac{GM}{d(L+d)}\right)\hat{x}$$

This is the same solution derived previously, as expected.

Example 2-10

Problem: A uniformly dense, thin rod with a total mass M is bent into a semicircular arc with radius R as shown in Figure 2.17.

The position P in Figure 2.17 is equidistant from all points on the arc. What is the gravitational potential at the position P? What is the gravitational field at the position P?

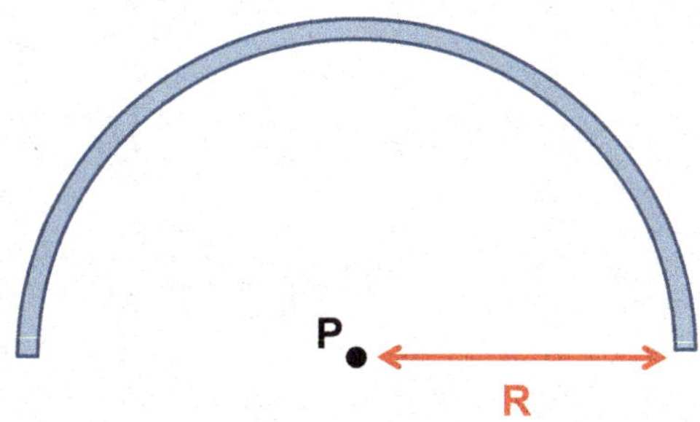

FIGURE 2.17: A uniformly dense semicircular arc.

Solution: The gravitational potential can be determined using Equation 2-3. For this calculation, let's define a reference frame for the system with an origin at the position P, as shown in Figure 2.18.

$$V_g=\int dV_g \rightarrow V_g=-G\int\frac{dm}{r} \rightarrow V_g=-\frac{G}{R}\int dm$$

The integral in this equation is simply the total mass of the arc.

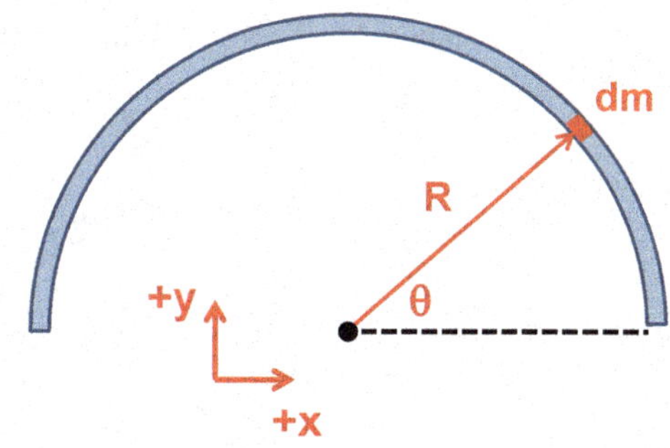

FIGURE 2.18: The location of an infinitesimal bit of mass that constitutes a uniformly dense semicircular arc.

$$\int dm = M \rightarrow V_g=-G\frac{M}{R}$$

This expression for the gravitational potential makes sense since all of the mass constituting the rod is at the same distance (the radius R) from the position P.

The gravitational field of the arc at the position P can be calculated using Equation 2-12. We can express the integral with respect to mass in this equation as an integral with respect to the circumference of the rod. We can further simplify this calculation by independently determining the x-axis and y-axis components of the gravitational field.

As shown in Figure 2.19, we can relate the magnitude of the x-axis component of the gravitational field to the magnitude of the gravitational field using trigonometry.

$$\cos\theta = \frac{(dg)_x}{dg} \rightarrow (dg)_x = dg\cos\theta \rightarrow (g)_x = \int dg\cos\theta$$

$$(g)_x = \int_0^{\pi}\left(G\frac{dm}{R^2}\right)\cos\theta \rightarrow (g)_x = \int_0^{\pi}\left(G\frac{\lambda R d\theta}{R^2}\right)\cos\theta$$

$$(g)_x = \frac{G\lambda}{R}\int_0^{\pi}\cos\theta\, d\theta \rightarrow (g)_x = \frac{G\lambda}{R}\left(\sin\theta\Big|_0^{\pi}\right) \rightarrow (g)_x = 0$$

Of course, we could have anticipated that the x-axis component of the gravitational field at the position P is zero, since this point lies along a line of symmetry of the arc. In other words, there is just as much mass in the rod to the left of position P as there is to the right of positon P.

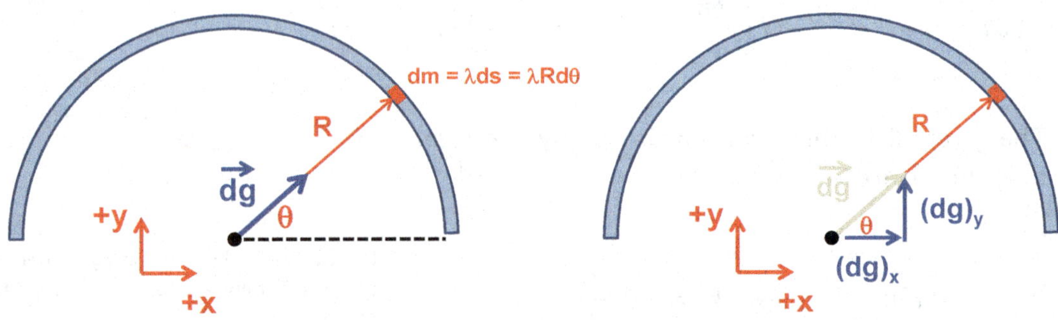

FIGURE 2.19: Determining the gravitational field for each bit of mass constituting the arc in Example 2-10.

We can use a similar approach to determine the y-axis component of the gravitational field.

$$\sin\theta = \frac{(dg)_y}{dg} \rightarrow (dg)_y = dg\sin\theta \rightarrow (g)_y = \int dg\sin\theta$$

$$(g)_y = \int_0^{\pi}\left(G\frac{dm}{R^2}\right)\sin\theta \rightarrow (g)_y = \int_0^{\pi}\left(G\frac{\lambda R d\theta}{R^2}\right)\sin\theta$$

$$(g)_y = \frac{G\lambda}{R}\int_0^{\pi}\sin\theta\, d\theta \rightarrow (g)_y = \frac{G\lambda}{R}\left(-\cos\theta\big|_0^{\pi}\right) \rightarrow (g)_y = \frac{2G\lambda}{R}$$

Substitution of the linear mass density of the rod gives us

$$\lambda = \frac{M}{\pi R} \rightarrow (g)_y = 2G\left(\frac{M}{\pi R}\right)$$

Hence, the gravitational field at the position P is

$$\vec{g} = \left(\frac{2GM}{\pi R^2}\right)\hat{y}$$

As anticipated, the gravitational field points in the positive y-axis direction (*i.e.*, toward the rod, as shown in Figure 2.19).

Finally, let's calculate the gravitational field of the Earth close to the surface of the Earth. Recall from classical mechanics that the gravitational potential energy of a point particle of mass m at a height y above the surface of the Earth was approximated by the following equation:

$$U_g = -G\frac{M_E m}{R_E} + G\frac{M_E m}{R_E^2}y$$

In this equation, the variables M_E and R_E are the mass and radius of the Earth, respectively. Thus, near the surface of the Earth the gravitational potential and gravitational field of the Earth are approximately

$$V_g = \frac{U_g}{m} \rightarrow V_g = -G\frac{M_E}{R_E} + G\frac{M_E}{R_E^2}y$$

Since this gravitational potential is expressed in terms of rectangular coordinates we will use Equation B-5 to determine the associated gravitational field using Equation 2-8.

$$\frac{\partial V_g}{\partial x} = \frac{\partial V_g}{\partial z} = 0 \rightarrow \vec{g} = -\left(\frac{\partial V_g}{\partial y}\right)\hat{y} \rightarrow \vec{g} = \left(-G\frac{M_E}{R_E^2}\right)\hat{y} \rightarrow \vec{g} = -g\hat{y}$$

In this equation, the variable g is commonly referred to as the acceleration of gravity.

$$g = G\frac{M_E}{R_E^2} = 9.8\frac{\text{m}}{\text{s}^2}$$

We see that near the surface of the Earth the gravitational field of Earth has a constant magnitude and always points down toward the surface of the Earth. Under this approximation we say that the Earth's gravitational field is ***uniform***.

> *Uniform field:* a field that is constant over a region of space. For a scalar field, the value of the scalar is constant. For a vector field, the magnitude and direction of the vector are constant.

2-7 The Bohr Model for the Hydrogen Atom, Part I

One of the earliest physical models for the atom, referred to as the Bohr model, consists of the electrons[17] in the atom orbiting the nucleus of the atom in the same way the planets in the solar system orbit the Sun[18]. For the hydrogen atom, there is

17 We will talk more about these particles in Chapter 3.

18 Of course we know from classical mechanics that all the objects in the system are actually orbiting the center of mass of the system.

only one electron and the nucleus consists of only one proton. Since the mass of the proton (1.67×10^{-27} kg) is so much larger than the mass of the electron is (9.11×10^{-31} kg), we can consider the proton to remain effectively stationary as the electron orbits it.

A net force must act on the electron to give rise to the centripetal acceleration required for the electron to orbit the proton. If this net force is the force of gravity acting between the two particles, we have

$$F_{g,electron} = m_{electron}\frac{v_{electron}^2}{R_{orbit}} \rightarrow G\frac{m_{electron}m_{proton}}{R_{orbit}^2} = m_{electron}\frac{v_{electron}^2}{R_{orbit}}$$

$$G\frac{m_{proton}}{R_{orbit}} = v_{electron}^2 \rightarrow v_{electron} = \left(G\frac{m_{proton}}{R_{orbit}}\right)^{\frac{1}{2}}$$

Let's assume the radius of the orbit of the electron (R_{orbit} in this expression) is 50 pm[19]. The speed of the electron would then be

$$v_{electron} = \left(\frac{\left(6.67 \times 10^{-11}\frac{\text{J m}}{\text{kg}^2}\right)\left(1.67 \times 10^{-27}\,\text{kg}\right)}{50 \times 10^{-12}\,\text{m}}\right)^{\frac{1}{2}} \rightarrow v_{electron} = \pm 4.7 \times 10^{-14}\,\frac{\text{m}}{\text{s}}$$

The (±) sign in our solution reminds us that the electron can orbit the proton clockwise or counterclockwise. We could equivalently determine the angular speed of the electron to be

$$\omega_{electron} = 9.4 \times 10^{-4}\,\frac{\text{rad}}{\text{s}}$$

Therefore, the period of the orbit (the time required for the electron to complete one orbit of the proton) is

19 At this point, we don't have any basis to make predictions/assumptions about the radius of an atom. We will introduce some experimental justification and associated theories for the radius of an atom in subsequent chapters.

$$T_{electron} = \frac{2\pi}{\omega_{electron}} \quad \rightarrow \quad T_{electron} = 6.7 \times 10^{3}\,\text{s}$$

According to this calculation it would take the electron almost two hours to orbit the proton. Somehow, intuitively, that seems a bit slow. We will therefore add modifications to the Bohr model in subsequent chapters in order to obtain a more "realistic" description of the motion of the electron in an atom.

Summary

- **Gravitational potential energy:** energy associated with the gravitational interaction between two objects. For two point particles, A and B, this energy is

$$U_g = -G\frac{m_A m_B}{r}$$

The gravitational potential energy of objects near the surface of the Earth is

$$U_g = \left(U_g\right)_0 + mgy$$

- **Scalar field:** a mathematical function or operation that assigns a scalar value to every point in a region of space.

- **Vector field:** a mathematical function or operation that assigns a vector value (*i.e.*, magnitude and direction) to every point in a region of space.

- **Gravitational potential:** an intrinsic property of mass that describes the alteration of space associated with mass. The detection of the gravitational potential of one object by another object is the origin of the gravitational interaction (potential energy and force) between the two objects. Gravitational potential is a scalar field and can be considered to be the gravitational potential energy per unit mass. The gravitational potential surrounding a point particle with mass m is

$$V_g = -G\frac{m}{r}$$

The gravitational potential for a continuous distribution of mass is given by

$$V_g = -G\int\frac{dm}{r}$$

- The gravitational potential energy associated with the gravitational interaction between two point particles is equal to the product of the mass of one of the particles and the gravitational potential of the other particle at that location.

$$U_g = mV_g$$

- The gravitational force acting on a point particle is equal to the product of the mass of the particle and the gradient of the gravitational potential experienced by the particle.

$$\vec{F}_g = -m\nabla V_g$$

- **Gravitational field:** an intrinsic property of mass that describes the alteration of space associated with mass. The gravitational field is a vector field obtained from the gradient of the gravitational potential.

$$\vec{g} = -\nabla V_g$$

The gravitational field surrounding a point particle is

$$\vec{g} = \left(-G\frac{m}{r^2}\right)\hat{r}$$

The gravitational field from a continuous distribution of mass is

$$\vec{g} = -G\left(\int \frac{dm}{r^2}\hat{r}\right)$$

The force of gravity acting on a point particle is equal to the product of the particle's mass and the gravitational field experienced by the particle.

$$\vec{F}_g = m\vec{g}$$

The gravitational field can thus be thought of as the gravitational force per unit mass.

- The gravitational potential and gravitational field of a system of point particles is the summation of the gravitational potential and gravitational field, respectively, of each of the point particles in the system.

- **Uniform field:** a field whose magnitude and direction and constant over a given region of space.

Problems

1. What is the gravitational potential of the Sun (1.99×10^{30} kg) measured at a distance corresponding to the orbit of the Earth? The distance between the Sun and Earth is 1.5×10^{11} m. You can assume that the Sun is effectively a point particle when viewed from this distance.

2. What is the gravitational potential energy of the interaction between the Sun (1.99×10^{30} kg) and the Earth (5.97×10^{24} kg)? The distance between the Sun and Earth is 1.5×10^{11} m. You can assume that both the Sun and the Earth are effectively point particles for this interaction.

3. A system consists of two point particles, A and B, separated by a distance of 3 m as shown in the figure. The mass of particle A is 8 kg and the mass particle B is 10 kg. What is the gravitational potential of this system at the position P?

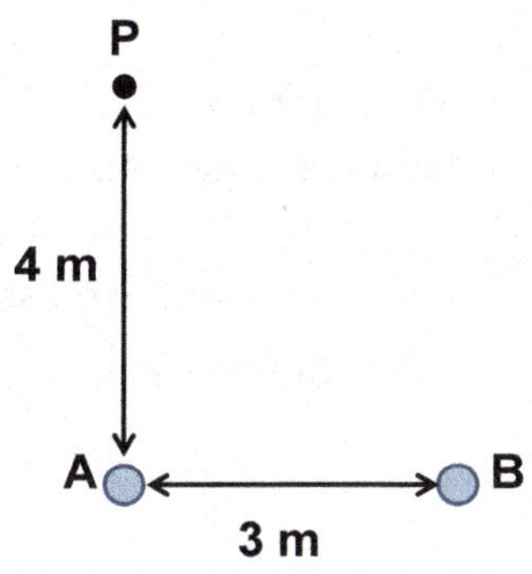

4. A system consists of three identical point particles, A, B, and C, as shown in the figure. The mass of each particle is 30 kg. What is the gravitational potential of this system at the position P?

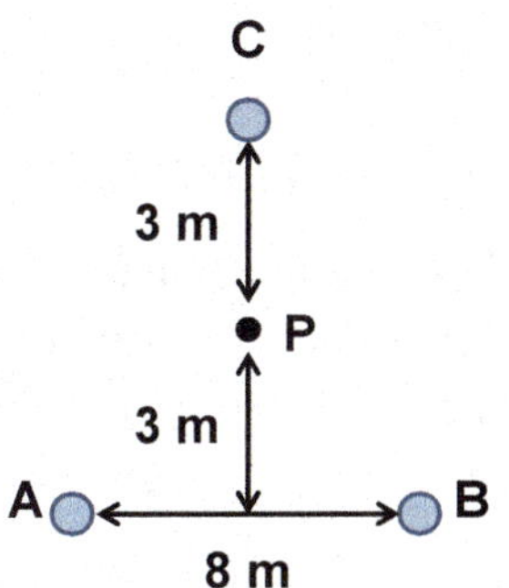

5. A system consists of a thin and uniformly dense rod of length L and mass M, as shown in the figure. What is the gravitational potential of this system at the position P that is a distance y "above" one end of the rod?

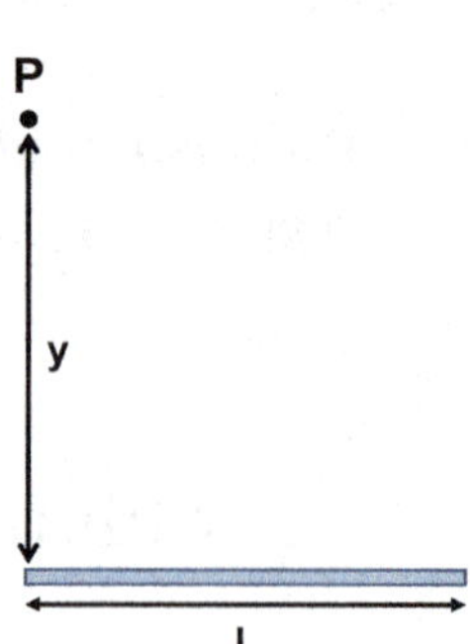

6. What is the gravitational potential at the center of a uniformly dense, thin ring of radius R and total mass M, as shown in the figure?

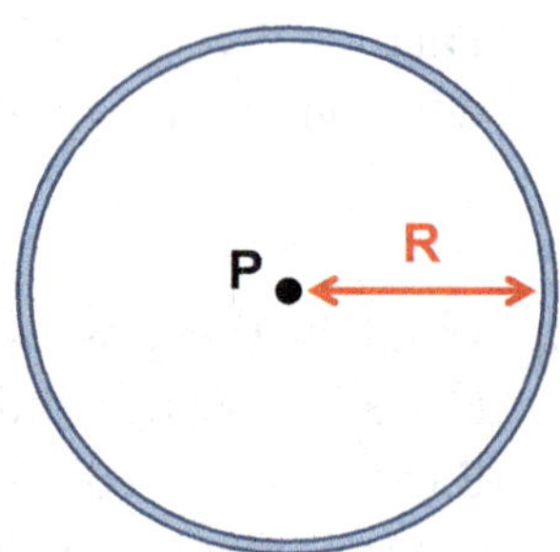

7. What is the gravitational potential at the position P, a distance z away from the center of a uniformly dense ring along the central axis of the ring, as shown in the figure? The total mass and radius of the ring are M and R, respectively.

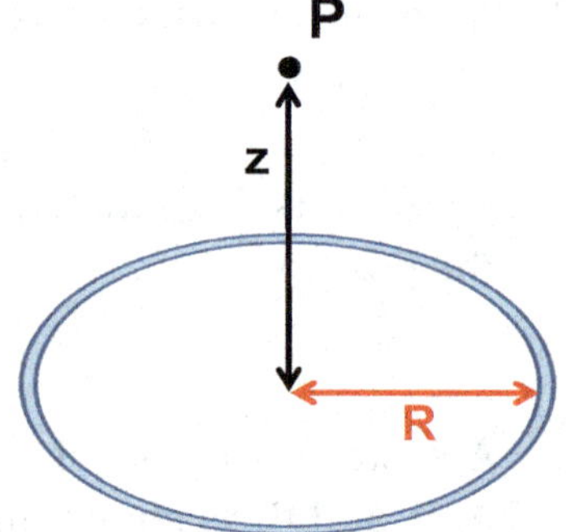

8. A system consists of two thin and uniformly dense parallel rods of length L and mass M, as shown in the figure. What is the gravitational potential energy of this system?

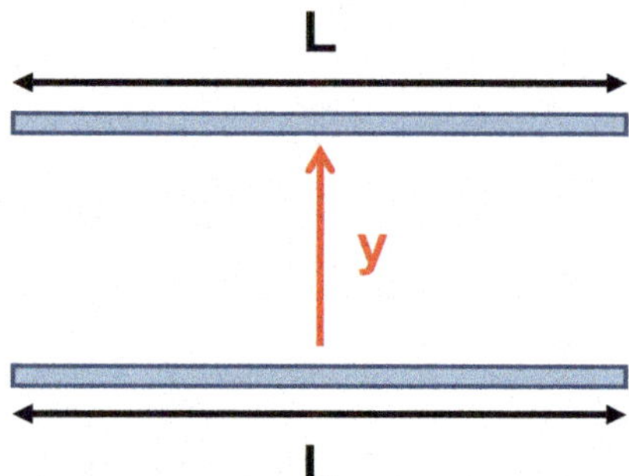

9. The gravitational potential a distance z away from a distribution of mass has the following equation:

$$V_g = -G\frac{M}{\left(R^2 + z^2\right)^{\frac{1}{2}}}$$

What is the gravitational field associated with this distribution of mass?

10. Two identical 10 kg point particles are located as shown in the figure. What is the gravitational field at the position P? Express your answer in terms of the x-axis and y-axis components of the gravitational field using the reference frame shown in the figure.

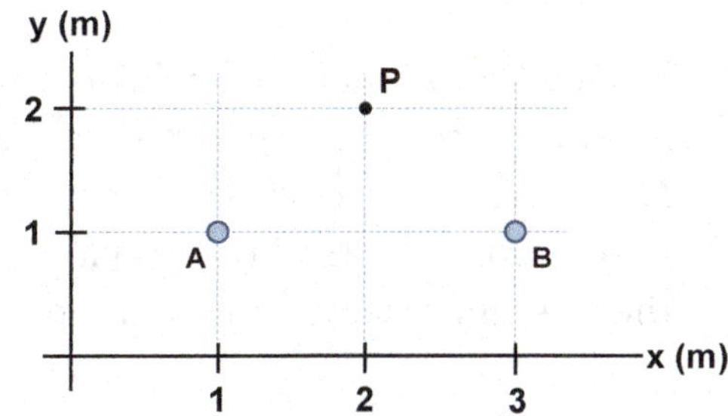

11. What is the gravitational field at a point P that lies on the perpendicular bisector between two identical point particles, as shown in the figure? The mass of each point particle is M. Express your answer in terms of the x-axis and y-axis components of the gravitational field using the reference frame shown in the figure.

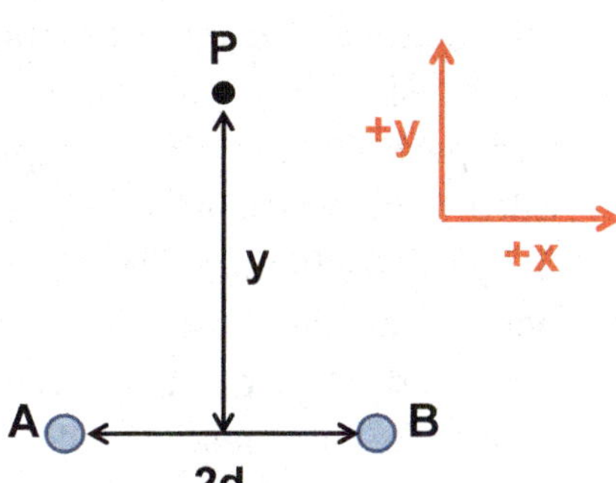

12. A system consists of a thin and uniformly dense rod of length L and mass M, as shown in the figure. What is the gravitational field of this system at the point P that is a distance y "above" one end of the rod? Express your answer in terms of the x-axis and y-axis components of the gravitational field using the reference frame shown in the figure.

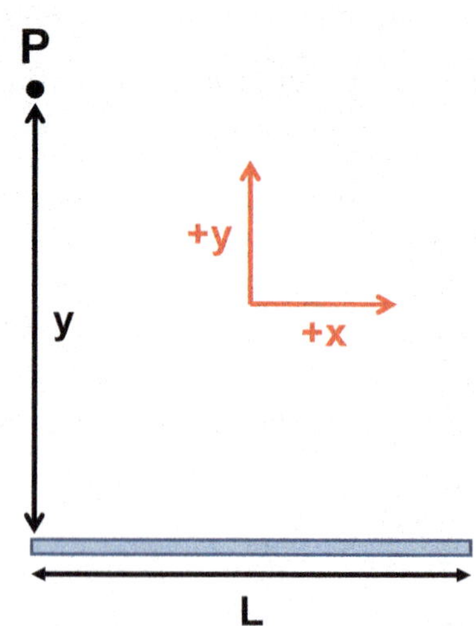

13. What is the gravitational field at the center of a uniformly dense thin ring of radius R and total mass M? Express your answer in terms of the x-axis and y-axis components of the gravitational field using the reference frame shown in the figure.

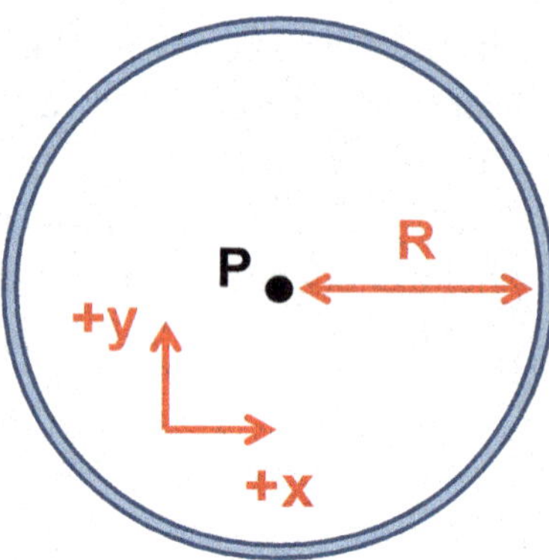

14. A system consists of two identical point particles, each with a mass of 10 kg, separated by a distance of 2 cm, as shown in the figure. These two particles are fixed in place. A third point particle, with a mass of 2 kg, is oscillating along the perpendicular bisector between the two fixed point particles. When the amplitude of these oscillations is small, this oscillation will be simple harmonic. What is the corresponding frequency?

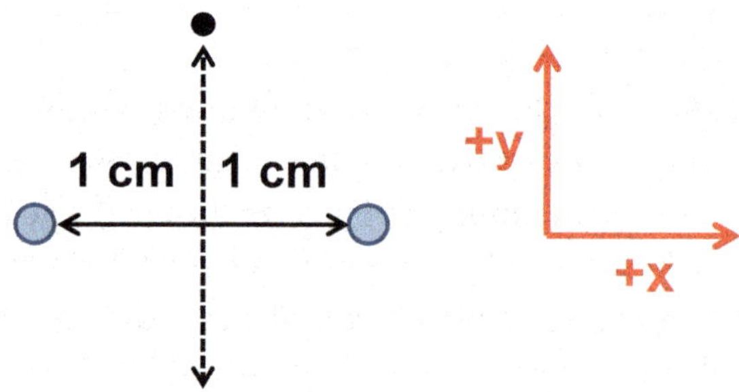

CHAPTER 3

Electric Interaction

3-1 Introduction

In this chapter we will introduce a new interaction, the electric interaction, which has an associated potential, potential energy, field, and force. While the gravitational interaction between two objects is a function of the masses of those objects, the electric interaction between two objects depends upon the electric charges of those objects.

3-2 Electric Charge

Electric charge, like mass, is an intrinsic property of matter; the SI unit for electric charge is the coulomb, denoted by C. Unlike mass, however, electric charge has two varieties: positive and negative. Furthermore, electric charge is *quantized*, meaning that it occurs only in integer multiples of a small unit, typically referred to as the fundamental electric charge, denoted by e.

$$e = 1.6 \times 10^{-19}\ \mathrm{C}$$

The electron has an electric charge of $-e$ and the proton[1] has an electric charge of $+e$. The *net electric charge* of an object is the sum of the positive and negative electric charges possessed by the object. Negatively electrically charged objects have a net negative electric charge, positively electrically charged objects have a net positive electric charge, and neutral objects have no net electric charge.

Example 3-1

Problem: What is the net electric charge of an object that has 3 protons and 4 electrons?

Solution: The net electric charge is the sum of the electric charges possessed by the object.

$$3(e)+4(-e)=-e$$

The object has a net electric charge of $-$e.

As demonstrated by the calculation in Example 3-1, we must always carefully distinguish between the electric charges possessed by an object and the *net* electric charge of that object. An object like the Earth might have as many as 10^{50} protons, but still be electrically neutral if it contains an equal number of electrons.

Furthermore, the terminology "positively electrically charged" or "negatively electrically charged" can be cumbersome. Instead, we will often simply refer to an object with a net electric charge as being "charged". Thus, an object with a net positive electric charge will be referred to as a positively charged object, and an object with a net negative electric charge (such as the object in Example 3-1) will be referred to as a negatively charged object.

1 It is worth noting that protons, as well as many other subatomic particles, are composed of smaller particles called quarks that possess electric charges that are a fraction of e.

Another fundamental property of electric charge is that it can neither be created nor destroyed. This is referred to as the ***conservation of electric charge***.

> *Conservation of electric charge:* electric charge can neither be created nor destroyed. Thus, the total electric charge for an isolated system is constant.

It is important to remember that electric charge is not a substance itself, but rather a property of matter. And although we do not know exactly what electric charge is, we can nevertheless explain how it behaves.

3-3 Electric Potential

In the same way that we say every bit of mass is associated with a gravitational potential that permeates the space around it, we say that every electric charge is associated with an ***electric potential*** that permeates the space around it.

> *Electric potential:* an intrinsic property of electric charge that describes the alteration of space associated with electric charge. The detection of the electric potential of one charged object by another charged object is the origin of the electric interaction (potential energy and force) between the two objects. Electric potential is a scalar field.

The electric potential surrounding a charged point particle, denoted by V_e, is defined in Equation 3-1[2].

$$V_e = \frac{q}{4\pi\varepsilon_0 r} \tag{3-1}$$

In equation 3-1, q denotes the net electric charge of the point particle, r is the distance from the point particle, and the constant ε_0 is the permittivity of free space (also called the vacuum permittivity), which is

$$\varepsilon_0 = 8.85 \times 10^{-12} \frac{\mathrm{C}^2}{\mathrm{J\,m}}$$

2 We will discuss the origin of this equation in Section 3-6.

The units of electric potential are energy divided by charge (J/C), which is also called a volt, denoted by *V*.

$$1\,\text{V} = 1\frac{\text{J}}{\text{C}} = 1\frac{\text{Nm}}{\text{C}}$$

The electric potential can be considered to be a measure of the electric potential energy per unit charge similar to how the gravitational potential is the gravitational potential energy per unit mass.

Example 3-2

Problem: What is the electric potential of a point particle with a net electric charge of −2 nC at a distance of 5 m from the particle?

Solution: According to Equation 3-1 the electric potential is

$$V_e = \frac{-2\times10^{-9}\,\text{C}}{4\pi\left(8.85\times10^{-12}\,\frac{\text{C}^2}{\text{J m}}\right)(5\text{m})}$$

$$V_e = -3.6\,\text{V}$$

Electric potential is different from gravitational potential in that while there is only one type of mass, there are two types of electric charge (positive and negative). Thus, while the gravitational potential is always negative, the electric potential can be positive or negative. Furthermore, as with gravitational potential, the electric potential of a system of point particles is the summation of the electric potential from each of the point particles.

Example 3-3

Problem: Two charged point particles, A and B, are separated by a distance of 30 cm as shown in Figure 3.1.

The net electric charge of particle A is +2 nC and the net electric charge of particle B is −4 nC. At what position in-between the two point particles is the electric potential of this system zero?

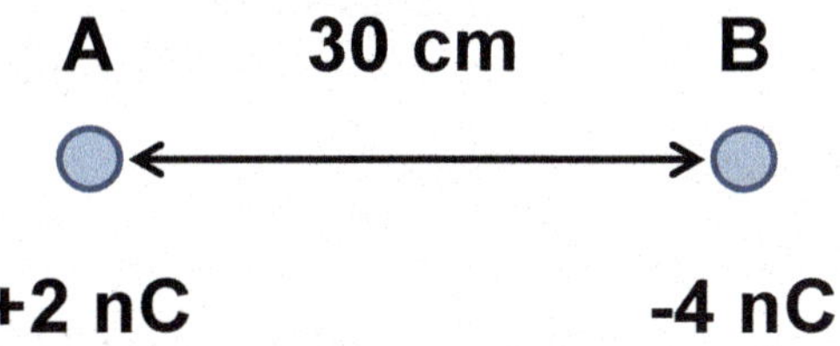

FIGURE 3.1: The system of two charged point particles in Example 3-3.

Solution: The net electric potential of this system is the sum of the electric potential of each of the point particles. Thus, from Equation 3-1 we have

$$\left(V_e\right)_{net} = \left(V_e\right)_A + \left(V_e\right)_B \quad \rightarrow \quad \left(V_e\right)_{net} = \frac{q_A}{4\pi\varepsilon_0 r_A} + \frac{q_B}{4\pi\varepsilon_0 r_B}$$

The variables r_A and r_B in this equation denote the distances from particle A and particle B, respectively, to the position where the electric potential is observed or measured. As shown in Figure 3.2, for all positions in-between the two particles the distances r_A and r_B are related by the total separation of the two particles.

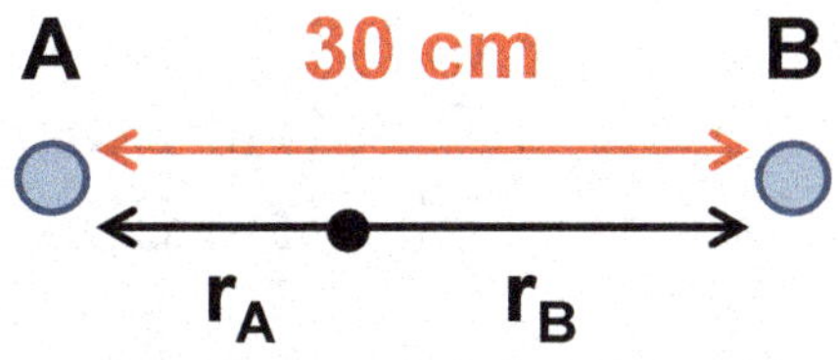

FIGURE 3.2: Determining the electric potential for a position (black circle) in-between the two charged particles constituting the system in Example 3-3.

$$r_A + r_B = 30\,\text{cm} \quad \rightarrow \quad r_B = 30\,\text{cm} - r_A$$

Substitution of this relationship then gives us

$$\left(V_e\right)_{net} = \frac{q_A}{4\pi\varepsilon_0 r_A} + \frac{q_B}{4\pi\varepsilon_0 \left(30\,\text{cm} - r_A\right)}$$

We can now solve for the value of r_A corresponding to a net electric potential of zero.

$$\left(V_e\right)_{net} = 0 \quad \rightarrow \quad 0 = \frac{q_A}{4\pi\varepsilon_0 r_A} + \frac{q_B}{4\pi\varepsilon_0 \left(30\,\text{cm} - r_A\right)} \quad \rightarrow \quad \frac{q_A}{r_A} = -\frac{q_B}{30\,\text{cm} - r_A}$$

$$r_A = -\left(30\,\text{cm}\right)\left(\frac{q_A}{q_B - q_A}\right) \quad \rightarrow \quad r_A = -\left(30\,\text{cm}\right)\left(\frac{2\times10^{-9}\text{C}}{-4\times10^{-9}\text{C} - 2\times10^{-9}\text{C}}\right)$$

$$r_A = 10\,\text{cm}$$

The system will have a net electric potential of zero in-between the two particles at a distance of 10 cm from particle A.

According to Equation 3-1, an electrically neutral object will have an electric potential of zero. This occurs because the electric potentials of all of the positive and negative electric charges constituting the object cancel each other out. This is similar to the calculation in Example 3-3 in which an electric potential of zero existed at a specific position in-between the two charged point particles. For neutral objects, this cancellation occurs at all locations so the net electric potential is zero everywhere. This is also why it is only the net electric charge of an object that is used for calculating the net electric potential surrounding that object.

3-4 Electric Potential for Continuous Distributions of Electric Charge

The electric potential for a system consisting of a continuous distribution of net electric charge is determined from the electric potential of each infinitesimal bit of net electric charge that constitute the system. *Each infinitesimal bit of net electric charge can be modeled as a charged point particle whose electric potential is calculated using Equation 3-1.* Specifically, if the net electric charge of one of these infinitesimal "bits" is dq, then the electric potential associated with this bit is

$$dV_e = \frac{dq}{4\pi\varepsilon_0 r}$$

The variable r in this equation is the distance between dq and the location where the electric potential is measured. The total electric potential from the distribution of net electric charge is thus determined by summing the electric potential of each infinitesimal bit of net electric charge; since these bits of net electric charge are so small, this summation is an integral.

$$V_e = \int dV_e \quad \rightarrow \quad V_e = \int \frac{dq}{4\pi\varepsilon_0 r} \quad \rightarrow \quad V_e = \frac{1}{4\pi\varepsilon_0}\int \frac{dq}{r} \qquad \textbf{(3-2)}$$

TABLE 3-1. Variable substitutions for integrals involving differential electric charge

Type of integral	Density used	Relation to dq	Units
linear integral	λ (linear net electric charge density)	$dq = \lambda \cdot dx$	C/m
area integral	σ (area net electric charge density)	$dq = \sigma \cdot dA$	C/m^2
volume integral	ρ (volume net electric charge density)	$dq = \rho \cdot dV$	C/m^3

The integral in Equation 3-2 is typically simplified by relating dq to an infinitesimal measure of distance, area, or volume using the appropriate net electric charge density (Table 3-1)[3].

Example 3-4

Problem: The system shown in Figure 3.3 consists of a thin and uniformly charged rod of length L and net electric charge Q. What is the electric potential of the rod at a distance d from one end (*i.e.*, at the position P in Figure 3.3)?

FIGURE 3.3: Determining the electric potential from a uniformly charged rod.

Solution: Since the rod is uniformly charged, the net electric charge density of the rod is constant throughout the rod. This allows us to express the infinitesimal amount of net electric charge in Equation 3-2 in terms of the linear net electric charge density of the rod and an infinitesimal distance along the rod, dx, as shown in Figure 3.4.

In Figure 3.4, we have defined an x-axis to be parallel to the length of the rod, with an origin at one end of the rod and a positive direction pointing from the rod to the position P. With these definitions, the integral with respect to net electric charge in Equation 3-2 can be written as an integral with respect to the length of the rod.

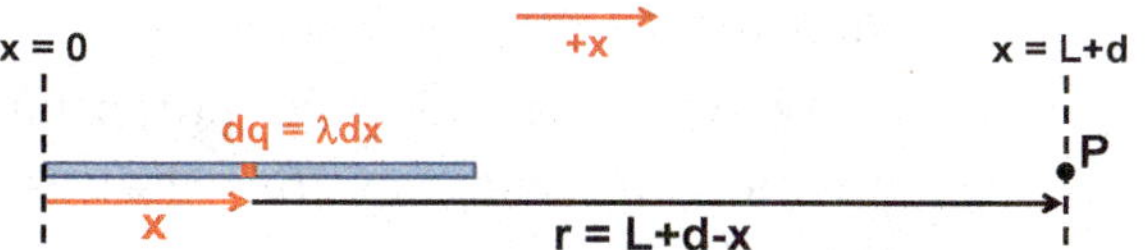

FIGURE 3.4: Determining the electric potential from each bit of net electric charge constituting the rod in Figure 3.3.

3 This is exactly how the integral in Equation 2-3 was simplified.

$$V_e = \frac{1}{4\pi\varepsilon_0}\int\frac{dq}{r} \rightarrow V_e = \frac{1}{4\pi\varepsilon_0}\int_0^L\frac{\lambda dx}{L+d-x} \rightarrow V_e = \frac{\lambda}{4\pi\varepsilon_0}\int_0^L\frac{dx}{L+d-x}$$

Performing the integration gives us

$$V_e = \frac{-\lambda}{4\pi\varepsilon_0}\left(\ln\left(L+d-x\right)\Big|_0^L\right)$$

We then arrive at the final answer by substituting for the net electric charge density.

$$\lambda = \frac{Q}{L} \rightarrow V_e = -\frac{Q}{4\pi\varepsilon_0 L}\ln\left(\frac{d}{L+d}\right) \rightarrow V_e = \frac{Q}{4\pi\varepsilon_0 L}\ln\left(\frac{L+d}{d}\right)$$

The solution to Example 3-4 demonstrates that calculations of the electric potential for uniformly charged objects are basically identical to calculations of the gravitational potential for uniformly dense objects. This is a consequence, of course, of the similarity between Equation 2-3 and Equation 3-2. Indeed, we can convert back and forth between solutions for these calculations by replacing the total mass of the distribution with the net electric charge of the distribution and the constant $-G$ with $\frac{1}{4\pi\varepsilon_0}$[4].

Example 3-5

Problem: A system consists of a thin and uniformly charged rod of length L and net electric charge $-Q$, as shown in Figure 3.5.

What is the electric potential of this system at the position P that lies on the perpendicular bisector of the rod?

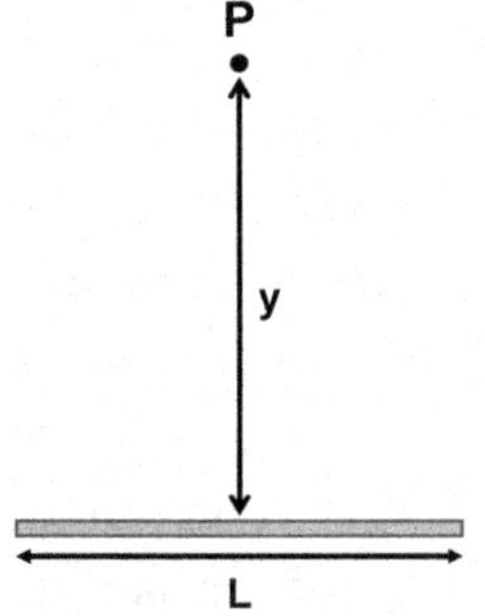

FIGURE 3.5: The uniformly charged rod in Example 3-5.

4 You might wonder why the constant $\frac{1}{4\pi\varepsilon_0}$ for the electric interaction isn't written more simply, as the constant "$-G$" is written for the gravitational interaction. In addition to historical considerations, the main reason why this is done is to simplify other formulas, some of which we will encounter later in this book.

Solution: From comparison to the solution to Example 2-4 we know that the answer is

$$V_e = \frac{-Q}{2\pi\varepsilon_0 L}\ln\left(\frac{2y}{\left(L^2+4y^2\right)^{\frac{1}{2}}-L}\right)$$

Example 3-6

Problem: A system consists of a thin and uniformly charged circular ring of radius *R* and net electric charge *Q*, as shown in Figure 3.6.

What is the electric potential at the position P, a distance *z* away from the center of the ring along the normal axis through the center of the ring?

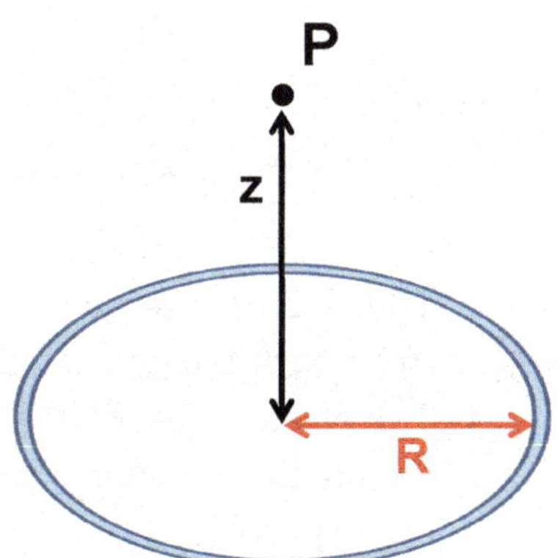

FIGURE 3.6: The uniformly charged ring in Example 3-6.

Solution: Since the ring is uniformly charged, we can divide the ring into infinitesimal bits of net electric charge, *dq*, and determine the total electric potential of the ring from the sum of the electric potential from each of these bits of net electric charge using Equation 3-2. This calculation is further simplified since each infinitesimal bit of net electric charge is the same distance from the position P, as shown in Figure 3.7.

$$V_e = \frac{1}{4\pi\varepsilon_0}\int\frac{dq}{r} \rightarrow V_e = \frac{1}{4\pi\varepsilon_0\left(R^2+z^2\right)^{\frac{1}{2}}}\int dq$$

The integral in this equation is simply the net electric charge of the ring.

$$\int dq = Q \rightarrow V_e = \frac{Q}{4\pi\varepsilon_0\left(R^2+z^2\right)^{\frac{1}{2}}}$$

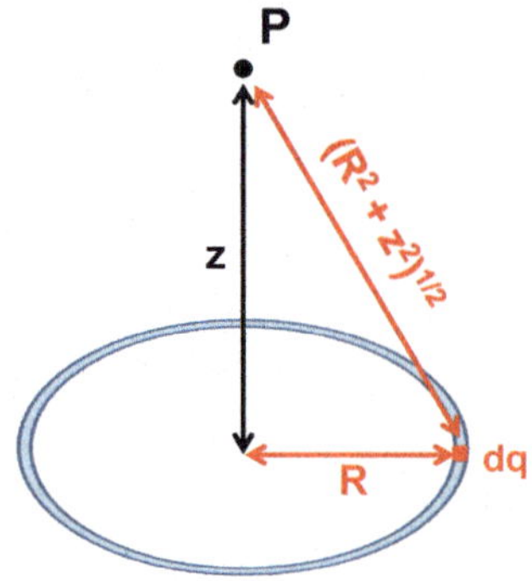

FIGURE 3.7: Determining the electric potential from the system in Figure 3.6.

This result gives a reasonable prediction in the limit of large values of z.

$$z >> R \;\rightarrow\; V_e \approx \frac{Q}{4\pi\varepsilon_0 \left(z^2\right)^{\frac{1}{2}}} \;\rightarrow\; V_e \approx \frac{Q}{4\pi\varepsilon_0 |z|}$$

In other words, when viewed from large distances, the ring appears to be a point particle with net electric charge Q, as expected.

3-5 Electric Potential and Electric Potential Energy

The electric interaction between two charged point particles can be considered to arise from each particle sensing (or reacting to) the electric potential of the other particle. The electric potential energy of this interaction between two charged point particles is simply the product of the electric charge of one particle and the electric potential of the other particle.

$$U_e = qV_e \tag{3-3}$$

Thus, the electric potential energy for a system of two charged point particles, A and B, with net electric charges q_A and q_B is

$$U_e = \frac{q_A q_B}{4\pi\varepsilon_0 r} \tag{3-4}$$

It is important to emphasize that Equation 3-4 is valid only when the electric potential is constant over the entire sizes of the objects, as in the case of the interaction between two charged point particles[5].

5 The same caution applies to Equation 2-4 for gravitational potential and gravitational potential energy.

Example 3-7

Problem: What is the electric potential energy of the system of charged point particles in Figure 3.1?

Solution: From Equation 3-4 we have

$$U_e = \frac{\left(2\times10^{-9}\,\text{C}\right)\left(-4\times10^{-9}\,\text{C}\right)}{4\pi\left(8.85\times10^{-12}\,\frac{\text{C}^2}{\text{J m}}\right)\left(30\,\text{cm}\right)} \quad\rightarrow\quad U_e = -2.4\times10^{-7}\,\text{J}$$

Let's consider a system consisting of two charged point particles, A and B, initially held in place a distance *r* away from each other before being released. The electric potential energy of this system is given by Equation 3-4. Differentiating this equation with respect to the variable *r* gives us

$$\frac{dU_e}{dr} = -\frac{q_A q_B}{4\pi\varepsilon_0 r^2} \quad\rightarrow\quad dU_e = -\frac{q_A q_B}{4\pi\varepsilon_0 r^2}\,dr$$

When released from rest the two charged particles will move in order to minimize their total potential energy[6]. For decreases in the electric potential energy, we have

$$dU_e < 0 \quad\rightarrow\quad -\frac{q_A q_B}{4\pi\varepsilon_0 r^2}\,dr < 0 \quad\rightarrow\quad \frac{q_A q_B}{4\pi\varepsilon_0 r^2}\,dr > 0$$

Therefore, if the net electric charges of the two particles have the same ($\pm$) sign, $dr > 0$ and the separation between the particles must increase for the electric potential energy of the system to decrease. If the net electric charges of the two particles have opposite ($\pm$) signs, $dr < 0$ and the separation between the particles must decrease for the electric potential energy of the system to decrease. Hence, like electric charges repel each other, and opposite electric charges attract each other.

6 This follows from the principle of minimum potential energy.

Example 3-8

Problem: Two identical 20 g charged point particles have the same net electric charge and are suspended from the same point by massless ropes of the equal length L as shown in Figure 3.8. The entire system is in static equilibrium. What is the net electric charge on each particle if $L = 10$ cm and $\theta = 30°$?

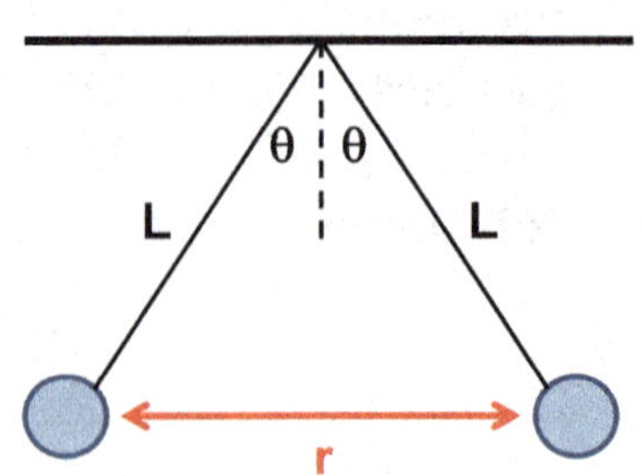

FIGURE 3.8: The system in Example 3-8.

Solution: At equilibrium, the total potential energy of the system must be a minimum. The total potential energy of the system is

$$U_{total} = U_e + U_g \quad \rightarrow \quad U_{total} = \frac{q^2}{4\pi\varepsilon_0 r} + 2mgy$$

The variable q in this equation denotes the net electric charge on each particle and the variable m denotes the mass of each particle. The horizontal separation (r) and vertical position (y) of the particles can be expressed in terms of the θ and L using trigonometry. Let's define the zero point of the gravitational potential energy of the system to be at the horizontal surface to which the strings are attached. Then, the total energy of the system can be written as

$$U_{total} = \frac{q^2}{4\pi\varepsilon_0 2L\sin\theta} + 2mg\left(-L\cos\theta\right)$$

We can then find the minimum of this function through differentiation

$$\frac{dU_{total}}{d\theta} = 0 \quad \rightarrow \quad \frac{-q^2\cos\theta}{4\pi\varepsilon_0 2L\sin^2\theta} + 2mg\left(L\sin\theta\right) = 0$$

Solving for q gives us

$$\frac{q^2\cos\theta}{4\pi\varepsilon_0 2L\sin^2\theta} = 2mg\left(L\sin\theta\right) \quad \rightarrow \quad q = \left(\frac{16\pi\varepsilon_0 mgL^2\sin^3\theta}{\cos\theta}\right)^{\frac{1}{2}}$$

Substitution then yields

$$q = \left(\frac{16\pi \left(8.85 \times 10^{-12} \frac{\mathrm{C}^2}{\mathrm{J\,m}} \right) (0.02\,\mathrm{kg}) \left(9.8 \frac{\mathrm{m}}{\mathrm{s}^2} \right) (0.1\,\mathrm{m})^2 \sin^3(30°)}{\cos 30°} \right)^{\frac{1}{2}}$$

$$q = \pm 354\,\mathrm{nC}$$

We cannot determine whether the net electric charge on each particle is positive or negative.

If the two charged point particles in Figure 3.1 were released from rest, they would move toward each other because their net electric charges have opposite (±) sign. What is the change in the electric potential energy of this system associated with a change in the separation distance of the charged point particles from 30 cm to 10 cm? We can determine the answer using Equation 3-4.

$$\Delta U_e = \left(\frac{q_A q_B}{4\pi\varepsilon_0} \right) \left(\frac{1}{r_f} - \frac{1}{r_i} \right) \rightarrow \Delta U_e = \left(\frac{(2 \times 10^{-9}\,\mathrm{C})(-4 \times 10^{-9}\,\mathrm{C})}{4\pi \left(8.85 \times 10^{-12} \frac{\mathrm{C}^2}{\mathrm{J\,m}} \right)} \right) \left(\frac{1}{10\,\mathrm{cm}} - \frac{1}{30\,\mathrm{cm}} \right)$$

$$\Delta U_e = -4.8 \times 10^{-9}\,\mathrm{J}$$

The electric potential has decreased, as expected, since the net electric charges of the two point particles have opposite sign and, thus, attract each other.

The equation for the change in the electric potential energy of this system can also be written as

$$\Delta U_e = q_B \left(\frac{q_A}{4\pi\varepsilon_0 r_f} - \frac{q_A}{4\pi\varepsilon_0 r_i} \right) \rightarrow \Delta U_e = q_B (\Delta V_e)_A$$

Or, equivalently, as

$$\Delta U_e = q_A\left(\frac{q_B}{4\pi\varepsilon_0 r_f} - \frac{q_B}{4\pi\varepsilon_0 r_i}\right) \quad \rightarrow \quad \Delta U_e = q_A\left(\Delta V\right)_B$$

Therefore, the change in the electric potential energy of this system can be expressed in terms of charged point particle B moving through the electric potential of particle A, or in terms of charged point particle A moving through the electric potential of particle B[7].

$$\Delta U_e = q\Delta V_e \tag{3-5}$$

Of course, this change in the electric potential energy refers to a change in the electric potential energy of the entire system consisting of both the point particle and the net electric charge(s) that is/are the source of the electric potential.

It follows from Equation 3-5 that positively charged particles will always move toward regions of lower electric potential so as to minimize their electric potential energy[8]. Similarly, negatively charged particles will always move toward regions of higher electric potential so as to minimize their electric potential energy.

3-6 Electric Fields

The electric force[9] between two charged point particles, A and B, with net electric charges q_A and q_B is found using Equation 1-1; because electric potential energy (Equation 3-4) is spherically symmetric we will use Equation B-7 for the gradient.

$$\vec{F}_e = -\nabla U_e \quad \rightarrow \quad \vec{F}_e = \left(-\frac{\partial U_e}{\partial r}\right)\hat{r} + \left(-\frac{1}{r}\frac{\partial U_e}{\partial \theta}\right)\hat{\theta} + \left(-\frac{1}{r\sin\theta}\frac{\partial U_e}{\partial \varphi}\right)\hat{\varphi}$$

$$\vec{F}_e = \left(-\frac{\partial}{\partial r}\left(\frac{q_A q_B}{4\pi\varepsilon_0 r}\right)\right)\hat{r} + \left(-\frac{1}{r}\frac{\partial}{\partial \theta}\left(\frac{q_A q_B}{4\pi\varepsilon_0 r}\right)\right)\hat{\theta} + \left(-\frac{1}{r\sin\theta}\frac{\partial}{\partial \varphi}\left(\frac{q_A q_B}{4\pi\varepsilon_0 r}\right)\right)\hat{\varphi}$$

---

7 Thanks to Galilean transformations, we cannot say which point particle is moving, or if both are moving. The mathematical description is equivalent, which is another reason why the change in the electric potential energy of the system can be written in terms of either description.

8 This is the principle of minimum potential energy.

9 This force is also known as the Coulomb force.

$$\vec{F}_e = \frac{q_A q_B}{4\pi\varepsilon_0 r^2}\hat{r} \qquad \textbf{(3-6)}$$

The unit vector $\hat{r}$ in Equation 3-6 points along an axis extending through the two particles and for each particle away from the other particle (see Figure 3.9). Therefore, if the two point particles have the same (±) signs for their net electric charges, the electric force will push the particles apart; but if they have opposite (±) signs for their net electric charges, the electric force will pull the particles together.

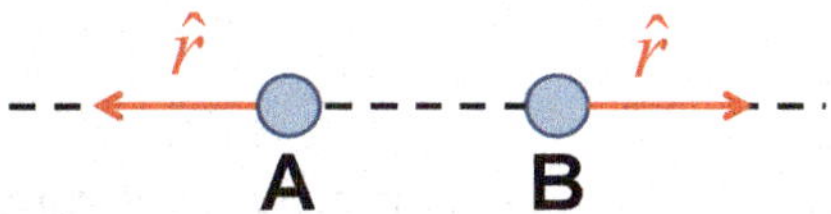

FIGURE 3.9: Definition for the unit vector in Equation 3-6.

Example 3-9

Problem: Determine the solution to Example 3-8 using a description of the forces acting on the particles.

Solution: Let's define the *x*-axis of the reference frame for this system to be parallel to the horizontal surface from which the particles are hanging in Figure 3.8. We can then define the *y*-axis to be perpendicular to the *x*-axis with a positive direction pointing up from the horizontal surface. In the free body diagram for each particle, we will denote the magnitude of the electric force as F_e, the magnitude of the tension as *T*, and the magnitude of the force of gravity as F_g. With these definitions, the free body diagram for the particle on the left side of Figure 3.8 is shown in Figure 3.10.

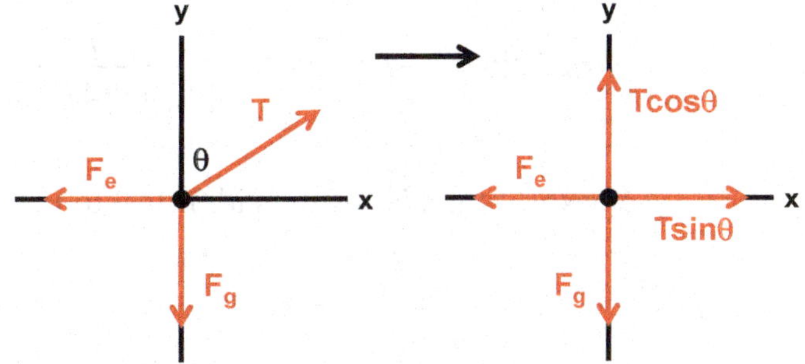

FIGURE 3.10: The free body diagram for the ball on the left side of Figure 3.8.

Applying Newton's 2nd law to each of the coordinate axes in this free body diagram yields the following equations:

$$\left(F_{net}\right)_x = ma_x \quad \rightarrow \quad T\sin\theta - F_e = ma_x \quad \rightarrow \quad T\sin\theta - \frac{q^2}{4\pi\varepsilon_0 r^2} = ma_x$$

$$\left(F_{net}\right)_y = ma_y \quad \rightarrow \quad T\cos\theta - F_g = ma_y \quad \rightarrow \quad T\cos\theta - mg = ma_y$$

Since the ball is in static equilibrium, the components of its acceleration along the x and y axes will both be zero. Substitution of $a_x = 0$ and $a_y = 0$ into the equations above yields

$$a_x = 0 \quad \rightarrow \quad T\sin\theta = \frac{q^2}{4\pi\varepsilon_0 r^2}$$

$$a_y = 0 \quad \rightarrow \quad T\cos\theta = mg$$

We can now algebraically eliminate T from these equations.

$$\frac{T\sin\theta}{T\cos\theta} = \frac{\frac{q^2}{4\pi\varepsilon_0 r^2}}{mg} \quad \rightarrow \quad \tan\theta = \frac{q^2}{4\pi\varepsilon_0 mgr^2}$$

Hence,

$$r = 2L\sin\theta \quad \rightarrow \quad \tan\theta = \frac{q^2}{4\pi\varepsilon_0 mg 4L^2 \sin^2\theta} \quad \rightarrow \quad q = \left(\frac{16\pi\varepsilon_0 L^2 mg \sin^3\theta}{\cos\theta}\right)^{\frac{1}{2}}$$

As expected, this result is identical to the solution derived before.

The electric force in Equation 3-6 can also be written in terms of the electric potential.

$$\vec{F}_e = -\nabla U_e \quad \rightarrow \quad \vec{F}_e = -\nabla(qV_e) \quad \rightarrow \quad \vec{F}_e = -q\nabla V_e \tag{3-7}$$

The variable q in Equation 3-7 is the net electric charge of the particle that is the object of the electric force (*i.e.*, the particle that experiences the electric potential). The gradient of the electric potential in Equation 3-7 is referred to as the ***electric field***.

Electric field: an intrinsic property of electric charge that describes the alteration of space associated with electric charge. The electric field is a vector field obtained from the gradient of the electric potential.

The electric field, denoted by $\vec{E}$, can be expressed[10] as

$$\vec{E} = -\nabla V_e \tag{3-8}$$

For example, the electric field surrounding a point particle with a net electric charge q can be found through the substitution of Equation 3-1 into Equation 3-8; because electric potential (Equation 3-1) is spherically symmetric we will use Equation B-7 for the gradient.

$$\vec{E} = -\nabla V_e \quad \rightarrow \quad \vec{E} = \left(-\frac{\partial V_e}{\partial r}\right)\hat{r} + \left(-\frac{1}{r}\frac{\partial V_e}{\partial \theta}\right)\hat{\theta} + \left(-\frac{1}{r\sin\theta}\frac{\partial V_e}{\partial \varphi}\right)\hat{\varphi}$$

$$\vec{E} = \left(-\frac{\partial}{\partial r}\left(\frac{q}{4\pi\varepsilon_0 r}\right)\right)\hat{r} + \left(-\frac{1}{r}\frac{\partial}{\partial \theta}\left(\frac{q}{4\pi\varepsilon_0 r}\right)\right)\hat{\theta} + \left(-\frac{1}{r\sin\theta}\frac{\partial}{\partial \varphi}\left(\frac{q}{4\pi\varepsilon_0 r}\right)\right)\hat{\varphi}$$

$$\vec{E} = \left(\frac{q}{4\pi\varepsilon_0 r^2}\right)\hat{r} \tag{3-9}$$

As shown in Equation 3-9, the magnitude of the electric field for a charged point particle is inversely proportional to the square of the distance away from the particle; the units of the magnitude of the electric field are typically expressed as N/C[11] or V/m. Furthermore, since the unit vector $\hat{r}$ in Equation 3-9 points away from the point particle, the electric field always points away from positive electric charges and toward negative electric charges.

The electric fields for two charged point particles are shown in Figure 3.11. In this representation the direction of the electric field at several positions is indicated by the direction of an arrow and the magnitude of the electric field by the length of the arrow. As shown in Figure 3.11, and as follows from Equation 3-8, the direction of the electric field at any position is always

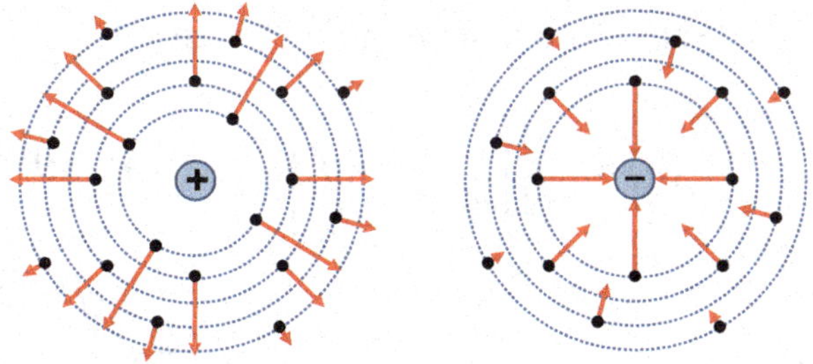

FIGURE 3.11: The electric field at various positions (black circles) around a positively charged point particle (left) and a negatively charged point particle (right). All positions on the same dashed line are at the same electric potential.

10 We will learn a modified equation for the electric field in Chapter 6 that is more universally valid.
11 Thus, the magnitude of the electric field is measure of the electric force per unit charge.

perpendicular to the equipotential surface associated with that position and points from high electric potential to low electric potential. It follows that the electric field always points toward objects with net negative electric charge and away from objects with net positive electric charge.

Since the electric field is determined from the gradient of the electric potential, many electric potentials can be associated with the same electric field. Indeed, any constant scalar field can be added to or subtracted from the electric potential without changing the associated electric field[12]. Therefore, our definition of the electric potential defined in Equation 3-1 represents only one possible electric potential.

To explore this further, let's rewrite the relationship between electric field and electric potential in Equation 3-8 as Equation 3-10[13].

$$\Delta V_e = -\int \vec{E} \bullet d\vec{s} \tag{3-10}$$

We can now use Equation 3-10 to define the electric potential of a point particle. Specifically, we define the electric potential of a point particle to be zero when measured infinitely far away from the point particle.

$$-\int_{\infty}^{r} \vec{E} \bullet d\vec{s} = V_e(r) - V_e(\infty) \quad \rightarrow \quad -\int_{\infty}^{r} \vec{E} \bullet d\vec{s} = V_e(r) - 0$$

$$V_e(r) = -\int_{\infty}^{r} \vec{E} \bullet d\vec{s}$$

Substitution of Equation 3-9 into this expression gives us[14]

$$V_e(r) = -\int_{\infty}^{r} \left[\left(\frac{q}{4\pi\varepsilon_0 r^2}\right)\hat{r}\right] \bullet \left[(dr)\hat{r} + (rd\theta)\hat{\theta} + (r\sin\theta d\varphi)\hat{\varphi}\right] \rightarrow V_e(r) = -\left(\frac{q}{4\pi\varepsilon_0}\right)\int_{\infty}^{r}\left(\frac{dr}{r^2}\right)$$

$$V_e(r) = -\left(\frac{q}{4\pi\varepsilon_0}\right)\left(-\frac{1}{r}\Big|_{\infty}^{r}\right) \quad \rightarrow \quad V_e(r) = \left(\frac{q}{4\pi\varepsilon_0 r}\right)$$

As anticipated, this result is identical to Equation 3-1.

---

12 See Section 2-5 for the similar discussion with gravitational potential and gravitational field.

13 The derivation of Equation 3-10 is similar to the derivation of Equation 2-11.

14 Since Equation 3-9 is spherically symmetric we will use spherical coordinates to define the differential displacement (Section B-5).

Example 3-10

Problem: What is the average magnitude of the electric field between the two equipotential surfaces of the electric potential in Figure 3.12?

Solution: Although the magnitude of the electric field at all positions in-between the two equipotential surfaces of the electric potential cannot be determined from the information given in the problem, we can nevertheless determine an average magnitude for the electric field across the entire separation distance between the two surfaces using Equation 3-10.

$$\Delta V_e = -\int \vec{E} \bullet d\vec{s} \quad \rightarrow \quad \Delta V_e = -\left(\vec{E}_{avg}\right) \bullet \left(\int d\vec{s}\right) \quad \rightarrow \quad \Delta V_e = -\left(\vec{E}_{avg}\right) \bullet \left(\Delta \vec{s}\right)$$

Let's define the origin of the s-axis to be at the 200 V equipotential surface. The positive direction for the s-axis will point toward the 100 V equipotential surface. Let's further assume that the direction of the electric field is the same as the direction of the s-axis. Then,

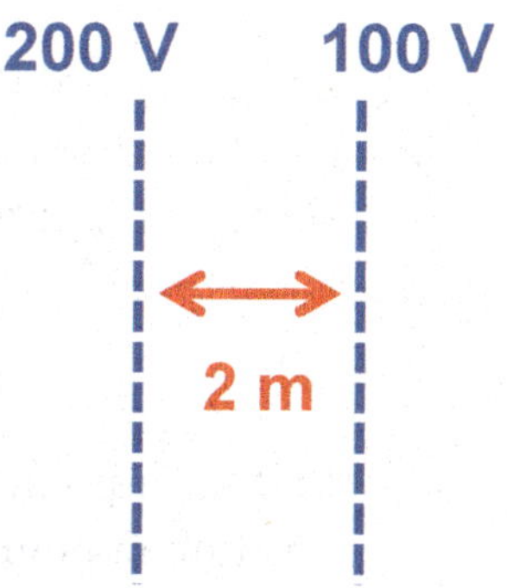

FIGURE 3.12: The electric field between two equipotential surfaces in Example 3-10.

$$\left(\vec{E}_{avg}\right) \bullet \left(\Delta \vec{s}\right) = E_{avg} \Delta s \quad \rightarrow \quad E_{avg} = -\frac{\Delta V_e}{\Delta s}$$

Substitution gives us

$$E_{avg} = -\frac{100\,\text{V} - 200\,\text{V}}{2\,\text{m}} \quad \rightarrow \quad E_{avg} = 50\frac{\text{V}}{\text{m}}$$

The positive direction in our result indicates that the direction of the electric field is indeed the same as the direction of the s-axis. Therefore, the electric field points from high electric potential to low electric potential, as expected.

The electric force acting on a point particle is equal to the product of the particle's net electric charge and the electric field experienced by the particle.

$$\vec{F}_e = q\vec{E} \qquad \textbf{(3-11)}$$

Example 3-11

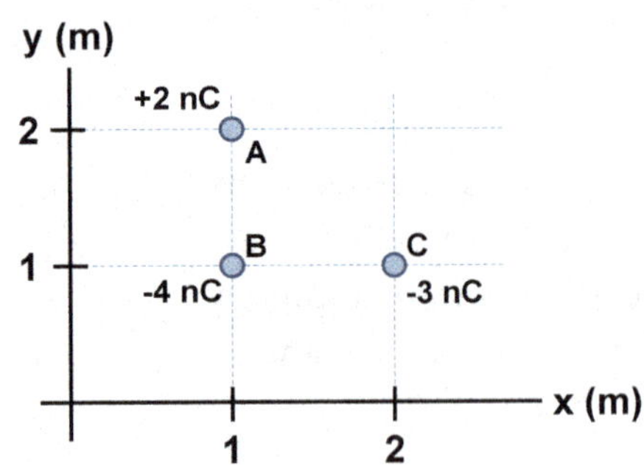

FIGURE 3.13: The system of three charged point particles in Example 3-11.

Problem: Three charged point particles, A, B, and C, with net electric charges +2 nC, −4 nC, and −3 nC, respectively, are arranged as shown in Figure 3.13. What is the electric field experienced by point particle C? What is the electric force acting on this particle?

Solution: The net electric field experienced by the point particle C is the sum of the electric fields from each of the other point particles at that position. Furthermore, from Equation 3-9 we know that the electric field for a negatively charged point particle points toward the point particle, and the electric field for a positively charged point particle points away from the point particle. Thus, for the reference frame shown in Figure 3.13, the electric field experience by point particle C is

$$\vec{E}_{net,C} = \vec{E}_{A,C} + \vec{E}_{B,C}$$

In this equation, $\vec{E}_{A,C}$ is the electric field from point particle A at the position of point particle C, $\vec{E}_{B,C}$ is the electric field from point particle B at the position of point particle C, and $\vec{E}_{net,C}$ is the net electric field at the position of point particle C. Using Equation 3-9 gives us

$$\vec{E}_{A,C} = \left(\frac{2 \times 10^{-9}\,\mathrm{C}}{4\pi\left(8.85 \times 10^{-12}\,\dfrac{\mathrm{C}^2}{\mathrm{J\,m}}\right)\left((1\,\mathrm{m})^2 + (1\,\mathrm{m})^2\right)} \right)\left((\cos 45°)\hat{x} - (\sin 45°)\hat{y}\right)$$

$$\vec{E}_{A,C} = \left(6.36\frac{\mathrm{N}}{\mathrm{C}}\right)\hat{x} + \left(-6.36\frac{\mathrm{N}}{\mathrm{C}}\right)\hat{y}$$

$$\vec{E}_{B,C} = \left(\frac{-4 \times 10^{-9}\,\mathrm{C}}{4\pi\left(8.85 \times 10^{-12}\,\dfrac{\mathrm{C}^2}{\mathrm{J\,m}}\right)(1\,\mathrm{m})^2} \right)\hat{x} \;\rightarrow\; \vec{E}_{B,C} = \left(-35.97\frac{\mathrm{N}}{\mathrm{C}}\right)\hat{x}$$

Thus, the net electric field experienced by point particle C is

$$\vec{E}_{net,C} = \left(6.36\frac{\text{N}}{\text{C}}\right)\hat{x} + \left(-6.36\frac{\text{N}}{\text{C}}\right)\hat{y} + \left(-35.97\frac{\text{N}}{\text{C}}\right)\hat{x}$$

$$\vec{E}_{net,C} = \left(-29.61\frac{\text{N}}{\text{C}}\right)\hat{x} + \left(-6.36\frac{\text{N}}{\text{C}}\right)\hat{y}$$

The electric force acting on point particle C can then be determined using Equation 3-11.

$$\vec{F}_e = \left(-3\times10^{-9}\text{C}\right)\left(\left(-29.61\frac{\text{N}}{\text{C}}\right)\hat{x} + \left(-6.36\frac{\text{N}}{\text{C}}\right)\hat{y}\right)$$

$$\vec{F}_e = \left(8.88\times10^{-8}\text{N}\right)\hat{x} + \left(1.91\times10^{-8}\text{N}\right)\hat{y}$$

3-7 The Electric Field for Continuous Distributions of Charge

The electric field for a system consisting of a continuous distribution of net electric charge is determined from the electric field of each infinitesimal bit of net electric charge that constitutes the system. *Each infinitesimal bit of net electric charge can be modeled as a charged point particle whose electric field is calculated using Equation 3-9.* Specifically, if the net electric charge of one of these infinitesimal "bits" is dq, then the electric field associated with this bit is

$$d\vec{E} = \left(\frac{dq}{4\pi\varepsilon_0 r^2}\right)\hat{r}$$

The variable r in this equation is the distance between dq and the location where the electric field is measured. The total electric field from the distribution of net electric charge is thus determined by summing the electric field of each infinitesimal bit of net electric charge; since these bits of net electric charge are so small, this summation is an integral.

$$\vec{E} = \int d\vec{E} \quad \rightarrow \quad \vec{E} = \left(\frac{1}{4\pi\varepsilon_0}\right)\int \frac{dq}{r^2}\hat{r} \tag{3-12}$$

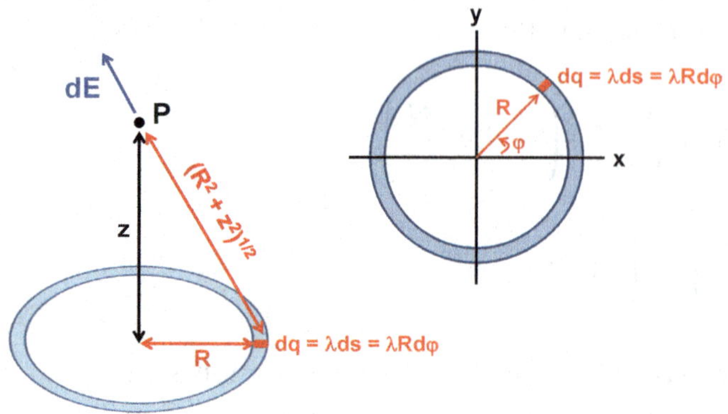

FIGURE 3.14: Determining the electric field from a uniformly charged ring. The electric field will be directed along the *z*-axis only.

Let's revisit the system shown in Figure 3.6 that consists of a thin and uniformly charged circular ring of radius R and net electric charge Q. We can now use Equation 3-12 to determine the electric field at the position P, a distance z away from the center of the ring along the normal axis through the center of the ring. Since the ring is uniformly charged, we can divide the ring into infinitesimal bits of net electric charge, dq, and determine the ring's total electric field from the sum of the electric field from each bit of net electric charge using Equation 3-12. Furthermore, as shown in Figure 3.14, we can express these infinitesimal bits of net electric charge in terms of the linear net electric charge density of the ring (Table 3-1) and an infinitesimal distance (*i.e.*, an infinitesimal arc length) along the ring. To simplify our calculation, let's use cylindrical coordinates[15] as a reference frame for this system. In this reference frame, the infinitesimal arc length can be expressed in terms of an infinitesimal change in the angle φ (Figure 3.14). The electric field vector can be decomposed into its r-axis and z-axis components (in terms of R and z) using trigonometry.

$$\vec{E}=\left(\frac{1}{4\pi\varepsilon_0}\right)\int\left(\frac{\lambda R d\varphi}{\left(R^2+z^2\right)^2}\left(\left(-\frac{R}{\left(R^2+z^2\right)^{\frac{1}{2}}}\right)\hat{\rho}+\left(\frac{z}{\left(R^2+z^2\right)^{\frac{1}{2}}}\right)\hat{z}\right)\right)$$

$$\vec{E}=\left(\frac{1}{4\pi\varepsilon_0}\right)\int\left(\frac{\lambda R d\varphi}{\left(R^2+z^2\right)^{\frac{3}{2}}}\left(\left(-R\right)\hat{\rho}+\left(z\right)\hat{z}\right)\right)$$

As noted in Section B-5, $\hat{\rho}$ points out from the origin of the reference frame. Thus, the location of the position P along the ρ-axis is "$-R$" with respect to each infinitesimal arc length. The integral in this equation is with respect to φ. Since the direction of $\hat{\rho}$ changes as φ changes[16], it is best to rewrite this integral in terms of its x-axis and y-axis components (Figure 3.14).

15 See Section B-5.
16 See Section B-5.

$$\hat{\rho} = (\cos\varphi)\hat{x} + (\sin\varphi)\hat{y}$$

$$\vec{E} = \left(\frac{1}{4\pi\varepsilon_0}\right)\int\left(\frac{\lambda R d\varphi}{\left(R^2+z^2\right)^{\frac{3}{2}}}\left((-R)\left((\cos\varphi)\hat{x}+(\sin\varphi)\hat{y}\right)+(z)\hat{z}\right)\right)$$

Let's determine the integral for the component of the electric field along each coordinate axis separately. For the *x*-axis component, we have

$$E_x = -\frac{1}{4\pi\varepsilon_0}\int\frac{\lambda R^2\cos\varphi}{\left(R^2+z^2\right)^{\frac{3}{2}}}d\varphi \;\rightarrow\; E_x = -\frac{\lambda R^2}{4\pi\varepsilon_0\left(R^2+z^2\right)^{\frac{3}{2}}}\int_0^{2\pi}\cos\varphi\, d\varphi$$

$$E_x = -\frac{\lambda R^2}{4\pi\varepsilon_0\left(R^2+z^2\right)^{\frac{3}{2}}}\left(\sin\varphi\Big|_0^{2\pi}\right) \;\rightarrow\; E_x = 0$$

Similarly,

$$E_y = -\frac{\lambda R^2}{4\pi\varepsilon_0\left(R^2+z^2\right)^{\frac{3}{2}}}\int_0^{2\pi}\sin\varphi\, d\varphi \;\rightarrow\; E_y = 0$$

This result makes sense intuitively from the symmetry of the ring. The *x*-axis and *y*-axis components of the electric field from any bit of net electric charge will always be exactly cancelled by the *x*-axis and *y*-axis components of the electric field from a diametrically opposite bit of net electric charge. Thus, the net electric field for the entire ring will have only a component directed perpendicular to the plane of ring (*i.e.*, parallel to the *z*-axis).

$$\vec{E} = \left(\frac{1}{4\pi\varepsilon_0}\int_0^{2\pi}\frac{\lambda zR}{\left(R^2+z^2\right)^{\frac{3}{2}}}d\varphi\right)\hat{z} \;\rightarrow\; \vec{E} = \left(\frac{\lambda zR}{4\pi\varepsilon_0\left(R^2+z^2\right)^{\frac{3}{2}}}\int_0^{2\pi}d\varphi\right)\hat{z}$$

$$\vec{E} = \left(\frac{\lambda zR}{4\pi\varepsilon_0\left(R^2+z^2\right)^{\frac{3}{2}}}(2\pi)\right)\hat{z} \;\rightarrow\; \vec{E} = \left(\frac{\lambda zR}{2\varepsilon_0\left(R^2+z^2\right)^{\frac{3}{2}}}\right)\hat{z}$$

Substituting for the linear net electric charge density gives us

$$\lambda = \frac{Q}{2\pi R} \rightarrow \vec{E} = \left(\frac{Qz}{4\pi\varepsilon_0 \left(R^2 + z^2\right)^{\frac{3}{2}}} \right) \hat{z}$$

Of course, we could have also arrived at this result from the electric potential of the ring that we derived earlier. Since the electric potential of the ring is expressed in terms of rectangular coordinates (Example 3-6) we will use Equation B-5 to determine the electric field.

$$\vec{E} = -\nabla V_e \rightarrow \vec{E} = -\left[\left(\frac{\partial}{\partial x} V_e \right) \hat{x} + \left(\frac{\partial}{\partial y} V_e \right) \hat{y} + \left(\frac{\partial}{\partial z} V_e \right) \hat{z} \right]$$

$$\frac{\partial}{\partial x} V_e = \frac{\partial}{\partial y} V_e = 0 \rightarrow \vec{E} = \left(-\frac{\partial}{\partial z} \left(\frac{Q}{2\pi\varepsilon_0 R^2} \left(\left(R^2 + z^2\right)^{\frac{1}{2}} - |z| \right) \right) \right) \hat{z}$$

Example 3-12

Problem: What are the electric potential and electric field for a uniformly charged disk of radius R and net electric charge Q at a position P, a distance z away from the center of the disk along the perpendicular bisector of the disk?

Solution: As shown in Figure 3.15, we can divide the disk into a series of infinitesimal narrow rings[17]. The electric potential of the disk is then the sum of the electric potential from each of these rings.

From Example 3-6 we have

$$V_e = \int dV_e \rightarrow V_e = \int \frac{dq}{4\pi\varepsilon_0 \left(r^2 + z^2\right)^{\frac{1}{2}}}$$

The variable r in this equation is the radius of an infinitesimal ring, as shown in Figure 3.15, and dq is the infinitesimal net electric charge associated with this

17 See Section 1-3.

ring. This charge can be related with the infinitesimal area of the ring (dA) using the area net electric charge density of the ring (Table 3-1).

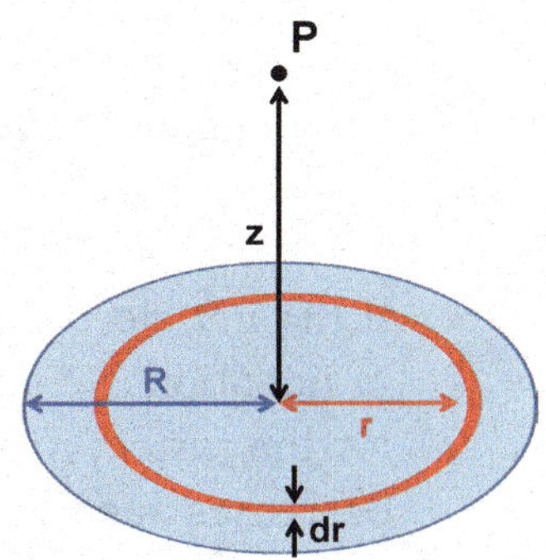

FIGURE 3.15: The uniformly charged disk in Example 3-12.

$$dq = \sigma dA \quad \rightarrow \quad dq = \sigma\left(2\pi r\, dr\right)$$

In this equation, the variable dr is the width of the ring, as shown in Figure 3.15.

$$V_e = \int \frac{\sigma\left(2\pi r\, dr\right)}{4\pi\varepsilon_0\left(r^2+z^2\right)^{\frac{1}{2}}} \quad \rightarrow \quad V_e = \frac{\sigma}{2\varepsilon_0}\int_0^R \frac{r\, dr}{\left(r^2+z^2\right)^{\frac{1}{2}}}$$

$$V_e = \frac{\sigma}{2\varepsilon_0}\left(r^2+z^2\right)^{\frac{1}{2}}\Bigg|_0^R \quad \rightarrow V_e = \frac{\sigma}{2\varepsilon_0}\left(\left(R^2+z^2\right)^{\frac{1}{2}} - |z|\right)$$

Substitution then gives us

$$\sigma = \frac{Q}{\pi R^2} \quad \rightarrow \quad V_e = \frac{Q}{2\pi\varepsilon_0 R^2}\left(\left(R^2+z^2\right)^{\frac{1}{2}} - |z|\right)$$

The electric field at the position P can then be determined using Equation 3-8. Since the electric potential is expressed in terms of rectangular coordinates we will use Equation B-5 to determine the electric field[18].

$$\vec{E} = -\nabla V_e \quad \rightarrow \quad \vec{E} = -\left[\left(\frac{\partial}{\partial x}V_e\right)\hat{x} + \left(\frac{\partial}{\partial y}V_e\right)\hat{y} + \left(\frac{\partial}{\partial z}V_e\right)\hat{z}\right]$$

$$\frac{\partial}{\partial x}V_e = \frac{\partial}{\partial y}V_e = 0 \quad \rightarrow \quad \vec{E} = \left(-\frac{\partial}{\partial z}\left(\frac{Q}{2\pi\varepsilon_0 R^2}\left(\left(R^2+z^2\right)^{\frac{1}{2}} - |z|\right)\right)\right)\hat{z}$$

$$\vec{E} = \frac{Q}{2\pi\varepsilon_0 R^2}\left(1 - \frac{z}{\left(R^2+z^2\right)^{\frac{1}{2}}}\right)\hat{z}$$

18 The same result can also be obtained assuming cylindrical symmetry and using Equation B-6.

The same expression for the electric field can also be derived using Equation 3-12.

If the net electric charge density is not uniform it must be included in the integrals in Equation 3-2 and Equation 3-12. For example, let's calculate the electric potential for the system in Example 3-12 when the area net electric charge density of the disk decreases linearly with distance from the center of the ring according to the following equation[19]:

$$\sigma = \sigma_0 - \gamma r$$

The variable r in this equation is the radial distance from the center of the disk, the variable σ_0 is the area net electric charge density when $r = 0$, and the variable γ is the rate of change of the area net electric charge density with respect to radial distance from the center of the disk. Following the solution of Example 3-12 and Example 1-2 we have

$$V_e = \int dV_e \quad \rightarrow \quad V_e = \int \frac{dq}{4\pi\varepsilon_0 \left(r^2 + z^2\right)^{\frac{1}{2}}}$$

$$dq = \sigma dA \quad \rightarrow \quad dq = \left(\sigma_0 - \gamma r\right)\left(2\pi r\, dr\right) \quad \rightarrow \quad dq = 2\pi\left(\sigma_0 r - \gamma r^2\right) dr$$

$$V_e = \int \frac{2\pi\left(\sigma_0 r - \gamma r^2\right) dr}{4\pi\varepsilon_0 \left(r^2 + z^2\right)^{\frac{1}{2}}} \quad \rightarrow \quad V_e = \frac{1}{2\varepsilon_0} \int_0^R \frac{\left(\sigma_0 r - \gamma r^2\right) dr}{\left(r^2 + z^2\right)^{\frac{1}{2}}}$$

$$V_e = \frac{1}{2\varepsilon_0}\left(\left(\sigma_0 - \frac{\gamma r}{2}\right)\left(r^2 + z^2\right)^{\frac{1}{2}} + \frac{1}{2}\gamma z^2 \ln\left(r + \left(r^2 + z^2\right)^{\frac{1}{2}}\right)\right)\Bigg|_0^R$$

$$V_e = \frac{1}{2\varepsilon_0}\left(\left(\left(\sigma_0 - \frac{\gamma R}{2}\right)\left(R^2 + z^2\right)^{\frac{1}{2}} + \frac{1}{2}\gamma z^2 \ln\left(R + \left(R^2 + z^2\right)^{\frac{1}{2}}\right)\right) - \left(\sigma_0 |z| + \frac{1}{2}\gamma z^2 \ln(z)\right)\right)$$

19 See Example 1-2.

$$V_e = \frac{1}{2\varepsilon_0}\left(\left(\sigma_0 - \frac{\gamma R}{2}\right)\left(R^2+z^2\right)^{\frac{1}{2}} - \sigma_0|z| + \frac{1}{2}\gamma z^2 \ln\left(\frac{R+\left(R^2+z^2\right)^{\frac{1}{2}}}{z}\right)\right)$$

This expression does not depend explicitly on the total net electric charge of the rod. In order to rewrite our result in terms of this net electric charge, let's determine an expression for σ_0 in terms of γ and the total net electric charge of the rod, Q using integration[20].

$$Q = \int dq \;\rightarrow\; Q = \int_0^R 2\pi\left(\sigma_0 r - \gamma r^2\right)dr \;\rightarrow\; Q = 2\pi\left(\left(\frac{\sigma_0 r^2}{2} - \frac{\gamma r^3}{3}\right)\Bigg|_0^R\right)$$

$$Q = 2\pi\left(\frac{\sigma_0 R^2}{2} - \frac{\gamma R^3}{3}\right) \;\rightarrow\; Q = \frac{\pi R^2}{3}\left(3\sigma_0 - 2\gamma R\right) \;\rightarrow\; \sigma_0 = \frac{Q}{\pi R^2} + \frac{2\gamma R}{3}$$

Substitution then gives us

$$V_e = \frac{1}{2\varepsilon_0}\left(\left(\frac{Q}{\pi R^2} + \frac{2\gamma R}{3} - \frac{\gamma R}{2}\right)\left(R^2+z^2\right)^{\frac{1}{2}} - \left(\frac{Q}{\pi R^2} + \frac{2\gamma R}{3}\right)|z| + \frac{1}{2}\gamma z^2 \ln\left(\frac{R+\left(R^2+z^2\right)^{\frac{1}{2}}}{z}\right)\right)$$

$$V_e = \frac{Q}{2\pi\varepsilon_0 R^2}\left(\left(R^2+z^2\right)^{\frac{1}{2}} - |z|\right) + \frac{\gamma}{2\varepsilon_0}\left(\frac{1}{2}z^2 \ln\left(\frac{R+\left(R^2+z^2\right)^{\frac{1}{2}}}{z}\right) - \frac{R}{6}\left(R^2+z^2\right)^{\frac{1}{2}} - \frac{2R}{3}|z|\right)$$

Note that when the area net electric charge density of the disk is constant (*i.e.*, when $\gamma = 0$) we have

$$\gamma = 0 \;\rightarrow\; V_e = \frac{Q}{2\pi\varepsilon_0 R^2}\left(\left(R^2+z^2\right)^{\frac{1}{2}} - |z|\right)$$

As expected.

20 See Section 1-3 and Section 2-3.

3-8 Electric Dipoles

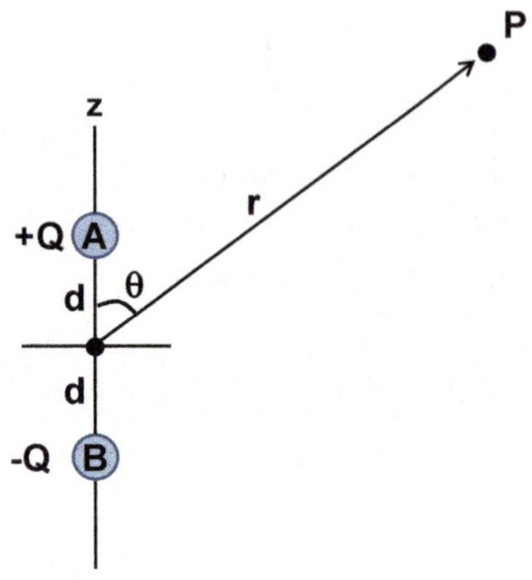

FIGURE 3.16: An electric dipole.

A system consisting of two net electric charges of opposite (±) sign separated by a small distance is called an ***electric dipole;*** the prefix *di* in dipole refers to the fact that this system has two objects[21] with different net electric charges[22].

Electric dipole: two net electric charges of opposite (±) sign separated by a small distance.

Consider the electric dipole shown in Figure 3.16 that consists of two charged point particles, A and B, separated by a distance 2*d*, and centered around the origin of a spherical coordinate system.

The net electric charge of particle A is +Q, and the net electric charge of particle B is -Q. The electric potential of this system at the position P in Figure 3.16 is

$$V_e = \frac{q_A}{4\pi\varepsilon_0 r_A} + \frac{q_B}{4\pi\varepsilon_0 r_B} \quad \rightarrow \quad V_e = \frac{1}{4\pi\varepsilon_0}\left(\frac{q_A}{r_A} + \frac{q_B}{r_B}\right)$$

In this equation r_A and r_B are the distances from the point particles A and B, respectively, to the position P, as shown in Figure 3.17.

Substitution of the net electric charge of the point particles gives us

$$V_e = \frac{1}{4\pi\varepsilon_0}\left(\frac{Q}{r_A} + \frac{-Q}{r_B}\right) \quad \rightarrow \quad V_e = \frac{Q}{4\pi\varepsilon_0}\left(\frac{1}{r_A} - \frac{1}{r_B}\right)$$

Our next step is to relate the variables r_A and r_B to the variable *r*, which denotes the location of the position *P* relative to the origin of the reference frame in Figure 3.16. The associated trigonometry[23] is shown in Figure 3.17.

21 For our calculation, we will assume that the objects are point particles, but similar results would be obtained with larger objects as long as the electric potential and electric field of the dipole are measured at a distance much larger than the distance separating the objects. See Section 3-10 for expressions for these quantitates in different limits.

22 This is something new. There is no gravitational dipole since there is only one type (or variety) of mass.

23 The law of cosines

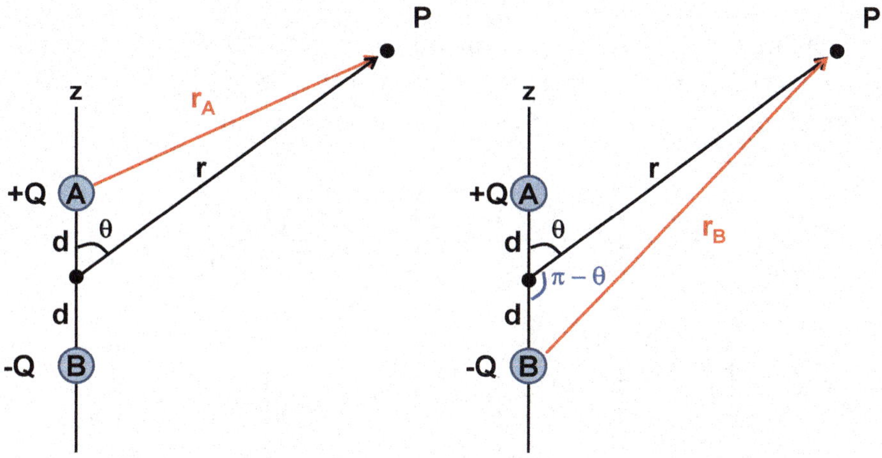

FIGURE 3.17: Determining the electric potential from the electric dipole in Figure 3.16.

$$r_A^2 = r^2 + d^2 - 2rd\cos\theta \quad \rightarrow \quad r_A = \left(r^2 + d^2 - 2rd\cos\theta\right)^{\frac{1}{2}}$$

$$r_A = r\left(1 + \left(\frac{d}{r}\right)^2 - 2\left(\frac{d}{r}\right)\cos\theta\right)^{\frac{1}{2}}$$

Similarly

$$r_B^2 = r^2 + d^2 - 2rd\cos\left(\pi - \theta\right) \quad \rightarrow \quad r_B = \left(r^2 + d^2 + 2rd\cos\theta\right)^{\frac{1}{2}}$$

$$r_B = r\left(1 + \left(\frac{d}{r}\right)^2 + 2\left(\frac{d}{r}\right)\cos\theta\right)^{\frac{1}{2}}$$

Substitution then gives us

$$V_e = \frac{Q}{4\pi\varepsilon_0 r}\left(\frac{1}{\left(1 + \left(\frac{d}{r}\right)^2 - 2\left(\frac{d}{r}\right)\cos\theta\right)^{\frac{1}{2}}} - \frac{1}{\left(1 + \left(\frac{d}{r}\right)^2 + 2\left(\frac{d}{r}\right)\cos\theta\right)^{\frac{1}{2}}}\right)$$

When r is much larger than d (*i.e.*, when the distance to the position where the electric potential is measured is much larger than the separation distance of the two charges) we can use the binomial expansion[24] to simplify this expression.

$$r >> d \;\rightarrow\; \left(1+\left(\frac{d}{r}\right)^2-2\left(\frac{d}{r}\right)\cos\theta\right)^{-\frac{1}{2}} \approx 1-\frac{1}{2}\left(\left(\frac{d}{r}\right)^2-2\left(\frac{d}{r}\right)\cos\theta\right)$$

$$r >> d \;\rightarrow\; \left(1+\left(\frac{d}{r}\right)^2+2\left(\frac{d}{r}\right)\cos\theta\right)^{-\frac{1}{2}} \approx 1-\frac{1}{2}\left(\left(\frac{d}{r}\right)^2+2\left(\frac{d}{r}\right)\cos\theta\right)$$

Hence, in this limit

$$V_e \approx \frac{Q}{4\pi\varepsilon_0 r}\left(\left(1-\frac{1}{2}\left(\left(\frac{d}{r}\right)^2-2\left(\frac{d}{r}\right)\cos\theta\right)\right)-\left(1-\frac{1}{2}\left(\left(\frac{d}{r}\right)^2+2\left(\frac{d}{r}\right)\cos\theta\right)\right)\right)$$

$$V_e \approx \frac{Q}{4\pi\varepsilon_0 r}\left(1-\frac{1}{2}\left(\frac{d}{r}\right)^2+\left(\frac{d}{r}\right)\cos\theta-1+\frac{1}{2}\left(\frac{d}{r}\right)^2+\left(\frac{d}{r}\right)\cos\theta\right)$$

$$V_e \approx \frac{Q}{4\pi\varepsilon_0 r}\left(2\frac{d}{r}\right)\cos\theta \;\rightarrow\; V_e \approx \frac{(Qd)}{4\pi\varepsilon_0 r^2}(2\cos\theta)$$

It is common to define the *electric dipole moment* for an electric dipole as a vector whose magnitude is equal to the product of the separation distance between the two net electric charges in the electric dipole and the magnitude of net electric charges, and whose direction points *from* the net negative electric charge *to* the net positive electric charge. For the electric dipole in Figure 3.16, the electric dipole moment, denoted[25] by $\vec{\mu}_e$, is

$$\vec{\mu}_e = \left(Q(2d)\right)\hat{z} \quad \textbf{(3-13)}$$

24 See Section A-3.

25 Most textbooks use the variable $\vec{p}$ for the electric dipole moment. I chose a different variable to avoid confusion with the variable for linear momentum.

With this definition we can also express the electric potential of the electric dipole as

$$V_e \approx \frac{2\vec{\mu}_e \cdot \hat{r}}{4\pi\varepsilon_0 r^2} \quad \rightarrow \quad V_e \approx \frac{\vec{\mu}_e \cdot \hat{r}}{4\pi\varepsilon_0 r^2} \tag{3-14}$$

Example 3-13

Problem: What is the electric field for the electric dipole in Figure 3.16 at the position *P*?

Solution: The electric field can be found from the electric potential of the electric dipole using Equation 3-8; because the electric potential is spherically symmetric we will use Equation B-7 for the gradient.

$$\vec{E} = -\nabla V_e \quad \rightarrow \quad \vec{E} = \left(-\frac{\partial V_e}{\partial r}\right)\hat{r} + \left(-\frac{1}{r}\frac{\partial V_e}{\partial \theta}\right)\hat{\theta} + \left(-\frac{1}{r\sin\theta}\frac{\partial V_e}{\partial \varphi}\right)\hat{\varphi}$$

$$\frac{\partial V_e}{\partial \varphi} \quad \rightarrow \quad \vec{E} = \left(-\frac{\partial}{\partial r}\left(\frac{(Qd)}{4\pi\varepsilon_0 r^2}(2\cos\theta)\right)\right)\hat{r} + \left(-\frac{1}{r}\frac{\partial}{\partial \theta}\left(\frac{(Qd)}{4\pi\varepsilon_0 r^2}(2\cos\theta)\right)\right)\hat{\theta}$$

$$\vec{E} = \left(\frac{Qd}{\pi\varepsilon_0 r^3}\cos\theta\right)\hat{r} + \left(\frac{Qd}{2\pi\varepsilon_0 r^3}\sin\theta\right)\hat{\theta}$$

The electric field is symmetric around the *z*-axis since it does not depend upon φ.

When an electric dipole is placed in a uniform electric field, the resulting electric forces exerted on the net positive and net negative electric charges of the electric dipole will pull on the electric dipole in opposite directions. As shown in Figure 3.18, this will result in a net torque acting on the electric dipole that will cause the electric dipole to rotate around its center of mass, which, for simplicity, we will assume is exactly half-way between the two net electric charges.

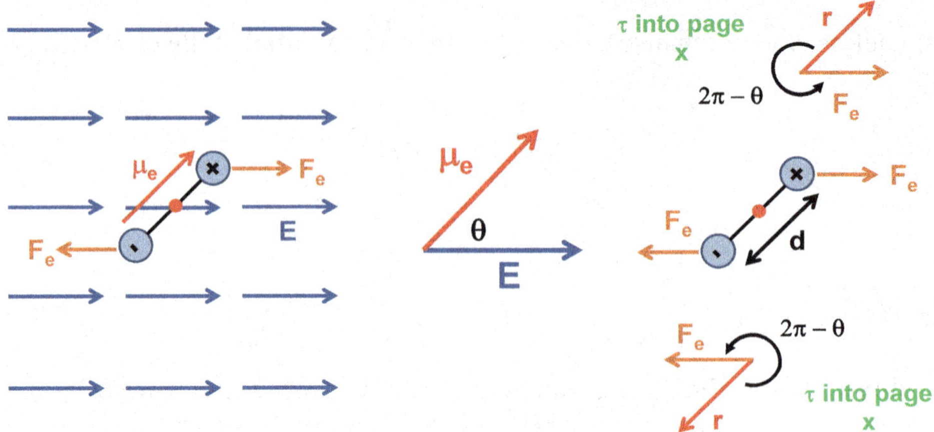

FIGURE 3.18: A uniform electric field exerts a torque on an electric dipole[26].

Let's define the angle θ for this system to be the angle between $\vec{\mu}_e$ and $\vec{E}$. The net torque acting on the electric dipole would then be

$$\tau_{net} = rF_e \sin(2\pi - \theta) + rF_e \sin(2\pi - \theta) \;\; \rightarrow \;\; \tau_{net} = -2rF_e \sin\theta$$

$$\tau_{net} = -2\left(\frac{d}{2}\right)(QE)\sin\theta \;\; \rightarrow \;\; \tau_{net} = -(Qd)E\sin\theta \;\; \rightarrow \;\; \tau_{net} = -\mu_e E\sin\theta$$

We can write this expression as

$$\vec{\tau}_{net} = \vec{\mu}_e \times \vec{E} \tag{3-15}$$

This net torque can be associated with a potential energy, which we can denote as $U_{e,dipole}$.

$$\tau_{net} = -\frac{dU_{e,dipole}}{d\theta} \;\; \rightarrow \;\; U_{e,dipole} = -\mu_e E\cos\theta$$

This expression for $U_{e,dipole}$ can be written in vector form as

$$U_{e,dipole} = -\vec{\mu}_e \bullet \vec{E} \tag{3-16}$$

As expected, this potential energy is a minimum when $\vec{\mu}_e$ is parallel to $\vec{E}$.

26 You can review vector notation in Section B-3.

In contrast, when an electric dipole is placed in a non-uniform electric field, the difference between the magnitudes of the electric forces acting on the two net electric charges will result in a net electric force acting on the electric dipole.

Example 3-14

Problem: What is the net electric force acting on the electric dipole in Figure 3.19?

The distance r in Figure 3.19 is the distance from the external negatively charged point particle to the center of the electric dipole. The three charged particles in Figure 3.19 all lie along the same line.

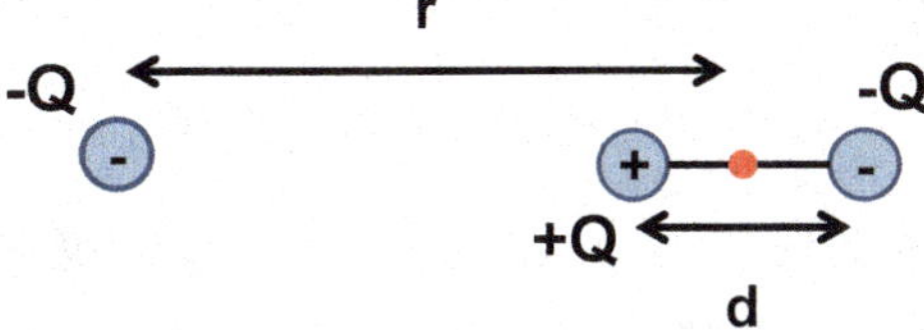

FIGURE 3.19: The system in Example 3-14.

Solution: The electric force acting on each net electric charge in the electric dipole can be found using Equation 3-6[27]. Let's denote as the x-axis the line connecting all three charged particles in this system; the origin of the x-axis will be at the external negatively charged particle, and the positive direction for the x-axis will point from this origin toward the electric dipole. With these definitions, the net electric force on the electric dipole is

$$\vec{F}_{e,net} = \frac{(-Q)(+Q)}{4\pi\varepsilon_0\left(r - \frac{d}{2}\right)^2}\hat{x} + \frac{(-Q)(-Q)}{4\pi\varepsilon_0\left(r + \frac{d}{2}\right)^2}\hat{x}$$

$$\vec{F}_{e,net} = \left(\frac{Q^2}{4\pi\varepsilon_0}\left(-\frac{1}{\left(r - \frac{d}{2}\right)^2} + \frac{1}{\left(r + \frac{d}{2}\right)^2}\right)\right)\hat{x} \rightarrow \vec{F}_{e,net} = \left(\frac{Q^2}{4\pi\varepsilon_0}\left(-\frac{32rd}{\left(d^2 - 4r^2\right)^2}\right)\right)\hat{x}$$

$$\vec{F}_{e,net} = \left(-\frac{8rQ^2d}{\pi\varepsilon_0\left(d^2 - 4r^2\right)^2}\right)\hat{x}$$

27 Or using Equation 3-9 and Equation 3-10

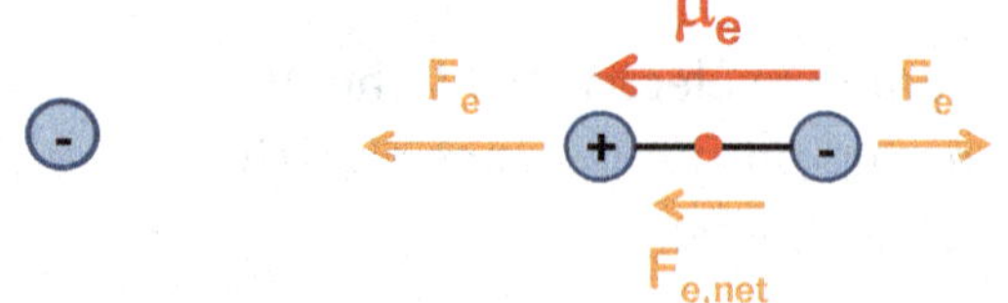

FIGURE 3.20: In a non-uniform electric field, the difference in the magnitude of the electric force acting on the two net electric charges of an electric dipole results in a net force electric force acting on the electric dipole.

The net electric force acting on the electric dipole is directed toward the external negatively charged particle, as shown in Figure 3.20. In the limit that $r >> d$ this force becomes

$$r >> d \quad \rightarrow \quad \vec{F}_{e,net} = \left(-\frac{8rQ^2d}{\pi\varepsilon_0 16r^4} \right) x$$

$$\vec{F}_{e,net} = \left(-\frac{Q^2d}{2\pi\varepsilon_0 r^3} \right) x$$

We could also express the net electric force acting on the electric dipole as

$$\vec{F}_{e,dipole} = -\nabla U_{e,dipole} \quad \rightarrow \quad \vec{F}_{e,dipole} = -\nabla\left(-\vec{\mu}_e \bullet \vec{E}\right)$$

$$\vec{F}_{e,dipole} = \nabla\left(\vec{\mu}_e \bullet \vec{E}\right) \tag{3-17}$$

The electric field in Equation 3-17 is the electric field experienced by the electric dipole (*i.e.*, the electric field produced by electric charges not part of the electric dipole). For the system shown in Figure 3.20, the electric field in Equation 3-17 is produced by the negatively charged particle to the left in the figure.

We can confirm that Equation 3-17 recapitulates the result of Example 3-14 within the dipole approximation limit (*i.e.*, for $r >> d$) using Equation 3-9 and Equation 3-13. Since the direction of the electric dipole moment is parallel to the direction of the electric field from the external negatively charged particle, the dot product in Equation 3-17 is simply the product of the magnitudes of the electric dipole moment and the electric field.

$$\vec{\mu}_e \bullet \vec{E} = \mu_e E \quad \rightarrow \quad \vec{F}_{e,dipole} = \nabla\left((Qd)\left(\frac{Q}{4\pi\varepsilon_0 r^2} \right) \right) \quad \rightarrow \quad \vec{F}_{e,dipole} = \nabla\left(\frac{Q^2d}{4\pi\varepsilon_0 r^2} \right)$$

We will use Equation B-7 for the gradient because the electric field is spherically symmetric around the external negatively charged point particle.

$$\vec{F}_{e,dipole} = \left(\frac{\partial}{\partial r}\left(\frac{Q^2d}{4\pi\varepsilon_0 r^2} \right) \right)\hat{r} + \left(\frac{1}{r}\frac{\partial}{\partial\theta}\left(\frac{Q^2d}{4\pi\varepsilon_0 r^2} \right) \right)\hat{\theta} + \left(\frac{1}{r\sin\theta}\frac{\partial}{\partial\varphi}\left(\frac{Q^2d}{4\pi\varepsilon_0 r^2} \right) \right)\hat{\varphi}$$

$$\vec{F}_{e,dipole} = \left(-\frac{2Q^2 d}{4\pi\varepsilon_0 r^3}\right)\hat{r} \rightarrow \vec{F}_{e,dipole} = \left(-\frac{Q^2 d}{2\pi\varepsilon_0 r^3}\right)\hat{r}$$

This result is identical to what we derived earlier (*N.B.*, $\hat{x} = \hat{r}$ for this comparison).

3-9 The Bohr Model for the Hydrogen Atom, Part II

We can now modify the Bohr model of the hydrogen atom introduced in Section 2-7 to include the electric force between the electron and the proton.

$$F_{g,electron} + F_{e,electron} = m_{electron}\frac{v_{electron}^2}{R_{orbit}}$$

$$G\frac{m_{electron}m_{proton}}{R_{orbit}^2} + \frac{e^2}{4\pi\varepsilon_0 R_{orbit}^2} = m_{electron}\frac{v_{electron}^2}{R_{orbit}}$$

$$v_{electron}^2 = \frac{1}{R_{orbit}}\left(Gm_{proton} + \frac{e^2}{4\pi\varepsilon_0 m_{electron}}\right)$$

$$v_{electron}^2 = \left(\frac{1}{50\times10^{-12}\,\mathrm{m}}\right) \times\left(\left(6.67\times10^{-11}\frac{\mathrm{J\,m}}{\mathrm{kg}^2}\right)\left(1.67\times10^{-27}\,\mathrm{kg}\right) + \frac{\left(1.6\times10^{-19}\,\mathrm{C}\right)^2}{4\pi\left(8.85\times10^{-12}\frac{\mathrm{C}^2}{\mathrm{J\,m}}\right)\left(9.11\times10^{-31}\,\mathrm{kg}\right)}\right)$$

$$v_{electron}^2 = \left(\frac{1}{50\times10^{-12}\,\mathrm{m}}\right)\left(1.11\times10^{-37}\frac{\mathrm{J\,m}}{\mathrm{kg}} + 2.53\times10^{2}\frac{\mathrm{J\,m}}{\mathrm{kg}}\right)$$

We see that the contribution of the electric force to the net force acting on the electron is 39 orders of magnitude larger that the contribution of the gravitational force acting on the electron. Solving this equation give us

$$v_{electron}^2 = 5.05\times10^{12}\frac{\mathrm{m}^2}{\mathrm{s}^2} \rightarrow v_{electron} = \pm2.25\times10^{6}\frac{\mathrm{m}}{\mathrm{s}}$$

Now the angular frequency and period of the electron orbiting the proton are

$$\omega_{electron} = 4.5\times10^{16}\,\frac{\text{rad}}{\text{s}} \qquad T_{electron} = 1.4\times10^{-16}\,\text{s}$$

This much shorter period seems more "realistic".

Furthermore, now that we have established that the magnitude of the gravitational interaction between the electron and proton in the Bohr atom is insignificant, we can omit it from our calculations. This will enable us, for example, to determine the radius of the electron's orbit in the Bohr atom. We begin this calculation by noting that the amount of energy required to remove the electron from a hydrogen atom is experimentally determined to be

$$E = 2.17\times10^{-18}\,\text{J}$$

This energy is commonly referred to as the *ionization energy* of the atom. The ionization energy is the energy required to remove the electron from its orbit around the proton. Thus, it is the sum of the kinetic and potential energies of the electron.

$$E = K + U_e \quad\rightarrow\quad E = \frac{1}{2}m_{electron}v_{electron}^2 + \frac{e(-e)}{4\pi\varepsilon_0 R_{orbit}}$$

$$E = \frac{1}{2}m_{electron}v_{electron}^2 - \frac{e^2}{4\pi\varepsilon_0 R_{orbit}}$$

As before, we can determine the speed of the electron from the centripetal acceleration of the electron (only now we can leave out the gravitational force).

$$\frac{e^2}{4\pi\varepsilon_0 R_{orbit}^2} = m_{electron}\frac{v_{electron}^2}{R_{orbit}} \quad\rightarrow\quad \frac{e^2}{4\pi\varepsilon_0 R_{orbit}} = m_{electron}v_{electron}^2$$

Substitution gives us

$$E = \frac{1}{2}\left(\frac{e^2}{4\pi\varepsilon_0 R_{orbit}}\right) - \frac{e^2}{4\pi\varepsilon_0 R_{orbit}} \quad\rightarrow\quad E = -\frac{1}{2}\left(\frac{e^2}{4\pi\varepsilon_0 R_{orbit}}\right)$$

The negative sign in this equation indicates that energy must be added to the system to remove the electron. Substitution of the ionization energy as the magnitude of this added energy allows us to determine the radius of the electron's orbit in the Bohr atom.

$$2.17\times10^{-18}\,\text{J} = \frac{1}{2}\left(\frac{\left(1.6\times10^{-19}\,\text{C}\right)^2}{4\pi\left(8.85\times10^{-12}\,\dfrac{\text{C}^2}{\text{J m}}\right)R_{orbit}}\right) \rightarrow R_{orbit} = 5.3\times10^{-11}\,\text{m}$$

The corresponding speed of the electron is therefore

$$v_{electron} = \left(\frac{e^2}{4\pi\varepsilon_0 R_{orbit} m_{electron}}\right)^{\frac{1}{2}}$$

$$v_{electron} = \left(\frac{\left(1.6\times10^{-19}\,\text{C}\right)^2}{4\pi\left(8.85\times10^{-12}\,\dfrac{\text{C}^2}{\text{J m}}\right)\left(5.3\times10^{-11}\,\text{m}\right)\left(9.11\times10^{-31}\,\text{kg}\right)}\right)^{\frac{1}{2}}$$

$$v_{electron} = \pm 2.18\times10^{6}\,\frac{\text{m}}{\text{s}}$$

3-10 Applications of Legendre Polynomials (Optional)

The expression for the electric potential of an electric dipole derived in Section 3-7 is valid only when the electric potential is measured at a distance from the electric dipole that is much larger than the separation distance between the net electric charges in the electric dipole. We could have obtained a more generally valid expression, of course, by retaining more terms in the binomial expansion of the r_A and r_B variables in our derivation, but a more mathematically elegant and simple approach is to use Legendre polynomials[28], which are the coefficients of the Taylor series expansion of the expressions for r_A and r_B. Indeed, the distances r_A and r_B in Figure 3.17 can be expressed in terms of the variables r, d, and θ in Figure 3.17 as

28 See Section A-4.

$$\frac{r}{r_A} = \frac{1}{\left(1+\left(\frac{d}{r}\right)^2 - 2\left(\frac{d}{r}\right)\cos\theta\right)^{\frac{1}{2}}} \rightarrow \frac{r}{r_A} = \sum_{n=0}^{\infty}\left(\frac{d}{r}\right)^n P_n(\cos\theta)$$

$$\frac{r}{r_B} = \frac{1}{\left(1+\left(\frac{d}{r}\right)^2 - 2\left(\frac{d}{r}\right)\cos(\pi-\theta)\right)^{\frac{1}{2}}} \rightarrow \frac{r}{r_B} = \sum_{n=0}^{\infty}\left(\frac{d}{r}\right)^n P_n(\cos(\pi-\theta))$$

The electric potential of the electric dipole in Figure 3.17 is therefore

$$V_e = \frac{Q}{4\pi\varepsilon_0 r}\left(\left(\sum_{n=0}^{\infty}\left(\frac{d}{2r}\right)^n P_n(\cos\theta)\right) - \left(\sum_{n=0}^{\infty}\left(\frac{d}{r}\right)^n P_n(\cos(\pi-\theta))\right)\right)$$

$$V_e = \frac{Q}{4\pi\varepsilon_0 r}\left(\sum_{n=0}^{\infty}\left(\frac{d}{r}\right)^n \left(P_n(\cos\theta) - P_n(\cos(\pi-\theta))\right)\right)$$

And since $\cos(\pi-\theta) = -\cos(\theta)$ we have

$$V_e = \frac{Q}{4\pi\varepsilon_0 r}\left(\sum_{n=0}^{\infty}\left(\frac{d}{r}\right)^n \left(P_n(\cos\theta) - P_n(-\cos(\theta))\right)\right)$$

We arrive at our final expression through the identity $P_n(-x) = (-1)^n P_n(x)$

$$V_e = \frac{Q}{4\pi\varepsilon_0 r}\left(\sum_{n=0}^{\infty}\left(\frac{d}{r}\right)^n P_n(\cos\theta)\left(1-(-1)^n\right)\right)$$

This expression for the electric potential of the electric dipole in Figure 3.17 is valid for all values of r and d, as well as all values of θ. We see immediately that all terms in the summation corresponding to even values of n are zero. Indeed, the fact that the leading term in the summation is zero (*i.e.*, the $n = 0$ term is zero) indicates that the electric dipole has no net electric charge. Hence,

$$V_e = \frac{Q}{4\pi\varepsilon_0 r}\left(2\left(\frac{d}{r}\right)P_1(\cos\theta) + 2\left(\frac{d}{r}\right)^3 P_3(\cos\theta) + 2\left(\frac{d}{r}\right)^5 P_5(\cos\theta) + \ldots\right)$$

The first term in this series (with $n = 1$) is the dipole term that we have derived previously

$$V_{e,dipole} = \frac{2Q}{4\pi\varepsilon_0 r}\left(\frac{d}{r}\right)P_1(\cos\theta) \rightarrow V_{e,dipole} = \frac{Q(2d)}{4\pi\varepsilon_0 r^2}\cos\theta$$

This is the term that dominates when $r >> d$ since the remaining terms in the series decrease in magnitude much more quickly as r increases.

The use of Legendre polynomials provides a convenient shortcut notation to simplify the series expansions associated with calculations of the electric potential for more complicated systems of charged particles.

Example 3-15

Problem: The system of charged point particles in Figure 3.21 is referred to as a linear electric *quadrupole*. What is the electric potential for this system at the position P in Figure 3.21?

Solution: The electric potential is the sum of the electric potentials from each of the charged particles in the system.

$$V_e = \frac{q_A}{4\pi\varepsilon_0 r_A} + \frac{q_B}{4\pi\varepsilon_0 r_B} + \frac{q_C}{4\pi\varepsilon_0 r_C}$$

$$\rightarrow V_e = \frac{1}{4\pi\varepsilon_0}\left(\frac{Q}{r_A} + \frac{-2Q}{r} + \frac{Q}{r_C}\right)$$

$$V_e = \frac{Q}{4\pi\varepsilon_0 r}\left(\frac{r}{r_A} - 2 + \frac{r}{r_C}\right)$$

As shown in Figure 3.22, r_A and r_C can be related to r, d, and θ using Legendre polynomials.

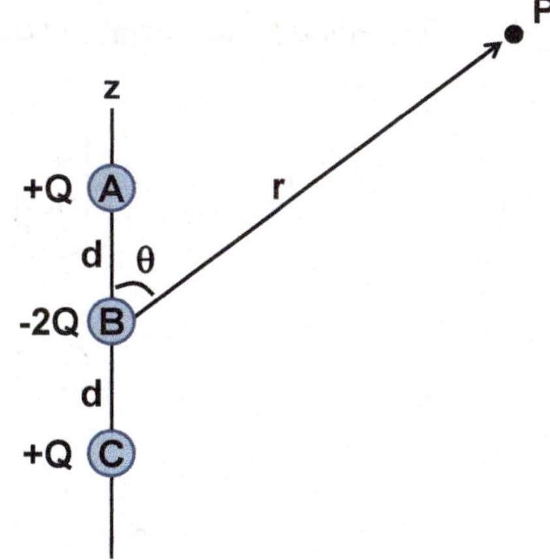

FIGURE 3.21: The linear electric quadrupole in Example 3-15.

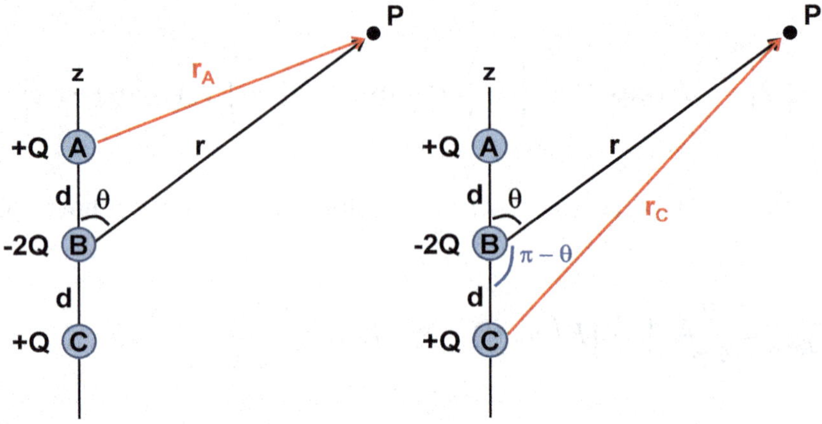

FIGURE 3.22: Determining the electric potential from the linear electric quadrupole in Figure 3.21.

$$\frac{r}{r_A} = \frac{1}{\left(1+\left(\frac{d}{r}\right)^2 - 2\left(\frac{d}{r}\right)\cos\theta\right)^{\frac{1}{2}}} \rightarrow \frac{r}{r_A} = \sum_{n=0}^{\infty}\left(\frac{d}{r}\right)^n P_n(\cos\theta)$$

$$\frac{r}{r_C} = \frac{1}{\left(1+\left(\frac{d}{r}\right)^2 + 2\left(\frac{d}{r}\right)\cos(\pi-\theta)\right)^{\frac{1}{2}}} \rightarrow \frac{r}{r_C} = \sum_{n=0}^{\infty}\left(\frac{d}{r}\right)^n P_n(\cos(\pi-\theta))$$

The electric potential for the system is therefore

$$V_e = \frac{Q}{4\pi\varepsilon_0 r}\left(\left(\sum_{n=0}^{\infty}\left(\frac{d}{r}\right)^n P_n(\cos\theta)\right) - 2 + \left(\sum_{n=0}^{\infty}\left(\frac{d}{r}\right)^n P_n(\cos(\pi-\theta))\right)\right)$$

$$V_e = \frac{Q}{4\pi\varepsilon_0 r}\left(-2+\sum_{n=0}^{\infty}\left(\frac{d}{r}\right)^n \left(P_n(\cos\theta) + P_n(\cos(\pi-\theta))\right)\right)$$

Following the same simplification that we used for the electric dipole gives us

$$V_e = \frac{Q}{4\pi\varepsilon_0 r}\left(-2+\sum_{n=0}^{\infty}\left(\frac{d}{r}\right)^n P_n(\cos\theta)\left(1+(-1)^n\right)\right)$$

For this expression, both the $n = 0$ and $n = 1$ terms are equal to zero. Furthermore, all higher order terms in the series where n is odd are also zero.

$$V_e = \frac{Q}{4\pi\varepsilon_0 r}\left(2\left(\frac{d}{r}\right)^2 P_2(\cos\theta) + 2\left(\frac{d}{r}\right)^4 P_4(\cos\theta) + 2\left(\frac{d}{r}\right)^6 P_6(\cos\theta) + \dots\right)$$

The first term in this series (with $n = 2$) is referred to as the *quadrupole term.*

$$V_{e.quadrupole} = \frac{Q}{4\pi\varepsilon_0 r}\left(2\left(\frac{d}{r}\right)^2 P_2(\cos\theta)\right)$$

$$V_{e.quadrupole} = \frac{Q}{4\pi\varepsilon_0 r}\left(2\left(\frac{d}{r}\right)^2\right)\left(\frac{1}{2}\left(3(\cos\theta)^2 - 1\right)\right)$$

This is also the dominate term in the series when $r >> d$.

$$V_e \approx \frac{Qd^2}{4\pi\varepsilon_0 r^3}\left(3(\cos\theta)^2 - 1\right)$$

The electric potential for any system of charged particles can be expressed as an infinite series using Legendre polynomials; the first term in this series ($n = 0$ in the summation) reflects the net electric charge of the system[29], the second term in this series ($n = 1$ in the summation) is the dipole term, the third term in this series ($n = 2$ in the summation) is the quadrupole term, and the remaining terms represent *higher order poles* of the electric potential. This is the basis for the *multipole expansion* of the electric potential (and by extension the electric field) a system of charged point particles.

$$V_e = V_{e,monopole} + V_{e,dipole} + V_{e,quadrupole} + \dots$$

29 This is also referred to as the monopole term.

Example 3-16

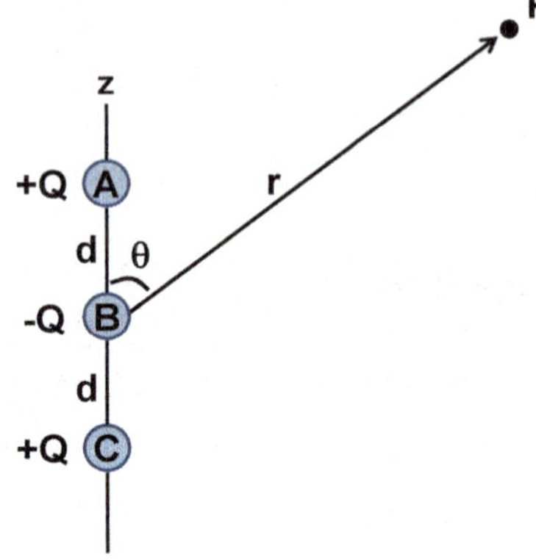

FIGURE 3.23: The system of charged point particles in Example 3-16.

Problem: What is the electric potential for this system of charged point particles at the position P in Figure 3.23?

Solution: The electric potential is the sum of the electric potentials from each of the charged particles in the system. Following the solution to Example 3-15, we have

$$V_e = \frac{q_A}{4\pi\varepsilon_0 r_A} + \frac{q_B}{4\pi\varepsilon_0 r_B} + \frac{q_C}{4\pi\varepsilon_0 r_C}$$

$$\rightarrow \quad V_e = \frac{1}{4\pi\varepsilon_0}\left(\frac{Q}{r_A} + \frac{-Q}{r} + \frac{Q}{r_C}\right)$$

$$V_e = \frac{Q}{4\pi\varepsilon_0 r}\left(\frac{r}{r_A} - 1 + \frac{r}{r_C}\right)$$

As shown in Figure 3.22 and Example 3-15, r_A and r_C can be related to r, d, and θ using Legendre polynomials.

$$V_e = \frac{Q}{4\pi\varepsilon_0 r}\left(\left(\sum_{n=0}^{\infty}\left(\frac{d}{r}\right)^n P_n(\cos\theta)\right) - 1 + \left(\sum_{n=0}^{\infty}\left(\frac{d}{r}\right)^n P_n(\cos(\pi-\theta))\right)\right)$$

$$V_e = \frac{Q}{4\pi\varepsilon_0 r}\left(-1 + \sum_{n=0}^{\infty}\left(\frac{d}{r}\right)^n \left(P_n(\cos\theta) + P_n(\cos(\pi-\theta))\right)\right)$$

$$V_e = \frac{Q}{4\pi\varepsilon_0 r}\left(-1 + \sum_{n=0}^{\infty}\left(\frac{d}{r}\right)^n P_n(\cos\theta)\left(1+(-1)^n\right)\right)$$

Let's examine the first few terms in this series.

$$V_e = \frac{Q}{4\pi\varepsilon_0 r}\left(-1 + 2 + 0 + 2\left(\frac{d}{r}\right)^2 \cos\theta + \ldots\right)$$

$$V_e = \frac{Q}{4\pi\varepsilon_0 r} + \frac{Qd^2}{2\pi\varepsilon_0 r^3}\cos\theta + \ldots$$

The mono-pole, dipole, and quadrupole terms of the electric potential of this system are

$$V_{e,monopole} = \frac{Q}{4\pi\varepsilon_0 r} \quad \text{(the net electric charge of the system is +Q)}$$

$$V_{e,dipole} = 0 \quad \text{(there is no separation of net opposite charge)}$$

$$V_{e,quadrupole} = \frac{Qd^2}{2\pi\varepsilon_0 r^3}\cos\theta$$

A similar multipole expansion is possible for the gravitational potential, but since there is only one type (or variety) of mass, there cannot be a gravitational dipole. It is possible, however, to have a gravitational quadrupole as well as higher order poles.

Example 3-17

Problem: What is the gravitational potential of the system of identical point particles at position P in Figure 3.24?

Solution: The gravitational potential is the sum of the gravitational potentials from each of the particles in the system.

$$V_g = -G\frac{m}{r_A} - G\frac{m}{r_B}$$

$$V_g = -Gm\left(\frac{1}{r_A} + \frac{1}{r_B}\right)$$

$$V_g = -\frac{Gm}{r}\left(\frac{r}{r_A} + \frac{r}{r_B}\right)$$

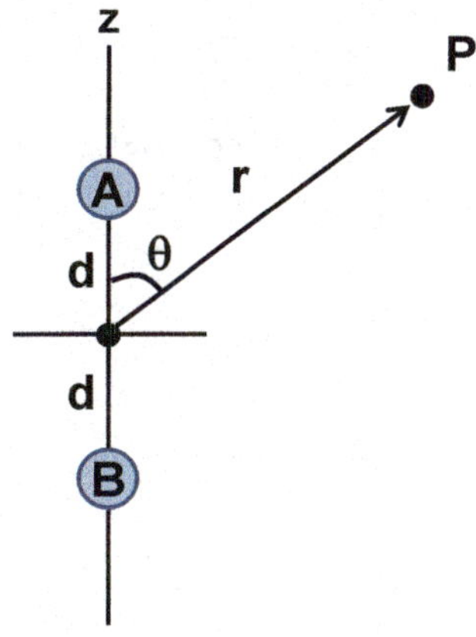

FIGURE 3.24: The system of point particles in Example 3-17.

The variable m in this equation is the mass of each point particle. The variables r_A and r_B in this equation are the distances between position P and particle A and particle B, respectively.

$$\frac{r}{r_A} = \frac{1}{\left(1+\left(\frac{d}{r}\right)^2 - 2\left(\frac{d}{r}\right)\cos\theta\right)^{\frac{1}{2}}} \rightarrow \frac{r}{r_A} = \sum_{n=0}^{\infty}\left(\frac{d}{r}\right)^n P_n(\cos\theta)$$

$$\frac{r}{r_B} = \frac{1}{\left(1+\left(\frac{d}{r}\right)^2 - 2\left(\frac{d}{r}\right)\cos(\pi-\theta)\right)^{\frac{1}{2}}} \rightarrow \frac{r}{r_B} = \sum_{n=0}^{\infty}\left(\frac{d}{r}\right)^n P_n(\cos(\pi-\theta))$$

The gravitational potential is therefore

$$V_g = -\frac{Gm}{r}\left(\sum_{n=0}^{\infty}\left(\frac{d}{r}\right)^n P_n(\cos\theta) + \sum_{n=0}^{\infty}\left(\frac{d}{r}\right)^n P_n(\cos(\pi-\theta))\right)$$

$$V_g = -\frac{Gm}{r}\left(\sum_{n=0}^{\infty}\left(\frac{d}{r}\right)^n \left(P_n(\cos\theta) + P_n(\cos(\pi-\theta))\right)\right)$$

Since $\cos(\pi-\theta) = -\cos(\theta)$ and $P_n(-x) = (-1)^n P_n(x)$ we have

$$V_g = -\frac{Gm}{r}\left(\sum_{n=0}^{\infty}\left(\frac{d}{r}\right)^n P_n(\cos\theta)\left(1+(-1)^n\right)\right)$$

$$V_g = -\frac{Gm}{r}\left(2+0+\left(\frac{d}{r}\right)^2 2\cos\theta + \ldots\right)$$

Therefore,

$$V_{g,monopole} = -\frac{2Gm}{r} \quad \text{(the net mass of the system is } 2m\text{)}$$

$$V_{g,dipole} = 0 \quad \text{(gravitational dipoles are impossible)}$$

$$V_{g,quadrupole} = -\frac{2Gmd^2}{r^3}\cos\theta$$

Summary

- **Electric charge:** an intrinsic property of matter. The SI unit for electric charge is the coulomb (C). Electric charges can be positive or negative and occur only in integer multiples of the fundamental electric charge, denoted by *e*.

$$e = 1.6 \times 10^{-23}\ \text{C}$$

- **Conservation of electric charge:** electric charge can neither be created nor destroyed. Thus, the total electric charge for an isolated system is constant.

- **Electric potential:** an intrinsic property of an electric charge that describes the alteration of space associated with that electric charge. The detection of the electric potential of one charged object by another charged object is the origin of the electric interaction (potential energy and force) between the two objects. Electric potential is a scalar field and can be considered to be the electric potential energy per unit charge. The electric potential surrounding a charged point particle with net electric charge q is

$$V_e = \frac{q}{4\pi\varepsilon_0 r}$$

The electric potential of a distribution of net electric charge is

$$V_e = \frac{1}{4\pi\varepsilon_0}\int \frac{dq}{r}$$

- The electric potential energy associated with the electric interaction between two charged point particles is equal to the product of the net electric charge of one of the particles and the electric potential of the other particle at that location.

$$U_e = qV_e$$

- The electric force acting on a charged point particle is equal to the product of the net electric charge of the particle and the gradient of the electric potential experienced by the particle.

$$\vec{F}_e = -q\nabla V_e$$

- **Electric field:** an intrinsic property of an electric charge that describes the alteration of space associated with that electric charge. The electric field is a vector field obtained from the gradient of the electric potential.

  $$\vec{E} = -\nabla V_e$$

  The electric field surrounding a point particle is

  $$\vec{E} = \left(\frac{q}{4\pi\varepsilon_0 r^2}\right)\hat{r}$$

  The electric field from a continuous distribution of net electric charge is

  $$\vec{E} = \left(\frac{1}{4\pi\varepsilon_0}\right)\int \frac{dq}{r^2}\hat{r}$$

  The electric force acting on a charged point particle is equal to the product of the particle's net electric charge and the electric field experienced by the particle.

  $$\vec{F}_e = q\vec{E}$$

  The electric field can thus be thought of as the electric force per unit charge.

- **Electric dipole:** two net electric charges of opposite ($\pm$) sign separated by a small distance.

- **Electric dipole moment:** a vector whose magnitude is equal to the product of the separation distance between the two electric charges in the electric dipole and the magnitude of the electric charge of the point particles in the dipole (both charges are assumed to have the same magnitude of electric charge), and whose direction points from the negative electric charge to the positive electric charge (defined to be the *z*-axis in the equation below).

  $$\vec{\mu}_e = (Qd)\hat{z}$$

  When observed from a large distance the electric potential of the electric dipole is

  $$V_{e,dipole} \approx \frac{\vec{\mu}_e \bullet \hat{r}}{4\pi\varepsilon_0 r^2}$$

The electric potential energy of an electric dipole in a uniform electric field is

$$U_{e,dipole} = -\vec{\mu}_e \bullet \vec{E}$$

Problems

1. What is the electric potential of a 3 kg point particle with a net electric charge of 5 nC at a distance of 4 m from the particle?

2. An isolated system consists of two identical charged point particles that each have a mass of 2 g and a net electric charge of 2 nC. The particles are initially held in place a distance of 1 cm and then released. After they are released, the particles move away from each other. What will be the speed of each particle when their separation distance is 10 cm?

3. A system consists of two charged point particles, A and B, separated by a distance of 3 m as shown in the figure. The net electric charge of particle A is 4 nC, and the net electric charge of particle B is -4 nC. What is the electric potential of this system at the position P?

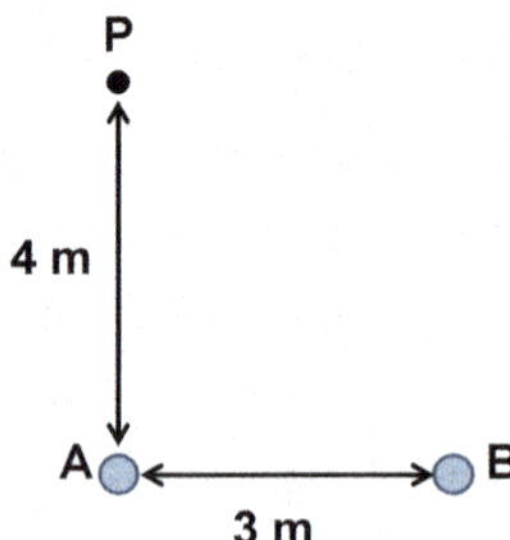

4. A system consists of three point particles, A, B, and C, as shown in the figure. The net electric charge of particle A is -3 nC, the net electric charge of particle B is +4 nC, and the net electric charge of particle C is -5 nC. What is the electric potential of this system at the position P?

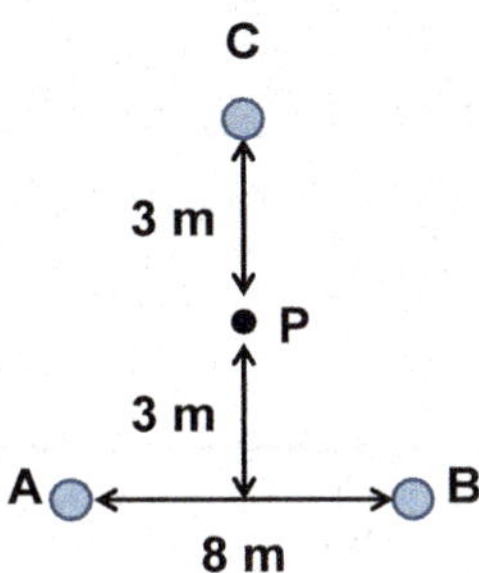

5. What is the electric potential at the center of a uniformly charged thin ring of radius R and net electric charge $-Q$?

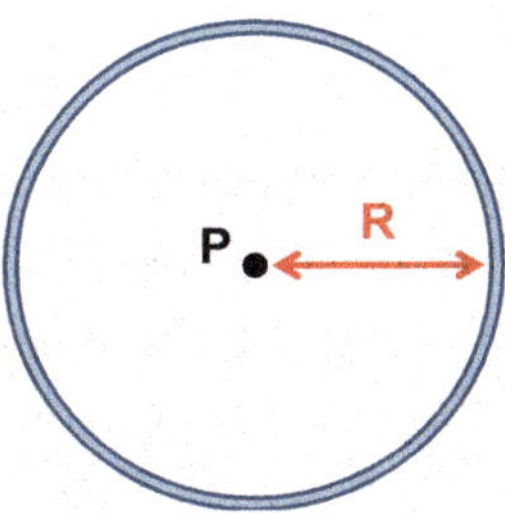

6. A system consists of a thin and uniformly charged rod of length L and net charge $+Q$, as shown in the figure. What is the electric potential of this system at the position P that is a distance y "*above*" one end of the rod?

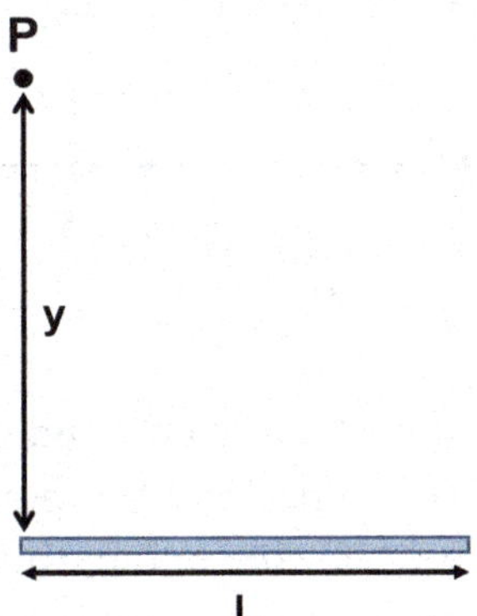

7. A 5 g point particle with a net electric charge of +4 μC is suspended by a massless rope, as shown in the figure. A uniform electric field with a magnitude of 1.2×10^4 N/C is also acts on the particle causing it to hang from the ceiling at an angle θ. What is the angle θ?

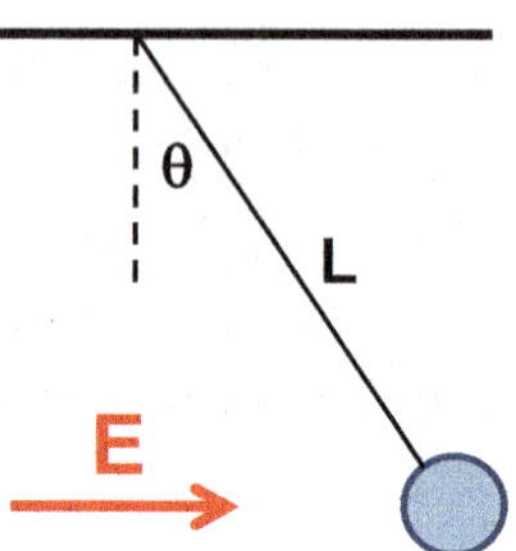

8. A system consists of two charged point particles, A and B, arranged as shown in the figure. The net electric charge of particle A is −5 nC, and the net electric charge of particle B is +4 nC. What is the electric field at position P? Express your answer in terms of the *x*-axis and *y*-axis components of the electric field using the reference frame shown in the figure.

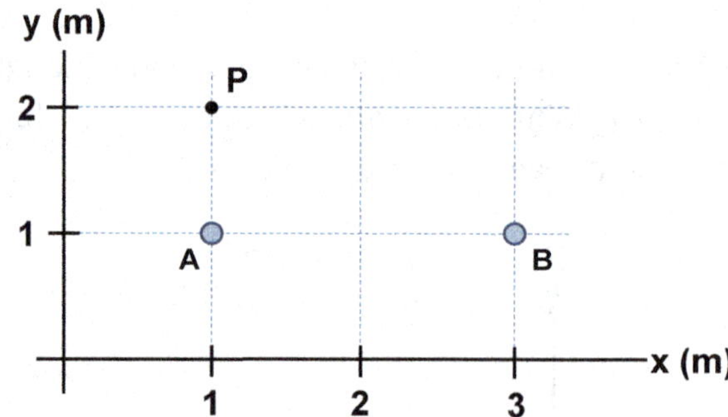

9. A system consists of a thin and uniformly charged circular ring of radius R and net electric charge $+Q$. The electric field along the perpendicular bisector of the ring always points directly away from the ring and parallel to the perpendicular bisector of the ring. At what distance from the center of the ring is the magnitude of this electric field a maximum? Express your answer in terms of the radius R.

10. What is the electric field from a uniformly charged annulus, with an inner radius a and outer radius b, a distance z away from the center of the annulus along the perpendicular bisector of the annulus (*i.e.*, at the position P in the figure)? The net electric charge on the annulus is $-Q$. Express your answer in terms of the z-axis component of the electric field using the reference frame shown in the figure.

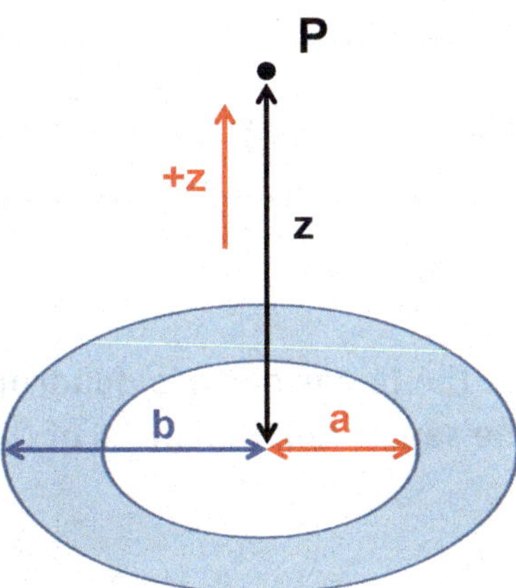

11. A system consists of a thin and uniformly charged rod of length L and net electric charge $-Q$. What is the electric field of this system at the position P that is a distance y "above" one end of the rod? Express your answer in terms of the x-axis and y-axis components of the electric field using the reference frame shown in the figure.

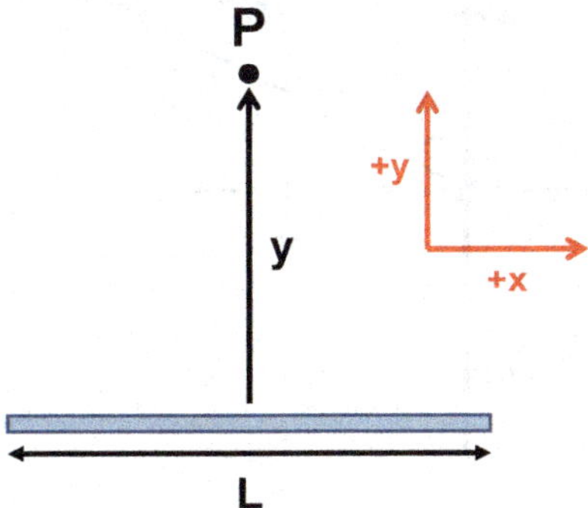

12. Two oppositely charged point particles are connected by a massless rod of length 2 cm to form an electric dipole; the magnitude of the electric charge on each point particle is 5 nC. This electric dipole is then placed in a uniform electric field of

magnitude 20 N/C, as shown in the figure. What is the electric potential energy of the electric dipole? What is the magnitude of the torque exerted on the electric dipole by the electric field?

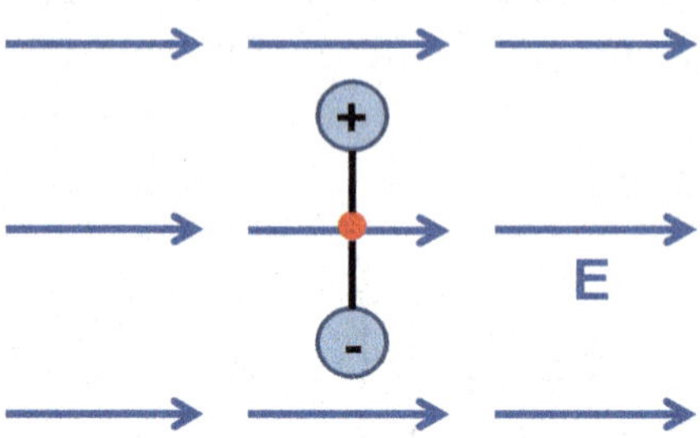

13. What is the electric field of the linear electric quadrupole when measured at distances much larger than the separation distances of the net electric charges in the quadrupole?

14. What is the electric potential for the electric quadrupole shown in the figure at the position P? The magnitude of the electric charge on each point particle in the quadrupole is Q.

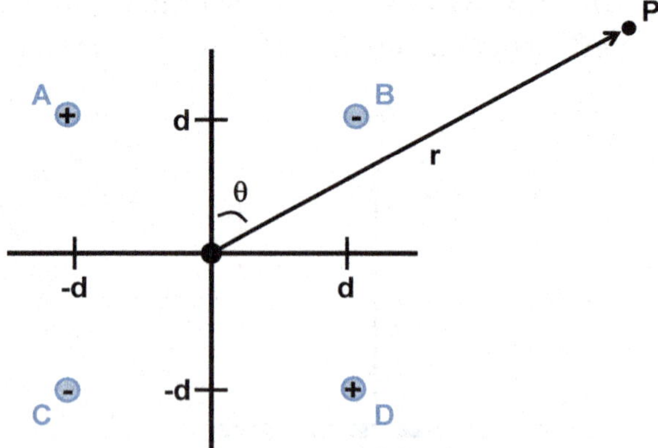

CHAPTER 4

Magnetic Interaction

4-1 Introduction

Gravitational fields produce a gravitational force on objects that have mass and electric fields produce an electric force on charged objects. In this chapter, we begin our discussion of magnetic fields, which are produced by moving charged objects and also produce a magnetic force on moving charged objects. We are also familiar with the magnetic properties of permanent magnets, such as those we stick on refrigerators, or a compass needle reacting to the magnetic field of the Earth. Permanent magnets result from the intrinsic magnetic properties of the matter that constitute them, as we will discuss near the end of this chapter. A more thorough discussion of the electric and magnetic properties of matter will resume in Chapter 7.

4-2 Magnetic Field Produced by a Moving Charged Particle

The magnetic field[1], denoted by $\vec{B}$, is a vector field produced by a charged object with net electric charge q moving with velocity $\vec{v}$. The magnetic field of a charged point particle moving through a vacuum is given by Equation 4-1[2].

$$\vec{B} = \left(\frac{\mu_0 q}{4\pi}\right)\frac{\vec{v}\times\vec{r}}{r^3} \tag{4-1}$$

Equation 4-1 is commonly known as the Law of Biot and Savart[3]. In Equation 4-1, the vector $\vec{r}$ points from the point particle to the location where the magnetic field is measured. The constant μ_0 in Equation 4-1 is called the vacuum permeability[4] and is

$$\mu_0 = 4\pi \times 10^{-7}\,\frac{\text{kgm}}{\text{C}^2}$$

The SI unit for the magnitude of the magnetic field is the tesla, denoted by T, which is defined as

$$1\,\text{T} = 1\frac{\text{kg}}{\text{Cs}} = 1\frac{\text{Ns}}{\text{Cm}} = 1\frac{\text{Vs}}{\text{m}^2}$$

Another common, but not SI, unit for the magnetic field is the gauss, denoted by G.

$$1\,\text{G} = 10^{-4}\,\text{T}$$

A magnetic field with a magnitude (or strength) of 1 T is a strong magnetic field. The magnitude of the Earth's magnetic field[5] varies with latitude, but is approximately 0.5 G or 50 μT. The magnitude of the magnetic field associated with a refrigerator magnet is around 5 μT, and the magnitude of the strongest magnetic fields routinely produced in a laboratory are less than 100 T[6].

---

1 This field is also called the magnetic flux density, the magnetic induction, or the B-field.

2 Strictly speaking, Equation 4-1 is valid only when the speed of the object is much less than the speed of light. You should feel inspired to take more advanced physics courses to learn how this equation changes when objects are moving close to the speed of light.

3 These French names rhyme with "leo" and "bazaar."

4 This constant is also called the permeability of free space or the magnetic constant.

5 Magnetic fields not in vacuum will be discussed in Chapter 7.

6 This limit is for a sustained magnetic field; transient magnetic fields of larger strength have been produced (sometimes by accident).

Example 4-1

Problem: A point particle with a net electric charge of +10 nC is moving with velocity $\vec{v} = \left(5\frac{\text{m}}{\text{s}}\right)\hat{x}$. What is the magnetic field of this point particle at a distance $\vec{r} = (0.2\,\text{m})\hat{y}$ away from the particle?

Solution: From Equation 4-1 we have

$$\vec{B} = \left(\frac{\left(4\pi \times 10^{-7}\frac{\text{kgm}}{\text{C}^2}\right)\left(10 \times 10^{-9}\text{C}\right)}{4\pi}\right)\frac{\left[\left(5\frac{\text{m}}{\text{s}}\right)\hat{x}\right] \times \left[(0.2\,\text{m})\hat{y}\right]}{(0.2\,\text{m})^3}$$

$$\vec{B} = \left(1.3 \times 10^{-13}\frac{\text{kg}}{\text{Cs}}\right)(\hat{x} \times \hat{y}) \quad \rightarrow \quad \vec{B} = \left(1.3 \times 10^{-13}\text{T}\right)\hat{z}$$

It follows from the cross product in Equation 4-1 that the direction of the magnetic field associated with a moving charged particle is perpendicular to the direction of the velocity of the charged particle and depends upon the (±) sign of the net electric charge of the particle (see Figure 4.1). The direction of the magnetic field associated with a moving charged particle can also be found using a right-hand rule. For a positively charged moving object, point the thumb of your right hand in the direction of the object's velocity and the fingers of your right hand will curl in the direction of the magnetic field created by the object. The direction of the magnetic field is opposite if the net electric charge is negative[7]. Indeed, the magnetic field created by a positively charge moving object is identical to the magnetic field

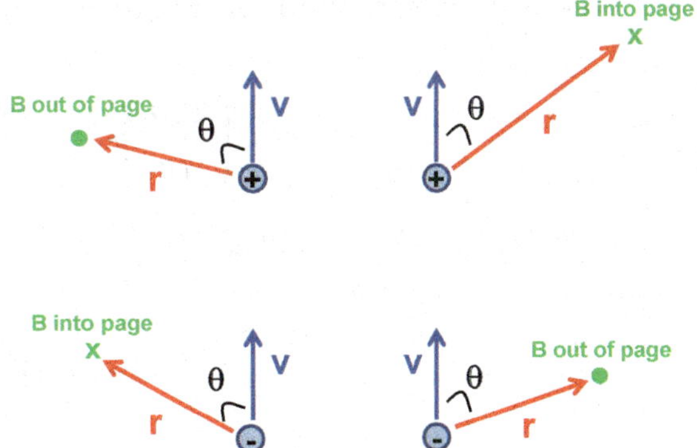

FIGURE 4.1: The direction of the magnetic field created by a moving particle with a net electric charge.

7 You can obtain this result by using your left hand.

created by a negatively charged object moving in the opposite direction with the same speed.

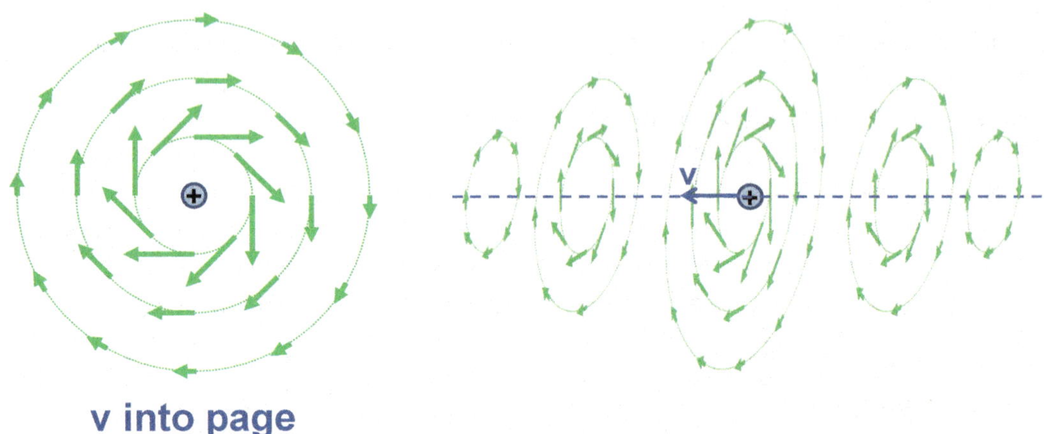

FIGURE 4.2: The magnetic field (green) created by a moving charged particle is circularly symmetric around the direction of the object's velocity. The direction of the magnetic field at several positions is indicated by the direction of an arrow and the magnitude of the magnetic field by the length of the arrow.

As shown in Figure 4.2, the magnetic field created by the moving charged object is circularly symmetric around the direction of the object's velocity. This cylindrical symmetry of the magnetic field motivates us to use cylindrical coordinates[8] as the reference frame for the calculation in Equation 4-1. For example, if we define the velocity of the point particle to be along the positive z-axis in a reference frame using cylindrical coordinates, then the magnetic field of the point particle is

$$\vec{B}=\left(\frac{\mu_0 q}{4\pi}\right)\frac{\left((v)\hat{z}\right)\times\vec{r}}{r^3} \;\rightarrow\; \vec{B}=\left(\frac{\mu_0 qv}{4\pi r^3}\right)\left(\hat{z}\times\vec{r}\right)$$

$$\vec{r}=(\rho)\hat{\rho}+(z)\hat{z} \;\rightarrow\; \hat{z}\times\vec{r}=\hat{z}\times\left[(\rho)\hat{\rho}+(z)\hat{z}\right] \;\rightarrow\; \hat{z}\times\vec{r}=(\rho)\left(\hat{z}\times\hat{\rho}\right)$$

$$\hat{z}\times\vec{r}=(\rho)\hat{\varphi} \;\rightarrow\; \vec{B}=\left(\frac{\mu_0 qv\rho}{4\pi r^3}\right)\hat{\varphi}$$

$$r^2=\rho^2+z^2 \;\rightarrow\; \vec{B}=\left(\frac{\mu_0 qv\rho}{4\pi\left(\rho^2+z^2\right)^{\frac{3}{2}}}\right)\hat{\varphi}$$

8 See Section B-5 for review.

This expression for the magnetic field is convenient since its description of the direction of the magnetic field does not require us to use the right-hand rule.

Since the magnetic field forms "closed loops" around the particle's velocity there is neither an origin (source) nor a termination (sink) for the magnetic field. This is different from the electric field associated with a stationary charged point particle (Section 3-4); the electric field points away from positively charged particles and toward negatively charged particles. We will discuss this difference between electricity and magnetism further in Section 4-7.

Finally, the magnetic field from a system of moving charged point particles is the sum of the magnetic field from each of the point particles.

Example 4-2

Problem: A system consists of two identical charged point particles, each with net electric charge q, connected by a massless rigid rod of length $2d$. This system is rotating at angular speed ω around an axis that passes through its center of mass and is perpendicular to the rigid rod, as shown in Figure 4.3. What is the magnetic field of these two charged particles at distance z along the axis of rotation (*i.e.*, at the position P in Figure 4.3)?

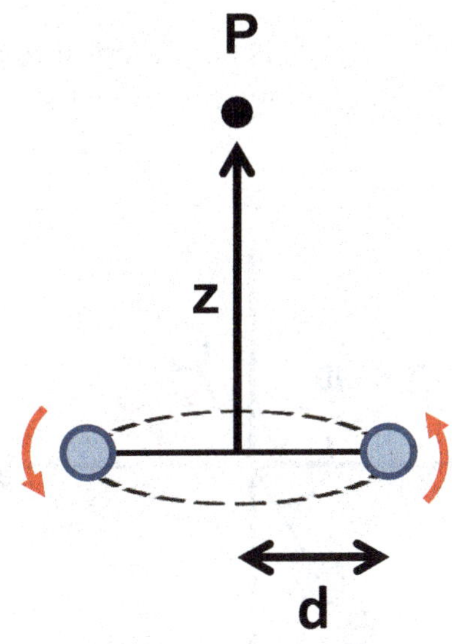

FIGURE 4.3: The moving charged particles in Example 4-2.

The magnetic field at the position P from each of the point particles can be determined using Equation 4-1; the velocity in this equation can be written in terms of the angular speed of the particles using cylindrical coordinates. The origin of the reference frame for this system will be at the center of mass of the system and the positive direction for the z-axis will be toward the position P. With these definitions, we have

$$\vec{v} = (\omega d)\hat{\varphi} \quad \rightarrow \quad \vec{B} = \left(\frac{\mu_0 q}{4\pi}\right)\frac{((\omega d)\hat{\varphi}) \times \vec{r}}{r^3} \quad \rightarrow \quad \vec{B} = \left(\frac{\mu_0 q \omega d}{4\pi r^3}\right)(\hat{\varphi} \times \vec{r})$$

The variable d in this equation is the radius of the circular motion of each particle, as shown in Figure 4.3. The position vector $\vec{r}$ for each particle in this equation can also be written in cylindrical coordinates as[9]

$$\vec{r} = (-d)\hat{\rho} + (z)\hat{z}$$

Hence,

$$\vec{B} = \left(\frac{\mu_0 q\omega d}{4\pi r^3}\right)\left(\hat{\varphi}\times\left[(-d)\hat{\rho}+(z)\hat{z}\right]\right) \rightarrow \vec{B} = \left(\frac{\mu_0 q\omega d}{4\pi r^3}\right)\left[(-d)(\hat{\varphi}\times\hat{\rho})+(z)(\hat{\varphi}\times\hat{z})\right]$$

$$\vec{B} = \left(\frac{\mu_0 q\omega d}{4\pi r^3}\right)\left[(-d)(-\hat{z})+(z)\hat{\rho}\right] \rightarrow \vec{B} = \left(\frac{\mu_0 q\omega d}{4\pi r^3}\right)\left[(d)\hat{z}+(z)\hat{\rho}\right]$$

It is not practical to use this expression for the magnetic field for this system, however, since the unit vector $\hat{\rho}$ will point in a different direction for each of the point particles and will also constantly change direction as the system rotates[10]. We can simplify our calculation by rewriting this expression in terms of the x-axis and y-axis components of the magnetic field (*i.e.*, the x-axis and y-axis positions of the particles), as shown in Figure 4.4.

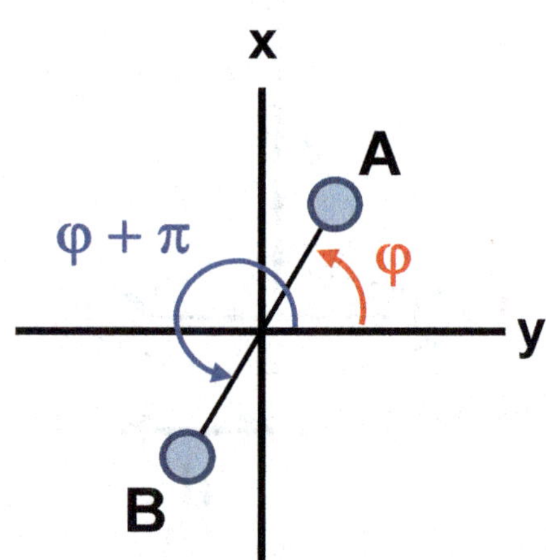

FIGURE 4.4: The x-axis and y-axis positions of the particles in Example 4-2.

$$\vec{B} = \left(\frac{\mu_0 q\omega d}{4\pi r^3}\right)\left[(d)\hat{z}+(z)\left((\cos\varphi)\hat{x}+(\sin\varphi)\hat{y}\right)\right]$$

Let's label the two charged point particles A and B. The magnetic field from particle A is

$$\vec{B}_A = \left(\frac{\mu_0 q\omega d}{4\pi r^3}\right)\left[(z\cos\varphi)\hat{x}+(z\sin\varphi)\hat{y}+(d)\hat{z}\right]$$

9 As noted in Section B-5, $\hat{\rho}$ points out from the origin of the reference frame. Thus, the component of the position P along the ρ-axis is "−d" with respect to each point particle.

10 We encountered a similar situation when calculating the electric field for the uniformly charged ring. The solution presented here for calculating the magnetic field for the system in Figure 4.3 follows the approach discussed in Section 3-7 for determining the electric field of the ring of net electric charge in Figure 3.14.

Since particle B is on the other side of the circle from particle A (see Figure 4.4), the magnetic field from particle B is

$$\vec{B}_B = \left(\frac{\mu_0 q\omega d}{4\pi r^3}\right)\left[\left(z\cos(\varphi+\pi)\right)\hat{x} + \left(z\sin(\varphi+\pi)\right)\hat{y} + (d)\hat{z}\right]$$

$$\vec{B}_B = \left(\frac{\mu_0 q\omega d}{4\pi r^3}\right)\left[(-z\cos\varphi)\hat{x} + (-z\sin\varphi)\hat{y} + (d)\hat{z}\right]$$

The net magnetic field of the system is the sum of the magnetic fields from the two point particles.

$$\vec{B} = \vec{B}_A + \vec{B}_B$$

$$\vec{B} = \left(\frac{\mu_0 q\omega d}{4\pi r^3}\right)\left[(z\cos\varphi - z\cos\varphi)\hat{x} + (z\sin\varphi - z\sin\varphi)\hat{y} + (d+d)\hat{z}\right]$$

$$\vec{B} = \left(\frac{\mu_0 q\omega d^2}{2\pi r^3}\right)\hat{z}$$

$$r^2 = d^2 + z^2 \quad \rightarrow \quad \vec{B} = \left(\frac{\mu_0 q\omega d^2}{2\pi\left(d^2+z^2\right)^{\frac{3}{2}}}\right)\hat{z}$$

Of course, we could have anticipated from the symmetry of the system that the magnetic field would have a component along only the *z*-axis.

4-3 Magnetic Force on a Moving Charged Object

Recall from Chapter 2 and Chapter 3 that the gravitational and electric fields are measures of gravitational force per unit mass and electric force per unit charge, respectively; the unit of the magnitude of the gravitational field is N/kg and the unit of the magnitude of the electric field is N/C. The unit of the magnitude of the magnetic field can be written as $\frac{\text{N}}{\text{C}\frac{\text{m}}{\text{s}}}$, suggesting that magnetic fields produce a magnetic force on moving charged particles (*i.e.*, on something which has the unit $\mathrm{C}\frac{\mathrm{m}}{\mathrm{s}}$ and is thus described by the product of electric charge and velocity). This magnetic force is defined in Equation 4-2.

$$\vec{F}_m = q\left(\vec{v} \times \vec{B}\right) \qquad \textbf{(4-2)}$$

In Equation 4-2, q is the net electric charge of the object, $\vec{v}$ is the object's velocity, and $\vec{B}$ is the magnetic field experienced by the object.

Example 4-3

Problem: A point particle with a net electric charge of -5 nC is moving with a velocity of $\vec{v} = \left(5\frac{\mathrm{m}}{\mathrm{s}}\right)\hat{x}$ through a region where there is a uniform magnetic field of $\vec{B} = \left(2\,\mathrm{T}\right)\hat{z}$. What is the magnetic force that acts on the particle?

Solution: From Equation 4-2 we have

$$\vec{F}_m = \left(-5\times10^{-9}\mathrm{C}\right)\left[\left(5\frac{\mathrm{m}}{\mathrm{s}}\right)\hat{x}\right]\times\left[\left(2\mathrm{T}\right)\hat{z}\right] \rightarrow \vec{F}_m = \left(-5\times10^{-9}\mathrm{C}\right)\left(5\frac{\mathrm{m}}{\mathrm{s}}\right)\left(2\frac{\mathrm{Ns}}{\mathrm{Cm}}\right)\left(\hat{x}\times\hat{z}\right)$$

$$\vec{F}_m = \left(-5\times10^{-8}\mathrm{N}\right)\left(-\hat{y}\right) \rightarrow \vec{F}_m = \left(5\times10^{-8}\mathrm{N}\right)\hat{y}$$

It follows from Equation 4-2 that the power supplied by the magnetic field to a moving charged object is zero.

$$P = \vec{F}\bullet\vec{v} \rightarrow P = \left(q\left(\vec{v}\times\vec{B}\right)\right)\bullet\vec{v} \rightarrow P = q\left(\vec{v}\times\vec{B}\right)\bullet\vec{v} \rightarrow P = 0$$

Since the power supplied by the magnetic force is zero, the magnetic field does no work on the object.

A magnetic field cannot do work on a moving charged object.

Furthermore, since the magnetic force exists only when the charged object is moving (Equation 4-2), we can refine this statement to be

A magnetic field cannot do work on a charged object.

Indeed, since the direction of the magnetic force acting on a moving charged object is perpendicular to the velocity of the object, the acceleration associated with the magnetic force can change the direction of the object's velocity, but not its speed. Imagine a point particle with a net electric charge $+q$ moving with speed v in the presence of a uniform magnetic field with magnitude B. Let's further assume that the direction of the magnetic field and the direction of particle's velocity are perpendicular, as shown in Figure 4.5.

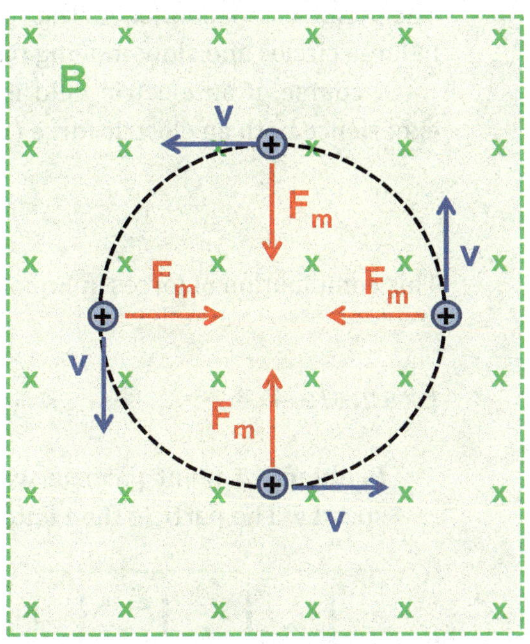

FIGURE 4.5: A point particle with a net positive electric charge moving in the presence of a magnetic field. The direction of the magnetic field is into the page.

If the net force[11] acting on the point particle is the magnetic force, then we can relate the magnitude of this net force to the magnitude of the object's acceleration using Newton's 2nd law.

$$F_{net} = F_m \quad \rightarrow \quad F_{net} = |q| vB \quad \rightarrow \quad ma = |q| vB$$

As shown in Figure 4.5, this net force will cause the point particle to move in a circle and, therefore, the acceleration associated with this net force will be the centripetal acceleration.

$$a = \frac{v^2}{r} \quad \rightarrow \quad m\frac{v^2}{r} = |q| vB$$

The period of this circular motion is therefore

$$T = \frac{2\pi r}{v} \quad \rightarrow \quad T = \frac{2\pi r}{\frac{r|q| B}{m}} \quad \rightarrow \quad T = \frac{2\pi m}{|q| B}$$

11 We will ignore other forces, such as gravity and air resistance.

The period is independent of the speed of the point particle. Fast-moving particles move in large circles and slow-moving particles move in small circles.

Of course, if an electric field is also present then the moving charged object will experience both an electric force (Equation 3-11) and a magnetic force (Equation 4-2).

$$\vec{F} = q\left(\vec{E} + \vec{v} \times \vec{B}\right) \tag{4-3}$$

This combination of forces in Equation 4-3 is commonly referred to as the *Lorentz force*.

Example 4-4

Problem: A point particle with a net electric charge of $+Q$ is moving with speed v. The particle then enters a region of space with perpendicular electric and magnetic fields, as shown in Figure 4.6. What speed must the particle have to avoid any deflection from either field?

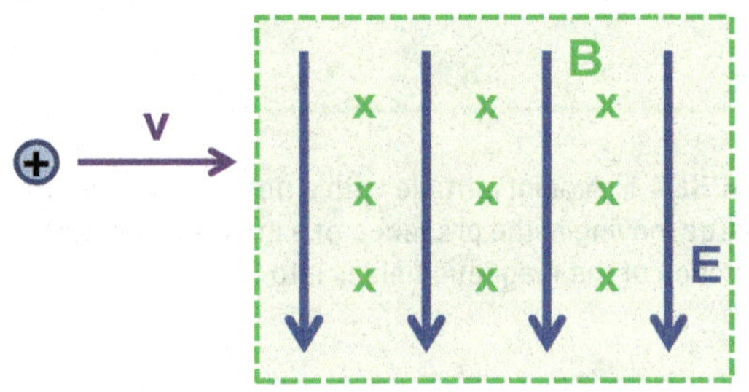

FIGURE 4.6: A moving charged point particle interacting with crossed electric and magnetic fields.

Solution: If the particle is not deflected, its velocity must be constant. This requires that the acceleration of the particle must be zero which from Newton's 2nd law requires that the net force acting on the particle must be zero. Substitution of $\vec{F} = 0$ into Equation 4-3 gives us

$$\vec{F} = 0 \quad \rightarrow \quad q\left(\vec{E} + \vec{v} \times \vec{B}\right) = 0 \quad \rightarrow \quad \vec{E} + \vec{v} \times \vec{B} = 0$$

We see from Figure 4.6 that the direction of $\vec{v} \times \vec{B}$ is opposite the direction of $\vec{E}$. Thus, the magnitudes of these two terms must be equal for the net force to be zero. Since $\vec{v}$ is perpendicular to $\vec{B}$ we have

$$E = vB \quad \rightarrow \quad v = \frac{E}{B}$$

Only particles whose speed matches this condition will not be deflected by the crossed fields in Figure 4.6. Interestingly, this result is independent of the magnitude and the (±) sign of the net electric charge of the particle, as well as the particle's mass.

4-4 Electric Current

An electric current is as a flow of charged objects. More specifically, electric current is defined to be a net flow of electric charge through a surface[12]. We can define current mathematically using Equation 4-4.

$$I = \frac{dq}{dt} \tag{4-4}$$

The units of electric current is electric charge per time (C/s), which we define as an ampere[13] (A).

$$1\,\mathrm{A} = 1\frac{\mathrm{C}}{\mathrm{s}}$$

It follows from Equation 4-4 that a positive current is associated with the flow of net positive electric charge through a surface, and a negative current is associated with the flow of net negative electric charge through a surface.

A very dramatic example of current is lightning, in which case the charged objects are flowing through the air. Although many other types of current exist, we are most familiar with the current flowing through wires. In this case, it is the negatively charged electrons that are moving in response to the electric field (or difference in electric potential) that exists in the wire. However, *we nevertheless define the current in the wire as though the moving charged objects were positively charged*. Thus, the direction of the current in a wire is opposite the direction of motion of the electrons (*i.e.*, is opposite the direction of the moving charged objects) that constitute the current. This annoying phenomenon is due to Benjamin Franklin's assignment of the negative charge (rather than the positive charge) to the electron. However, we forgive him this inconvenience for all his help with the founding of our country.

12 We will discuss this further in Chapter 7.

13 The abbreviation "amp" is commonly used for ampere.

4-5 Magnetic Field Produced by an Electric Current

Since an electric current is a flow of charged objects, we know from Section 4-3 that electric currents will produce magnetic fields. The equation for the magnetic field created by an electric current can be derived from Equation 4-1. Let's assume that the current consists of infinitesimal bits of net electric charge, dq, moving at a velocity $\vec{v}$. The magnetic field from each of these bits of net electric charge would be

$$d\vec{B} = \left(\frac{\mu_0}{4\pi}\right)\frac{(dq)\vec{v}\times\vec{r}}{r^3}$$

The velocity of this bit of net electric charge can be expressed in terms of the distance it travels. Let's define the axis for this motion to be the s-axis.

$$(dq)\vec{v} = (dq)\frac{d\vec{s}}{dt} \rightarrow (dq)\vec{v} = \left(\frac{dq}{dt}\right)d\vec{s} \rightarrow (dq)\vec{v} = Id\vec{s}$$

Thus, the magnetic field created by each bit of the current is

$$d\vec{B} = \left(\frac{\mu_0}{4\pi}\right)\frac{I(d\vec{s}\times\vec{r})}{r^3} \qquad \textbf{(4-5)}$$

The total magnetic field produced by the current is found by integrating Equation 4-5 with respect to the path of the current (*e.g.*, the length of a wire).

Example 4-5

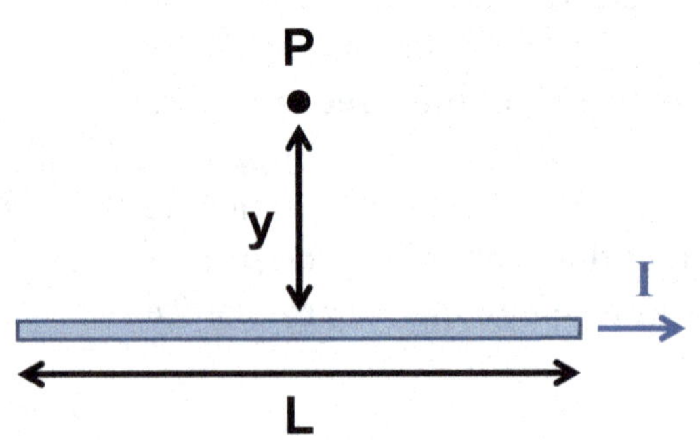

FIGURE 4.7: The system in Example 4-5.

Problem: A wire of length L carries a current I, as shown in Figure 4.7. What is the magnetic field produced by this current at the position P that lies along the perpendicular bisector of the wire?

Solution: Let's define an s-axis for this system that is parallel to the wire with an origin at the wire's perpendicular bisector. The positive direction for the s-axis will be the same as the direction of the current, as shown in

Figure 4.8. We will also define a y-axis for this system with an origin also at the wire's perpendicular bisector. The positive direction for the y-axis will point from the wire to the position P.

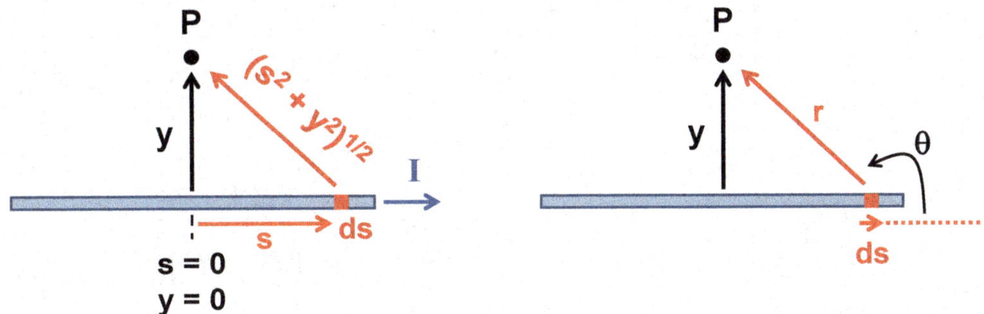

FIGURE 4.8: Determining the magnetic field from a segment of the wire in Figure 4.7.

We can divide the wire in Figure 4.7 into a series of infinitesimally small segments of length *ds*, and determine the magnetic field from each of these segments using Equation 4-5[14]. Let's first determine the magnitude of the magnetic field.

$$d\vec{B} = \left(\frac{\mu_0}{4\pi}\right)\frac{I\left(d\vec{s} \times \vec{r}\right)}{r^3} \rightarrow dB = \left(\frac{\mu_0}{4\pi}\right)\frac{I\left(ds\right)r}{r^3}\sin\theta \rightarrow dB = \left(\frac{\mu_0 I}{4\pi}\right)\frac{ds}{r^2}\sin\theta$$

The angle θ in this equation is the angle between $d\vec{s}$ and $\vec{r}$, as shown in Figure 4.8.

$$\sin\theta = \frac{y}{r} \rightarrow dB = \left(\frac{I\mu_0 y}{4\pi}\right)\frac{ds}{r^3}$$

We can then use the Pythagorean theorem to related r and y to the variable s, which denotes the distance of the infinitesimally small segment from the origin (Figure 4.8).

$$r^2 = s^2 + y^2 \rightarrow dB = \left(\frac{I\mu_0 y}{4\pi}\right)\frac{ds}{\left(s^2 + y^2\right)^{\frac{3}{2}}}$$

14 This is the same approach we have used previously to determine the gravitational field from a continuous distribution of mass and the electric field from a continuous distribution of net electric charge.

The total magnitude of the magnetic field from the current is found by integrating this expression with respect to s over the entire length of the wire.

$$B = \int dB \quad \rightarrow \quad B = \left(\frac{I\mu_0 y}{4\pi}\right)\int_{-\frac{L}{2}}^{\frac{L}{2}} \frac{ds}{\left(s^2 + y^2\right)^{\frac{3}{2}}} \quad \rightarrow \quad B = \left(\frac{I\mu_0 y}{4\pi}\right)\left(\left.\frac{s}{y^2\left(y^2 + s^2\right)^{\frac{1}{2}}}\right|_{-\frac{L}{2}}^{\frac{L}{2}}\right)$$

$$B = \left(\frac{I\mu_0 y}{4\pi}\right)\frac{2L}{y^2\left(4y^2 + L^2\right)^{\frac{1}{2}}} \quad \rightarrow \quad B = \frac{I\mu_0 L}{2\pi y\left(4y^2 + L^2\right)^{\frac{1}{2}}}$$

The direction of the magnetic field can be found from the cross-product in Equation 4-5 using the right-hand rule. The direction of the magnetic field is out of the page at the position P for each segment ds of the wire. Thus, the net magnetic field at the position P is also directed out of the page.

We can also solve for the direction of the magnetic field using an extension of the right-hand rule introduced in Section 4-2. Point the thumb of your right hand in the direction of the current and the fingers of your right hand will curl in the direction of the magnetic field, as shown in Figure 4.9.

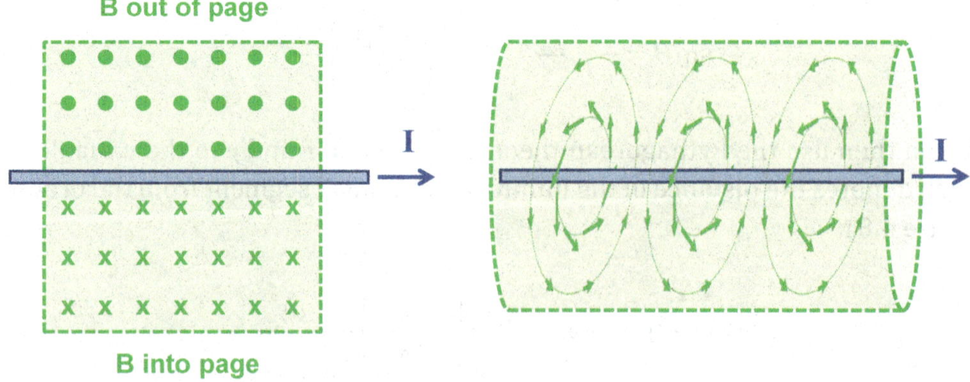

FIGURE 4.9: The direction of the magnetic field around a current carrying wire.

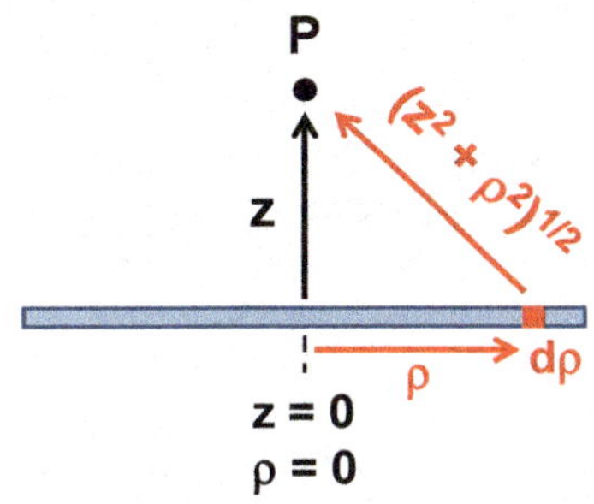

FIGURE 4.10: Determining the magnetic field from a segment of the wire in Figure 4.7 using cylindrical coordinates.

The cylindrical symmetry of the magnetic field from a current carrying wire shown in Figure 4.9 results from the fact that the current in the wire consists of moving charged particles whose magnetic fields are cylindrically symmetric. To explore this further, let's recalculate the magnetic field of the current-carrying wire using a reference frame with cylindrical coordinates, as shown in Figure 4.10. In this reference frame, the z-axis is aligned parallel to the direction of the current in the wire, and the ρ-axis is perpendicular to the z-axis; the origin of both of these axes is at the perpendicular bisector of the wire.

The infinitesimal bit of length $d\vec{s}$ in Equation 4-5 can then be written as an infinitesimal bit of length $d\vec{z}$ along the z-axis.

$$d\vec{s} = d\vec{z} = (dz)\hat{z} \rightarrow d\vec{B} = \left(\frac{\mu_0}{4\pi}\right)\frac{I\left(\left((dz)\hat{z}\right)\times\vec{r}\right)}{r^3} \rightarrow d\vec{B} = \left(\frac{\mu_0}{4\pi}\right)\frac{Idz}{r^3}(\hat{z}\times\vec{r})$$

$$\vec{r} = (\rho)\hat{\rho} + (z)\hat{z} \rightarrow \hat{z}\times\vec{r} = \hat{z}\times\left[(\rho)\hat{\rho} + (z)\hat{z}\right] \rightarrow \hat{z}\times\vec{r} = (\rho)(\hat{z}\times\hat{\rho})$$

$$\hat{z}\times\vec{r} = (\rho)\hat{\varphi} \rightarrow d\vec{B} = \left(\frac{\mu_0}{4\pi}\right)\frac{I\rho dz}{r^3}\hat{\varphi}$$

The length of the position vector can be related to ρ and z using trigonometry, as shown in Figure 4.10.

$$r = \left(z^2 + \rho^2\right)^{\frac{1}{2}} \rightarrow d\vec{B} = \left(\frac{\mu_0}{4\pi}\right)\frac{I\rho dz}{\left(z^2 + \rho^2\right)^{\frac{3}{2}}}\hat{\varphi}$$

We now have an expression for the infinitesimal bit of magnetic field (magnitude and direction) from each infinitesimal bit of length of wire through which the current is flowing. The total magnetic field from the wire is found by summing (*i.e.*, integrating) all of these infinitesimal bits of magnetic field along the entire length of the wire.

$$\vec{B} = \int d\vec{B} \rightarrow \vec{B} = \left(\left(\frac{I\mu_0\rho}{4\pi}\right)\int_{-\frac{L}{2}}^{\frac{L}{2}}\frac{dz}{\left(z^2 + \rho^2\right)^{\frac{3}{2}}}\right)\hat{\varphi}$$

$$\vec{B} = \left(\left(\frac{I\mu_0\rho}{4\pi} \right) \left(\frac{z}{\rho^2\left(\rho^2 + z^2\right)^{\frac{1}{2}}} \Bigg|_{-\frac{L}{2}}^{\frac{L}{2}} \right) \right) \hat{\varphi} \rightarrow \vec{B} = \left(\left(\frac{I\mu_0\rho}{4\pi} \right) \frac{2L}{\rho^2\left(4\rho^2 + L^2\right)^{\frac{1}{2}}} \right) \hat{\varphi}$$

$$\vec{B} = \left(\frac{I\mu_0 L}{2\pi\rho\left(4\rho^2 + L^2\right)^{\frac{1}{2}}} \right) \hat{\varphi}$$

The magnitude of the magnetic field derived here is the same as that obtained in the solution of Example 4-5. The advantage of this approach, however, is that we have simultaneously also calculated the direction of the magnetic field without having to separately use the right-hand rule.

The magnetic field of a system of currents is the summation of the magnetic field from each of the currents.

Example 4-6

Problem: Two parallel current carrying wires of length L are separated by a distance d as shown in Figure 4.11. The magnitude of the current in each wire is the same and equal to I, but the directions of the currents are opposite.

What is the magnetic field produced by this current at the position P that lies halfway between the wires along the perpendicular bisectors of the wires?

Solution: The magnitude of the magnetic field from each current will be identical at the position P, since that point is at the same distance from each wire. The direction of the magnetic field from each current can be found using the right-hand rule or through direct calculation in cylindrical coordinates, as discussed previously. Through either calculation, we see that the direction of the magnetic field at the position P from the top wire in Figure 4.11 will point out of the page, and the direction of the magnetic field at the position P from the bottom wire in Figure 4.11 will also

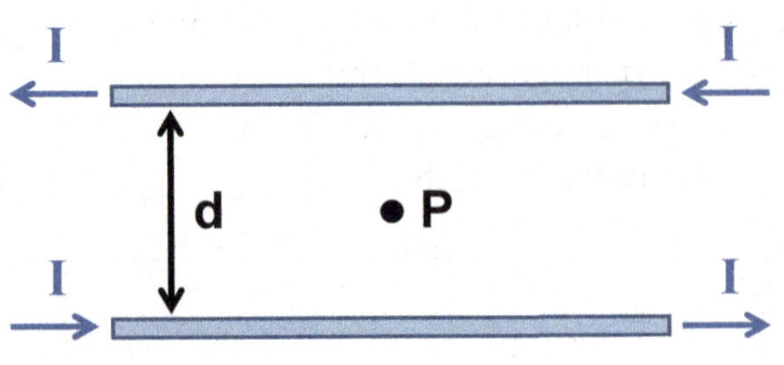

FIGURE 4.11: The system of two wires in Example 4-6.

point out of the page. Therefore, the magnetic field at the position P is directed out of the page with a magnitude of

$$B = 2\left(\frac{I\mu_0 L}{2\pi d\left(4\left(\frac{d}{2}\right)^2 + L^2\right)^{\frac{1}{2}}}\right) \rightarrow B = \frac{I\mu_0 L}{\pi d\left(d^2 + L^2\right)^{\frac{1}{2}}}$$

Example 4-7

Problem: The wire in Figure 4.12 consists of two straight segments and one segment in the shape of a semicircular arc with radius *R*.

The wire carries a current *I*. What is the magnetic field of the wire at the position P in Figure 4.12 that lies at the center of the semicircular arc?

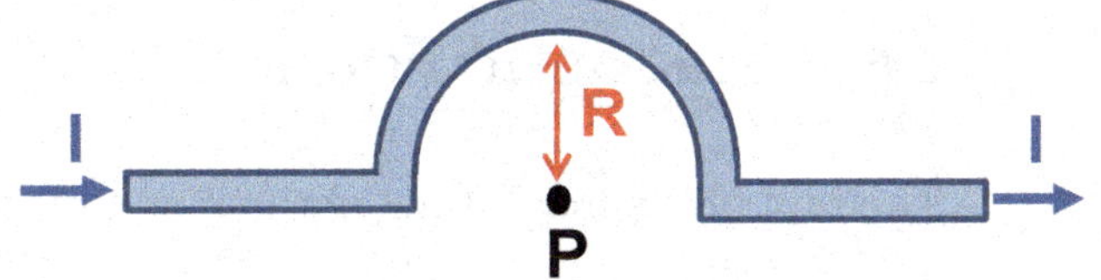

FIGURE 4.12: The wire in Example 4-7.

Solution: It is easiest to determine the magnetic field by dividing the wire into three segments, labeled A, B, and C in Figure 4.13.

The magnetic field at position P from the current in wire segments A and C is zero since the angles between $d\vec{s}$ and $\vec{r}$ for those segments are 0° and 180°, respectively; in other words, the cross product in Equation 4-5 for the current in those segments is zero. For the current in segment B, $d\vec{s}$ is always perpendicular to $\vec{r}$.

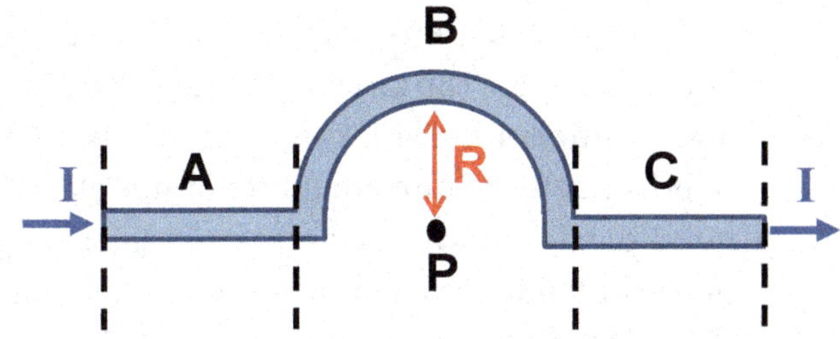

FIGURE 4.13: Wire segments for calculating the magnetic field in Example 4-7.

Thus, the magnitude of the magnetic field at the position P from the current in segment B is

$$B = \int dB \quad \rightarrow \quad B = \int \left(\frac{\mu_0}{4\pi} \right) \frac{I(ds)r}{r^3} \quad \rightarrow \quad B = \left(\frac{\mu_0}{4\pi} \right) \int \frac{I\,ds}{r^2} \quad \rightarrow \quad B = \left(\frac{\mu_0}{4\pi} \right) \frac{I}{R^2} \int ds$$

The integral in this equation is simply the circumference of the arc.

$$\int ds = \pi R \quad \rightarrow \quad B = \left(\frac{\mu_0}{4\pi} \right) \frac{I}{R^2} (\pi R) \quad \rightarrow \quad B = \frac{\mu_0 I}{4R}$$

The direction of the magnetic field can be found from the right-hand rule. In this case, the magnetic field is directed into the page.

4-6 Magnetic Force Acting on a Current Carrying Wire

The unit of the magnitude of the magnetic field can also be written as

$$1\,\mathrm{T} = 1 \frac{\mathrm{N}}{\mathrm{C}\frac{\mathrm{m}}{\mathrm{s}}} \quad \rightarrow \quad 1\,\mathrm{T} = 1 \frac{\mathrm{N}}{\left(\frac{\mathrm{C}}{\mathrm{s}} \right) \mathrm{m}} \quad \rightarrow \quad 1\,\mathrm{T} = 1 \frac{\mathrm{N}}{\mathrm{Am}}$$

Therefore, we would expect by analogy to the gravitational fields and electric fields that the magnetic field will exert a magnetic force on a length of wire carrying current (*i.e.*, on something which has the unit Am and is thus described by the product of current and length).

Indeed, a current-carrying wire in a magnetic field will experience a magnetic force, because each of the moving charged particles that constitute the current will experience a magnetic force from the magnetic field. If we consider the current to be composed of effectively infinitesimal bits of net electric charge (*i.e.*, infinitesimally small charged point particles), we can determine the net magnetic force on the wire, $\vec{F}_m$, from the sum of the individual magnetic forces exerted on these point particles, $d\vec{F}_m$. In the limit that these individual magnetic forces are infinitesimally small, we can replace the summation with an integral.

$$\vec{F}_m = \int d\vec{F}_m$$

We can use Equation 4-2 to determine the infinitesimal magnetic force associated with an infinitesimally small charged point particle with net electric charge dq moving through a magnetic field.

$$d\vec{F}_m = dq\left(\vec{v} \times \vec{B}\right)$$

We can then use the same substitutions that we used in Section 4-5 to express this force in terms of the current in the wire.

$$d\vec{F}_m = dq\left(\frac{d\vec{s}}{dt} \times \vec{B}\right) \rightarrow d\vec{F}_m = \frac{dq}{dt}\left(d\vec{s} \times \vec{B}\right)$$

$$d\vec{F}_m = I\left(d\vec{s} \times \vec{B}\right) \tag{4-6}$$

Example 4-8

Problem: An infinitely long wire carries a current I. A segment of that wire, with a length L, is in the presence of a uniform magnetic field of magnitude B as shown in Figure 4.14.

What is the magnetic force exerted on the segment of the wire by the magnetic field?

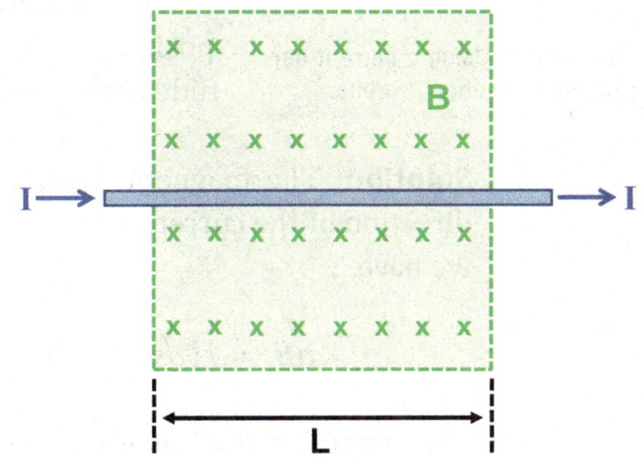

FIGURE 4.14: The wire in Example 4-8.

Solution: The magnitude of the force can be determined using Equation 4-6. Since the direction of the current is perpendicular to the direction of the magnetic field, we have

$$dF_m = I\left(ds\right)B \rightarrow F_m = \int I\left(ds\right)B \rightarrow F_m = IB\int ds$$

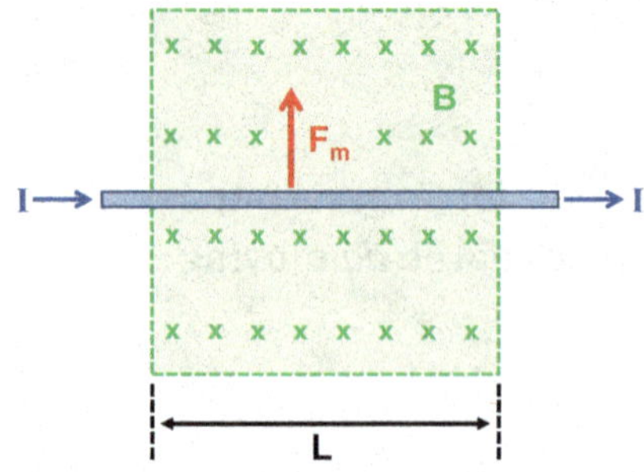

FIGURE 4.15: The magnetic force exerted on the wire in Example 4-8.

The integral in this equation is simply the length of the wire segment.

$$\int ds = L \quad \rightarrow \quad F_m = IBL$$

The direction of the force can be found using the right-hand rule. For the system shown in Figure 4.14, the direction of the magnetic field is upwards, as shown in Figure 4.15.

Example 4-9

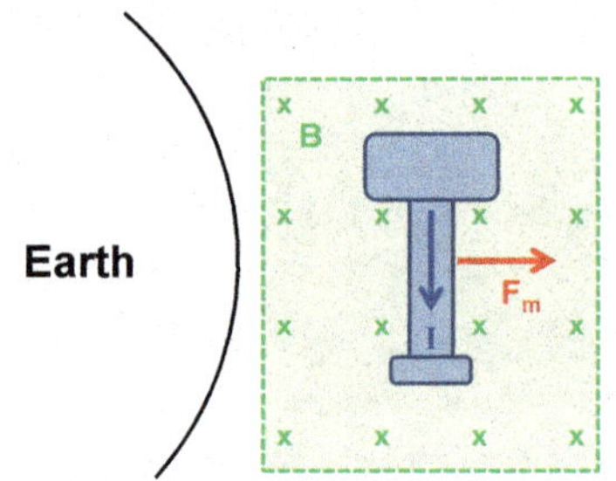

FIGURE 4.16: Using a current carrying tether to change orbits.

Problem: Spacecraft and satellites in orbit around the Earth can use electric currents in tethers to change their orbits (*i.e.*, their distance from the Earth). Let's imagine that a spacecraft generates a current of 2 A in a 15 m tether aligned perpendicular to the Earth's magnetic field, as shown in Figure 4.16.

The magnitude of the magnetic field of the Earth at the position of the satellite's orbit is 20 μT. What is the magnitude of the magnetic force acting on the tether/satellite?

Solution: The magnetic force is determined using Equation 4-6. Since the direction of the current is perpendicular to the direction of the magnetic field, we have

$$dF_m = I(ds)B \quad \rightarrow \quad F_m = \int I(ds)B \quad \rightarrow \quad F_m = IB\int ds$$

The integral in this equation is simply the length of the wire.

$$\int ds = L \quad \rightarrow \quad F_m = IBL \quad \rightarrow \quad F_m = (2\,\mathrm{A})(20\times10^{-6}\,\mathrm{T})(15\,\mathrm{m})$$

$$F_m = 0.6\,\mathrm{mN}$$

Now let's re-examine the situation shown in Example 4-6, in which two parallel wires carry current. The current in each wire will create a magnetic field that will then exert a force on the current in the other wire. From the application of the right-hand rule to the vector cross-product in Equation 4-6, we see that the direction of the magnetic forces exerted by the wires on each other will cause the wires to move apart, as shown in Figure 4.17.

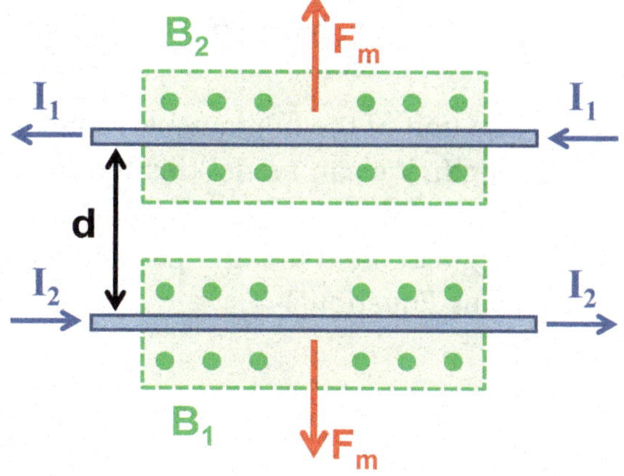

FIGURE 4.17: The magnetic forces exerted on parallel wires carrying current in opposite direction cause the wires to repel each other. The magnetic field B_1 is generated by the current I_1 and acts on the current I_2. The magnetic field B_2 is generated by the current I_2 and acts on the current I_1.

Similarly, if the wires were carrying current in the same direction the magnetic forces would cause the wires to be attracted to each other.

These results are interesting from a thermodynamics point of view. Specifically, we know from the 2nd law of thermodynamics that systems will always arrange themselves in order to minimize their total potential energy. Therefore, the current in the wires appears to generate some kind of potential energy that exists between the two wires, and, furthermore, this *magnetic potential energy* depends upon the relative directions of the currents in the wires.

Finally, it is clear from the examples in this section that the magnetic force exerted by a magnetic field on a current carrying wire can cause displacement of the wire and therefore appears to do work on the current in the wire. In order to reconcile this observation with the previous statement that "a magnetic field cannot do work on a moving charged object" we recognize that the motions of the charged objects constituting the current in a wire are constrained by the wire itself. Indeed, some of these charged objects are unable to undergo the circular motion illustrated in Figure 4.5 because of collisions with the boundaries of the wire. The impulse associated with the momentum transfer of these collisions contributes to the magnetic force acting on the wire and hence to the work done on the wire (and by extension on the current in the wire). In this way, we argue that the work reflects a transfer of kinetic energy from the charged objects in the current to the wire. Therefore, the ultimate source of the energy responsible for the work done on these systems can be considered to be the electric field (or electric potential) that drives the current in the wire and thereby provides the charged objects in the current with their kinetic energy[15].

15 We will return to this discussion again in Section 6-6.

4-7 Magnetic Dipoles

The unit of the fundamental source of the magnetic field is the product of current and length. We can model this fundamental source of magnetism as a closed loop of current. Such an idealized loop of current is shown in Figure 4.18. We can determine the magnetic field along a perpendicular bisector of the loop (the z-axis in Figure 4.18) using Equation 4-5.

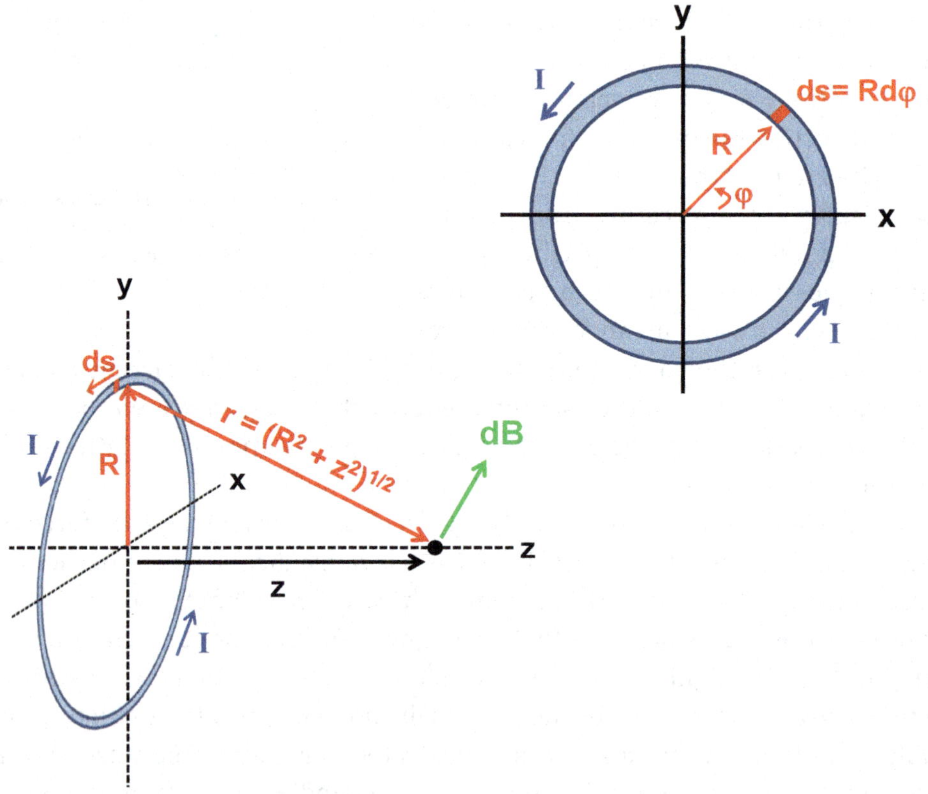

FIGURE 4.18: The magnetic field from a circular loop of current.

Let's model our solution to this problem on the calculation of the electric field of a uniformly charged ring (Section 3-7), since the two problems are so similar[16].

16 As noted in the solution to Example 4-2, $\hat{\rho}$ points out from the origin of the reference frame. Thus, the ρ-axis component of the location of the perpendicular bisector along is "–R" with respect to each infinitesimal bit of current in the loop.

$$d\vec{s} = (Rd\varphi)\hat{\varphi} \rightarrow d\vec{B} = \left(\frac{\mu_0 I}{4\pi r^3}\right)\left(\left[(Rd\varphi)\hat{\varphi}\right]\times\left[(-R)\hat{\rho}+(z)\hat{z}\right]\right)$$

$$d\vec{B} = \left(\frac{\mu_0 I}{4\pi r^3}\right)\left[(R^2)\hat{z}+(Rz)\hat{\rho}\right]d\varphi \rightarrow d\vec{B} = \left(\frac{\mu_0 I}{4\pi\left(R^2+z^2\right)^{\frac{3}{2}}}\right)\left[(R^2)\hat{z}+(Rz)\hat{\rho}\right]d\varphi$$

$$\vec{B} = \frac{\mu_0 I}{4\pi}\int\left(\frac{(R^2)\hat{z}+(Rz)\hat{\rho}}{\left(R^2+z^2\right)^{\frac{3}{2}}}\right)d\varphi$$

The integral in this equation is with respect to ϕ. Since the direction of $\hat{\rho}$ changes as ϕ changes, it is best to rewrite this integral in terms of its x-axis and y-axis components (Figure 4.18).

$$\vec{B} = \frac{\mu_0 I}{4\pi}\int_0^{2\pi}\left(\frac{(R^2)\hat{z}+(Rz)\left((\cos\varphi)\hat{x}+(\sin\varphi)\hat{y}\right)}{\left(R^2+z^2\right)^{\frac{3}{2}}}\right)d\varphi$$

As shown previously (Section 3-7), the x-axis and y-axis components of this integral will be zero. As in the case of the uniformly charged ring, we argue here that the x-axis and y-axis components of the magnetic field from any bit of electric current will always be exactly cancelled by the x-axis and y-axis components of the magnetic field from a diametrically opposite bit of electric current. Thus, the net magnetic field for the entire loop will have a component along only the z-axis.

$$\vec{B} = \left[\frac{\mu_0 IR^2}{4\pi\left(R^2+z^2\right)^{\frac{3}{2}}}\int_0^{2\pi}d\varphi\right]\hat{z} \rightarrow \vec{B} = \left[\frac{\mu_0 IR^2 2\pi}{4\pi\left(R^2+z^2\right)^{\frac{3}{2}}}\right]\hat{z}$$

$$\vec{B} = \left[\frac{\mu_0 IR^2}{2\left(R^2+z^2\right)^{\frac{3}{2}}}\right]\hat{z}$$

In the limit of $z >> R$ (*i.e.*, when the magnetic field of the current loop is measured at a distance much larger than the radius of the loop), the magnitude of the magnetic field is

$$\lim_{z>>R} \vec{B} = \left(\frac{\mu_0 I R^2}{2z^3} \right) \hat{z}$$

In this limit, the magnitude of the magnetic field is inversely proportional to the cube of the distance away from the current loop. This is the same dependence on separation distance that we previously derived for an electric dipole[17] (Section 3-8). Indeed, we can rewrite our expression for this magnetic field as

$$\lim_{z>>R} \vec{B} = \left(\frac{\mu_0 \left(\pi I R^2 \right)}{2\pi z^3} \right) \hat{z} \quad \rightarrow \quad \lim_{z>>R} \vec{B} = \left(\frac{\mu_0 \mu_m}{2\pi z^3} \right) \hat{z}$$

By analogy to Equation 3-14, the magnetic dipole moment of the current loop, denoted by μ_m, is

$$\vec{\mu}_m = \left(\pi R^2 I \right) \hat{z}$$

It follows from this comparison that the fundamental source of the magnetic field is a magnetic dipole, either created by a moving electric charge or, equivalently, by current flowing in a wire. However, whereas the magnitude of the electric dipole moment is the product of charge and distance (Section 3-8), the magnitude of the magnetic dipole moment is the product of current and area, specifically the area enclosed by the current "loop" generating the magnetic dipole[18].

$$\mu_m = IA \qquad \textbf{(4-7)}$$

We might have anticipated this, of course, since we know that the magnitude of the magnetic field can be written as the product of current and length. Thus, by analogy to the electric dipole, the unit of the magnetic dipole must be the product of current and length and length, or current and area. The direction of the magnetic dipole moment associated with a closed loop of current can be determined using a right-hand rule. If you curl the fingers of your right hand in the direction of the current, the thumb

17 The magnetic field for this system can be written as a multipole expansion similar to the multipole expansion for the electric field (Section 3-10). See Section C-4 for more details.

18 This equation is valid for all closed paths of current, whether they are circular, square, or any other shape.

of your right hand will point in the direction of the magnetic dipole moment.

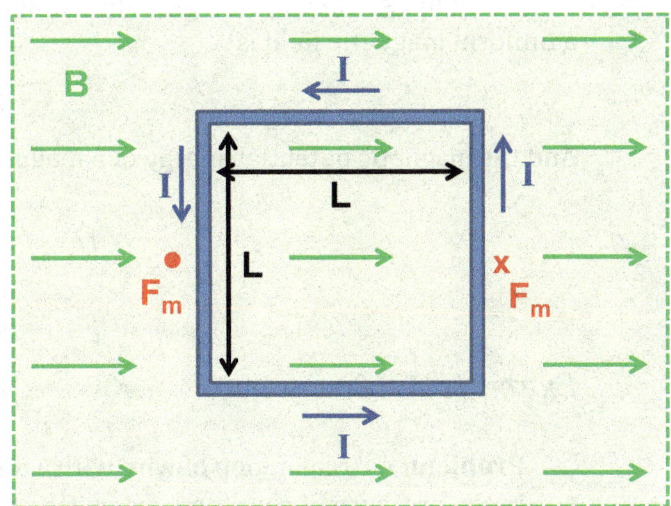

FIGURE 4.19: A square loop of current in a uniform magnetic field. The net magnetic force acting on the current carrying wires generates a net torque.

Now let's imagine a square "loop" of current placed in a uniform magnetic field, as shown in Figure 4.19. The magnetic force exerted on each segment of the current- carrying wire in this system can be determined using Equation 4-6.

The magnetic force is zero for the wire segments that are parallel or anti-parallel to the direction of the magnetic field (*i.e.*, on the top and bottom horizontal segments in Figure 4.19). The magnetic forces acting on the remaining two wire segments have the same magnitude, but point in opposite directions, and thus generate a net torque on the current loop. Using the result of Example 4-8, we see that the magnitude of this net torque is

$$\tau_{net} = 2\left(\frac{L}{2}F_m\right) \rightarrow \tau_{net} = 2\left(\frac{L}{2}\right)(IBL) \rightarrow \tau_{net} = IL^2 B$$

This expression can also be written in terms of the area of the current loop

$$L^2 = A \rightarrow \tau_{net} = IAB$$

We recognize immediately that the quantity IA in this expression is the magnetic dipole moment of the current loop (Equation 4-7).

$$\tau_{net} = \mu_m B$$

From Newton's 2nd law we know that this net torque will cause the current loop to rotate. As it starts to rotate, the angle between the direction of the current and the direction of the magnetic field (*i.e.*, the angle between $d\vec{s}$ and $\vec{B}$ in Equation 4-6) will change, and the associated torque will also change. This can be easily viewed as a change in the angle between $\vec{\mu}_m$ and $\vec{B}$. Indeed, by analogy to our discussion of electric dipoles in Section 3-8 we know that the torque acting on a magnetic dipole in a uniform magnetic field is

$$\vec{\tau} = \vec{\mu}_m \times \vec{B} \tag{4-8}$$

And the magnetic potential energy of a magnetic dipole in a magnetic field is[19]

$$U_{m,dipole} = -\vec{\mu}_m \bullet \vec{B} \tag{4-9}$$

Example 4-10

Problem: A circular loop of wire with a radius of 4 cm carries a current of 2 mA in the presence of a uniform magnetic field with a magnitude of 0.5 T. Initially the direction of magnetic dipole moment of the loop is parallel to the direction of the magnetic field. How much work would be required to rotate the current loop so that the direction of its magnetic dipole moment is anti-parallel to the direction of the magnetic field?

Solution: The work required is equal to the change in the magnetic potential energy of the current loop (*i.e.*, the change in the magnetic potential energy of the magnetic dipole associated with the current loop). This change in potential energy can be determined using Equation 4-9.

$$U_{m,dipole} = -\vec{\mu}_m \bullet \vec{B} \;\rightarrow\; U_{m,dipole} = -\mu_m B\cos\theta$$

$$\Delta U_{m,dipole} = \left(-\mu_m B\cos\theta_f\right) - \left(-\mu_m B\cos\theta_i\right)$$

$$\Delta U_{m,dipole} = -\mu_m B\left(\cos\theta_f - \cos\theta_i\right)$$

$$\Delta U_{m,dipole} = -\left(\pi\left(0.04\,\text{m}\right)^2\left(0.002\,\text{A}\right)\right)\left(0.5\,\text{T}\right)\left(\cos\left(180°\right) - \cos\left(0°\right)\right)$$

$$\Delta U_{m,dipole} = 10\;\mu\text{J} \;\rightarrow\; W = 10\;\mu\text{J}$$

19 The ultimate source of this potential energy is the electric field or electric potential that drives the current in the loop. This is analogous to the electric field or electric potential being the source of the energy associated with the work done by a magnetic field on a current carrying wire.

When placed in a uniform magnetic field, a magnetic dipole will align itself so that the direction of its magnetic dipole moment is the same as the direction of the magnetic field. This is the same behavior seen when an electric dipole is placed in a uniform electric field (Section 3-8). In both of this instances, this alignment of the dipole corresponds to the minimum potential energy of the system.

When a magnetic dipole is placed in a non-uniform magnetic field, the dipole will also experience a net magnetic force; similarly, an electric dipole will experience a net electric force when placed in a non-uniform electric field (Section 3-8). This net magnetic force can be determined from the magnetic potential energy of the dipole.

$$\vec{F}_{m,dipole} = -\nabla U_{m,dipole} \quad \rightarrow \quad \vec{F}_{m,dipole} = -\nabla\left(-\vec{\mu}_m \bullet \vec{B}\right)$$

$$\vec{F}_{m,dipole} = \nabla\left(\vec{\mu}_m \bullet \vec{B}\right) \qquad \textbf{(4-10)}$$

Thus, as shown in Figure 4.20, if the magnetic dipole is aligned with a non-uniform external magnetic field, it will experience a net magnetic force in the direction of increasing magnetic field intensity. Similarly, if the magnetic dipole is aligned opposite the direction of a non-uniform magnetic field, it will experience a net magnetic force in the direction of decreasing magnetic field intensity.

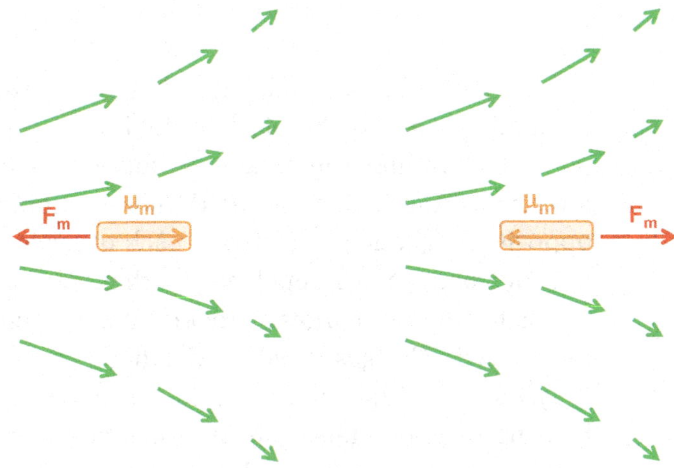

FIGURE 4.20: Directions of the net magnetic force acting on a magnetic dipole in a non-uniform magnetic field.

The net magnetic force acting on a magnetic dipole in a non-uniform magnetic field can also be understood by analogy to electric forces. Rather than positive and negative charges, we refer to north and south *poles*[20] of a magnetic dipole. As shown in Figure 4.21, *outside* of magnetic dipole the magnetic field always points away from the north pole and toward the south pole. Inside of the magnetic dipole, however, the magnetic field points from the south pole to the north pole. Similarly, the magnetic dipole moment always points from the south pole to the north pole. As discussed before, the magnetic field is continuous, forming closed "loops", and is therefore different

20 This terminology reminds us of the multipole expansion for distribution of charged objects in Section 3-10.

from the electric field which "originates" at positive electric charges and "terminates" at negative electric charges.

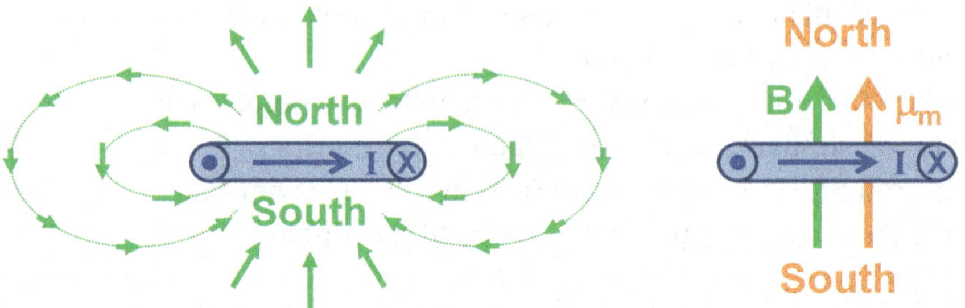

FIGURE 4.21: The north and south poles of a closed loop of current. The direction of the current is indicated by the blue arrow and the direction of the magnetic dipole moment by the orange arrow.

Nevertheless, an analogy between the magnetic field and the electric field is helpful in understanding the magnetic force associated with the magnetic field. For example, opposite magnetic poles attract each other and like magnetic poles repel each other, just as opposite electric charges attract each other and like electric charges repel each other (Section 3-5). As shown in Figure 4.22, this property can help us to understand the net magnetic force acting on a magnetic dipole in a non-uniform magnetic field. The two poles of the magnetic dipole will each experience a magnetic force in the presence of an external magnetic field, and these forces will pull on the dipole in opposite directions. If the external magnetic field is uniform, these two forces will have the same magnitude, and the net magnetic force will be zero. However, if the magnetic field is non-uniform, the magnetic force will have a larger magnitude where the magnitude of the magnetic field is larger. This will result in a net force acting on the magnetic dipole. If the dipole moment is aligned with the magnetic field, this net magnetic force

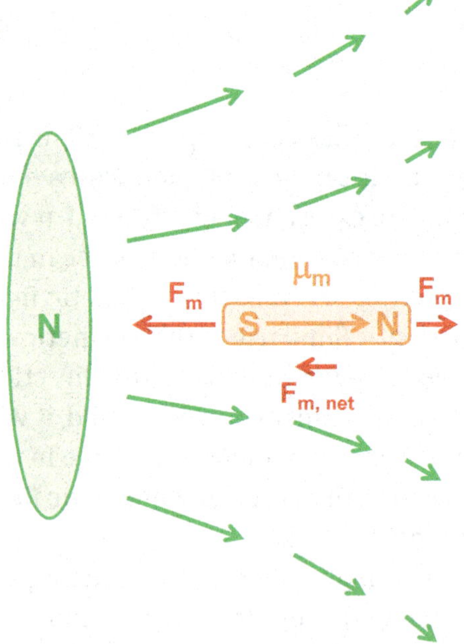

FIGURE 4.22: In a non-uniform magnetic field, the difference in the magnitude of the magnetic force acting on the two poles of a magnetic dipole results in a net force magnetic force acting on the magnetic dipole.

will pull the magnetic dipole toward the region where the magnitude of the magnetic field is larger.

Finally, we can use the formulism of magnetic poles to determine the net magnetic force acting between current loops. *If you curl the fingers of your right hand in the direction of the current, the thumb of your right hand will be the north pole of the magnetic dipole moment associated with the current.* Thus, as shown in Figure 4.23, if the direction of the current is the same in the two loops, the two loops will be attracted to each other. In contrast, if the current in the two loops are in opposite directions, the two loops will repel each other.

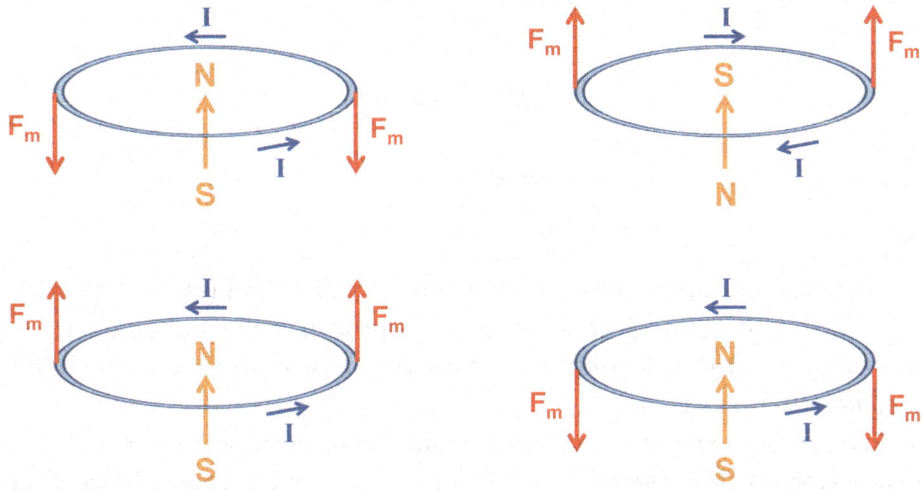

FIGURE 4.23: Net magnetic force acting between two current loops can be understood from the interactions between the associated magnetic dipoles.

This is, of course, exactly the same conclusion that that we derived from the forces acting between two current carrying wires.

4-8 The Bohr Model for the Atom, Part III

In the Bohr model of the hydrogen atom, the electron is orbiting the proton. Since the electron is moving, it will generate a magnetic field. In Section 3-9 we determined the speed of the electron to be

$$v_{electron} = 2.25 \times 10^6 \frac{\text{m}}{\text{s}}$$

The magnitude of the magnetic field of the electron at the center of its orbit (*i.e.*, at the position of the proton) can then be calculated using Equation 4-1. If the orbit of the electron is perfectly circular, we have

$$\|\vec{v}\times\vec{r}\| = vr \quad \rightarrow \quad B = \left(\frac{\mu_0 q}{4\pi}\right)\frac{vr}{r^3} \quad \rightarrow \quad B = \frac{\mu_0 q v}{4\pi r^2}$$

Substitution then gives us

$$B = \frac{\left(4\pi \times 10^{-7}\,\frac{\text{kgm}}{\text{C}^2}\right)\left(1.6\times 10^{-19}\,\text{C}\right)\left(2.18\times 10^{6}\,\frac{\text{m}}{\text{s}}\right)}{4\pi\left(5.3\times 10^{-11}\,\text{m}\right)^2}$$

$$B = 12.4\,\text{T}$$

The direction of the magnetic field depends upon the direction (clockwise *vs.* counter-clockwise) of the orbit of the electron around the proton. A magnetic field that results from the movement of the electrons in atoms is said to result from the *orbital magnetic dipole moment* of these electrons.

We can determine the magnitude of the orbital magnetic dipole moment by comparing our expression for the magnetic field of the electron at the center of its orbit to the expression for the magnetic field of a closed loop at the center of the loop (Section 4-7). If we denote the radius of the orbit and loop to be *R*, we have

$$\frac{\mu_0 q v}{4\pi R^2} = \frac{\mu_0 I}{2R} \quad \rightarrow \quad I = \frac{qv}{2\pi R}$$

$$\mu_m = IA \quad \rightarrow \quad \mu_m = \frac{qv}{2\pi R}\left(\pi R^2\right) \quad \rightarrow \quad \mu_m = \frac{qvR}{2}$$

Substitution gives us

$$\mu_m = \frac{\left(1.6\times 10^{-19}\,\text{C}\right)\left(2.18\times 10^{6}\,\frac{\text{m}}{\text{s}}\right)\left(5.3\times 10^{-11}\,\text{m}\right)}{2} \quad \rightarrow \quad \mu_m = 9.24\times 10^{-24}\,\text{Am}^2$$

The magnetic potential energy of a magnetic dipole in a uniform magnetic field depends upon the directions of the magnetic dipole moment and the magnetic field (Equation 4-9). Therefore, hydrogen atoms with electrons orbiting in different directions will have different magnetic potential energies when placed in a uniform magnetic field.

4-9 Permanent Magnets

The magnetic properties of permanent magnets, such as refrigerator magnets, arise from the intrinsic magnetic properties of the subatomic particles that constitute these materials. Both the proton and the electron have a magnetic dipole moment, called a *spin magnetic dipole moment*[21], which, along with charge and mass, is an intrinsic property of these particles. The net magnetic dipole moment of an atom is the sum of the spin magnetic dipole moments of the protons and electrons, together with the magnetic dipole moments associated with the movements of the electrons within the atom[22]; the latter quantities include the orbital magnetic dipole moments of the electrons discussed in Section 4-8. If the sum of these magnetic dipole moments produces a net magnetic dipole moment, then the material is magnetic. There are three types of magnetism: ferromagnetism, paramagnetism, and diamagnetism. The basic properties of these types of magnetism are summarized here, and we will return to a more thorough discussion of them in Chapter 7.

Ferromagnetism is a property of materials containing iron, nickel, and certain other elements. The net magnetic dipole moments of groups of atoms in these materials align with one another so that regions (called domains) of the material have a strong magnetic dipole moment. These domains are typically randomly aligned, but when an external magnetic field is then applied, all of the magnetic dipole moments of these domains will align with this magnetic field. When this magnetic field is then removed, this alignment remains, and the material has a strong permanent magnetic dipole. Refrigerator magnetics are made from ferromagnetic material.

Paramagnetism is a property typically exhibited by materials containing actinide elements, rare earth elements, and transition elements. The atoms of these materials contain a permanent dipole moment, but unlike ferromagnetic materials, these dipoles are all randomly aligned; *i.e.*, there are no domains of strong magnetic dipole moment

---

21 The term spin evokes the idea that the movement of charge (in this case spinning of the particle) causes a magnetic dipole moment. Unfortunately, this is not a realistic picture of how these particles behave or what gives rise to their intrinsic magnetic dipole moments.

22 As discussed in Section 2-7, the motion of the protons within the atom is insignificant compared to the motions of the electrons.

as in a ferromagnetic material. When placed in an external magnetic field, the magnetic dipole moments of these atoms will align with the field to create a net magnetic dipole moment inside the material. However, this alignment and the associated net magnetic dipole moment disappear once the material is removed from the magnetic field.

Diamagnetism is a very weak magnetic property exhibited by all materials. When placed in an external magnetic field, a diamagnetic material will develop a net magnetic dipole moment that is aligned opposite the direction of the external magnetic field. As with paramagnetism, this alignment is temporary and disappears once the material is removed from the external magnetic field.

Finally, the fact that the magnetic properties of electrons are the ultimate source of magnetism in matter explains another interesting property of magnets. Namely, that you are not able to isolate a north pole or south pole by breaking a magnet in half. Indeed, if you break a magnet in half you will produce two magnets, each with its own north and south pole (Figure 4.24). Since you can't break up the electron into constitute parts, you can't break up its magnetic dipole moment into separate north and south poles.

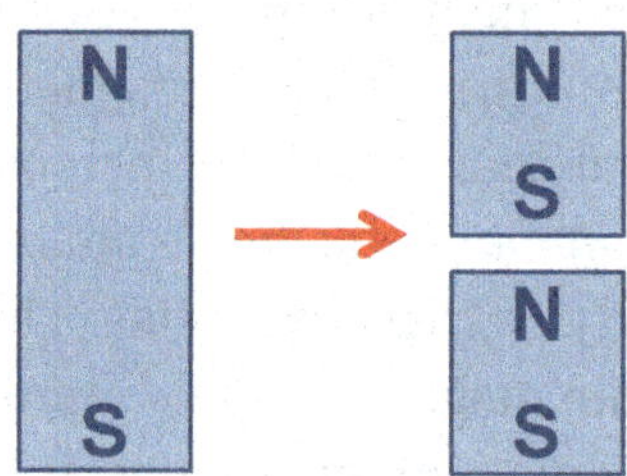

FIGURE 4.24: Breaking a magnet results in two new magnets, each with a north and south pole.

This fact is typically expressed as: magnetic monopoles do not exist.

4-10 Magnetic Potential

Based upon analogy to gravitational fields and electric fields, we would correctly assume that the magnetic field would also be associated with a potential. However, the differences between magnetic fields, electric fields, and gravitational fields suggest that the equation for this magnetic potential will be different from the equations for the gravitational or electric potentials. Gravitational fields always point toward objects with mass and terminate on these objects. Electric fields always point from positive electric charges, where they originate, to negative electric charges, where they terminate. Because of these properties, it is possible to define the relationship between these fields and their associated potentials through the gradient operations in Equation 2-7 and Equation 3-8[23]. However since the magnetic field forms closed loops, its potential

23 If this reasoning is not clear to you from your vector calculus class, you should feel inspired to take additional math and physics classes to understand why it is true.

cannot be described using the gradient[24]. Rather, the relationship between the magnetic field and the magnetic potential is given by Equation 4-11.

$$\vec{B} = \nabla \times \vec{V}_m \qquad \textbf{(4-11)}$$

The magnetic potential, denoted by $\vec{V}_m$ in Equation 4-11[25] is also commonly referred to as the vector potential, since it is a potential function that is also a vector[26]. The magnetic potential is thus a vector field, in contrast to the gravitational and electric potentials which are scalar fields. An equation for the magnetic potential[27] is

$$\vec{V}_m = \left(\frac{\mu_0}{4\pi}\right)\int \frac{Id\vec{s}}{r} \qquad \textbf{(4-12)}$$

Example 4-11

Problem: A wire of length L carries a current I, as shown in Figure 4.7. What is the magnetic potential produced by this current at the position P that lies along the perpendicular bisector of the wire segment?

Solution: As discussed in Section 4-5, the best reference frame for describing the magnetic field of a current-carrying wire uses cylindrical coordinates. Therefore, let's use cylindrical coordinates (Figure 4.8) for the integration in Equation 4-12 and define the positive direction for the z-axis to be the same as the direction of the current.

$$d\vec{s} = d\vec{z} = (dz)\hat{z} \quad \rightarrow \quad \vec{V}_m = \left(\left(\frac{\mu_0 I}{4\pi}\right)\int \frac{dz}{\left(z^2 + \rho^2\right)^{\frac{1}{2}}}\right)\hat{z}$$

24 This is the equivalent of saying that magnetic monopoles do not exist.

25 The magnetic potential is frequently represented by the variable $\vec{A}$. I have chosen to use $\vec{V}_m$ for the magnetic potential and $\vec{A}$ for the area vector, which we will encounter in the following chapters.

26 Similarly, the electric potential is commonly referred to as the scalar potential.

27 See Section C-5 for derivation.

$$\vec{V}_m = \left(\left(\frac{\mu_0 I}{4\pi} \right) \int_{-\frac{L}{2}}^{\frac{L}{2}} \frac{dz}{\left(z^2 + \rho^2 \right)^{\frac{1}{2}}} \right) \hat{z} \rightarrow \vec{V}_m = \left(\left(\frac{\mu_0 I}{4\pi} \right) \ln \left(z + \left(z^2 + \rho^2 \right)^{\frac{1}{2}} \right) \Bigg|_{-\frac{L}{2}}^{\frac{L}{2}} \right) \hat{z}$$

$$\vec{V}_m = \left(\left(\frac{\mu_0 I}{2\pi} \right) \ln \left(\frac{2\rho}{\left(L^2 + 4\rho^2 \right)^{\frac{1}{2}} - L} \right) \right) \hat{z}$$

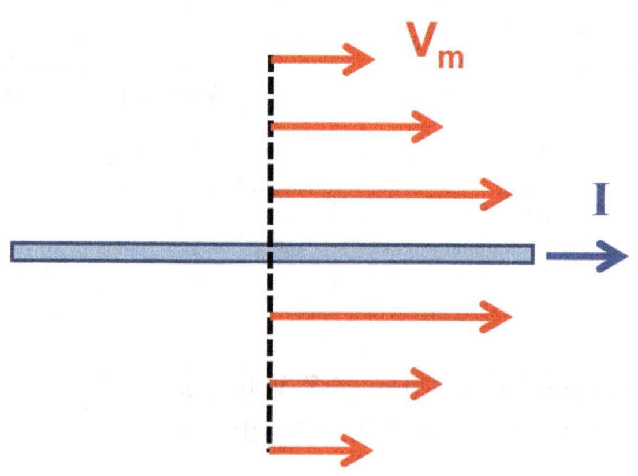

FIGURE 4.25: The magnetic potential for a straight current carrying wire.

The direction of the magnetic potential for the wire is thus parallel to the direction of the current in the wire, and has a magnitude that decreases with increasing distance away from the wire.

As shown in Figure 4.25 the magnitude of the magnetic potential decreases with increasing distance from the wire. Since the magnitude of this vector field is decreasing with respect to a direction perpendicular to the direction of the vector field (*i.e.*, the magnitude of the vector field along the z-axis is decreasing as the distance from the wire along the ρ-axis increases), we know that the curl of the vector field is non-zero[28].

We can examine the validity of our solution in Example 4-11 by calculating the associated magnetic field using Equation 4-11. We begin by recognizing that the magnetic potential for the wire has a component along the z-axis only.

$$\vec{V}_m = \left(V_m \right)_z \hat{z} \rightarrow \left(V_m \right)_z = \left(\frac{\mu_0 I}{2\pi} \right) \ln \left(\frac{2\rho}{\left(L^2 + 4\rho^2 \right)^{\frac{1}{2}} - L} \right)$$

$$\vec{B} = \nabla \times \vec{V}_m \rightarrow \vec{B} = \nabla \times \left(\left(V_m \right)_z \hat{z} \right)$$

28 See Section B-9.

Applying Equation B-11 gives us

$$\left(V_m\right)_\rho = \left(V_m\right)_\varphi = 0$$

$$\nabla \times \vec{V}_m = \left(\frac{1}{\rho}\frac{\partial \left(V_m\right)_z}{\partial \varphi}\right)\hat{\rho} + \left(-\frac{\partial \left(V_m\right)_z}{\partial \rho}\right)\hat{\varphi}$$

$$\nabla \times \vec{V}_m = \left(-\frac{\partial \left(V_m\right)_z}{\partial \rho}\right)\hat{\varphi}$$

Hence,

$$\vec{B} = \left(\frac{\mu_0 I}{2\pi}\right)\left(-\frac{\partial}{\partial \rho}\right)\left(\ln\left(\frac{2\rho}{\left(L^2+4\rho^2\right)^{\frac{1}{2}} - L}\right)\right)\hat{\varphi}$$

$$\vec{B} = \left(\frac{\mu_0 I}{4\pi}\right)\left(\frac{2L}{\rho\left(L^2+4\rho^2\right)^{\frac{1}{2}}}\right)\hat{\varphi} \rightarrow \vec{B} = \left(\frac{\mu_0 IL}{2\pi\rho\left(L^2+4\rho^2\right)^{\frac{1}{2}}}\right)\hat{\varphi}$$

This result is identical to the expression derived previously.

Example 4-12

Problem: What is the magnetic potential of a charged point particle, with net electric charge q, moving with velocity $\vec{v}$, measured a distance r away from the particle?

Solution: We can determine the solution through manipulation of Equation 4-12.

$$\vec{V}_m = \left(\frac{\mu_0}{4\pi}\right)\int \frac{I d\vec{s}}{r} \rightarrow d\vec{V}_m = \left(\frac{\mu_0}{4\pi}\right)\frac{I d\vec{s}}{r} \rightarrow d\vec{V}_m = \left(\frac{\mu_0}{4\pi}\right)\frac{dq}{dt}\frac{d\vec{s}}{r}$$

$$d\vec{V}_m = \left(\frac{\mu_0}{4\pi}\right)\frac{dq}{r}\frac{d\vec{s}}{dt} \rightarrow d\vec{V}_m = \left(\frac{\mu_0}{4\pi}\right)\frac{dq}{r}\vec{v}$$

Thus, for a point particle with net charge q,

$$\vec{V}_m = \int d\vec{V}_m \;\rightarrow\; \vec{V}_m = \int\left(\frac{\mu_0}{4\pi}\right)\frac{dq}{r}\vec{v} \;\rightarrow\; \vec{V}_m = \left(\frac{\mu_0}{4\pi r}\vec{v}\right)\int dq$$

$$\vec{V}_m = \left(\frac{\mu_0 q}{4\pi r}\right)\vec{v}$$

Although the magnetic field can be uniquely determined through experimental measurements, using Equation 4-1, using Equation 4-5, *etc.*, the magnetic potential is not uniquely defined by a distribution of currents or other moving charges. Indeed, there are multiple solutions for $\vec{V}_m$ in Equation 4-10 that generate the same value of $\vec{B}$; all of these possible $\vec{V}_m$ are equally valid. Consider the two magnetic potentials $\left(\vec{V}_m\right)_1$ and $\left(\vec{V}_m\right)_2$ defined as

$$\left(\vec{V}_m\right)_2 = \left(\vec{V}_m\right)_1 + \nabla\chi$$

In this expression, χ is an arbitrary scalar field. We see from Equation 4-10 that the magnetic fields associated with these two magnetic potentials are identical.

$$\nabla\times\left(\vec{V}_m\right)_2 = \nabla\times\left(\left(\vec{V}_m\right)_1 + \nabla\chi\right) \;\rightarrow\; \nabla\times\left(\vec{V}_m\right)_2 = \nabla\times\left(\vec{V}_m\right)_1 + \nabla\times\nabla\chi$$

$$\nabla\times\nabla\chi = 0 \;\rightarrow\; \nabla\times\left(\vec{V}_m\right)_2 = \nabla\times\left(\vec{V}_m\right)_1 \;\rightarrow\; \vec{B}_2 = \vec{B}_1$$

This result is reminiscent of our discussion of the ambiguity in the scalar fields for the gravitational potential (Section 2-5) and the electric potential (Section 3-6). Recall that we overcame this ambiguity by defining the gravitational potential (Equation 2-2) and the electric potential (Equation 3-1) to be zero infinitely far away from their source. Similarly, in the limit of large values of ρ, the previously derived equation for the magnetic potential for a current carrying wire (using Equation 4-12) is also zero.

$$\lim_{\rho>>L}\vec{V}_m = \left(\frac{\mu_0 I}{4\pi}\right)\left(\ln\left(\frac{2\rho}{2\rho}\right)\right)\hat{z} \;\rightarrow\; \lim_{\rho>>L}\vec{V}_m = \left(\frac{\mu_0 I}{4\pi}\right)\left(\ln(1)\right)\hat{z} \;\rightarrow\; \lim_{\rho>>L}\vec{V}_m = 0$$

The ambiguity in the definition of the magnetic potential for the wire is actually visible, however, in the limit of long lengths of wire.

$$\lim_{L\to\infty}\vec{V}_m=\left(\frac{\mu_0 I}{4\pi}\right)\left(\ln\left(\frac{L+L}{L-L}\right)\right)\hat{z} \;\rightarrow\; \lim_{L\to\infty}\vec{V}_m=\left(\frac{\mu_0 I}{4\pi}\right)\left(\ln(\infty)\right)\hat{z} \;\rightarrow\; \lim_{L\to\infty}\vec{V}_m=\infty$$

We typically circumvent this problem by defining a different *zero point* for the magnetic potential[29]; let's denote this zero point as ρ_0. With this definition, the magnetic potential for the infinite current carrying wire is

$$\vec{V}_m=\left(\frac{\mu_0 I}{2\pi}\right)\left(\ln\left(\frac{\rho_0}{\rho}\right)\right)\hat{z}$$

The fact that the vector and scalar potentials are not uniquely defined, but the magnetic, electric, and gravitational fields are uniquely defined (because we can measure the effects of the associated forces), motivates us to consider the fields, rather than the potentials, to be the physically *real* quantities.

Finally, the magnetic potential of a system of currents (or moving charged particles) is the sum of the magnetic potentials of each current (or moving charged particle).

Example 4-13

Problem: What is the magnetic potential at position P in Figure 4.11?

Solution: The magnetic potential at position P is the sum of the magnetic potentials from each of the wires at that point. Since the current in the wires is flowing in opposite directions, the directions of the magnetic potentials of each wire at the position P will also be opposite. The magnitude of the magnetic potentials from the wires will be the same at position P, since that position is equidistant from each wire. Therefore, the net magnetic potential at position P will be zero.

29 Recall from classical mechanics that the zero point of the potential energy did not affect the associated force or the kinematics of the system, since it is only changes in energy that affect kinematics. Similarly, it is only changes in potential that are significant since the associated fields are defined in terms of changes in potential.

4-11 Magnetic Potential Energy

The magnetic potential energy of a charged particle moving in a magnetic potential is

$$U_m = \vec{V}_m \cdot (q\vec{v}) \tag{4-13}$$

Similarly, the component of the magnetic potential energy of an infinitesimal bit of electric current is

$$dU_m = \vec{V}_m \cdot Id\vec{s} \tag{4-14}$$

The total magnetic potential energy of the system is thus found by integrating Equation 4-14 with respect to the path of the current (*e.g.*, the length of a wire).

Example 4-14

Problem: A charged point particle with net electric charge $+Q$ is moving at speed v a distance ρ away from a current carrying wire of length L, as shown in Figure 4.26.

The direction of the velocity of the point particle is the same as the direction of the current. What is the magnetic potential energy of the interaction of the point particle and wire?

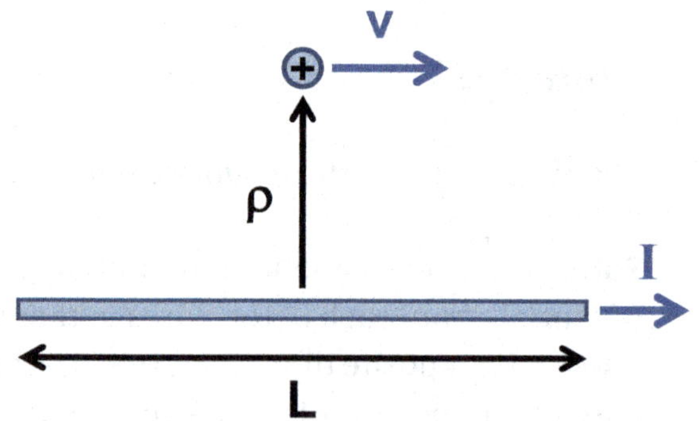

FIGURE 4.26: The system in Example 4-14.

Solution: The magnetic potential energy can be found using Equation 4-13 or Equation 4-14. Let's calculate the magnetic potential energy by from the point of view of the current in the wire interacting with the magnetic potential of the moving point particle. We will use cylindrical coordinates and define the direction of the velocity of the charged point particle and the current to be along the positive z-axis (Figure 4.27). The magnetic potential of a moving charged point particle was determined in Example 4-12.

$$\vec{V}_m = \left(\frac{\mu_0 q}{4\pi r}\right)\vec{v} \rightarrow \vec{V}_m = \left(\frac{\mu_0 q v}{4\pi\left(\rho^2 + z^2\right)^{\frac{1}{2}}}\right)\hat{z}$$

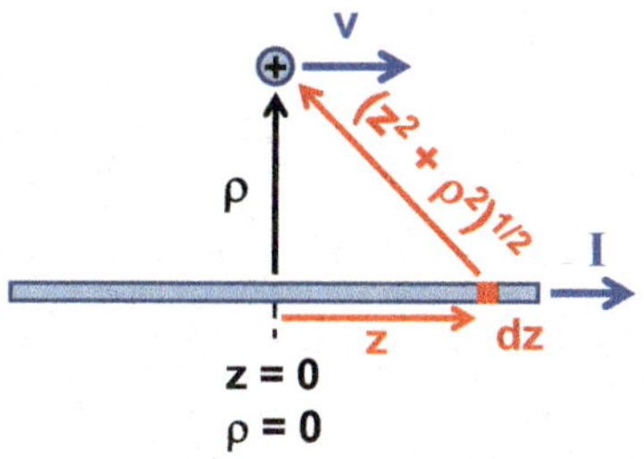

FIGURE 4.27: The variables for the integral in Example 4-14.

We can determine the total magnetic potential energy by integrating Equation 4-14 with respect to the length of the wire.

$$d\vec{s} = (dz)\hat{z} \rightarrow U_m = \int\left[\left(\frac{\mu_0 q v}{4\pi\left(\rho^2 + z^2\right)^{\frac{1}{2}}}\right)\hat{z}\right]\bullet\left[(I dz)\hat{z}\right]$$

$$U_m = \frac{\mu_0 q v I}{4\pi}\int_{-\frac{L}{2}}^{\frac{L}{2}}\frac{dz}{\left(\rho^2 + z^2\right)^{\frac{1}{2}}} \rightarrow U_m = \frac{\mu_0 q v I}{4\pi}\left(\ln\left(z + \left(\rho^2 + z^2\right)^{\frac{1}{2}}\right)\Big|_{-\frac{L}{2}}^{\frac{L}{2}}\right)$$

$$U_m = \frac{\mu_0 q v I}{4\pi}\ln\left(\frac{\left(\rho^2 + \left(\frac{L}{2}\right)^2\right)^{\frac{1}{2}} + \frac{L}{2}}{\left(\rho^2 + \left(-\frac{L}{2}\right)^2\right)^{\frac{1}{2}} - \frac{L}{2}}\right) \rightarrow U_m = \frac{\mu_0 q v I}{2\pi}\ln\left(\frac{2\rho}{\left(4\rho^2 + L^2\right)^{\frac{1}{2}} - L}\right)$$

We would naturally assume that we can use the magnetic potential energy determined in Example 4-14 to determine the magnetic force acting between the point particle and the wire with Equation 1-1.

$$\vec{F}_m = -\nabla U_m$$

The magnetic potential energy for this system is expressed in cylindrical coordinates and depends upon only the distance ρ separating the point particle from the wire; all of the other variables in the equation, such as the length of the rod, are constants. Hence, we can use Equation B-6 to determine the gradient.

$$\frac{\partial U_m}{\partial \varphi} = \frac{\partial U_m}{\partial z} = 0 \;\rightarrow\; \vec{F}_m = -\left(\frac{\partial}{\partial \rho}\left(\frac{\mu_0 q v I}{2\pi}\ln\left(\frac{2\rho}{\left(4\rho^2 + L^2\right)^{\frac{1}{2}} - L}\right)\right)\right)\hat{\rho}$$

$$\vec{F}_m = -\left(\frac{\mu_0 q v I}{2\pi}\left(-\frac{L}{\rho\left(4\rho^2 + L\right)^{\frac{1}{2}}}\right)\right)\hat{\rho} \;\rightarrow\; \vec{F}_m = \left(\frac{L\mu_0 q v I}{2\pi\rho\left(4\rho^2 + L\right)^{\frac{1}{2}}}\right)\hat{\rho}$$

However, this result is not identical to what we obtain using Equation 4-3[30].

$$\vec{F}_m = q\left(\vec{v} \times \vec{B}\right) \;\rightarrow\; \vec{F}_m = q\left[\left(v\right)\hat{z}\right] \times \left[\left(\frac{I\mu_0 L}{2\pi\rho\left(4\rho^2 + L^2\right)^{\frac{1}{2}}}\right)\hat{\varphi}\right]$$

$$\vec{F}_m = \left(\frac{qI\mu_0 Lv}{2\pi\rho\left(4\rho^2 + L^2\right)^{\frac{1}{2}}}\right)\left(\hat{z} \times \hat{\varphi}\right) \;\rightarrow\; \vec{F}_m = \left(\frac{qI\mu_0 Lv}{2\pi\rho\left(4\rho^2 + L^2\right)^{\frac{1}{2}}}\right)\left(-\hat{\rho}\right)$$

Why is there a difference in these predictions? The answer has to do with the fundamental interactions between magnetic and electric fields. As we will discuss in Chapter 6, moving the wire and the charged particle away from each other will change the current in the wire. Therefore, additional work (*i.e.*, additional energy) will be required if, as in this problem, we wish to maintain a constant current in the wire (and constant velocity for the charged particle). It is our neglecting of this additional energy that has given rise to the discrepancy here. We will return to this problem, and a more general discussion of the relationship between magnetic force and magnetic potential energy, in Section 6-8.

30 The magnetic field for the wire was determined in Section 4-5.

Summary

- The magnetic field produced by a charged object with net electric charge q moving through a vacuum with velocity $\vec{v}$ is

$$\vec{B} = \left(\frac{\mu_0 q}{4\pi}\right)\frac{\vec{v} \times \vec{r}}{r^3}$$

The constant μ_0 is called the vacuum permeability.

$$\mu_0 = 4\pi \times 10^{-7}\,\frac{\text{kgm}}{\text{C}^2}$$

The SI unit of the magnitude of the magnetic field is the tesla, denoted by T, which is defined as

$$1\text{T} = 1\frac{\text{kg}}{\text{Cs}} = 1\frac{\text{Ns}}{\text{Cm}} = 1\frac{\text{Vs}}{\text{m}^2}$$

The magnetic field is a vector field and can be considered to be the magnetic force per amp-meter.

- **Lorentz Force:** the combined electric and magnetic force acting on a charged object with net electric charge q moving with velocity $\vec{v}$ in the presence of both an electric and a magnetic field is

$$\vec{F} = q\left(\vec{E} + \vec{v} \times \vec{B}\right)$$

From the Lorentz force we also derive the equation for the magnetic force acting on a current-carrying wire in the presence of a magnetic field

$$\vec{F}_m = \int I d\vec{s} \times \vec{B}$$

- A magnetic field cannot do work on a charged object.

- **Electric current:** a net flow of electric charge through a surface.

$$I = \frac{dq}{dt}$$

The unit of electric current is charge per time (C/s), which we define as an ampere (A).

$$1\,\mathrm{A} = 1\frac{\mathrm{C}}{\mathrm{s}}$$

- The magnetic field created by a current I is

$$d\vec{B} = \left(\frac{\mu_0}{4\pi}\right)\frac{I\left(d\vec{s}\times\vec{r}\right)}{r^3}$$

- **Magnetic dipole:** a closed loop of current or a pair of magnetic poles in a permanent magnet.

- **Magnetic dipole moment:** for a closed loop of current, the magnitude of the magnetic dipole moment is product of the current in the loop and the cross-sectional area of the loop

$$\mu_m = IA$$

The direction of the magnetic dipole moment is determined using a right-hand rule. If you curl the fingers of your right hand in the direction of the current, the thumb of your right hand will point in the direction of the magnetic dipole moment. The magnetic potential energy of a magnetic dipole in a uniform magnetic field is

$$U_{m,dipole} = -\vec{\mu}_m \bullet \vec{B}$$

The net magnetic force acting on a magnetic dipole moment in a non-uniform magnetic field is

$$\vec{F}_{m,dipole} = \nabla\left(\vec{\mu}_m \bullet \vec{B}\right)$$

- **Ferromagnetism:** a ferromagnetic material will develop a net magnetic dipole moment in the presence of an external magnetic field, and will retain that net magnetic dipole moment when the external magnetic field is removed.

- **Paramagnetism:** a paramagnetic material will develop a net magnetic dipole moment in the presence of an external magnetic field, but that net magnetic dipole moment will not be retained when the external magnetic field is removed.

- **Diamagnetism:** a very weak magnetic property exhibited by all materials. When placed in an external magnetic field, a diamagnetic material will develop a net magnetic dipole moment that is aligned opposite the direction of the external magnetic field. This net magnetic dipole moment will not be retained when the external magnetic field is removed.

- **Magnetic Potential:** The potential associated with the magnetic field; it is also frequently referred to as the vector potential.

$$\vec{B} = \nabla \times \vec{V}_m$$

The magnetic potential for a current distribution can be found using

$$\vec{V}_m = \left(\frac{\mu_0}{4\pi}\right)\int \frac{Id\vec{s}}{r}$$

The magnetic potential is a vector field and can be considered to be the magnetic potential energy per amp-meter.

- **Magnetic Potential Energy:** the magnetic potential energy of a charged particle moving in a magnetic potential is

$$U_m = \vec{V}_m \bullet (q\vec{v})$$

Similarly, the component of the magnetic potential energy of an infinitesimal bit of electric current in a magnetic potential is

$$dU_m = \vec{V}_m \bullet Id\vec{s}$$

Problems

1. What is the magnetic dipole moment of the system in Example 4-2? What is the effective current for this system?

2. A thin and uniformly positively charged ring of radius R and net electric charge Q is lying in the x-y plane. The z-axis is perpendicular to the plane of the ring and passes through the center of the ring. If the ring is stationary, the magnetic field at position P is zero, but if the ring is rotating around the z-axis, the magnetic field at position P is not zero. What is the magnetic field at position P if the ring is rotating around the z-axis with angular frequency ω? What is the magnetic dipole moment of this system?

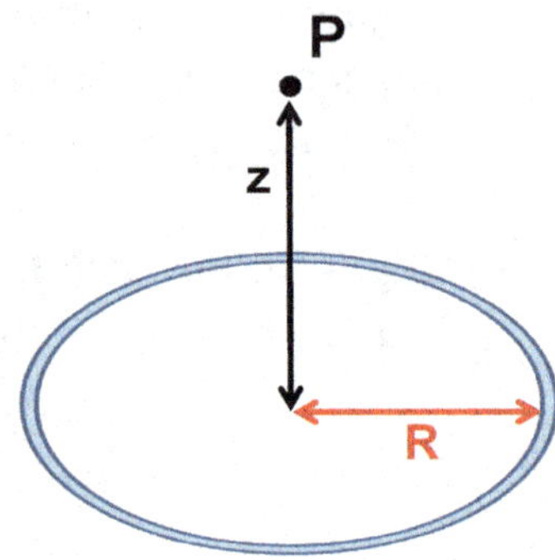

3. A point particle with a net electric charge of +4 nC is moving with a velocity of $\vec{v} = \left(2\frac{\text{m}}{\text{s}}\right)\hat{z}$ through a region where there is a uniform magnetic field of $\vec{B} = (2\text{T})\hat{\varphi}$. What is the magnetic force that acts on the particle?

4. A charged point particle is moving with a velocity $\vec{v} = \left(4\frac{\text{m}}{\text{s}}\right)\hat{z}$ through a region of space containing a uniform electric field $\vec{E} = \left(2\frac{\text{N}}{\text{C}}\right)\hat{x}$ and a uniform magnetic field $\vec{B} = (3\text{T})\hat{y}$. Due to its interactions with these fields, the particle experiences an acceleration $\vec{a} = \left(5\frac{\text{m}}{\text{s}}\right)\hat{x}$. What is the ratio of the net electric charge of the point particle to the mass of the point particle?

5. Calculate the magnitude and direction of the magnetic field for the current-carrying wire shown in the figure at the position P.

I
R
P θ
I

The wire consists of two straight portions and a semicircular arc of radius R, subtends an angle θ (measured in radians), and is centered on the position P.

6. Calculate the magnitude and direction of the magnetic field at the center of a circular wire of radius R carrying a current I (*i.e.*, at the position P shown in the figure).

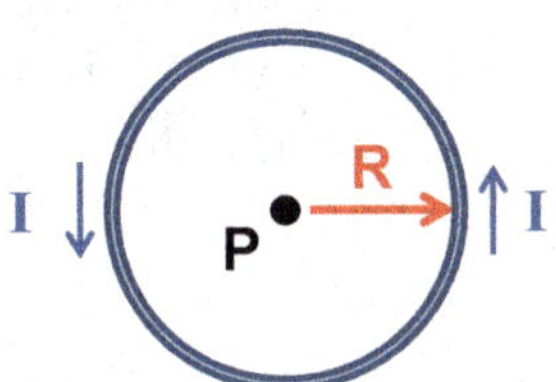

7. A long wire carries a current I. A right angle bend is made in the middle of the wire, forming a semicircle with radius R. What is the magnitude of the magnetic field at the position P? You should consider the two straight segments of the wire to be infinitely long.

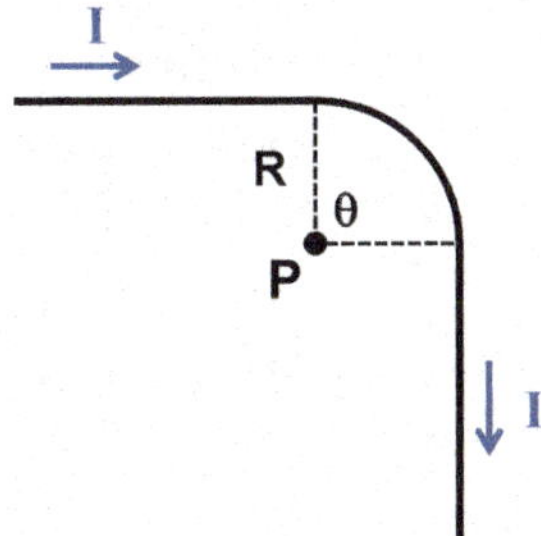

8. A wire has a linear mass density of 0.5 kg/m and carries a current of 2 A. What is the magnitude of the magnetic field that is required to levitate the wire (see figure)?

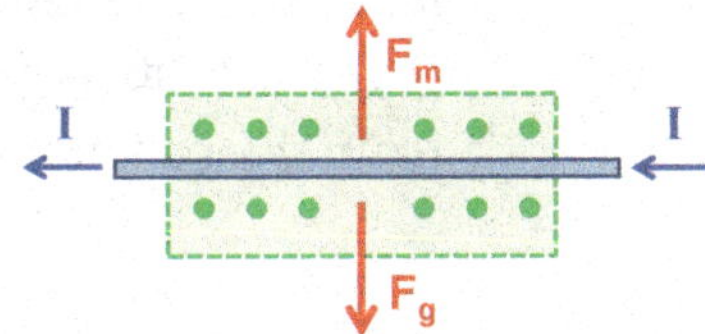

9. A wire of length L carries current I, as shown in the figure. What is the magnetic field at the position P that is a distance ρ "above" one end of the rod?

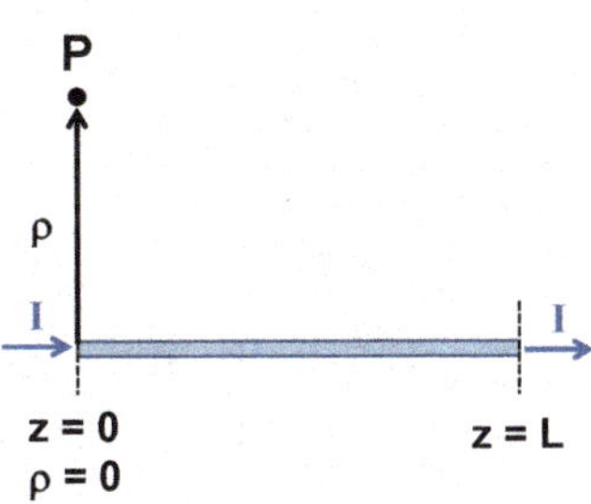

10. An annulus, with an inner radius a and outer radius b, carries a current I that is uniformly distributed throughout the annulus. What is the magnetic field from the annulus a distance z away from the center of the annulus along the perpendicular bisector of the annulus? What is the magnetic dipole moment for the annulus?

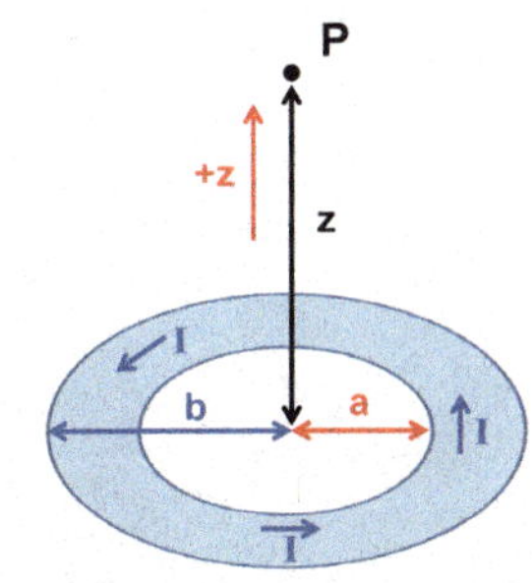

11. A magnetic dipole is placed a distance ρ away from an infinitely long current-carrying wire. If the direction of the current is along the positive z-axis and the magnetic dipole moment is $\vec{\mu}_m = (\mu_m)\hat{\rho}$, what is the magnetic torque acting on the magnetic dipole?

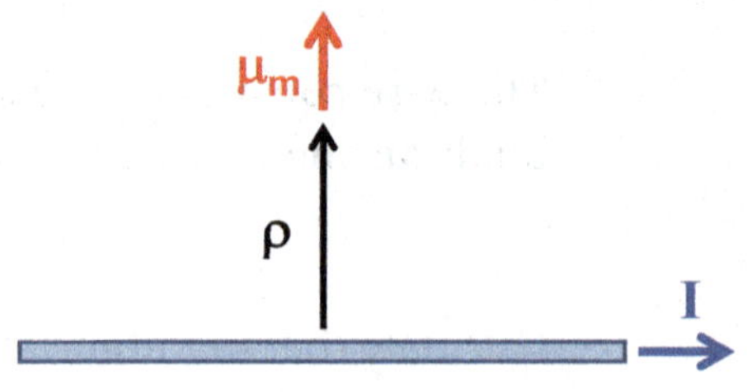

12. A magnetic dipole is placed a distance ρ away from an infinitely long current carrying wire. If the direction of the current is along the positive z-axis and the magnetic dipole moment is $\vec{\mu}_m = (\mu_m)\hat{\varphi}$, what is the magnetic force acting on the magnetic dipole?

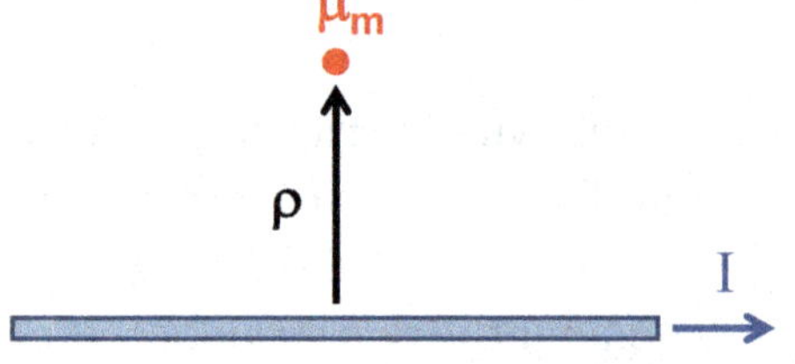

13. A thin and uniformly positively charged rigid rod is rotating around an axis through its center at angular frequency ω. The net electric charge and length of the rod are Q and L, respectively. What is the magnitude of the magnetic field a distance z away from the rod along the axis of rotation of the rod?

14. A uniformly positively charged disk of radius R is rotating with angular velocity ω around an axis that is perpendicular to the plane of the disk and passes through the center of the disk. The net electric charge of the disk is Q. What is the magnitude of the magnetic dipole moment of the disk?

15. Confirm that $\vec{V}_m = \frac{1}{2}\vec{B} \times \vec{r}$ is a possible magnetic potential for a uniform magnetic field.

CHAPTER 5

Symmetry Laws

5-1 Introduction

We argued in the solution to Example 2-10 that we could have guessed the *x*-axis component of the gravitational field for that system would be zero based on the fact that there were equal amounts of mass to the left and right of the position P along the *x*-axis (see Figure 2.17). In other words, the geometry of the distribution of mass should be reflected in the geometry of the gravitational field. Furthermore, since the direction of the gravitational field of a distribution of mass is related to the shape of the equipotential surfaces of the gravitational potential of the distribution of mass[1], it follows that the geometry of the distribution of mass should be reflected in the

1 We know from Equation 2-7 that the gravitational field is always perpendicular to the equipotential surfaces of the gravitational potential. Similarly, we know from Equation 3-8 that the electric field is always perpendicular to the equipotential surfaces of the electric potential.

geometry of the gravitational potential. Similarly, the geometry of the distribution of net electric charge and electric current should be reflected in the geometry of the electric and magnetic potentials, respectively. In this chapter, we will learn about several different fundamental laws that relate gravitational fields, electric fields, and magnetic fields to the distributions of mass, electric charge, and current that create them. Under certain circumstances, these laws allow us to more easily calculate these fields. More generally, however, these laws set the stage for the discussion of the deeper relationship between these fields in the next chapter.

5-2 Symmetry

Let's consider, for example, the gravitational potential of a uniformly dense, thin cylindrical rod. The gravitational potential for such a system was calculated in Example 2-4.

$$V_g = -2G\lambda \ln\left(\frac{2y}{\left(L^2+4y^2\right)^{\frac{1}{2}}-L}\right)$$

In this equation, the variable y denotes the distance from the rod along the perpendicular bisector of the rod. We can calculate the gravitational field from this gravitational potential using Equation 2-8.

$$\vec{g} = \left[2G\lambda\frac{\partial}{\partial y}\left(\ln\left(\frac{2y}{\left(L^2+4y^2\right)^{\frac{1}{2}}-L}\right)\right)\right]\hat{y} \rightarrow \vec{g} = \left[-\frac{2G\lambda L}{y\left(L^2+4y^2\right)^{\frac{1}{2}}}\right]\hat{y}$$

In the limit that the rod is infinitely long, we have

$$\lim_{L\to\infty}\vec{g} = \left[-\frac{2G\lambda}{y}\right]\hat{y}$$

As expected, the gravitational field has a component along the y-axis only and always points toward the rod[2] (see Figure 5.1).

2 The gravitational field always points toward the distribution of mass (Section 2-6).

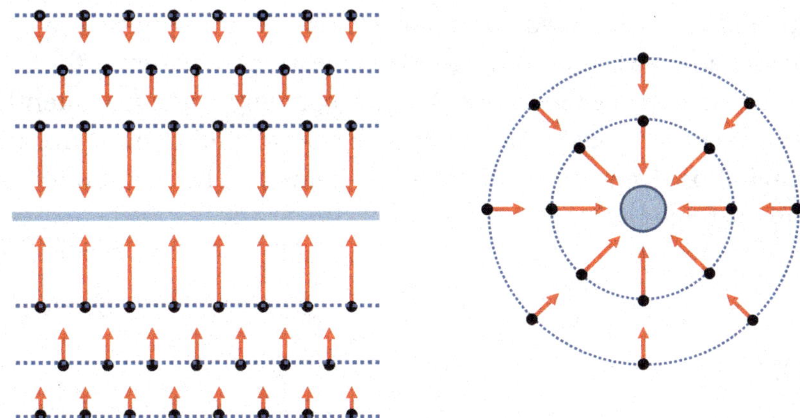

FIGURE 5.1: Gravitational field around an infinitely long and uniformly dense rigid rod. All locations on the same dashed line are at the same gravitational potential. The left panel is a view of the system from the side of the rod, and the right panel is a view of the system "head-on" from one end of the rod.

As shown in Figure 5.1, the equipotential surfaces of the gravitational potential associated with the rod share the symmetry of the rod. Specifically, they are all infinitely long cylindrical shells. Furthermore, since the gravitational field always points perpendicular to the equipotential surfaces of the gravitational potential, the gravitational field of the rod cannot have any component parallel to the surface of the rod, but must instead always point directly toward the rod. Similarly, as shown in Section 2-2, the equipotential surfaces of the gravitational potential around a point particle are concentric spherical shells, reflecting the spherical symmetry of the point particle.

Example 5-1:

Problem: Describe the gravitational field and the equipotential surfaces of the gravitational potential associated with an infinitely large and uniformly dense square slab.

Solution: The equipotential surfaces of the gravitational potential must have the same symmetry as the square slab and thus must be infinitely long square planes. The gravitational field must be perpendicular to the equipotential surfaces and must point directly toward the slab.

The electric potential of a charged point particle reflects the spherical symmetry of the point particle (Section 3-3), and the magnetic potential of a current reflects the cylindrical symmetry of the current (Section 4-10). Recognizing and subsequently exploiting the symmetry of distributions of mass, electric charge, and current in calculating the associated gravitational, electric, and magnetic fields will be the foundation of the remaining sections of this chapter.

5-3 Flux

Just as one can infer information about the gravitational potential and gravitational field of a distribution of mass from the symmetries of the distribution of mass, so too can one infer information about a distribution of mass from its gravitational potential and/or gravitational field. For example, we could infer that a spherically or cylindrically symmetric gravitational potential results from a spherically or cylindrically symmetric distribution of mass, respectively. Since the gravitational field is determined from the gravitational potential (Equation 2-8), it follows that a determination of the gravitational field (magnitude and direction) can tell us about the distribution of mass generating that field.

We can quantify measurements of the magnitude and direction of a vector field through its associated *flux*. The *gravitational flux* is a measure of the *component of the gravitational field perpendicular to a given surface*. Consider the situation shown in Figure 5.2, in which there is a uniform gravitational field. The gravitational flux through the square surface in Figure 5.2 depends upon the relative orientation of the surface and the gravitational field. The gravitational flux is largest when the direction of the gravitational field is perpendicular to the surface and zero when the direction of the gravitational field is parallel to the surface. We can use vector calculus to quantify this further. We can define a surface area vector (or simply an area vector), denoted as A in Figure 5.2, to be perpendicular to the surface. The magnitude of the area vector is equal to the associated surface area. With this definition, we see that the gravitational flux through a surface is maximum when the gravitational field and the area vector are parallel. Similarly, the gravitational flux is zero when the gravitational field and area vector are perpendicular and negative when the gravitational field and area vector are anti-parallel.

Gravitational flux, denoted by Φ_g, is defined using Equation 5-1.

$$\Phi_g = \int_{surface} \vec{g} \cdot d\vec{A} \tag{5-1}$$

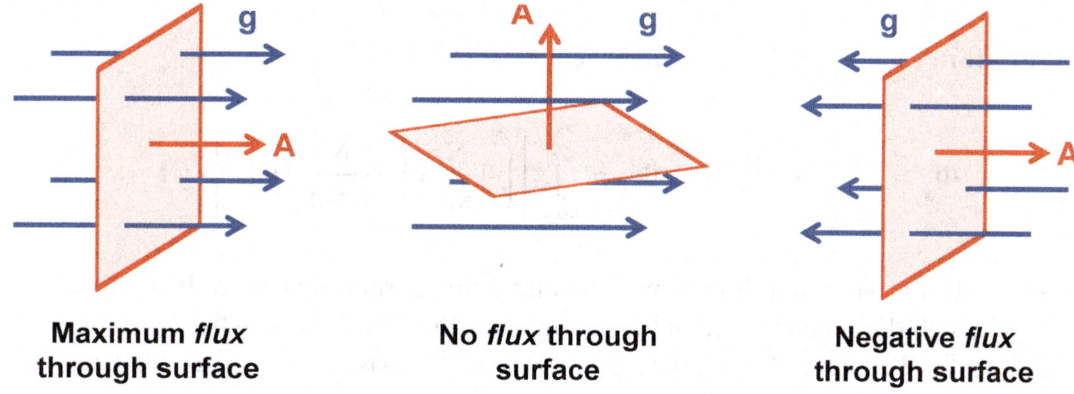

FIGURE 5.2. The gravitational flux through a surface is maximum when the direction of the gravitational field is perpendicular to the surface (*i.e.*, when it is parallel to the area vector for the surface). The gravitational flux through a surface is zero when the direction of the gravitational field is parallel to the surface (*i.e.*, when it is perpendicular to the area vector for the surface). The gravitational flux is negative when the gravitational field and the area vector are antiparallel.

The vector $d\vec{A}$ is a differential (*i.e.*, infinitesimally small) area vector; it has a magnitude equal to an infinitesimal area of, and a direction perpendicular to, the surface with respect to which the integration is performed. For a uniform gravitational field, the gravitational flux is therefore simply the dot product of the gravitational field and the area vector.

$$\Phi_g = \vec{g} \cdot \vec{A} \text{ (uniform gravitational field)}$$

Gravitational flux is thus a scalar quantity with the unit $\dfrac{\mathrm{Nm}^2}{\mathrm{kg}} = \dfrac{\mathrm{Jm}}{\mathrm{kg}}$.

Example 5-2

Problem: The gravitational field in a particular region of space can be described by the following equation:

$$\vec{g} = \left(4\frac{\mathrm{N}}{\mathrm{kg}} + \left(3\frac{\mathrm{N}}{\mathrm{kg\,m}}\right)y\right)\hat{z}$$

What is the flux of this gravitational field through the rectangular surface from $x = 0$ m to $x = 2$ m, and $y = 0$ m to $y = 4$ m?

Solution: From Equation 5-1, we have

$$\Phi_g = \int_{surface} \vec{g} \bullet d\vec{A} \rightarrow \Phi_g = \int_{surface} \left[\left(4\frac{\text{N}}{\text{kg}} + \left(3\frac{\text{N}}{\text{kg m}} \right) y \right) \hat{z} \right] \bullet d\vec{A}$$

The differential area in Equation 5-1 must point perpendicular to the surface over which the gravitational field is integrated. Since this surface lies in the *x*-*y* plane, the direction of $d\vec{A}$ must be parallel to the *z*-axis. For this calculation, we are free to choose the direction of $d\vec{A}$ to be $+\hat{z}$ or $-\hat{z}$; in later calculations we will introduce a rule for determining the direction of $d\vec{A}$ and similar differential area vectors.

For our calculation, let's assume that $d\vec{A}$ is pointed in the $+\hat{z}$ direction. We can further model these infinitesimally small bits of area as infinitesimally small rectangles (see Section 1-3). If the lengths of the sides of these rectangles are *dx* and *dy*, the magnitude of the area of each of these infinitesimally small rectangles (*i.e.*, the magnitude of $d\vec{A}$) is the product of *dx* and *dy*.

$$d\vec{A} = (dx\, dy)\hat{z}$$

Hence,

$$\Phi_g = \int_{surface} \left[\left(4\frac{\text{N}}{\text{kg}} + \left(3\frac{\text{N}}{\text{kg m}} \right) y \right) \hat{z} \right] \bullet \left[(dx\, dy)\hat{z} \right]$$

$$\Phi_g = \int_{0\,\text{m}}^{4\,\text{m}} \int_{0\,\text{m}}^{2\,\text{m}} \left(4\frac{\text{N}}{\text{kg}} + \left(3\frac{\text{N}}{\text{kg m}} \right) y \right) dx\, dy$$

$$\Phi_g = \int_{0\,\text{m}}^{4\,\text{m}} \left(4\frac{\text{N}}{\text{kg}} + \left(3\frac{\text{N}}{\text{kg m}} \right) y \right) dy \int_{0\,\text{m}}^{2\,\text{m}} dx \rightarrow \Phi_g = \int_{0\,\text{m}}^{4\,\text{m}} \left(4\frac{\text{N}}{\text{kg}} + \left(3\frac{\text{N}}{\text{kg m}} \right) y \right) (2\,\text{m})\, dy$$

$$\Phi_g = \int_{0\,\text{m}}^{4\,\text{m}} \left(8\frac{\text{Nm}}{\text{kg}} + \left(6\frac{\text{N}}{\text{kg}} \right) y \right) dy \rightarrow \Phi_g = \left(\left(8\frac{\text{Nm}}{\text{kg}} \right) y + \left(3\frac{\text{N}}{\text{kg}} \right) y^2 \right) \Big|_{0\,\text{m}}^{4\,\text{m}}$$

$$\Phi_g = \left(\left(8\frac{\text{Nm}}{\text{kg}} \right) (4\,\text{m}) + \left(3\frac{\text{N}}{\text{kg}} \right) (4\,\text{m})^2 \right)$$

$$\Phi_g = 80\frac{\text{Nm}^2}{\text{kg}} = 80\frac{\text{J m}}{\text{kg}}$$

The positive sign for the gravitational flux indicates that the *net direction* of the gravitational flux through the surface (*i.e.*, in the locations defined by the surface) is in the same direction as $d\vec{A}$ (*i.e.*, in the $+\hat{z}$ direction).

5-4 Gauss's Law for the Gravitational Field

Gauss's law[3] for the gravitational field states that the total gravitational flux through a closed surface is proportional to the total mass enclosed by that surface. We can express Gauss's law for the gravitational field mathematically using Equation 5-2.

$$\oint_{surface} \vec{g} \cdot d\vec{A} = -4\pi G M_{enclosed} \tag{5-2}$$

The vector $d\vec{A}$ in Equation 5-2 is directed *away* from the interior of the surface[4]. It is important to note that while Equation 5-2 is true for all surfaces, the integral in Equation 5-2 is simplest when the surface used for the integral matches the equipotential surfaces of the gravitational potential of the distribution of mass that generates the gravitational field. This is true since the direction of the gravitational field is always perpendicular to these surfaces[5], and the magnitude of the gravitational field is constant at all positions along these surfaces. The variable $M_{enclosed}$ in Equation 5-2 denotes the mass contained within the Gaussian surface.

Example 5-3

Problem: What is the magnitude of the gravitational field for an infinitely long, uniformly dense rod?

Solution: The equipotential surfaces of this system will reflect the symmetry of the distribution of mass and thus will be cylindrically symmetric. For this reason, it is best to choose the Gaussian surface for our application

3 Named for Johann Carl Friedrich Gauss.

4 The surfaces used in Equation 5-2 are frequently referred to as Gaussian surfaces.

5 Recall from Equation 2-8 that the gravitational field is always perpendicular to the equipotential surface.

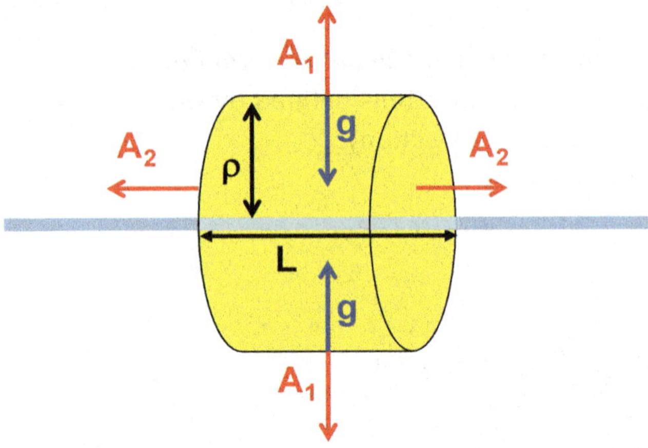

FIGURE 5.3. Determining the gravitational field for an infinitely long, uniformly dense rod. The Gaussian surface for this problem is a cylinder, shown in in yellow. The direction of the vectors associated with the surface areas and the gravitational field are indicated by the arrows.

of Equation 5-2 to be a cylinder, as shown in Figure 5.3. The radius and length of this cylinder are denoted by the variables y and L, respectively, as shown in Figure 5.3.

This Gaussian surface has two surface areas: the area around the cylinder, denoted as A_1 in Figure 5.3, and the areas of the top and bottom of the cylinder, denoted as A_2 in Figure 5.3. The vectors A_1 and A_2 in Figure 5.3 point perpendicular to their associated surfaces, and out from the inside of the cylinder. The direction of the gravitational field is toward the rigid rod (Section 2-6).

Since the direction of the gravitational field is perpendicular to the direction of the surface area A_2, there is no gravitational flux through this surface. Furthermore, since the surface A_1 is an equipotential surface, the magnitude of the gravitational field is constant at all locations along that surface. Therefore,

$$\oint_{surface} \vec{g} \bullet d\vec{A} = gA_1 \cos(180^\circ) \rightarrow \oint_{surface} \vec{g} \bullet d\vec{A} = -gA_1$$

$$A_1 = 2\pi yL \rightarrow \oint_{surface} \vec{g} \bullet d\vec{A} = -g(2\pi yL)$$

Hence, from Equation 5-2 we have

$$-g(2\pi yL) = -4\pi GM_{enclosed} \rightarrow g = \frac{4\pi GM_{enclosed}}{2\pi yL} \rightarrow g = \frac{2GM_{enclosed}}{yL}$$

We can simplify this result by expressing $M_{enclosed}$ in terms of the linear mass density of the rod.

$$M_{enclosed} = \lambda L \rightarrow g = \frac{2G\lambda L}{yL} \rightarrow g = \frac{2G\lambda}{y}$$

As expected, this expression of the magnitude of the gravitational field is identical to the expression derived in Section 5-2. The direction of gravitational field is indicated by the directions of the associated arrow in Figure 5.3. Incidentally, had we drawn Figure 5.3 with the arrow for the gravitational field pointing in the opposite direction (*i.e.*, in the same direction as A_1), the resulting magnitude for the gravitational field would have been negative. Obtaining a negative magnitude thus indicates that the relative orientation of the gravitational field and area vectors in our calculation is incorrect.

Example 5-4

Problem: What is the gravitational field a distance r away from the center of a uniformly dense, solid sphere of mass M and radius R?

Solution: The equipotential surfaces of the gravitational potential of this system will reflect the symmetry of the distribution of mass and thus will be spherically symmetric. For this reason, it is best to choose the Gaussian surface for our application of Equation 5-2 to be a spherical shell, as shown in Figure 5.4.

The magnitude of the gravitational field is constant at all points along this Gaussian surface since it is an equipotential surface for the gravitational potential. We don't necessarily know the direction of the gravitational field since we are interested in determining the gravitational field both inside and outside of the sphere. However, we recall from the solution to Example 5-3 that an incorrect guess for the direction of the gravitational field will result in a negative sign in our calculation of the magnitude of the gravitational field using Equation 5-2. Should this occur in our calculation, we will know that our guess for the direction of the gravitational field was incorrect. Thus, we assume that the gravitational field is always directed toward the center of the sphere, as shown in Figure 5.4.

$$\oint_{surface} \vec{g} \bullet d\vec{A} = gA\cos(180°) \quad \rightarrow \quad \oint_{surface} \vec{g} \bullet d\vec{A} = -g4\pi r^2$$

$$\oint_{surface} \vec{g} \bullet d\vec{A} = -4\pi G M_{enclosed} \quad \rightarrow \quad -g4\pi r^2 = -4\pi G M_{enclosed}$$

$$g = \frac{G}{r^2} M_{enclosed}$$

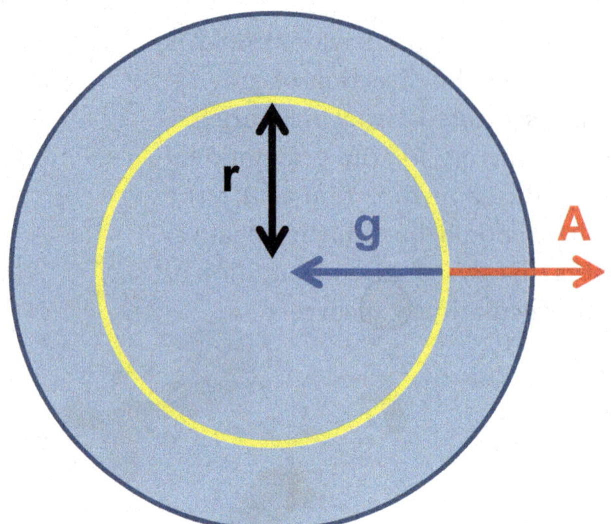

FIGURE 5.4. Determining the gravitational field for a uniformly dense solid sphere. The Gaussian surface for this problem is a spherical shell, shown in cross-section in yellow. The direction of the vectors associated with the area of this surface and the gravitational field are indicated by the arrows.

Since the magnitude of the gravitational field is positive (*i.e.*, all of the variables on the right hand side of this expression are positive), our choice for the direction of the gravitational field in Figure 5.4 is correct for all values of the variable r.

The next step in our calculation is to determine $M_{enclosed}$. First, we'll consider the case where $r < R$. In this case, $M_{enclosed}$ is determined from the radius of the Gaussian surface and the volume mass density of the sphere (Table 2-1).

$$M_{enclosed} = \rho V_{enclosed} \quad \rightarrow \quad M_{enclosed} = \left(\frac{4}{3}\pi r^3\right)\rho$$

The mass density can, similarly, be determined from the radius and total mass of the sphere.

$$M = \rho\left(\frac{4}{3}\pi R^3\right) \quad \rightarrow \quad \rho = \frac{3M}{4\pi R^3} \quad \rightarrow \quad M_{enclosed} = \left(\frac{4}{3}\pi r^3\right)\left(\frac{3M}{4\pi R^3}\right)$$

$$M_{enclosed} = M\left(\frac{r^3}{R^3}\right)$$

Substitution gives us

$$g = \frac{G}{r^2}\left(M\left(\frac{r^3}{R^3}\right)\right) \quad \rightarrow \quad g = \left(\frac{GM}{R^3}\right)r$$

The magnitude of the gravitational field increases linearly with increasing distance away from the center of the sphere.

If the radius of the Gaussian surface is larger than the radius of the sphere, then $M_{enclosed}$ is simply the total mass of the sphere.

$$M_{enclosed} = M \quad \rightarrow \quad g = \frac{GM}{r^2}$$

It is interesting to note that this expression for the magnitude of the gravitational field is identical to the magnitude of the gravitational field for a point particle (Equation 2-9). Therefore, when viewed from the outside, it is impossible to distinguish between a point particle and a uniformly dense solid sphere, regardless of how close you may be to the surface of the sphere. Finally, it is also worth noting that the two expressions for the gravitational field (inside and outside the sphere) agree at the surface of the sphere (*i.e.*, the two expressions are identical when $r = R$).

Of course, for more complicated (non-symmetric, *e.g.*) distributions of mass, the integral in Equation 5-2 cannot be as readily simplified as we have done in Example 5-3 and Example 5-4. Nevertheless, the resulting more complicated integral would still have equaled $-4\pi GM_{enclosed}$.

Example 5-5

Problem: A spherical Gaussian surface contains 3 point particles with masses 3 kg, 5 kg, and 7 kg. What is the gravitational flux through this surface?

Solution: The gravitational flux through the surface is found using Equation 5-2.

$$\Phi_g = -4\pi GM_{enclosed} \quad \rightarrow \quad \Phi_g = -4\pi\left(6.67\times10^{-11}\,\frac{\text{J m}}{\text{kg}^2}\right)(3\,\text{kg}+5\,\text{kg}+7\,\text{kg})$$

$$\Phi_g = -1.26\times10^{-8}\,\frac{\text{J m}}{\text{kg}}$$

5-5 Gauss's Law for the Electric Field

Expressions similar to Equation 5-1 and Equation 5-2 exist for electric fields.

$$\Phi_E = \int_{surface} \vec{E}\bullet d\vec{A} \qquad \textbf{(5-3)}$$

$$\oint_{surface} \vec{E} \bullet d\vec{A} = \frac{q_{enclosed}}{\varepsilon_0} \tag{5-4}$$

The variable $q_{enclosed}$ in Equation 5-4 denotes the net electric charge contained within the Gaussian surface[6]. Consider, for example a system that contains two charge point particles with net electric charges of equal magnitude, but opposite sign[7]. The total electric flux through a Gaussian surface containing these particles would be zero since the net electric charge contained within the surface. More specifically, the electric flux associated with the positively charged particle would be positive and the electric flux associated with the negatively charged particle would be negative, and these two electric fluxes have the same magnitude. Thus, the summation of these electric fluxes (*i.e.*, the total or net flux) is zero.

Example 5-6

Problem: A cylindrically symmetric electric field has a magnitude given by the following equation:

$$\vec{E} = \left[\left(2\frac{\text{N}}{\text{Cm}}\right)\rho\right]\hat{z}$$

What is the total electric flux through a circular cross-sectional area of radius 3 m lying in the *x-y* plane (*i.e.*, perpendicular to the *z*-axis)?

Solution: Since the electric field is cylindrically symmetric, the differential area for the integration in Equation 5-4 is a thin ring of circumference $2\pi\rho$ and width $d\rho$ (see Section 1-3).

$$d\vec{A} = \left[2\pi\rho\, d\rho\right]\hat{z}$$

Thus,

$$\Phi_E = \int_{surface} \left(\left[\left(2\frac{\text{N}}{\text{Cm}}\right)\rho\right]\hat{z}\right) \bullet \left(\left[2\pi\rho\, d\rho\right]\hat{z}\right) \rightarrow \Phi_E = \int_{0\text{m}}^{3\text{m}} \left(4\pi\frac{\text{N}}{\text{Cm}}\right)\rho^2 d\rho$$

---

6 Upon comparing Equations 2-9, 3-9, 5-2, and 5-4 we see how 4π either appears in the equation for the vector field (electric field) or the equation for Gauss's law (gravitational field). In either case, its ultimate origin is an integration over a surface area (*i.e.*, over a solid angle). You should definitely take some more advanced physics and math courses to understand exactly how and why this occurs.

7 An example of such a system is the electric dipole in Figure 3.16.

$$\Phi_E = \left(4\pi \frac{\text{N}}{\text{Cm}}\right)\int_{0\text{m}}^{3\text{m}} \rho^2 dr \rightarrow \Phi_E = \left(4\pi \frac{\text{N}}{\text{Cm}}\right)\left(\frac{\rho^3}{3}\Bigg|_{0\text{m}}^{3\text{m}}\right) \rightarrow \Phi_E = \left(4\pi \frac{\text{N}}{\text{Cm}}\right)\left(\frac{(3\,\text{m})^3}{3}\right)$$

$$\Phi_E = 36\pi \frac{\text{Nm}^2}{\text{C}} = 36\pi\,\text{Vm}$$

Example 5-7

Problem: A solid sphere of radius R and net electric charge $+Q$ has the following volumetric net electric charge density:

$$\rho = \left(\frac{5Q}{4\pi R^5}\right) r^2$$

The variable r in this equation is the distance from the center of the sphere. What is the magnitude of the electric field as a function of r?

Solution: The electric potential and field will both be spherically symmetric, so we should choose a spherical shell as the area of integration in Equation 5-4 (see Example 5-4). The sphere has a net positive electric charge density, so the direction of the electric field will be the same as the direction as the area vector for this surface, and the magnitude of the electric field will be constant at all locations along this surface. Thus,

$$\oint_{surface} \vec{E} \bullet d\vec{A} = E\left(4\pi r^2\right) \rightarrow E\left(4\pi r^2\right) = \frac{q_{enclosed}}{\varepsilon_0} \rightarrow E = \frac{q_{enclosed}}{4\pi\varepsilon_0 r^2}$$

For all values of $r < R$ the net electric charge enclosed by this surface can be found through the summation (*i.e.*, integration) of the infinitesimal bits of net electric charge that constitute the sphere using the volume net electric charge density.

$$dq_{enclosed} = \rho dV$$

The differential volume in this equation is the differential volume of a spherical shell (Section B-5).

$$q_{enclosed} = \int dq_{enclosed} \quad \rightarrow \quad q_{enclosed} = \int \rho\, dV \quad \rightarrow \quad q_{enclosed} = \int \rho\left(4\pi r^2\right) dr$$

$$q_{enclosed} = \int_0^r \left(\left(\frac{5Q}{4\pi R^5}\right) r^2\right)\left(4\pi r^2\right) dr \quad \rightarrow \quad q_{enclosed} = \left(\frac{5Q}{R^5}\right)\int_0^r r^4\, dr$$

$$q_{enclosed} = \left(\frac{5Q}{R^5}\right)\frac{r^5}{5} \quad \rightarrow \quad q_{enclosed} = \frac{Q}{R^5} r^5$$

Hence,

$$E = \left(\frac{1}{4\pi\varepsilon_0 r^2}\right)\left(\frac{Q}{R^5} r^5\right) \quad \rightarrow \quad E = \frac{Qr^3}{4\pi\varepsilon_0 R^5}$$

For all values of $r > R$ the net electric charge enclosed by this surface is equal to the net electric charge of the sphere.

$$q_{enclosed} = Q \quad \rightarrow \quad E = \frac{Q}{4\pi\varepsilon_0 r^2}$$

As expected, these two expressions agree when $r = R$. Furthermore, it is worth noting that this expression for the magnitude of the electric field is identical to the magnitude of the electric field for a point particle (Equation 3-9). Therefore, when viewed from the outside it is impossible to distinguish between a point particle and a solid sphere, regardless of how close you may be to the surface of the sphere[8].

Example 5-8

Problem: What is the electric field produced by an infinite plane with uniform positive net electric charge density?

8 A similar result was obtained in Example 5-4.

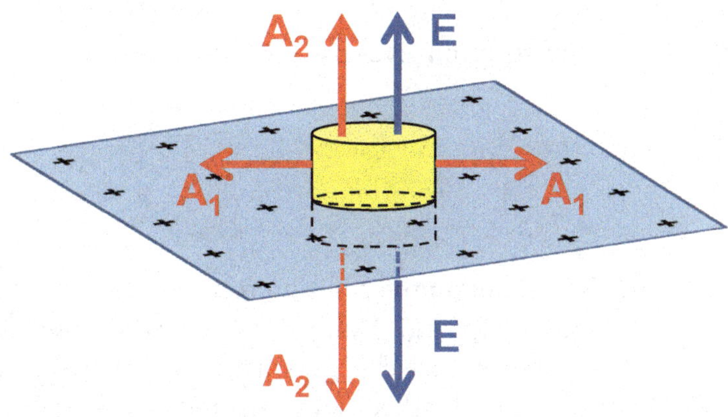

FIGURE 5.5. Determining the electric field for an infinite plane with uniform net positive electric charge density. The Gaussian surface for this problem is a cylinder. The direction of the vectors associated with the area of this surface and the electric field are indicated by the arrows.

Solution: The equipotential surfaces of the electric potential of this system will reflect the symmetry of the net electric charge distribution and thus will be infinite planes. The electric field, which is always perpendicular to the equipotential surfaces of the electric potential, will everywhere be aligned perpendicular to the charged plane; furthermore, since the plane is positively charged, the electric field will point away from the plane. Finally, the magnitude of the electric field will be constant everywhere on the equipotential surface of the electric potential.

We are not limited in our choice of a Gaussian surface for this net electric charge distribution as long as it has a surface that is parallel to the charged plane. For simplicity, let's choose a cylinder, as shown in Figure 5.5.

This Gaussian surface has two surface areas: the area around the cylinder, denoted as A_1 in Figure 5.5, and the areas of the top and bottom of the cylinder, denoted as A_2 in Figure 5.5[9]. The vectors A_1 and A_2 in Figure 5.5 point perpendicular to their associated surfaces, and out from the inside of the cylinder.

Since the magnitude of the electric field is constant at all points along the surface area A_2 and is perpendicular to the surface area A_1, we have

$$\oint_{surface} \vec{E} \bullet d\vec{A} = 2\left(EA_2 \cos\left(0°\right)\right) \quad \rightarrow \quad \oint_{surface} \vec{E} \bullet d\vec{A} = 2EA_2$$

$$2EA_2 = \frac{q_{enclosed}}{\varepsilon_0} \quad \rightarrow \quad E = \frac{q_{enclosed}}{2A_2\varepsilon_0}$$

The next step in our calculation is to determine $q_{enclosed}$. We can express this quantity in terms of the surface net electric charge density, σ, and the area A_2.

$$q_{enclosed} = \sigma A_2$$

9 These are identical to definitions in Example 5-3.

Substitution gives us

$$E = \frac{\sigma A_2}{2A_2\varepsilon_0} \rightarrow E = \frac{\sigma}{2\varepsilon_0}$$

Let's build upon the solution to Example 5-8 by considering a system that consists of two infinitely large planes with opposite uniform net electric charge density; we can denote the surface charge densities of the two planes as $+\sigma$ and $-\sigma$. The two planes are parallel to each other and separated by a distance *d*, as shown in Figure 5.6.

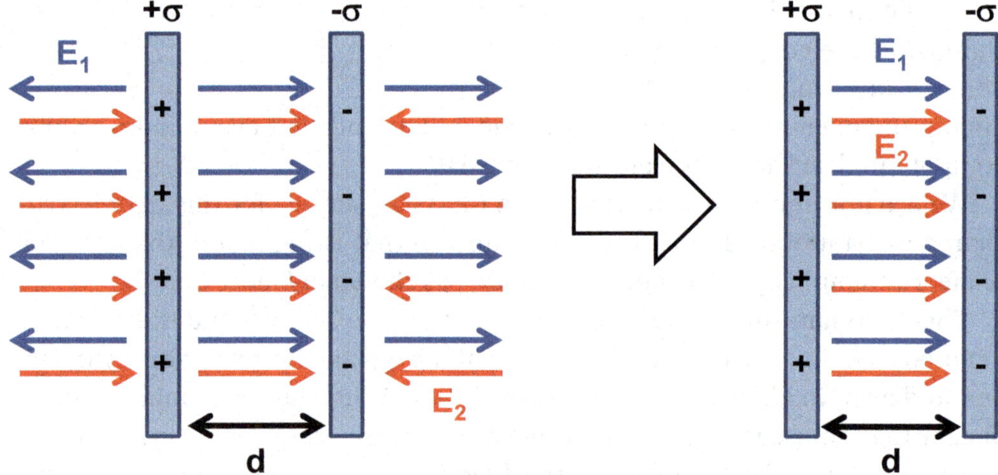

FIGURE 5.6. The electric field of two infinite parallel planes of uniform, but opposite, net electric charge density, is uniform between the two planes and zero everywhere else.

As shown in Figure 5.6, the electric fields of the two planes point in the same direction in-between the two planes, and in opposite directions elsewhere. Since the magnitudes of these two electric fields are the same, we see that the electric fields of the two planes will cancel everywhere except in-between the two planes. The electric field in-between the two planes points from the positively charged plane to the negatively charged plane, as shown in Figure 5.6, and has a magnitude that is simply the sum of the magnitudes of the electric field from each plane (Example 5-8).

$$E = \frac{\sigma}{2\varepsilon_0} + \frac{\sigma}{2\varepsilon_0} \rightarrow E = \frac{\sigma}{\varepsilon_0}$$

Although this result was determined from a system consisting of infinite planes, it is nevertheless still a good approximation of the result for non-infinite planes under the condition that the separation distance between the two planes (the distance d in Figure 5.6) is much smaller than the length and width of the plane.

Now let's consider a system consisting of two parallel planes of uniform, but opposite, charge density separated by a distance d. Each plane is a square with surface area A. Under the condition that $d \ll A^{1/2}$ the magnitude of the electric field between the two planes is

$$E = \frac{q}{A\varepsilon_0} \tag{5-5}$$

The variable q in Equation 5-5 denotes the net electric charge on the planes; one plane has a net electric charge of $+q$, and the other plane has a net electric charge of $-q$. The separation of charges in this system results not only in an electric field in-between the two planes but also in a difference in electric potential between the two planes. This electric potential difference can be determined from the electric field using Equation 3-10.

$$\Delta V_e = -\int \vec{E} \bullet d\vec{s}$$

Let's define the origin for the s-axis to be the positively charged plane in Figure 5.6 and the positive direction for the s-axis to point from the positively charged plane to the negatively charged plane (*i.e.*, in the same direction as the electric field). With these definitions we have

$$\vec{E} \bullet d\vec{s} = \left(\frac{q}{A\varepsilon_0}\right)(ds)\cos(0°) \;\rightarrow\; \Delta V_e = -\int_0^d \left(\frac{q}{A\varepsilon_0}\right) ds \cos(0°)$$

$$\Delta V_e = -\frac{q}{A\varepsilon_0}\int_0^d ds \;\rightarrow\; \Delta V_e = -\frac{q}{A\varepsilon_0} d$$

We can now define a new quantity called the capacitance, denoted by C.

$$C = \frac{A\varepsilon_0}{d} \tag{5-6}$$

The capacitance of a system is a measure of how much net electric charge must be separated to produce a certain difference in electric potential between the separated electric

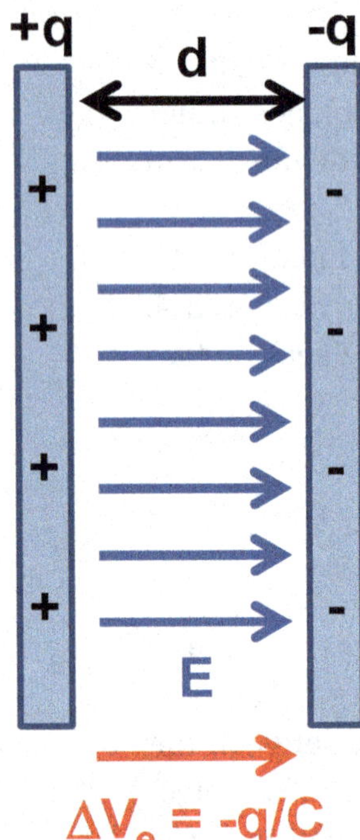

FIGURE 5.7: The electric field and electric potential of a parallel plate capacitor.

charges; the larger the capacitance, the more net electric charge that is required. For the system of two parallel planes, the capacitance depends upon only the surface area of the planes and the distance between them. For other systems, the capacitance is similarly defined by geometric parameters (length, width, *etc.*) associated with the system. The unit of capacitance is the farad, denoted by *F*.

$$1\,\mathrm{F} = 1\frac{\mathrm{A}^2\mathrm{s}^4}{\mathrm{kg\,m}^2} = 1\frac{\mathrm{C}}{\mathrm{V}}$$

The electric potential difference between two separated net electric charges (or electric charge distributions) can be expressed in terms of the associated capacitance as

$$\Delta V_e = -\frac{q}{C} \tag{5-7}$$

Furthermore, this system of uniform distributions of net electric charge density of the same magnitude but opposite (±) sign separated by a distance can also be referred to as a capacitor[10]. The particular capacitor shown in Figure 5.6 and Figure 5.7 is a parallel plate capacitor.

The separation of charges in the parallel plate capacitor in Figure 5.7 results in an electric potential energy for that system (Section 3-5). We can determine the total electric potential energy associated with this system by calculating the energy required to move an infinitesimal charge dq from the negatively charged plate to the positively charged plate *against* the electric field existing between the two plates. From Equation 3-5, we see that this change in energy would be

$$dU_e = (dq)\Delta V_e \quad \rightarrow \quad dU_e = (dq)\left(\frac{q}{C}\right)$$

So the total energy required to form the charge separation[11] in Figure 5.7 would be

$$\Delta U_e = \int dU_e \quad \rightarrow \quad \Delta U_e = \int_0^q \frac{q}{C}dq \quad \rightarrow \quad \Delta U_e = \left(\left.\frac{q^2}{2C}\right|_0^q\right)$$

10 We will discuss capacitors in more detail in later chapters.

11 See Section C-6.

$$\Delta U_e = \frac{q^2}{2C} \quad \textbf{(5-8)}$$

Of course, since the energy of the system is a state function, Equation 5-8 is valid regardless of how the charged plates came to be separated. In other words, Equation 5-8 is valid regardless of how the charge separation was created. Furthermore, we see from Equation 5-6, Equation 5-7, and Equation 5-8 that a system with infinite capacitance would be a system for which $d = 0$, $\Delta V_e = 0$, and $\Delta U_e = 0$. In other words, there is no charge separation for a system with infinite capacitance.

Example 5-9

Problem: The spherical capacitor in Figure 5.8 consists of a spherical shell of radius r_1 and uniform net positive electric charge density with net electric charge $+Q$, surrounded by a spherical shell of radius r_2 with uniform negative electric charge density and net electric charge $-Q$.

What is the capacitance of this capacitor?

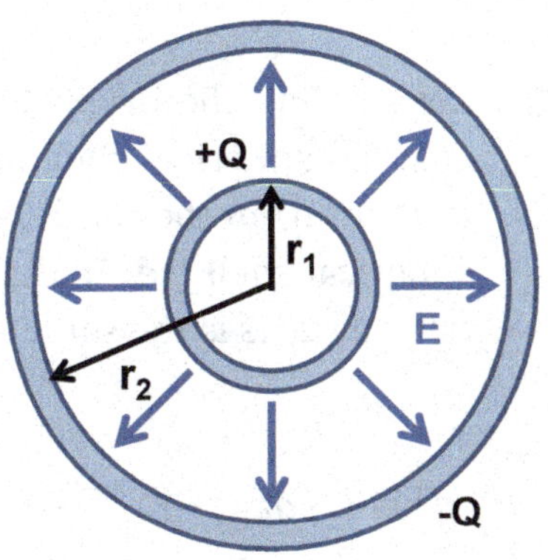

FIGURE 5.8: The spherical capacitor in Example 5-9.

Solution: The first step is to determine the electric field in-between the two shells of this capacitor. The equipotential surfaces of the electric potential of this system will reflect the symmetry of the electric charge distribution and thus will be spherically symmetric. For this reason, it is best to choose the Gaussian surface for our application of Equation 5-4 to be a spherical shell, as shown in Figure 5.9. Furthermore, as discussed in Section 3-6, the electric field will point from the high electric potential surface (the positively charged surface) to the low electric potential surface (the negatively charged surface).

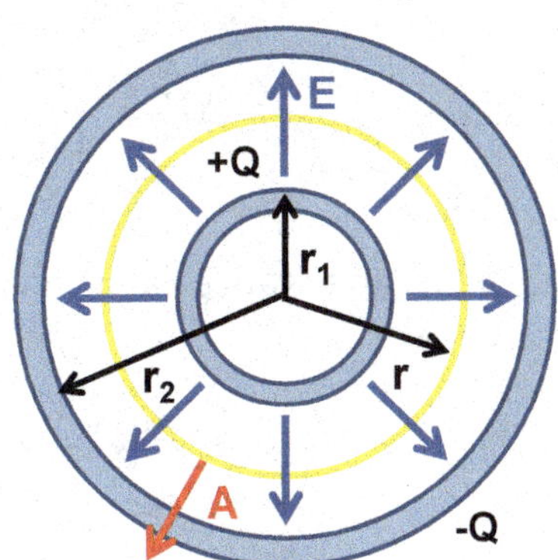

FIGURE 5.9: The Gaussian surface for determining the electric field inside the spherical capacitor in Example 5-9 is a spherical shell, shown in cross-section in yellow. The direction of the vectors associated with the area of this surface and the electric field are indicated by the arrows.

Building upon the solution to Example 5-4 we have

$$\oint_{surface} \vec{E} \bullet d\vec{A} = E 4\pi r^2$$

The charge enclosed by the Gaussian surface is the total charge on the positively charged sphere in the capacitor.

$$q_{enclosed} = Q \rightarrow E\left(4\pi r^2\right) = \frac{Q}{\varepsilon_0} \rightarrow E = \frac{Q}{4\pi r^2 \varepsilon_0}$$

The electric potential between the two spheres in the capacitor can be found using Equation 3-10.

$$\Delta V_e = -\int \vec{E} \bullet d\vec{s}$$

Let's define the origin for the s-axis in this integration to be the positively charged spherical shell in Figure 5.9 and the positive direction for the s-axis to point from the positively charged spherical shell to the negatively charged spherical shell (*i.e.*, in the same direction as the electric field). This definition for the s-axis is the same as the definition for the r-axis.

$$\vec{E} \bullet d\vec{s} = \vec{E} \bullet d\vec{r} \rightarrow \vec{E} \bullet d\vec{r} = \left(\frac{Q}{4\pi r^2 \varepsilon_0}\right)(dr)\cos\left(0°\right) \rightarrow \vec{E} \bullet d\vec{r} = \left(\frac{Q}{4\pi r^2 \varepsilon_0}\right)dr$$

$$\Delta V_e = -\int_{r_1}^{r_2} \left(\frac{Q}{4\pi r^2 \varepsilon_0}\right)dr \rightarrow \Delta V_e = -\frac{Q}{4\pi\varepsilon_0}\int_{r_1}^{r_2} \frac{dr}{r^2} \rightarrow \Delta V_e = -\frac{Q}{4\pi\varepsilon_0}\left(-\frac{1}{r}\Bigg|_{r_1}^{r_2}\right)$$

$$\Delta V_e = \frac{Q}{4\pi\varepsilon_0}\left(\frac{1}{r_2} - \frac{1}{r_1}\right) \rightarrow \Delta V_e = \frac{Q}{4\pi\varepsilon_0}\left(\frac{r_1 - r_2}{r_2 r_1}\right)$$

The capacitance can then be found from Equation 5-7.

$$C = -\frac{Q}{\Delta V_e} \rightarrow C = -\frac{Q}{\frac{Q}{4\pi\varepsilon_0}\left(\frac{r_1 - r_2}{r_2 r_1}\right)}$$

$$C = 4\pi\varepsilon_0\left(\frac{r_2 r_1}{r_2 - r_1}\right)$$

The net electric force between the two plates of a capacitor is determined using Equation 1-1 and Equation 5-8. For this calculation let's define the perpendicular axis between the two plates of the capacitor to be the x-axis.

$$F_e = -\frac{dU_e}{dx} \;\rightarrow\; F_e = -\left(-\frac{q^2}{2C^2}\frac{dC}{dx}\right) \;\rightarrow\; F_e = \left(\frac{q^2}{2C^2}\right)\frac{dC}{dx}$$

Thus, for a parallel plate capacitor whose plates are separated by a distance x we have

$$C = \frac{A\varepsilon_0}{x} \;\rightarrow\; F_e = \frac{q^2}{2C^2}\left(-\frac{A\varepsilon_0}{x^2}\right) \;\rightarrow\; F_e = -\frac{A\varepsilon_0 q^2}{2C^2x^2}$$

The negative sign for this force indicates that this force will pull the two plates together, as expected.

5-6 Ampère's Law and Gauss's Law for the Magnetic Field

Ampère's law[12], shown in Equation 5-9, is the expression for magnetic fields that can be considered to be somewhat equivalent to Equation 5-2 for gravitational fields and Equation 5-4 for electric fields.

$$\oint_{line} \vec{B} \cdot d\vec{s} = \mu_0 I_{enclosed} \tag{5-9}$$

The integral in Equation 5-9 is a *line integral* that is performed around the boundary of a surface through which the enclosed current *passes*; this boundary/line is frequently referred to as an Ampèrian loop. Furthermore, we define the direction for the vector associated with the area of this surface (the equivalent of the $d\vec{A}$ vector in Equation 5-2 and Equation 5-4) using a right-hand rule. Curl the fingers of your right hand in the direction of the integration around the Ampèrian loop and the thumb of your right hand points in the direction of the area vector, denoted as $\vec{A}$. As shown in Figure 5.10,

12 Named for André-Marie Ampère.

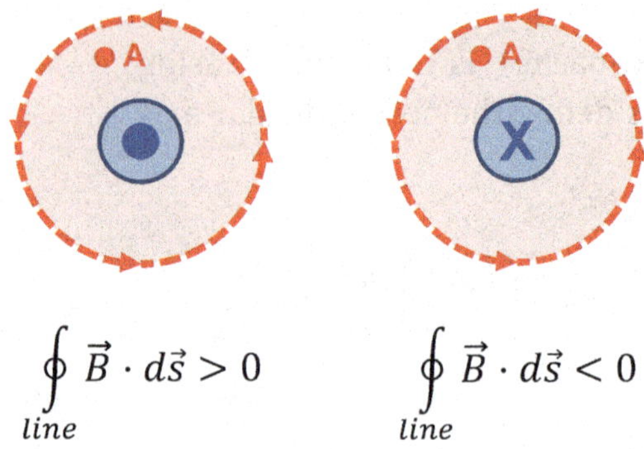

FIGURE 5.10: Applying Ampère's law to two different currents using the same Ampèrian loop. The dashed red line indicates the direction of integration. The corresponding $\vec{A}$ for the enclosed surface is directed out of the page. Thus, the current in the left panel is positive and the current in the right panel is negative.

any current in the same direction as $\vec{A}$ is a positive current in Equation 5-9, and any current in the opposite direction of $\vec{A}$ is a negative current.

Let's now consider the system shown in Figure 5.11, in which three currents are passing through the surface enclosed by the Ampèrian loop.

From the right-hand rule, we know that $\vec{A}$ for the enclosed area associated with the Ampèrian loop points out of the page. Therefore, applying Equation 5-9 gives us

$$\oint_{line} \vec{B} \bullet d\vec{s} = \left(4\pi \times 10^{-7} \frac{\text{kgm}}{\text{C}^2}\right)\left(2\,\mu\text{A} + 4\,\mu\text{A} - 3\mu\text{A}\right)$$

$$\oint_{line} \vec{B} \bullet d\vec{s} = \left(12\pi \times 10^{-13}\,\text{Tm}\right)$$

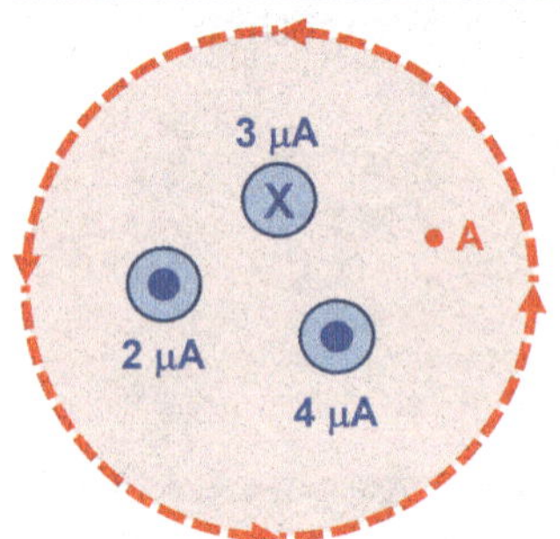

FIGURE 5.11: Three currents passing through a surface enclosed by an Ampèrian loop.

The magnitude of the magnetic field will, of course, not be constant at each point along the Ampèrian loop because of the lack of symmetry between the distribution of currents and the Ampèrian loop. Therefore, we cannot directly calculate the magnitude of the magnetic field at each position along the loop through this calculation.

Indeed, as with Gauss's law for gravitational and electric fields, Ampère's law is most useful when calculating the magnetic field associated with symmetric distributions of current.

Example 5-10

Problem: What is the magnetic field around an infinitely long wire carrying current I?

Solution: The symmetry of the magnetic field will be reflected in the symmetry of the current distribution that generates it. In this case, the magnetic field should exhibit the cylindrical symmetry of the wire carrying the current; this is, indeed, what we have previously calculated (Section 4-5). Therefore, as shown in Figure 5.12, let's choose an Ampèrian loop that is a circle centered on the wire/current.

By symmetry we know that the magnetic field is constant everywhere along the Ampèrian loop (*i.e.*, along the path of integration). If we further assume that the direction of the magnetic field is the same as the direction of the path of integration, we have

$$\oint_{line} \vec{B} \bullet d\vec{s} = \vec{B} \bullet \left(\oint_{line} d\vec{s} \right)$$

$$\oint_{line} \vec{B} \bullet d\vec{s} = B(2\pi r)\cos(0^\circ)$$

$$\oint_{line} \vec{B} \bullet d\vec{s} = B(2\pi r)$$

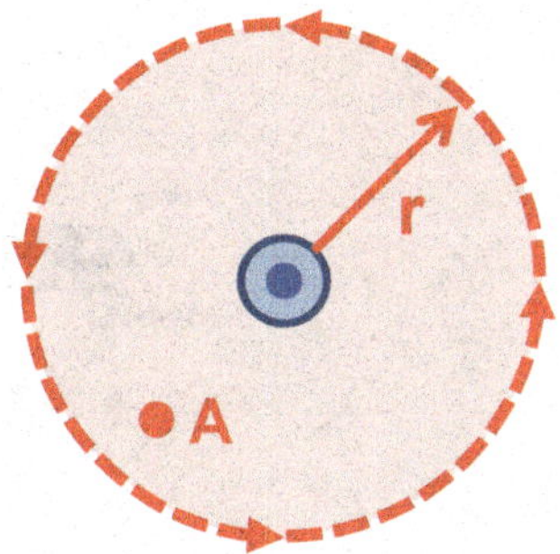

FIGURE 5.12: The Ampèrian loop for determining the magnetic field in Example 5-10.

In this equation the variable r is the radius of the circle corresponding to the Ampèrian loop, as shown in Figure 5.12. From the right-hand rule we know that the direction of $\vec{A}$ for the surface enclosed by the Ampèrian loop is out of the page. Since this is the same direction as the direction of the current we have

$$\oint_{line} \vec{B} \bullet d\vec{s} = \mu_0 I_{enclosed} \quad \rightarrow \quad B(2\pi r) = \mu_0 I \quad \rightarrow \quad B = \frac{\mu_0 I}{2\pi r}$$

The positive sign in our result indicates that the direction of the magnetic field is the same as the direction of the path of integration for the Ampèrian loop[13], consistent with the right-hand rule introduced in Section 4-5. Furthermore, the expression for the magnitude of the magnetic field agrees with the equation derived in Example 4-5 in the limit that the length of the wire is infinite.

A solenoid consists of a long, tightly wound helical coil of wire. We can model a solenoid as a series of tightly packed loops of current, each of which will generate a magnetic field (Section 4-7). As shown in Figure 5.13, the magnetic field from each of these coils will reinforce each other inside the solenoid and nearly completely cancel each other immediately outside the solenoid.

The more closely packed the loops of wire the more uniform and intense the magnetic field inside the solenoid and the smaller the intensity of the magnetic field immediately outside of the solenoid. As shown in Figure 5.14, a solenoid closely resembles a permanent magnet with a north pole and a south pole, whose position (at

13 See discussion in Example 5-3 and Example 5-4.

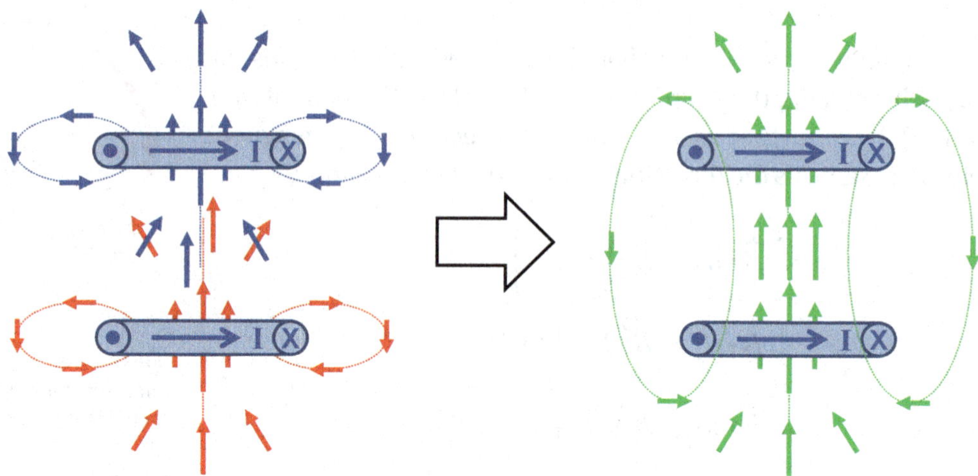

FIGURE 5.13: Determining the electric field for a solenoid from the contributions of each loop of current.

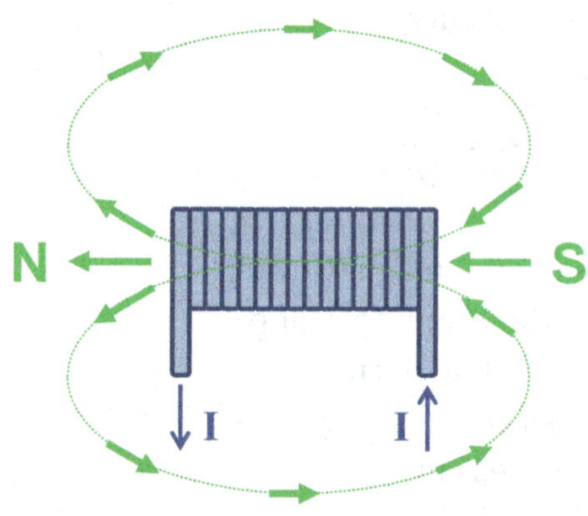

FIGURE 5.14: A solenoid is a simple example of an electromagnet.

either end of the solenoid) depends upon the direction of the current flowing through the solenoid. For this reason a solenoid is also referred to as an *electromagnet* (*i.e.*, a magnet whose magnetic field is generated by an electric current).

For a perfect solenoid, we assume that the magnetic field inside is completely uniform with a constant magnitude, and the magnetic field immediately outside is exactly zero. We can calculate the magnitude of the magnetic field inside the solenoid using Ampère's law. As shown in Figure 5.15, let's use a rectangle as the Ampèrian loop for this calculation; in Figure 5.15 the vertices of this rectangle are numbered 1 through 4.

From the right-hand rule we know that the direction of $\vec{A}$ is into the page, and thus all of the enclosed currents are positive in Equation 5-9. If the current in the wire is I and N loops of wire pass through this enclosed surface area, then

$$\mu_0 I_{enclosed} = \mu_0 NI$$

The line integral of the magnetic field can be calculated for each segment of the Ampèrian loop.

$$\oint_{line} \vec{B} \bullet d\vec{s} = \int_1^2 \vec{B} \bullet d\vec{s} + \int_2^3 \vec{B} \bullet d\vec{s} + \int_3^4 \vec{B} \bullet d\vec{s} + \int_4^1 \vec{B} \bullet d\vec{s}$$

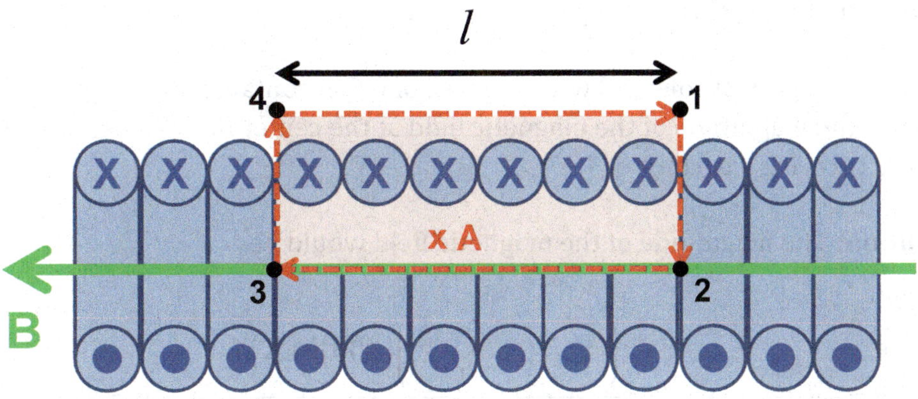

FIGURE 5.15: Determining the magnetic field inside a perfect solenoid using Ampère's law.

Since the magnetic field inside the perfect solenoid is uniform we have

$$\int_2^3 \vec{B} \bullet d\vec{s} = Bl$$

The first and third integrals in the expression for $\oint_{line} \vec{B} \bullet d\vec{s}$ are zero because $\vec{B} \bullet d\vec{s} = 0$ for those paths. The final integral in the expression for $\oint_{line} \vec{B} \bullet d\vec{s}$ are zero because the magnetic field is zero immediately outside of the perfect solenoid. Therefore, from Equation 5-9 we have

$$Bl = \mu_0 NI \quad \rightarrow \quad B = \frac{\mu_0 NI}{l}$$

It is common to express the magnitude of the magnetic field inside the perfect solenoid in terms of the number of loops of wire per unit length of the perfect solenoid; let's use the variable *n* to denote this *density of loops*.

$$n = \frac{N}{l} \quad \rightarrow \quad B = \mu_0 nI \qquad \textbf{(5-10)}$$

Example 5-11

Problem: A perfect solenoid with a length of 0.5 m contains 400 loops of wire. What is the magnitude of the magnetic field at the center of this solenoid if the wire is carrying a current of 4 A?

Solution: The magnitude of the magnetic field would be

$$B = \frac{\mu_0 NI}{l} \quad \rightarrow \quad B = \frac{\left(4\pi \times 10^{-7} \frac{\text{kgm}}{\text{C}^2}\right)(400)(4\,\text{A})}{(0.5\,\text{m})} \quad \rightarrow \quad B = 4\,\text{mT}$$

This magnetic field is several orders of magnitude more intense than the Earth's magnetic field.

In analogy to Equation 5-1 and Equation 5-3 we can define the magnetic flux through a surface using Equation 5-11.

$$\Phi_B = \int_{surface} \vec{B} \bullet d\vec{A} \tag{5-11}$$

The unit of magnetic flux is the weber (Wb).

$$1\,\text{Wb} = 1\,\text{Tm}^2$$

Other common expressions for this unit are

$$1\,\text{Wb} = 1\,\text{Vs} = 1\frac{\text{J}}{\text{A}}$$

Example 5-12

Problem: What is the magnetic flux through the rectangular area near a current carrying wire in Figure 5.16? Assume that the wire is infinitely long.

Solution: The magnetic flux is found using Equation 5-11. Let's define $\vec{A}$ for this surface to point out of the page (Figure 5.17). As shown in Figure 5.17, $\vec{A}$ and $\vec{B}$ will then point in the same direction.

Let's define a *y*-axis to point perpendicular to the wire as shown in Figure 5.18. With this definition we can write our equation for the magnetic flux[14] as

$$B = \frac{\mu_0 I}{2\pi y} \quad \rightarrow \quad \Phi_B = \int_{surface} \left(\frac{\mu_0 I}{2\pi}\right)\frac{dA}{y}$$

It's clear from this equation that the magnetic field has the same magnitude (and direction) along any horizontal line parallel to the wire. Therefore, the magnetic field will be the same for all locations within the red rectangle in Figure 5.18. We can thus use the area of this rectangle as the differential area in our integration.

$$dA = Ldy$$

$$\Phi_B = \int_r^{r+h} \left(\frac{\mu_0 I}{2\pi}\right)\frac{Ldy}{y}$$

$$\Phi_B = \left(\frac{\mu_0 IL}{2\pi}\right)\int_r^{r+h} \frac{dy}{y}$$

$$\Phi_B = \left(\frac{\mu_0 IL}{2\pi}\right)\left(\ln y\Big|_r^{r+h}\right)$$

$$\Phi_B = \left(\frac{\mu_0 IL}{2\pi}\right)\ln\left(\frac{r+h}{r}\right)$$

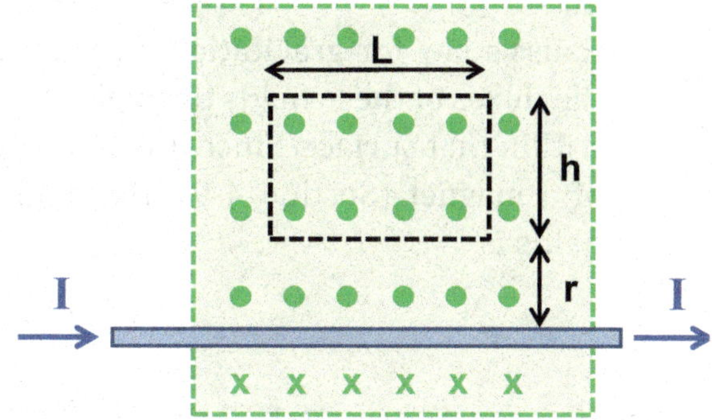

FIGURE 5.16: The system in Example 5-12.

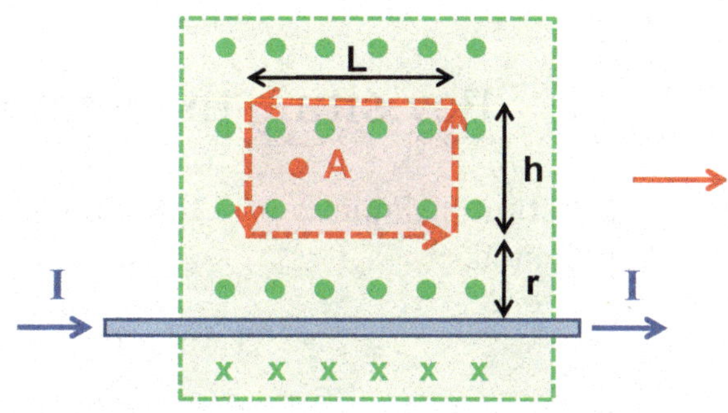

FIGURE 5.17: Defining the direction for the surface area vector for calculating the magnetic flux in Example 5-12.

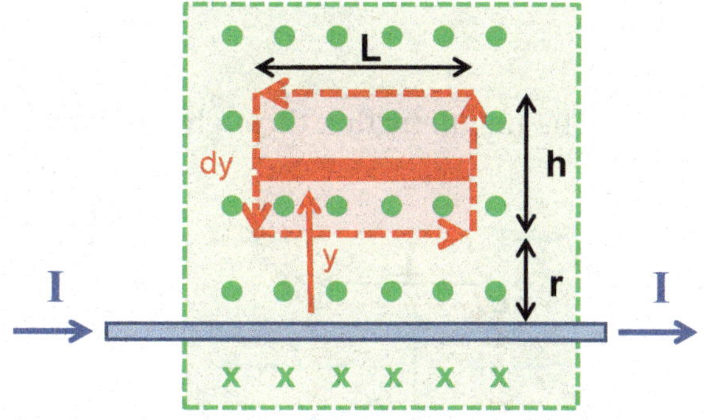

FIGURE 5.18: Defining the differential area for the integral of the magnetic flux.

14 The equation for the magnetic field of an infinite wire was determined in Example 5-10.

Gauss's law for gravitational fields (Equation 5-2) and electric fields (5-4) related the fluxes of these fields through a closed surface to the mass or net charge contained within that surface. Since it is impossible to isolate individual north and south poles of a magnet (Section 4-9), the equivalent expression of Gauss's law for magnetic fields is

$$\oint_{surface} \vec{B} \cdot d\vec{A} = 0 \tag{5-12}$$

In other words, magnetic monopoles (*i.e.*, individual magnetic poles) do not exist.

5-7 Revisiting the Vector Potential

Substituting Equation 4-11 into Equation 5-11 gives us another equation for the vector potential.

$$\vec{B} = \nabla \times \vec{V}_m \quad \rightarrow \quad \Phi_B = \int_{surface} \left(\nabla \times \vec{V}_m\right) \cdot d\vec{A}$$

This equation can then be simplified using Stoke's theorem (Equation B-15).

$$\Phi_B = \oint_{line} \vec{V}_m \cdot d\vec{s} \tag{5-13}$$

The magnetic flux through a surface is therefore equal to the line integral of the vector potential along the boundary of that surface.

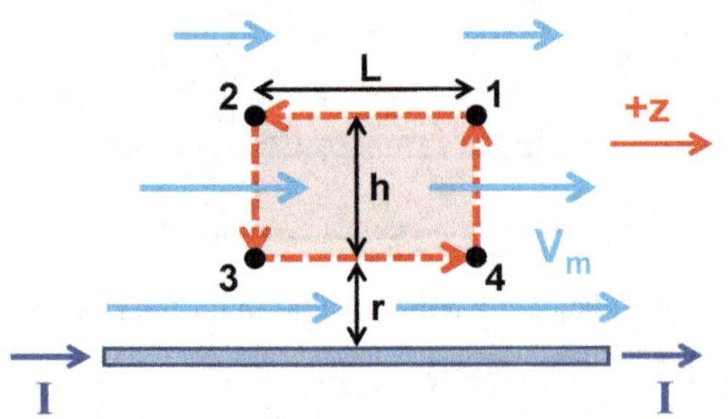

FIGURE 5.19: Defining the path for the integral of the magnetic potential to determine the magnetic flux.

Let's use Equation 5-13 to determine the solution to Example 5-12. The magnetic potential for an infinite wire was determined in Section 4-10.

$$\vec{V}_m = \left(\frac{\mu_0 I}{2\pi}\right)\left(\ln\left(\frac{\rho_0}{\rho}\right)\right)\hat{z}$$

In this equation the *z*-axis is parallel to the direction of the current in the wire, as shown in Figure 5.19.

The line integral of the magnetic potential can be calculated for each segment of the boundary of the surface in Figure 5.19.

$$\oint_{line} \vec{V}_m \bullet d\vec{s} = \int_1^2 \vec{V}_m \bullet d\vec{s} + \int_2^3 \vec{V}_m \bullet d\vec{s} + \int_3^4 \vec{V}_m \bullet d\vec{s} + \int_4^1 \vec{V}_m \bullet d\vec{s}$$

The first integral in this expression is

$$\int_1^2 \vec{V}_m \bullet d\vec{s} = \left(\frac{\mu_0 I}{2\pi}\right)\left(\ln\left(\frac{\rho_0}{r+h}\right)\right)(-L)$$

The negative sign indicates that the direction of $\vec{V}_m$ is opposite the direction of $d\vec{s}$ for this integral, as shown in Figure 5.19. The second and fourth integrals in the expression for $\oint_{line} \vec{V}_m \bullet d\vec{s}$ are zero because $\vec{V}_m \bullet d\vec{s} = 0$ for those paths. The final integral in the expression for $\oint_{line} \vec{V}_m \bullet d\vec{s}$ is

$$\int_4^1 \vec{V}_m \bullet d\vec{s} = \left(\frac{\mu_0 I}{2\pi}\right)\left(\ln\left(\frac{\rho_0}{r}\right)\right)(L)$$

Putting it all together gives us

$$\oint_{line} \vec{V}_m \bullet d\vec{s} = \left(\frac{\mu_0 IL}{2\pi}\right)\left(\ln\left(\frac{\rho_0}{r}\right)\right) - \left(\frac{\mu_0 IL}{2\pi}\right)\left(\ln\left(\frac{\rho_0}{r+h}\right)\right)$$

$$\oint_{line} \vec{V}_m \bullet d\vec{s} = \left(\frac{\mu_0 IL}{2\pi}\right)\ln\left(\frac{r+h}{r}\right) \quad \rightarrow \quad \Phi_B = \left(\frac{\mu_0 IL}{2\pi}\right)\ln\left(\frac{r+h}{r}\right)$$

The result is identical to that obtained in Example 5-12.

Example 5-13

Problem: What is the magnetic potential of the magnetic field of a perfect solenoid with a radius *R*, a density of loops *n*, carrying a current *I*?

Solution: We begin by defining a reference frame for this system using cylindrical coordinates and a *z*-axis parallel to the longitudinal axis of the solenoid. The positive direction of the *z*-axis will be the same as the direction of the solenoid's magnetic field. From the symmetry of the solenoid and the orientation of the current carrying wires in the solenoid we know that the magnetic potential will have a component along only the φ-axis and that this component will depend upon ρ, but not upon z or φ.

$$\vec{V}_m = \left(\left(V_m\right)_\varphi\right)\hat{\varphi}$$

The magnetic flux through a single loop of the solenoid can be calculated using Equation 5-13. The path of integration in this expression is around a single loop of the solenoid, which we will define using cylindrical coordinates.

$$d\vec{s} = (\rho d\varphi)\hat{\varphi} \rightarrow \Phi_B = \oint_{line} \left[\left(\left(V_m\right)_\varphi\right)\hat{\varphi}\right] \cdot \left[(\rho d\varphi)\hat{\varphi}\right] \rightarrow \Phi_B = \left(V_m\right)_\varphi \rho \oint_{line} d\varphi$$

$$\Phi_B = \left(V_m\right)_\varphi \rho(2\pi) \rightarrow \left(V_m\right)_\varphi = \frac{\Phi_B}{2\pi\rho}$$

Since the magnetic field inside the solenoid is uniform, the magnetic potential inside the solenoid ($\rho < R$) is

$$\Phi_B = B\left(\pi\rho^2\right) \rightarrow \left(V_m\right)_\varphi = \frac{B\left(\pi\rho^2\right)}{2\pi\rho} \rightarrow \left(V_m\right)_\varphi = \frac{B\rho}{2}$$

Outside of the solenoid ($\rho > R$) we have

$$\Phi_B = B\left(\pi R^2\right) \rightarrow \left(V_m\right)_\varphi = \frac{B\left(\pi R^2\right)}{2\pi\rho} \rightarrow \left(V_m\right)_\varphi = \frac{BR^2}{2\rho}$$

Substitution of Equation 5-10 gives us

$$\left(V_m\right)_\varphi = \frac{\mu_0 n I \rho}{2} \qquad \rho < R$$

$$\left(V_m\right)_\varphi = \frac{\mu_0 n I R^2}{2\rho} \qquad \rho > R$$

The magnetic potential for a perfect solenoid is illustrated in Figure 5.20.

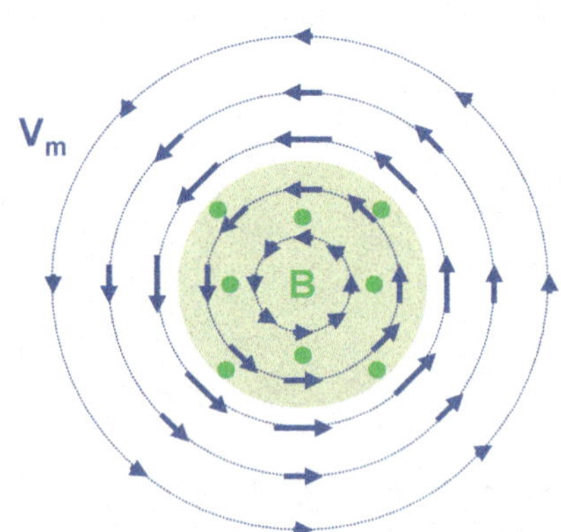

FIGURE 5.20: The magnetic potential for a perfect solenoid. The uniform magnetic field in the interior of the solenoid is shown in green.

Summary

- **Flux:** a quantitative measurement of the magnitude and direction of a vector field with respect to a surface. The gravitational, electric, and magnetic fluxes are defined to be

$$\Phi_g = \int_{surface} \vec{g} \bullet d\vec{A} \qquad \Phi_E = \int_{surface} \vec{E} \bullet d\vec{A} \qquad \Phi_B = \int_{surface} \vec{B} \bullet d\vec{A}$$

The magnetic flux can also be determined from the magnetic potential.

$$\Phi_B = \oint_{line} \vec{V}_m \bullet d\vec{s}$$

- **Gauss's Law for gravitational fields:** the total gravitational flux through a closed surface is proportional to the total mass enclosed by that surface

$$\oint_{surface} \vec{g} \bullet d\vec{A} = -4\pi G M_{enclosed}$$

- **Gauss's Law for electric fields:** the total electric flux through a closed surface is proportional to the net electric charge enclosed by that surface

$$\oint_{surface} \vec{E} \bullet d\vec{A} = \frac{q_{enclosed}}{\varepsilon_0}$$

- **Gauss's Law for magnetic fields:** the total magnetic flux through a closed surface is zero.

$$\oint_{surface} \vec{B} \bullet d\vec{A} = 0$$

This is equivalent to the statement that no magnetic monopoles exist.

- **Ampère's Law:** a mathematical relationship between a magnetic field and the distribution of currents that created it.

$$\oint_{line} \vec{B} \bullet d\vec{s} = \mu_0 I_{enclosed}$$

- **Capacitance:** a measure of how much net electric charge must be separated to produce a certain difference in electric between the separated net electric charges. The capacitance is defined by geometric parameters (length, width, *etc.*) associated with the system. The capacitance of a parallel plate capacitor is

$$C = \frac{A\varepsilon_0}{d}$$

The capacitance also determines the energy store in a capacitor.

$$\Delta U_e = \frac{q^2}{2C}$$

Problems

1. Imagine that a tunnel existed from one end of the moon to the other, directly through the center of the moon. In other words, the tunnel is along a line of symmetry through the moon. In this case, an object released from rest at the opening of one end of the tunnel would undergo simple harmonic motion through the center of the moon. What is the period of this oscillation? The mass and radius of the moon are 7.35×10^{22} kg and 1.74×10^6 m, respectively.

2. What is the magnitude of the net electric force between the positively and negatively charged surfaces of the spherical capacitor in Figure 5.8?

3. Two objects have net electric charges of +50 pC and −50 pC, which results in a 20 V electric potential difference between them. What is the capacitance of this system?

4. Three infinite planes with uniform net electric charge densities are separated as shown in the figure. The net electric charge densities are: $\sigma_1 = 6\ \mu C/m^2$, $\sigma_2 = -3\ \mu C/m^2$, and $\sigma_3 = -2\ \mu C/m^2$. What is the magnitude of the electric field at position P if $d = 5$ cm?

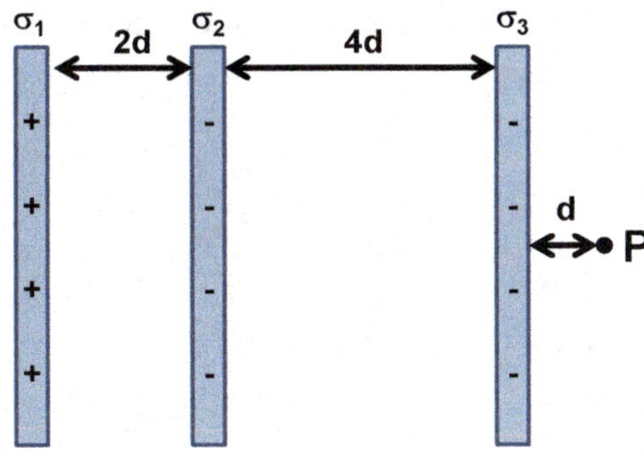

5. The infinitely long cylindrical capacitor in the figure consists of a cylindrical shell of radius r_1 and uniform net positive electric charge density (σ) surrounded by a cylindrical shell of radius r_2 with uniform negative electric charge density ($-\sigma$). What is the capacitance per unit length of this capacitor?

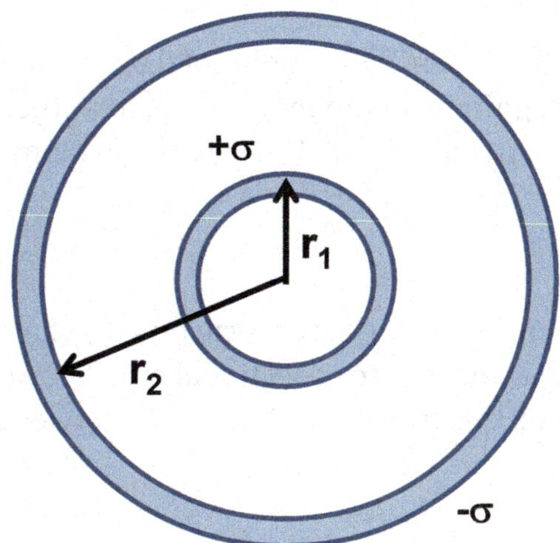

6. What is the value of $\oint_{line} \vec{B} \bullet d\vec{s}$ for the path and enclosed distribution of currents shown in the figure?

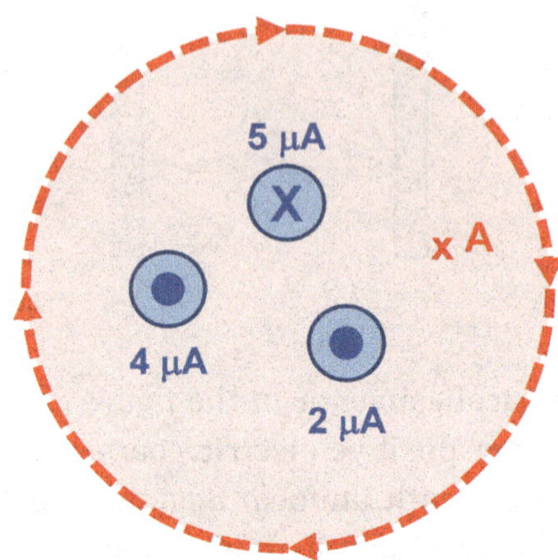

7. A solenoid with a length of 0.2 m and a radius of 0.02 m contains 1000 circular loops of wire. What is the magnitude of the magnetic field at the center of the solenoid if the wire is carrying a current of 2 A? What is the magnitude of the magnetic flux through the solenoid?

8. A closed loop of wire of radius R carries a current I. What is the magnitude of the magnetic flux through a smaller cross-sectional circular area of radius ρ within the loop? You can determine this from the relationship between the magnetic potential and the magnetic flux.

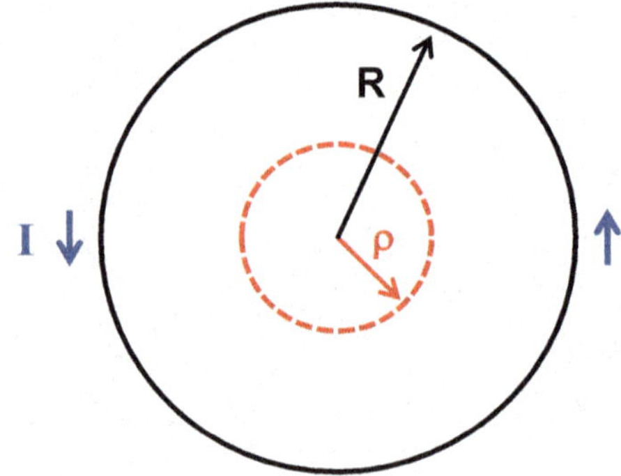

CHAPTER 6

Maxwell's Equations

6-1 Introduction

Recall that the (gravitational, electric, and magnetic) fields, rather than the associated potentials, can be considered to be the physically real entities since they give rise to forces, and through Newton's 2nd law to accelerations, that can be measured[1]. Indeed, we can detect and measure these fields based upon the forces associated with them. Furthermore, since Newton's laws are the same in all inertial reference frames, it must be the case that fields can be associated with forces measured (through the accelerations they cause, *e.g.*) in all inertial reference frames. However, since a magnetic field exerts a force on only a moving charge, the perception of whether a magnetic field or an electric field is causing an acceleration *will depend upon the reference frame applied to*

1 This also follows from the fact that the gravitational, electric, and magnetic potentials are not uniquely defined.

the system. This perception-based distinction between electric and magnetic fields is the first step into a deeper understanding of the relationships between electric and magnetic fields, which will be the focus of this chapter.

6-2 Galilean Transformations of Electric and Magnetic Fields

Consider a particle with a net positive charge moving in a uniform magnetic field, as shown in Figure 6.1. When viewed from a reference frame that is stationary with respect to the magnetic field (reference frame 1 in Figure 6.1) this particle will experience a magnetic force (Equation 4-2) directed along the positive *z*-axis. Let's denote this magnetic force as $\vec{F}_m$ and the associated magnetic field as $\vec{B}_1$; the subscript for the magnetic field denotes the reference frame in which the field is measured. However, when viewed from a reference frame that is moving with the particle (reference frame 2 in Figure 6.1) there cannot be a magnetic force, since the particle is not moving (Equation 4-2); there is still a magnetic field in reference frame 2, denoted as $\vec{B}_2$, and we will see shortly that $\vec{B}_1 \neq \vec{B}_2$. The velocity of reference frame 2 is denoted as $\vec{v}_{2,1}$ to indicate that it is the velocity of reference frame 2 relative to reference frame 1; $\vec{v}_{2,1}$ is therefore also the velocity of the particle as measured in reference frame 1.

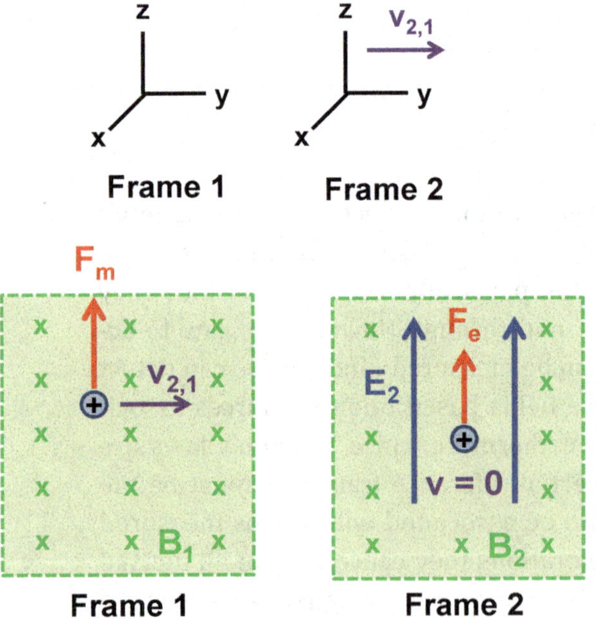

FIGURE 6.1: An electric field appears when a magnetic field is viewed from a moving reference frame.

The particle still feels a force in reference frame 2. Since this cannot be a magnetic force, we attribute it to an electric force[2].

The electric force associated with this electric field is

$$\vec{F}_e = q_2\vec{E}_2$$

2 If necessary, we could conduct additional experiments with particles of different net electric charges to convince us that it is an electric force rather than a gravitational force or some other force.

And since measurements of the force must be the same regardless of the reference frame we have

$$\vec{F}_m = \vec{F}_e \quad \rightarrow \quad q_2\left(\vec{v}_{2,1} \times \vec{B}_1\right) = q_2\vec{E}_2 \quad \rightarrow \quad \vec{E}_2 = \vec{v}_{2,1} \times \vec{B}_1$$

So whether you think an electric field or a magnetic field is present depends upon the motion of your reference frame relative to the reference frame of the source of the field. It follows that if an electric field were already present in reference frame 1, denoted as $\vec{E}_1$, then the total electric field measured in reference frame 2 would be

$$\vec{E}_2 = \vec{E}_1 + \vec{v}_{2,1} \times \vec{B}_1 \qquad \textbf{(6-1)}$$

Now let's consider the situation shown in Figure 6.2, in which a particle with a net positive charge is at rest in reference frame 1. The particle is moving, however, when viewed from reference frame 2 and therefore must produce a magnetic field in reference frame 2. Let's denote this magnetic field as $\vec{B}_2$. We can determine this magnetic field using Equation 4-1.

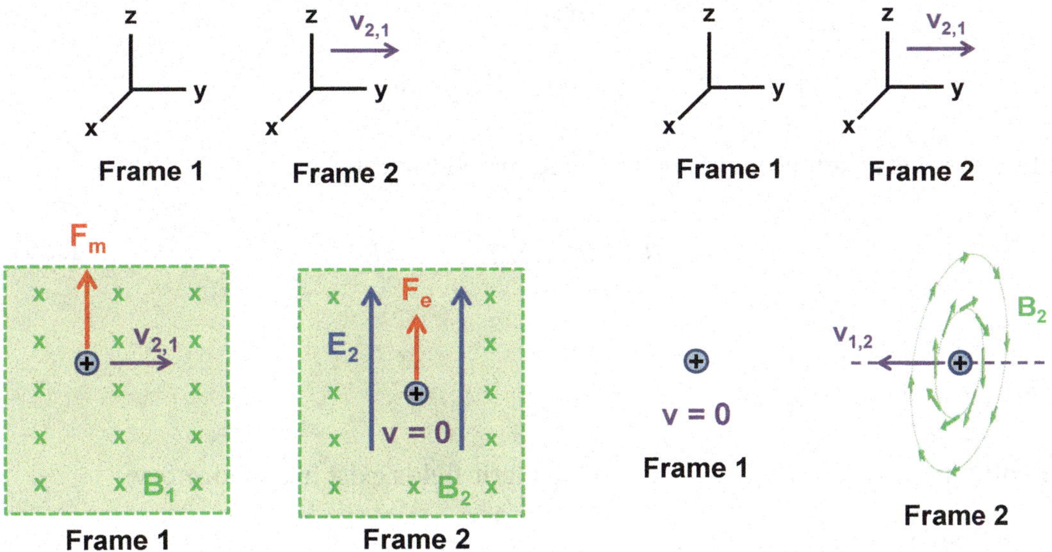

FIGURE 6.2: A charged particle appears to create a magnetic field when viewed from a moving reference frame.

$$\vec{B}_2 = \left(\frac{\mu_0 q_1}{4\pi}\right)\frac{\vec{v}_{1,2} \times \vec{r}}{r^3} \quad \rightarrow \quad \vec{B}_2 = \left(\mu_0 \varepsilon_0\right)\vec{v}_{1,2} \times \left(\frac{q_1 \vec{r}}{4\pi\varepsilon_0 r^3}\right)$$

$$\vec{B}_2 = \left(\mu_0 \varepsilon_0\right)\vec{v}_{1,2} \times \left(\frac{q_1 \hat{r}}{4\pi\varepsilon_0 r^2}\right)$$

We recognize the second term in parenthesis as the electric field of the particle (Equation 3-9).

$$\vec{B}_2 = \left(\mu_0 \varepsilon_0\right)\vec{v}_{1,2} \times \vec{E}_2$$

This is fairly profound. *The magnetic field of a moving point charge is the electric field of a stationary point charge transformed into a moving reference frame.* Furthermore, since there is no magnetic field in reference frame 1, we know from Equation 6-1 that

$$\vec{B}_1 = 0 \quad \rightarrow \quad \vec{E}_2 = \vec{E}_1$$

Therefore,

$$\vec{B}_2 = \left(\mu_0 \varepsilon_0\right)\vec{v}_{1,2} \times \vec{E}_1$$

$$\vec{v}_{1,2} = -\vec{v}_{2,1} \quad \rightarrow \quad \vec{B}_2 = -\left(\mu_0 \varepsilon_0\right)\vec{v}_{2,1} \times \vec{E}_1$$

Finally, if there is an additional magnetic field present in reference frame 1, then

$$\vec{B}_2 = \vec{B}_1 - \left(\mu_0 \varepsilon_0\right)\vec{v}_{2,1} \times \vec{E}_1 \tag{6-2}$$

Example 6-1

Problem: The following electric and magnetic fields exist in a laboratory:

$$\vec{E} = \left(4000\frac{\text{V}}{\text{m}}\right)\hat{y} \qquad \vec{B} = \left(0.1\text{T}\right)\hat{x}$$

A positively charged point particle is fired through these fields. What must be the velocity of the particle so that no electric field is perceived in the particle's reference frame? What is the magnetic field in the particle's reference frame?

Solution: We start with Equation 6-1.

$$\vec{E}_2 = \vec{E}_1 + \vec{v}_{2,1} \times \vec{B}_1 \quad \rightarrow \quad 0 = \left(4000\frac{\text{V}}{\text{m}}\right)\hat{y} + \vec{v} \times \left[(0.1\text{T})\hat{x}\right]$$

$$\vec{v} \times \left[(0.1\text{T})\hat{x}\right] = -\left(4000\frac{\text{V}}{\text{m}}\right)\hat{y}$$

The direction of the velocity must be in the negative z-axis direction. The speed is

$$\left[(v)(-\hat{z})\right] \times \left[(0.1\text{T})\hat{x}\right] = -\left(4000\frac{\text{V}}{\text{m}}\right)\hat{y} \quad \rightarrow \quad v = 4 \times 10^4 \frac{\text{m}}{\text{s}}$$

We can then determine the magnetic field using Equation 6-2.

$$\vec{B}_2 = \vec{B}_1 - (\mu_0 \varepsilon_0)\vec{v}_{2,1} \times \vec{E}_1$$

$$\vec{B}_2 = (0.1\text{T})\hat{x} - \left(4\pi \times 10^{-7}\frac{\text{kgm}}{\text{C}^2}\right)\left(8.85 \times 10^{-12}\frac{\text{C}^2}{\text{Jm}}\right)\left[\left(-4 \times 10^4 \frac{\text{m}}{\text{s}}\right)\hat{z}\right] \times \left[\left(4000\frac{\text{V}}{\text{m}}\right)\hat{y}\right]$$

$$\vec{B}_2 = (0.1\text{T})\hat{x} - (1.8 \times 10^{-9}\text{T})\hat{x} \quad \rightarrow \quad \vec{B}_2 \approx (0.1\text{T})\hat{x}$$

The magnetic field is nearly identical in the two reference frames.

We can conclude from Equation 6-1 and Equation 6-2 that it is not possible to distinguish between electric and magnetic fields[3]. Indeed, we instead argue that there is only a single electromagnetic field, and that our perception or measurement of it depends upon which reference frame we use for the measurement.

---

3 It is worth noting that Equation 6-1 and Equation 6-2 are valid only when $v_{2,1} << \frac{1}{\sqrt{\mu_0 \varepsilon_0}}$. For more information, see Section C-7.

6-3 A Changing Magnetic Field Creates an Electric Field, Part 1

Since electric and magnetic fields can be considered to be different manifestations of a single electromagnetic field, it is not surprising that we can associate a change in one field (electric or magnetic) with a corresponding change in the other field. For example, a changing magnetic field is said to create (or *induce*) an electric field. This phenomenon is commonly expressed mathematically as

$$\oint_{line} \vec{E} \bullet d\vec{s} = -\frac{d\Phi_B}{dt} \tag{6-3}$$

Equation 6-3 is called Faraday's law of induction, after Michael Faraday who empirically determined it. Equation 6-3 relates a change in magnetic flux through a surface to the electric field that is created (or *induced*) around the boundary of that surface; the integral in Equation 6-3 is performed along a path around the surface through which the magnetic flux is calculated. This is identical to how the surface and boundary were defined for the integral in Ampère's law (Section 5-6).

Consider the system shown in Figure 6.3 which consists of a uniform magnetic field contained within a cylindrical volume with radius R.

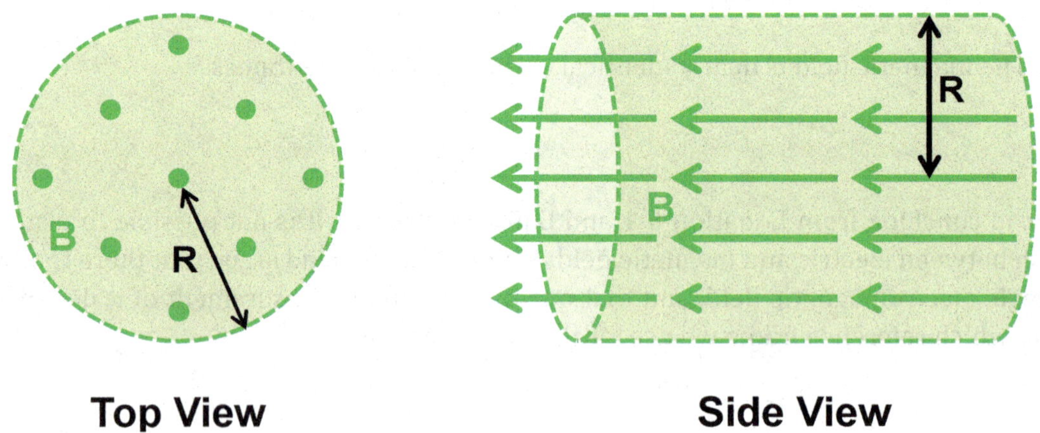

FIGURE 6.3: A uniform magnetic field contained in a cylindrical volume with radius R.

If the magnitude of the magnetic field is changing, then the magnetic flux through the cylinder is changing. This changing magnetic flux will then create an electric field. To

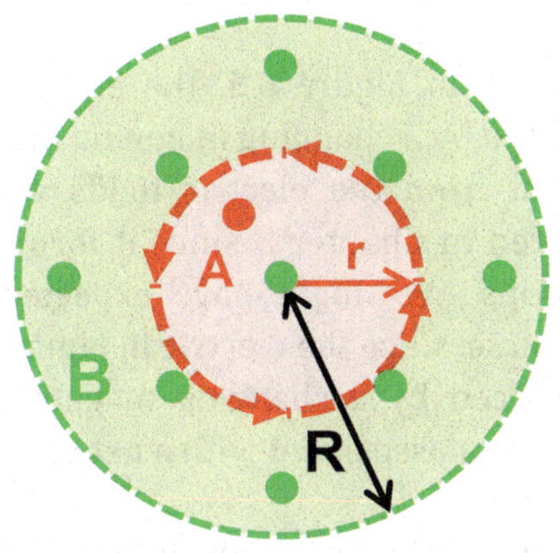

FIGURE 6.4: A uniform magnetic field contained in a cylindrical volume with radius *R*. The dashed red line indicates the path of integration for determining the electric field associated with a change in magnetic flux. For this path, the surface area vector points out of the page.

determine the magnitude of this electric field let's define the path for the integration in Equation 6-3 to be a circle of radius *r* centered on the longitudinal axis of the cylinder, as shown in Figure 6.4. It is the change in the magnetic flux through the surface area bounded by this path which will create the electric field along the path. The direction of the area vector for this surface is determined from the path of integration using the right-hand rule (Section 5-6).

Since the magnetic field is uniform and parallel to the surface area vector, we have

$$\frac{d\Phi_B}{dt} = \frac{d}{dt}\left(B\pi r^2\right) \rightarrow \frac{d\Phi_B}{dt} = \pi r^2 \frac{dB}{dt}$$

Furthermore, because the magnetic field is uniform and symmetric about the longitudinal axis of the cylinder, we can argue from symmetry that the electric field will have a constant magnitude for all locations along the path of integration. Thus,

$$\oint_{line} \vec{E} \bullet d\vec{s} = E\left(2\pi r\right)$$

Substitution of these results into Equation 6-3 gives us

$$E\left(2\pi r\right) = -\pi r^2 \frac{dB}{dt} \rightarrow E = -\frac{r}{2}\frac{dB}{dt}$$

If the magnitude of the magnetic field is increasing (*i.e.*, if $\frac{dB}{dt} > 0$), then $E < 0$. This result indicates that if the magnitude of the magnetic field is increasing, the direction of the induced electric field is opposite the direction of the integration (*i.e.*, the direction of the induced electric field is clockwise in Figure 64). Similarly, if the magnitude of the magnetic field is decreasing (*i.e.*, if $\frac{dB}{dt} < 0$), then $E > 0$ and the direction of the induced electric field is the same as the direction of the integration (*i.e.*, the direction of the induced electric field is counter-clockwise in Figure 64)[4].

4 See discussion in Example 5-10.

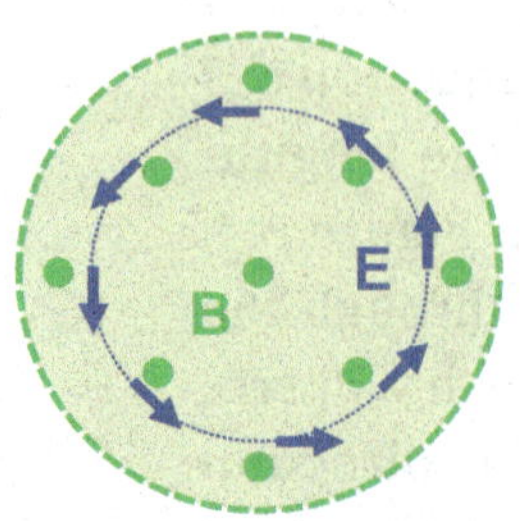

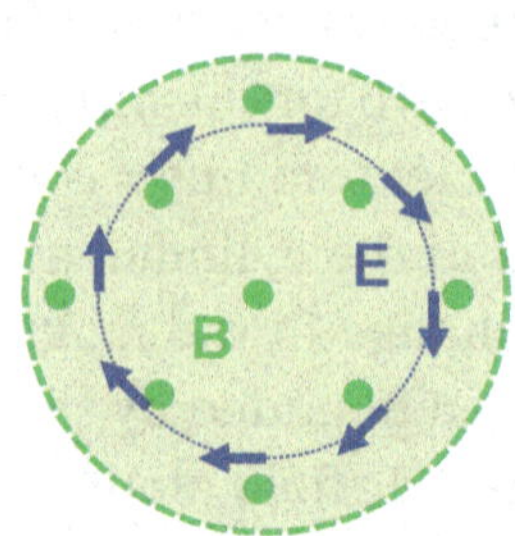

FIGURE 6.5: Orientation of the electric field induced by a changing magnetic flux.

As shown in Figure 6.5, the electric field induced by a changing magnetic flux is different from the electric fields we encountered in Chapter 3 since it forms closed loops. We might have expected this, of course, since the electric in Figure 6.5 is induced by a changing magnetic flux, and is not associated with a net electric charge.

An electric field will also be induced at locations outside of the cylinder in Figure 6.3. As shown in Figure 6.6, for these locations the magnetic flux is calculated using the area of the cylinder.

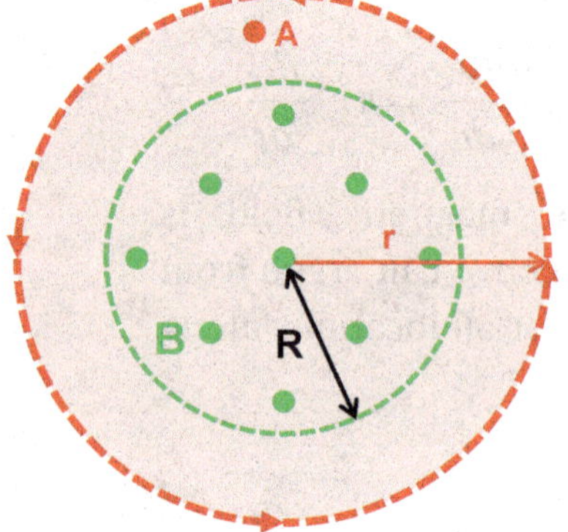

FIGURE 6.6: Determining the magnetic field for locations outside of the change in magnetic flux.

$$\frac{d\Phi_B}{dt} = \frac{d}{dt}\left(B\pi R^2\right) \rightarrow \frac{d\Phi_B}{dt} = \pi R^2 \frac{dB}{dt}$$

Since the path for the integration of the electric field is still centered on the longitudinal axis of the cylinder, the electric field will be constant at all points along the path.

$$\oint \vec{E} \bullet d\vec{s} = E(2\pi r)$$

Thus,

$$\oint \vec{E} \bullet d\vec{s} = -\frac{d\Phi_B}{dt} \rightarrow E(2\pi r) = -\pi R^2 \frac{dB}{dt} \rightarrow E = -\frac{R^2}{2r}\frac{dB}{dt}$$

It is interesting to note that the magnitude of this electric field decreases proportional to $1/r$, rather than $1/r^2$ as seen for electric fields created by a net electric charge (Equation 3-9).

If charged particles are present, an induced electric field will give rise to an induced current (Figure 6.7) that will be in the same direction as the induced electric field (Section 4-4).

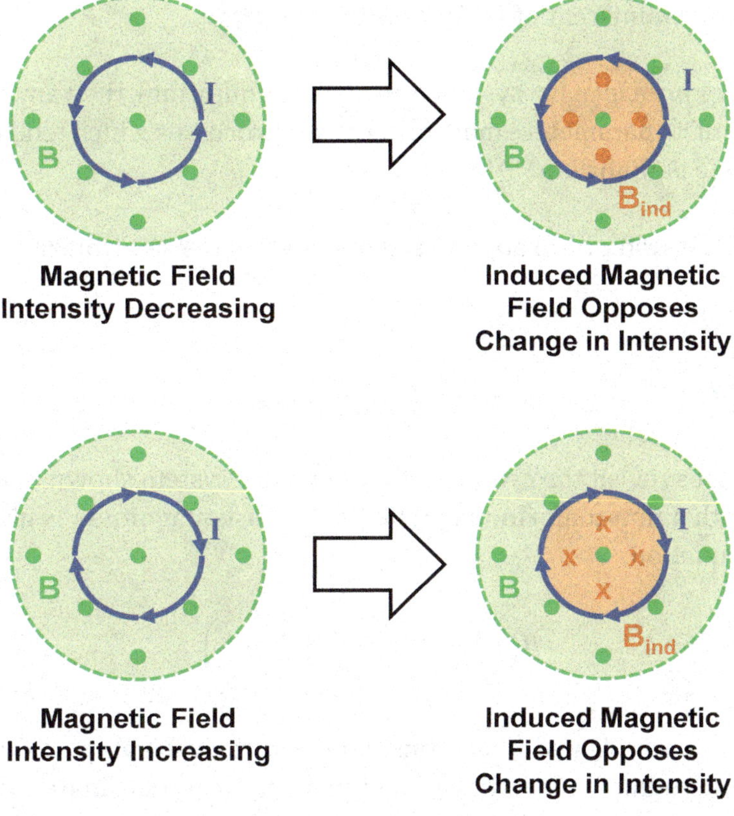

FIGURE 6.7: Lenz's law. The current induced by a changing magnetic flux will generate an induced magnetic field that opposes the change in that magnetic flux.

Any induced current associated with the induced electric field will itself give rise to a magnetic field (Section 4-5 and Section 4-7). Furthermore, as shown in Figure 6.7, the direction of this induced magnetic field always *counteracts* the change in magnetic flux that induced the current (or electric field). Thus, if the magnitude of the magnetic flux through a surface is decreasing, the induced magnetic field will be aligned with the direction of the original magnetic field. In contrast, if the magnitude of the magnetic flux through a surface is increasing, the induced magnetic field will be aligned opposite the direction of the original magnetic field. This phenomenon (which is succinctly expressed as the negative sign in Equation 6-3) is known as ***Lenz's law***.

> *Lenz's law*: The induced current in a loop is in the direction that creates a magnetic field that opposes the change in magnetic flux through the area enclosed by the loop. Hence, *systems always attempt to maintain a constant magnetic flux*.

Lenz's law is also reminiscent of ***Le Châtelier's principle***.

> *Le Châtelier's principle*: If a system is in stable equilibrium, then any spontaneous charge of its parameters must bring about processes which tend to restore the system to equilibrium.

In other words, the system will adjust itself to maintain the same magnetic flux through the surface.

Example 6-2

Problem: Let's revisit the cylindrically symmetric system shown in Figure 6.3, but now with a non-uniform magnetic field whose magnitude is given by the following equation:

$$B(\rho)=\frac{B_0}{\left(\eta+\rho^2\right)^2}\left(1-e^{-kt}\right)$$

In this equation B_0, k, and η are constants, and the variable ρ is the distance from the longitudinal axis of the cylinder in cylindrical coordinates. What is the magnitude of the induced electric field a distance r from the longitudinal axis of the cylinder?

Solution: Let's define a path of integration and enclosed surface area as shown in Figure 6.4; in other words, the direction of the area vector for the enclosed surface is the same as the direction of the magnetic field. Since the magnetic field is cylindrically symmetric, we can use a thin ring[5] as the differential area in the integration of the magnetic flux in Equation 5-11.

$$dA=\left(2\pi\rho\right)d\rho \quad\rightarrow\quad \Phi_B=\int_0^r\frac{B_0}{\left(\eta+\rho^2\right)^2}\left(1-e^{-kt}\right)\left(2\pi\rho\right)d\rho$$

$$\Phi_B=\left(2\pi B_0\right)\left(1-e^{-kt}\right)\int_0^r\frac{\rho}{\left(\eta+\rho^2\right)^2}d\rho \quad\rightarrow\quad \Phi_B=\left(2\pi B_0\right)\left(1-e^{-kt}\right)\left(-\frac{1}{2\left(\eta+\rho^2\right)}\Bigg|_0^r\right)$$

5 See Example 1-2, Example 3-12, and Example 5-6.

$$\Phi_B = \left(-\pi B_0\right)\left(1 - e^{-kt}\right)\left(\frac{1}{\eta + r^2} - \frac{1}{\eta}\right) \rightarrow \Phi_B = \left(\frac{\pi B_0}{\eta}\right)\left(\frac{r^2}{\eta + r^2}\right)\left(1 - e^{-kt}\right)$$

$$\frac{d\Phi_B}{dt} = \left(\frac{\pi B_0}{\eta}\right)\left(\frac{r^2}{\eta + r^2}\right)\left(ke^{-kt}\right)$$

Since the path for the integration of the electric field is still centered on the longitudinal axis of the cylinder and the magnitude field is cylindrically symmetric, we know that the electric field will be constant at all points along the path.

$$\oint \vec{E} \bullet d\vec{s} = E\left(2\pi r\right)$$

Putting everything together we have

$$E\left(2\pi r\right) = -\left(\frac{\pi B_0}{\eta}\right)\left(\frac{r^2}{\eta + r^2}\right)\left(ke^{-kt}\right)$$

$$E = -\frac{B_0 rke^{-kt}}{2\eta\left(\eta + r^2\right)}$$

The negative sign in our result indicates that the direction of the induced electric field is opposite the direction of the path of integration; specifically, the direction of the induced electric field is clockwise in Figure 6.4. This is in agreement with Lenz's law (Figure 6.5) since the magnitude of the magnetic field is increasing in this example.

Example 6-3

Problem: The electric current in a perfect solenoid (Section 5-6) is changing. What is the induced electric field in the wires of the solenoid?

Solution: Let's define the path for the integration in Equation 6-3 to be a circle centered on the longitudinal axis of the solenoid; the direction of the path will be the same as the direction of the current so that $\vec{A}$ is in the same direction as $\vec{B}$. The radius of this circle will be equal to the radius of the solenoid, which we can denote as R; this path corresponds to the location of the wires. Since the magnitude of the magnetic field is constant throughout the solenoid, we have

$$\frac{d\Phi_B}{dt} = \frac{d}{dt}\left(B\pi R^2\right) \quad \rightarrow \quad \frac{d\Phi_B}{dt} = \pi R^2 \frac{dB}{dt}$$

Substitution of the magnitude of this magnetic field (Equation 5-10) gives us

$$B = \mu_0 nI \quad \rightarrow \quad \frac{d\Phi_B}{dt} = \pi R^2 \mu_0 n \frac{dI}{dt}$$

From symmetry, we know that the magnitude of the induced electric field will be the same for all locations along this path of integration.

$$\oint \vec{E} \bullet d\vec{s} = E\left(2\pi R\right)$$

Putting everything together, we have

$$E\left(2\pi R\right) = -\pi R^2 \mu_0 n \frac{dI}{dt} \quad \rightarrow \quad E = -\left(\frac{R\mu_0 n}{2}\right)\frac{dI}{dt}$$

As shown in Figure 6.8, the direction of the electric field induced by the changing current will be opposite the direction of the change in the current. In other words, a change in the electric current will induce a new electric current that will oppose this change. This is yet another example of Lenz's law (or Le Châtelier's principle).

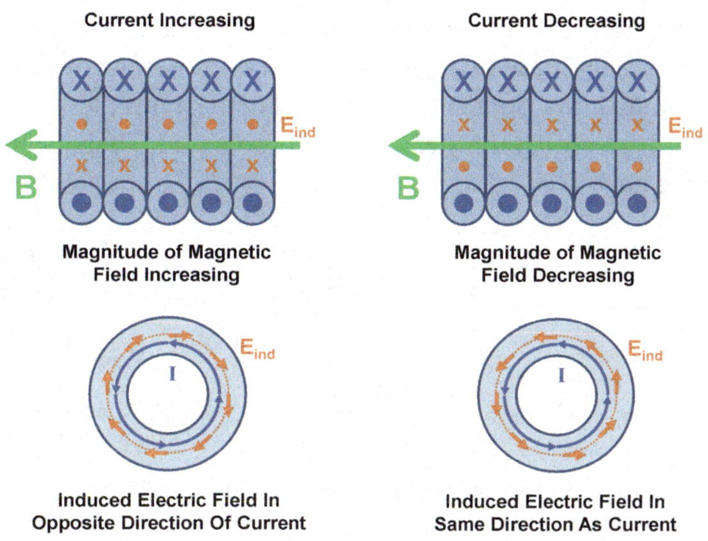

FIGURE 6.8: The electric field induced in a solenoid when the current in the solenoid is changing.

The induced electric field (or equivalently, the induced current) resulting from the change in the electric current in a solenoid will cause the charged particles (*e.g.*, the electrons) in the wires of the solenoid to move. These charged particles will continue to move until their separation builds up a potential difference that counteracts the induced electric field (Figure 6.9); this is analogous to the electric potential created by the separation of electric charges in a capacitor (Section 5-5). At this point, the system will once again be at equilibrium and no additional induced current will exist.

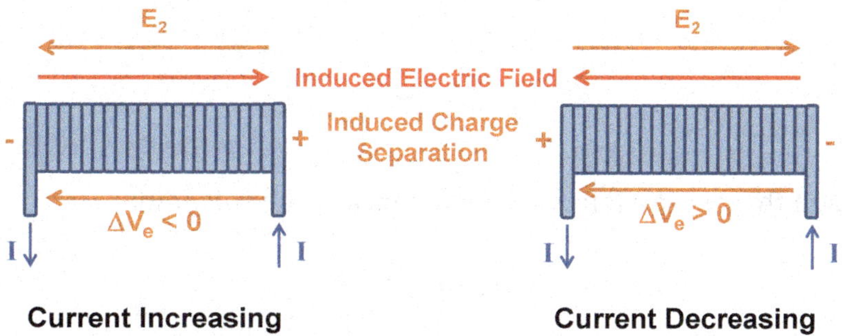

FIGURE 6.9: The electric potential induced in a solenoid when the current in the solenoid is changing.

Let's denote the electric field associated with this electric potential difference as $\vec{E}_2$ (see Figure 6.9). The magnitude of $\vec{E}_2$ must be equal to the magnitude of the induced electric field but point in the opposite direction. Furthermore, let's assume that both of these electric fields are constant throughout the solenoid. Then, from Equation 3-10 and Example 6-3 we have

$$E_2 = -E \rightarrow E_2 = -\left(-\left(\frac{R\mu_0 n}{2}\right)\frac{dI}{dt}\right) \rightarrow E_2 = \left(\frac{R\mu_0 n}{2}\right)\frac{dI}{dt}$$

$$\Delta V_e = -E_2 \Delta s \rightarrow \Delta V_e = -\left(\left(\frac{R\mu_0 n}{2}\right)\frac{dI}{dt}\right)\Delta s \rightarrow \Delta V_e = -\left(\frac{R\mu_0 n}{2}\right)\Delta s\frac{dI}{dt}$$

The length Δs is the entire length of wire in the solenoid. If the solenoid has N turns of wire, then

$$\Delta s = 2\pi RN \rightarrow \Delta V_e = -\left(\frac{R\mu_0 n}{2}\right)(2\pi RN)\frac{dI}{dt} \rightarrow \Delta V_e = -\pi R^2 \mu_0 nN\frac{dI}{dt}$$

Let's define the length of the solenoid to be l. Then,

$$n = \frac{N}{l} \rightarrow \Delta V_e = -\pi R^2 \mu_0 n(nl)\frac{dI}{dt} \rightarrow \Delta V_e = -\left(\pi R^2 l\right)\mu_0 n^2\frac{dI}{dt}$$

The term in parenthesis in this expression is the volume of the solenoid.

$$V = \pi R^2 l \rightarrow \Delta V_e = -\mu_0 n^2 V\frac{dI}{dt}$$

This equation is typically written as

$$\Delta V_e = -L\frac{dI}{dt} \tag{6-4}$$

The variable L in this expression is the *inductance* of the solenoid.

$$L = \mu_0 n^2 V$$

The inductance of a solenoid depends upon the geometry of the solenoid (*i.e.*, on the volume and the density of winding of the wire). The SI unit of inductance is the henry (H).

$$1\,\mathrm{H} = 1\frac{\mathrm{Vs}}{\mathrm{A}}$$

More generally, *the inductance of loop of wire is the ratio of the magnetic flux passing through the surface enclosed by the loop to the current in the wire.* Thus, the inductance for a combination of N stacked loops of wire (such as the loops of wire in a solenoid) would be

$$L = N\frac{\Phi_B}{I} \tag{6-5}$$

Example 6-4

Problem: A coaxial cable can be modeled as a thin cylindrical shell of radius r_2 surrounding a solid cylinder of radius r_1 (Figure 6.10). The cylinder and shell carry current of the same magnitude, but in opposite directions.

What is the inductance of a coaxial cable of length l?

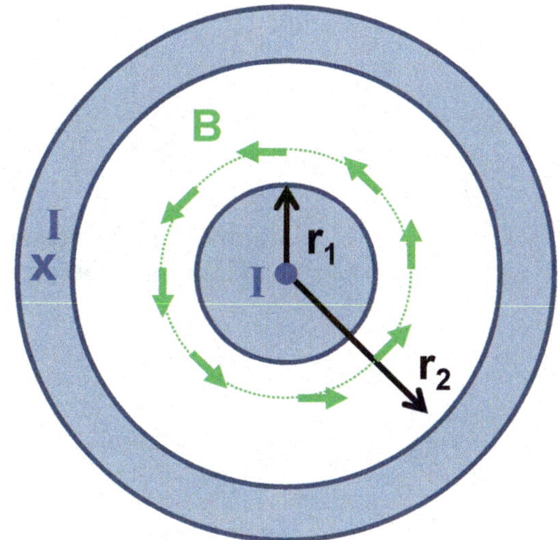

FIGURE 6.10: The coaxial cable in Example 6-4.

Solution: From Ampère's law, we know that the magnetic field inside the coaxial cable is due to the current flowing through the cylinder only. Furthermore, from Example 5-12, we know that the magnetic flux inside a coaxial cable of length l would be

$$\Phi_B = \left(\frac{\mu_0 Il}{2\pi}\right)\ln\left(\frac{r_2}{r_1}\right)$$

Thus, from Equation 6-5 we have

$$L = \frac{\Phi_B}{I} \rightarrow L = \frac{\left(\frac{\mu_0 Il}{2\pi}\right)\ln\left(\frac{r_2}{r_1}\right)}{I} \rightarrow L = \left(\frac{\mu_0 l}{2\pi}\right)\ln\left(\frac{r_2}{r_1}\right)$$

Now let's imagine that we have two loops of wire very close together, as shown in Figure 6.11. A change in the current in loop 1 will change the magnetic flux through loop 2 and thus induce and electric field and current in loop 2.

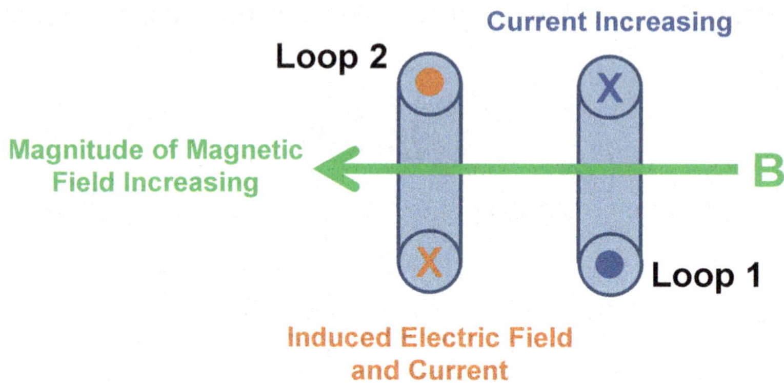

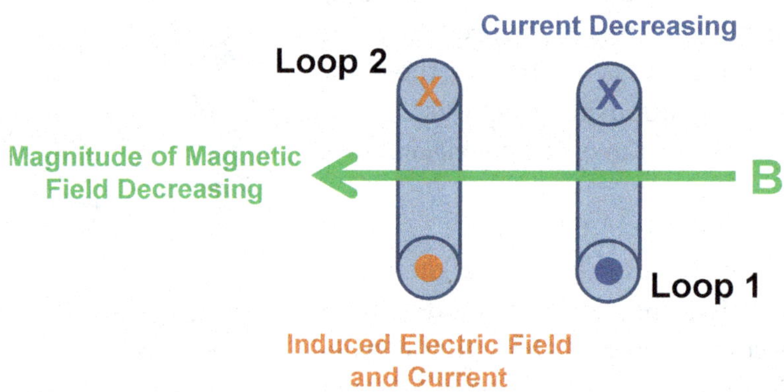

FIGURE 6.11: The mutual inductance of two loops of wire.

In analogy to Equation 6-5, we can define a mutual inductance (M_{21}) of loop 2 with respect to loop 1 as

$$M_{21} = \frac{(\Phi_B)_{21}}{I_1} \quad \textbf{(6-6)}$$

In Equation 6-6, I_1 is the current in loop 1 and $(\Phi_B)_{21}$ is the flux through loop 2 of the magnetic field from loop 1. Mutual inductance is reciprocal so that $M_{21} = M_{12}$.

Example 6-5

Problem: A coil of wire containing N_2 loops of wire (each with a radius R_2) is wrapped around a solenoid (with radius R_1 and loop density n_1) that is carrying current I_1. What is the mutual inductance of the coil with respect to the solenoid? What is the magnitude of the electric field in the coil if the current in the solenoid is changing?

Solution: Let's denote the coil with subscript 2 and the solenoid with subscript 1. Then the mutual inductance of the coil with respect to the solenoid would be

$$M_{21} = \frac{N_2 (\Phi_B)_{21}}{I_1}$$

The flux through the coil is contained completely within the area of the solenoid.

$$(\Phi_B)_{21} = B_1 \pi R_1^2 \quad \rightarrow \quad M_{21} = \frac{N_2 B_1 \pi R_1^2}{I_1}$$

Substitution of Equation 5-10 gives us

$$B_1 = \mu_0 n_1 I_1 \quad \rightarrow \quad M_{21} = \frac{N_2 \mu_0 n_1 I_1 \pi R_1^2}{I_1} \quad \rightarrow \quad M_{21} = N_2 \mu_0 n_1 \pi R_1^2$$

When the current in the solenoid is changing the flux through the coil changes and an electric field is induced in the coil. From Equation 6-3 and Equation 6-6, we have

$$(\Phi_B)_{21} = I_1 M_{21} \quad \rightarrow \quad \oint_{line} \vec{E}_2 \bullet d\vec{s} = -\frac{d}{dt}\left(\frac{M_{21} I_1}{N_2}\right)$$

$$\oint_{line} \vec{E}_2 \bullet d\vec{s} = -\left(\frac{M_{21}}{N_2}\right)\frac{dI_1}{dt}$$

From symmetry, we can assume that the induced electric field has the same magnitude at all points in the coil.

$$\oint_{line} \vec{E}_2 \bullet d\vec{s} = E_2 (2\pi R_2) \quad \rightarrow \quad E_2 (2\pi R_2) = -\left(\frac{M_{21}}{N_2}\right)\frac{dI_1}{dt}$$

$$E_2 = -\left(\frac{M_{21}}{2\pi R_2 N_2}\right)\frac{dI_1}{dt}$$

$$M_{21} = N_2\mu_0 n_1 \pi R_1^2 \quad \rightarrow \quad E_2 = -\left(\frac{N_2\mu_0 n_1 \pi R_1^2}{2\pi R_2 N_2}\right)\frac{dI_1}{dt}$$

$$E_2 = -\left(\frac{\mu_0 n_1 R_1^2}{2R_2}\right)\frac{dI_1}{dt}$$

When we compare the derivations of the inductance of a solenoid and the mutual inductance of two loops of wire (Example 6-5), we see that the inductance of the solenoid in Equation 6-4 is more accurately the *self-inductance* of the solenoid, since it is based upon the mutual inductance of the loops of wire within the solenoid.

For very simple systems, such as those discussed above, the self-inductance or mutual inductance can be calculated directly from the geometry of the loops in the system. However, it is often the case that these calculations cannot be done (or, at least cannot be done analytically). In those situations, the self-inductance or mutual inductance can be directly calculated by measuring the changing current and associated induced current in the system.

Finally, we see that substitution of Equation 5-13 into Equation 6-5 (with $N = 1$) allows us to express the inductance of a single loop of wire in terms of the magnetic potential associated with the current flowing in the wire.

$$L = \frac{\oint_{line} \vec{V}_m \bullet d\vec{s}}{I} \quad \rightarrow \quad L = \oint_{line}\left(\frac{\vec{V}_m}{I}\right)\bullet d\vec{s} \tag{6-7}$$

We can use Equation 6-7 together with the results of Example 5-13 to calculate the self-inductance of a solenoid. Let's begin by determining the self-inductance for a single loop of the solenoid. In that case, the path of integration is around one of the loops of wire in the solenoid, which we will define using cylindrical coordinates.

$$d\vec{s} = (\rho\, d\varphi)\hat{\varphi} \quad \rightarrow \quad L_{loop} = \oint_{line}\left[\left(\frac{\frac{\mu_0 nIR}{2}}{I}\right)\hat{\varphi}\right]\bullet\left[(R\,d\varphi)\hat{\varphi}\right] \quad \rightarrow \quad L_{loop} = \left(\frac{\mu_0 nR^2}{2}\right)\oint_{line} d\varphi$$

The integral in this expression is simply the total enclosed angle of a circle.

$$\oint_{line} d\varphi = 2\pi \quad \rightarrow \quad L_{loop} = \mu_0 n R^2 \pi$$

The self-inductance of the entire solenoid is thus

$$L = NL_{loop} \quad \rightarrow \quad L = N\mu_0 n R^2 \pi \quad \rightarrow \quad L = (nl)\mu_0 n R^2 \pi$$

$$L = \mu_0 n^2 (\pi R^2 l) \quad \rightarrow \quad L = \mu_0 n^2 V$$

As expected, this result is identical to that obtained previously.

6-4 A Changing Magnetic Field Creates an Electric Field, Part 2

In the examples of Section 6-3, a change in the magnitude of a magnetic field caused a change in magnetic flux which induced an electric field. Of course, the magnetic flux through a surface will also change if the size of the surface changes. Consider, for example, the system shown in Figure 6.12, in which a rectangular loop of wire is pulled at a constant speed v out of a region where a uniform magnetic field exists.

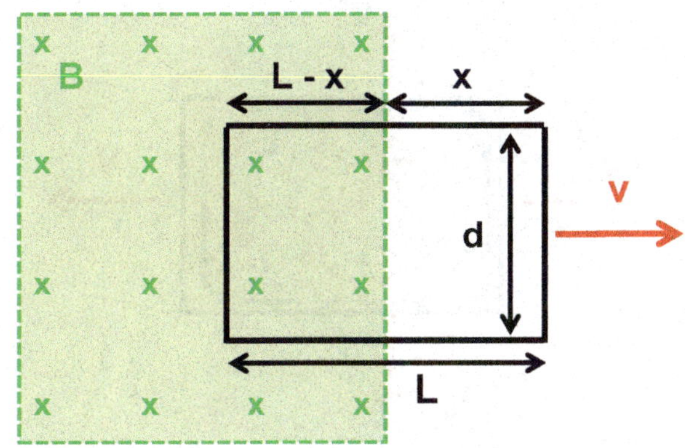

FIGURE 6.12: Changing the magnetic flux through a loop of wire by changing the area of the loop in contact with the magnetic field. The total length of the loop is L and the length of the loop outside the magnetic field is x.

If the area of the loop in the magnetic field is changing, then the magnetic flux through the loop is changing. This changing magnetic flux will create an electric field. To determine the magnitude of this electric field, let's define the path for the integration in Equation 6-3 to be counter-clockwise around the loop, as shown in Figure 6.13.

The magnetic flux through the loop is

$$\Phi_B = \int_{surface} \vec{B} \cdot d\vec{A} \quad \rightarrow \quad \Phi_B = B(L-x)d\cos(180°) \quad \rightarrow \quad \Phi_B = -B(L-x)d$$

This magnetic flux is changing as the loop is pulled out of the magnetic field.

$$\frac{d\Phi_B}{dt} = \frac{d}{dt}\big(B(x-L)d\big) \rightarrow \frac{d\Phi_B}{dt} = Bd\frac{d}{dt}(x-L) \rightarrow \frac{d\Phi_B}{dt} = Bd\left(\frac{dx}{dt} - \frac{dL}{dt}\right)$$

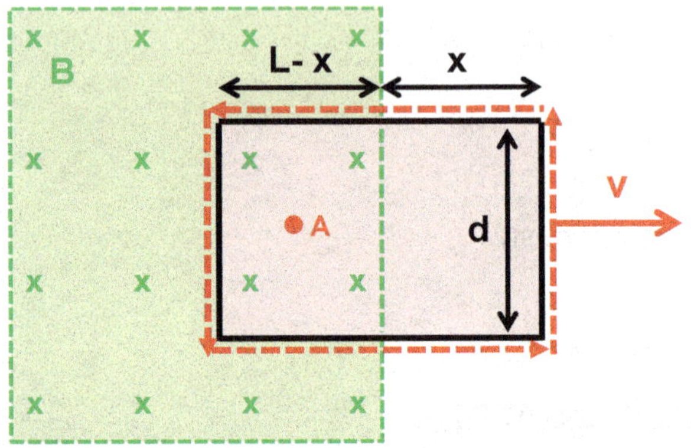

FIGURE 6.13: Determining the magnetic flux through the loop of wire in Figure 6.11. The length *x* increases as the rod is pulled.

The length of the loop is constant, and the length *x* increases as the loop is pulled to the right in Figure 6.12.

$$\frac{dL}{dt} = 0 \rightarrow \frac{d\Phi_B}{dt} = Bd\left(\frac{dx}{dt}\right)$$

$$\frac{dx}{dt} = v \rightarrow \frac{d\Phi_B}{dt} = Bdv$$

The induced electric field associated with this changing magnetic flux can then be found using Equation 6-3.

$$\oint_{line} \vec{E} \bullet d\vec{s} = -(Bdv) \rightarrow \oint_{line} \vec{E} \bullet d\vec{s} = -Bdv$$

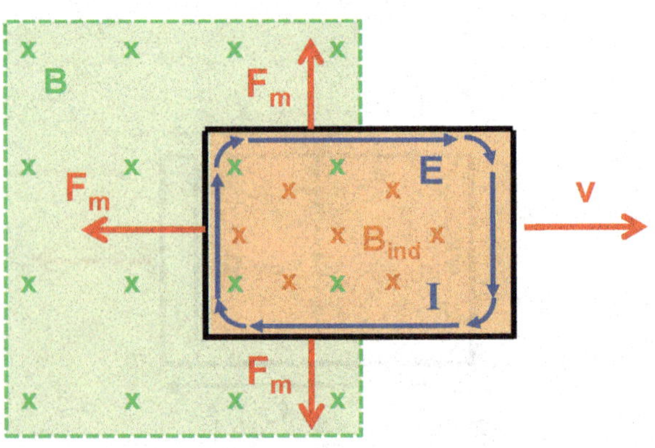

FIGURE 6.14: Induced current, magnetic force, and magnetic field for the movement of the loop in Figure 6.11.

The induced electric field is thus in a direction opposite the path of integration in Figure 6.13. As shown in Figure 6.14, this induced electric field will then give rise to a current, which will circulate through the loop in the same direction as the electric field, and which can then interact with the magnetic field to create magnetic forces acting on the loop (Section 4-6). These forces all act to increase the area of the loop that is within the magnetic field so as to increase the magnetic flux through the loop, in agreement with Lenz's law. If the loop is rigid, so that its area cannot change, the vertically oriented forces in Figure 6.14 will cancel each other out, leaving only the magnetic force directed opposite to the direction of the loop's velocity. This net magnetic force acts to pull the loop back into the magnetic field, and thus acts so as to maintain the magnetic flux through the loop.

Similarly, the direction of the induced magnetic field associated with the induced current is also in agreement with Lenz's law. As shown in Figure 6.14, the direction of the induced magnetic field is in the same direction as the original magnetic field, and thus acts to maintain the same magnetic flux passing through the loop.

Example 6-6

Problem: A circular loop of wire is located in a region of uniform magnetic field as shown in Figure 6.15.

The radius of the loop of wire now begins to decrease at a speed v. What is the magnitude of the electric field induced in the wire? What is the direction of the induced current associated with this induced electric field? What is the direction of the induced magnetic field associated with the induced current?

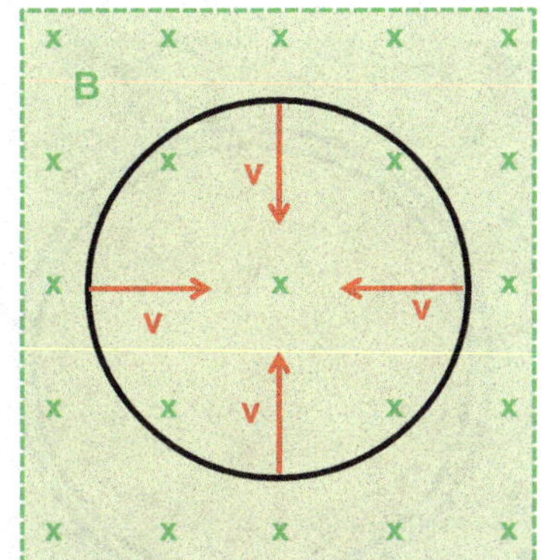

FIGURE 6.15: The circular loop of wire in Example 6-6.

Solution: If the area of the loop in the magnetic field is changing then the magnetic flux through the loop is changing. This changing magnetic flux will create an electric field. To determine the magnitude of this electric field, let's define the path for the integration in Equation 6-3 to be counter-clockwise around the loop, as shown in Figure 6.16.

The magnetic flux through the loop is therefore

$$\Phi_B = \int_{surface} \vec{B} \cdot d\vec{A} \quad \rightarrow \quad \Phi_B = B\pi r^2 \cos(180°) \quad \rightarrow \quad \Phi_B = -B\pi r^2$$

In this equation, the variable r denotes the radius of the loop at any instant. This magnetic flux changes as the area of the loop changes.

$$\frac{d\Phi_B}{dt} = \frac{d}{dt}\left(-B\pi r^2\right) \quad \rightarrow \quad \frac{d\Phi_B}{dt} = -2\pi Br\frac{dr}{dt}$$

The radius of the loop is decreasing with speed v. Thus,

$$\frac{dr}{dt} = -v \quad \rightarrow \quad \frac{d\Phi_B}{dt} = -2\pi Br(-v) \quad \rightarrow \quad \frac{d\Phi_B}{dt} = 2\pi Brv$$

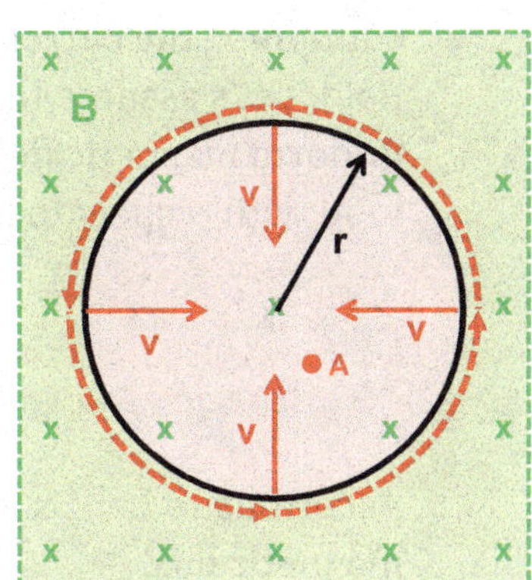

FIGURE 6.16: Determining the magnetic flux through the loop of wire in Example 6-6.

From the symmetry of the magnetic field and the loop, we know that the magnitude of the induced electric field will be the same for all locations along the path of its integration.

$$\oint \vec{E} \bullet d\vec{s} = E(2\pi r)$$

Substitution into Equation 6-3 gives us

$$E(2\pi r) = -(2\pi Brv) \quad \rightarrow \quad E = -Bv$$

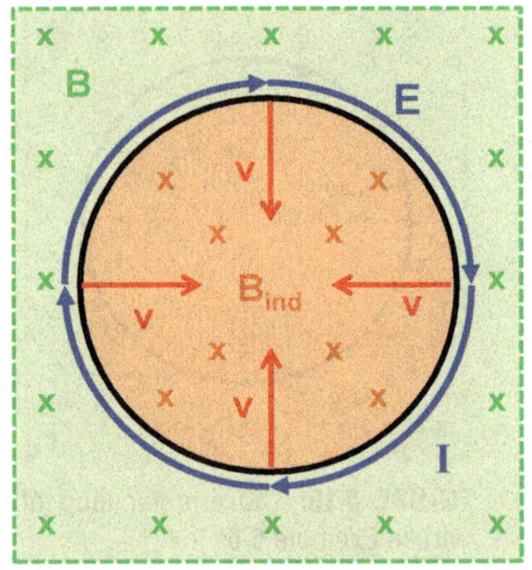

FIGURE 6.17: The induced electric field, current, and magnetic field in Example 6-6.

The negative sign in the solution indicates that the direction of the induced electric field is opposite the direction of the path of integration in Figure 6.16. Namely, the induced electric field and associated induced current are directed clockwise, as shown in Figure 6.17.

As shown in Figure 6.17, the direction of the induced magnetic field is the same as the direction of the original magnetic field. As the size of the loop decreases, this induced magnetic field acts to maintain the same magnetic flux passing through the loop, in agreement with Lenz's law.

Finally, we can always determine the direction of the magnetic force acting on the induced electric current in terms of how this force would affect the area of the loop of wire containing the current; specifically, the area of the loop of wire exposed to the magnetic field. Let's assume that we have a loop of wire enclosing an area A in the presence of a uniform magnetic field of magnitude B. The derivative of the magnetic flux through the loop with respect to time is

$$\Phi_B = BA \quad \rightarrow \quad \frac{d\Phi_B}{dt} = \frac{dB}{dt}A + B\frac{dA}{dt}$$

According to Lenz's law, a system will always adjust itself to try to maintain a constant magnetic flux.

$$\frac{d\Phi_B}{dt} = 0 \quad \rightarrow \quad \frac{dB}{dt}A = -B\frac{dA}{dt} \quad \rightarrow \quad \frac{dA}{dt} = -\frac{A}{B}\frac{dB}{dt}$$

Thus, in order to maintain a constant magnetic flux, the area of the loop exposed to the magnetic field must change when the magnitude of the magnetic field passing through the loop changes. If the magnitude of the magnetic field is increasing, the area of the loop exposed to the magnetic field will decrease.

$$\frac{dB}{dt} > 0 \quad \rightarrow \quad \frac{dA}{dt} < 0$$

Similarly, if the magnitude of the magnetic field passing through the loop is decreasing, the area of the loop exposed to the magnetic field will increase.

$$\frac{dB}{dt} < 0 \quad \rightarrow \quad \frac{dA}{dt} > 0$$

Consider the system shown in Figure 6.14. As the loop is pulled out of the region of magnetic field, the magnitude of the magnetic field passing through the loop is decreasing (*i.e.*, less of the enclosed area of the loop is exposed to the magnetic field). The magnetic force acting on the induced currents in the loop seeks to increase the area of the loop exposed to the magnetic field by pulling the loop back into the region of magnetic field.

6-5 Electric Potential Difference vs. Electromotive Force

The electric field driving (or causing) a current can result either from an electric potential difference (Section 3-6) or a changing magnetic field (Section 6-3 and Section 6-4). For simplicity, the term ***electromotive force (EMF)***, denoted by the symbol $\mathcal{E}$, is often used to describe the "source" driving any current, regardless of is origin. This terminology, while historical, is nevertheless unfortunate, however, since the word "force" doesn't mean a mechanical force, but rather an *effective* electric potential; thus, the unit of EMF is a volt and not a newton.

> *Electromotive force (EMF):* the effective electric potential for a current developed by any source of electrical energy. The unit of EMF is a volt.

Mathematically, we can define an EMF as the work per unit charge that would be done on a charged point particle as it travels along a path.

$$\mathcal{E} = \frac{W}{q} \tag{6-8}$$

In this equation, q is the net electric charge on the point particle. From classical mechanics we know that this work is simply the integral of the electric force acting on the particle over the path through which the particle moves.

$$\mathcal{E} = \frac{\int \vec{F}_e \cdot d\vec{s}}{q} \quad \rightarrow \quad \mathcal{E} = \int \left(\frac{\vec{F}_e}{q} \right) \cdot d\vec{s}$$

The EMF can then be expressed in terms of the associated electric field using Equation 3-11.

$$\mathcal{E} = \int \vec{E} \cdot d\vec{s} \tag{6-9}$$

Thus, for the electric field associated with an electric potential difference (Equation 3-10),

$$\mathcal{E} = -\Delta V_e \tag{6-10}$$

and for an induced electric field resulting from a changing magnetic field (Equation 6-3),

$$\mathcal{E} = -\frac{d\Phi_B}{dt} \tag{6-11}$$

Equation 6-11 is also the most commonly used expression of Faraday's law of induction. The unifying concept behind the common definition of EMF for describing current is that a current requires an input of energy. Consider, for example, the constant electric potential between the two plates of a parallel plate capacitor (Figure 5.7). If electric charges were able to flow between the two plates as a current[6], then the total electric charge on each plate, and the associated electric field and electric potential between the plates, would change. In order to maintain a constant EMF for this current (*i.e.*, in order to maintain a constant electric potential to drive the current), the original electric charge separation on the plates would have to be maintained through the constant removal or addition of the electric charges of the current interacting with the plates.

6 For example, the capacitor could be submerged in a solution of salt water and the salt ions could flow between the plates as a current.

This would require the input of energy (Equation 5-8). Similarly, the EMF associated with a changing magnetic flux (Section 6-3 and Section 6-4) also requires the input of energy. This energy could, for example, be the work required for the movement of a permanent magnet resulting in a changing magnetic field (Figure 6.7) or the movement of a loop of wire with respect to a constant magnetic field (Figure 6.13). Indeed, the induced electric field can be directly related to a change in magnetic potential (and by extension to a change in the magnetic potential energy of the system).

$$\oint_{line} \vec{E} \bullet d\vec{s} = -\frac{d}{dt}\left(\oint_{line} \vec{V}_m \bullet d\vec{s} \right)$$

If we assume that the path of integration does not change with time, we have

$$\oint_{line} \vec{E} \bullet d\vec{s} = -\left(\oint_{line} \frac{\partial \vec{V}_m}{\partial t} \bullet d\vec{s} \right) \quad \rightarrow \quad \oint_{line} \left(\vec{E} + \frac{\partial \vec{V}_m}{\partial t} \right) \bullet d\vec{s} = 0$$

$$\vec{E} = -\frac{\partial \vec{V}_m}{\partial t} \tag{6-12}$$

We can use Equation 6-12 to determine the electric field induced within the wires of a solenoid; the vector potential for a perfect solenoid was determined in Example 5-13. Substitution of the value of that expression at the location of the wires in the solenoid (*i.e.*, when $\rho = R$) gives us

$$\left(V_m\right)_\varphi = \frac{\mu_0 nIR}{2} \quad \rightarrow \quad \vec{E} = -\frac{\partial}{\partial t}\left[\left(\frac{\mu_0 nIR}{2} \right) \hat{\varphi} \right] \quad \rightarrow \quad \vec{E} = -\left(\frac{\mu_0 nR}{2} \frac{dI}{dt} \right) \hat{\varphi}$$

This result is identical to that obtained in Example 6-3.

6-6 Eddy Currents and Motional EMF

If some of the electrons in a material are free to move throughout the material (*i.e.*, are not bound to individual atoms), the material is referred to as an *electrical conductor*, or simply a conductor. If none of the electrons in a material are able to move throughout the material (*i.e.*, they are all bound to individual atoms), the material is referred to as an *electrical insulator*, or simply an insulator[7]. We will discuss the electrical and magnetic properties of conductors and insulators more thoroughly in Chapter 7 and for now will focus on just one aspect of the magnetic properties of conductors that is related to the induction of magnetic and electric fields within them.

If an electrical conductor moves through a uniform magnetic field, the free electrons in the conductor (which are obviously moving along with the conductor) will experience a magnetic force (Equation 4-2). As shown in Figure 6.18, the direction of the magnetic force is initially perpendicular to the direction of the velocity of the conductor (which follows from the cross product in Equation 4-2). As a result of this magnetic force, and the acceleration associated with it, the free electrons will gain a component of velocity in the same direction as the magnetic force. Of course, this component of velocity will, in turn, give rise to a magnetic force that is directed perpendicular to it. This process will continue until the electrons are moving in circular paths (Section 4-3). The currents associated with these moving electrons are referred to as *eddy currents*.

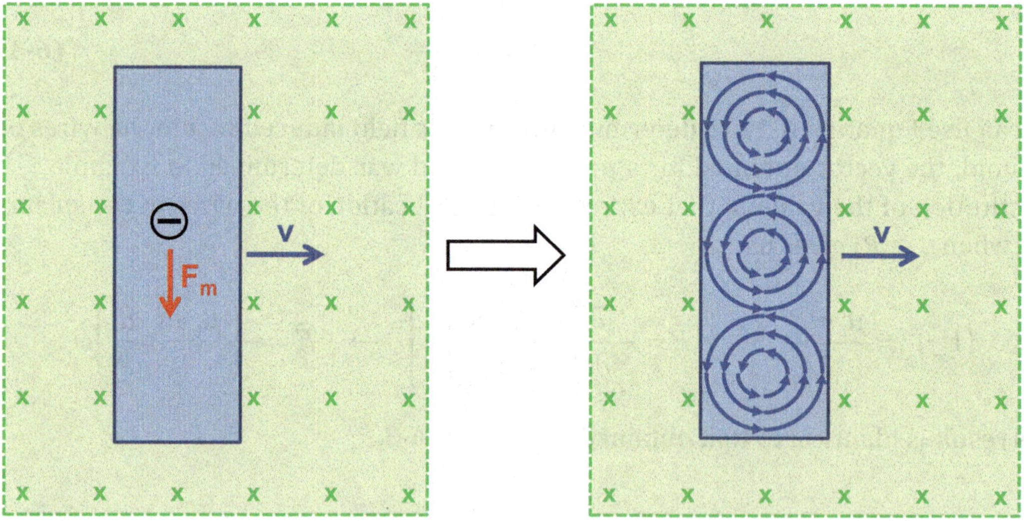

FIGURE 6.18: The magnetic force acting on free electrons in a conductor moving through a magnetic field (left panel) generates eddy currents within the conductor (right panel).

Of course, the free electrons in the conductor respond not only to the magnetic field through which they are passing but also to the changing electric fields associated with the movements of the other electrons in the conductor. In other words, it's best to think of eddy currents in terms of a bulk flow of free electrons, rather than the movement

7 Semiconductors are a third class of materials whose electrical properties are somewhere between those of conductors and insulators.

of individual electrons[8]. Furthermore, because it is not possible for the free electrons to leave the conductor, the shape and strength of the eddy currents will be affected by the boundaries of the conductor. In addition, the electrons will always have some component of their velocity parallel with the velocity of the conductor itself. Over time, these constraints on the bulk motion of the electrons will give rise to a net flow of current and associated charge separation within the conductor (Figure 6.19). This net current within the conductor will, of course, interact with the magnetic field to cause a force to act on the conductor (Figure 6.19). The direction of this magnetic force acting on the conductor will be in a direction opposite the direction of the velocity of the conductor, in agreement with Lenz's law.

As shown in Figure 6.19, the net current in the conductor causes a separation of charge within the conductor, which in turn creates an electric field inside the conductor; this is similar to the electric field created by the separation of charges in an inductor (Section 6-3). The bulk flow of electrons in the conductor will continue until the magnitude of the electric force on the free electrons associated with this *induced* electric field is identical to the magnitude of the magnetic force acting on the free electrons (Figure 6.19). At this point, there will be no net force acting on the free electrons and therefore no current (*i.e.*, no net bulk movement of the electrons within the conductor). In other words, the free electrons will be in steady-state equilibrium.

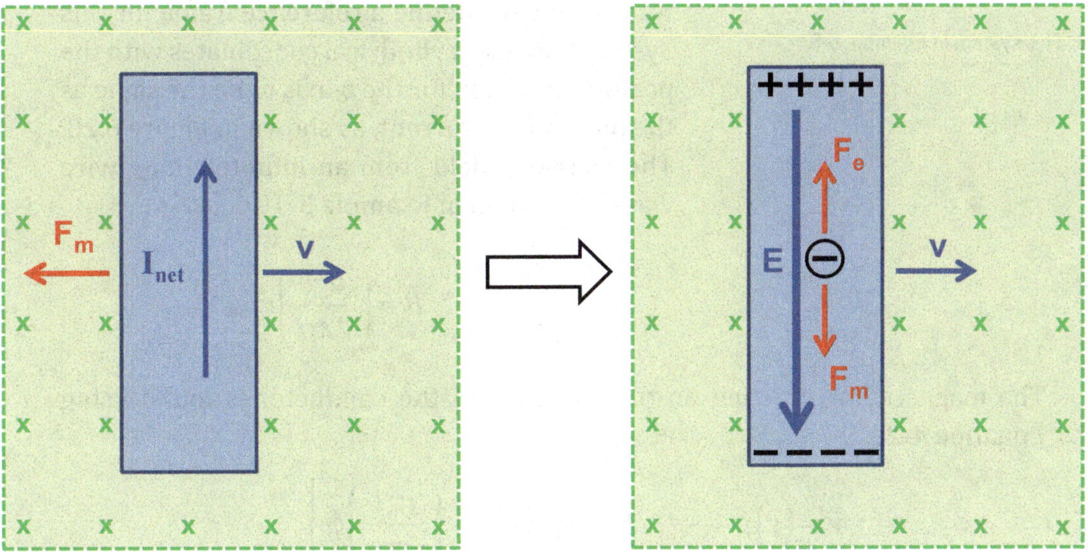

FIGURE 6.19: The net current associated with a conductor moving through a uniform magnetic field results in a charge separation and associate electric field within the conductor.

8 This is true for all currents, as we will discuss in Chapter 7.

We can solve for this steady-state induced electric field from the fact that the net force acting on the free electrons must be zero when the system is in equilibrium.

$$F_e = F_m \quad \rightarrow \quad qE = qvB \quad \rightarrow \quad E = vB$$

This electric potential associated with this induced electric field is historically referred to as motional electromotive force (or motional EMF), as it results from the motion of the conductor through the magnetic field. Not surprisingly, this result is also reminiscent of Equation 6-1.

Example 6-6

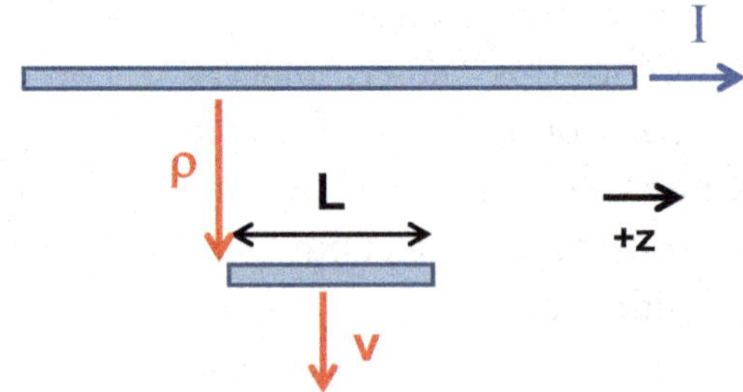

FIGURE 6.20: The system in Example 6-6.

Problem: A thin, linear conductor of length L is moving with constant speed in a direction perpendicular to a long, straight wire carrying current I, as shown in Figure 6.20. What is the magnitude of the motional EMF induced in the conductor?

Solution: Let's define a reference frame for this system that uses cylindrical coordinates with the positive direction for the z-axis to be the same as the direction of current, as shown in Figure 6.20. The magnetic field from an infinitely long wire was determined in Example 5-10[9].

$$\vec{B} = \left(\frac{\mu_0 I}{2\pi\rho}\right)\hat{\varphi}$$

The magnetic force acting on the electrons in the conductor is found using Equation 4-2.

$$\vec{v} = (v)\hat{\rho} \quad \rightarrow \quad \vec{F}_m = q\left[(v)\hat{\rho} \times \left(\frac{\mu_0 I}{2\pi\rho}\right)\hat{\varphi}\right]$$

$$\vec{F}_m = \left(\frac{\mu_0 qIv}{2\pi\rho}\right)\left[\hat{\rho} \times \hat{\varphi}\right] \quad \rightarrow \quad \vec{F}_m = \left(\frac{\mu_0 qIv}{2\pi\rho}\right)\hat{z}$$

9 See also Example 4-5.

The direction of the magnetic force is thus parallel to the length of the rod, and electrons[10] in the rod will be pushed in the direction opposite the direction of the current in the wire[11]. The movement of electrons in the rod will create an electric potential and electric field that will oppose the magnetic force; the direction of this induced electric field will be the same as the direction of the current in the wire. At equilibrium, the magnitude of the magnetic force and electric force from this electric field will be equal.

$$F_m = F_e \quad \rightarrow \quad \frac{q\mu_0 Iv}{2\pi\rho} = qE \quad \rightarrow \quad E = \frac{\mu_0 Iv}{2\pi\rho}$$

This explanation of the origin of the motional EMF is unsatisfying, however, since it relies upon a magnetic force to drive the current in the rod; we recall from Section 4-3 that magnetic fields can do no work on a charged particle. Perhaps a better explanation is that an electric field is induced in the rod as the rod moves through the changing magnetic potential of the rod (Equation 6-12), and that this induced electric field that creates the EMF in the rod that drives the current. The magnetic potential of an infinite wire was determined in Section 4-10; in this derivation, the direction of the current in the wire was defined to be the z-axis of the reference frame.

$$\vec{V}_m = \left[\left(\frac{\mu_0 I}{2\pi}\right)\ln\left(\frac{\rho_0}{\rho}\right)\right]\hat{z} \quad \rightarrow \quad \vec{E} = \left[-\frac{\partial}{\partial t}\left(\frac{\mu_0 I}{2\pi}\right)\left(\ln\left(\frac{\rho_0}{\rho}\right)\right)\right]\hat{z}$$

$$\vec{E} = \left(-\left(\frac{\mu_0 I}{2\pi}\right)\left(\frac{\rho}{\rho_0}\right)\left(-\frac{\rho_0\frac{d\rho}{dt}}{\rho^2}\right)\right)\hat{z} \quad \rightarrow \quad \vec{E} = \left(\left(\frac{\mu_0 I}{2\pi}\right)\left(\frac{\frac{d\rho}{dt}}{\rho}\right)\right)\hat{z}$$

$$\frac{d\rho}{dt} = v \quad \rightarrow \quad \vec{E} = \left(\frac{\mu_0 Iv}{2\pi\rho}\right)\hat{z}$$

As expected, this induced electric field will produce a force on the electrons in the rod that is identical to the magnetic force derived earlier. The induced EMF can then be determined using Equation 6-9 and integrating with respect to the

10 Recall that q is negative for an electron.
11 This is another example of Lenz's law.

length of the rod. Since the electric field is constant throughout the rod this integration yields

$$d\vec{s}=(dz)\hat{z} \rightarrow \mathcal{E}=\int\left[\left(\frac{\mu_0 I v}{2\pi\rho}\right)\hat{z}\right]\cdot\left[(dz)\hat{z}\right] \rightarrow \mathcal{E}=\left(\frac{\mu_0 I v}{2\pi\rho}\right)\int dz \rightarrow \mathcal{E}=\frac{\mu_0 I v L}{2\pi\rho}$$

Example 6-7

Problem: A thin conducting rod with length L is rotating around an axis at one end of the rod at angular frequency ω in the presence of a uniform magnetic field of magnitude B, as shown in Figure 6.21. What is the magnitude of the motional EMF across the length of the rod?

Solution: The magnitude of the magnetic force acting on an electron in the conductor can be determined using Equation 4-2. Since the magnetic field is perpendicular to the linear velocity of the electron, we have

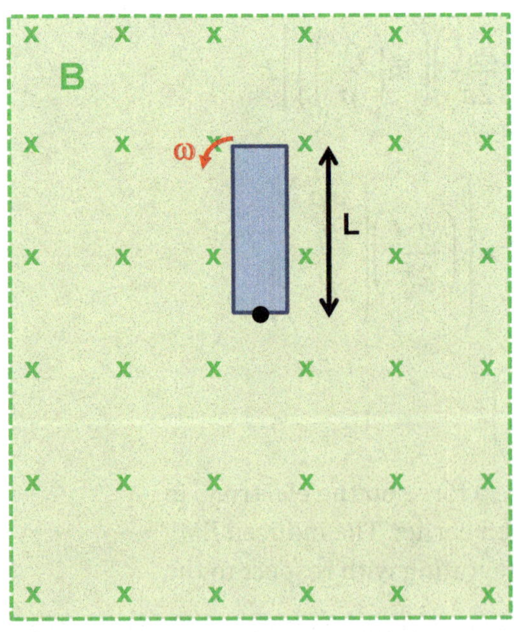

FIGURE 6.21: The rotating rod in Example 6-7.

$$F_m = qvB$$

The linear velocity of the electrons will depend upon their position from the axis of rotation. We can therefore rewrite our expression for the magnitude of the magnetic force in terms of the angular speed of the rotation of the rod and the distance of the electron from the axis of rotation, which we will denote by r.

$$v=\omega r \quad \rightarrow \quad F_m = q\omega rB$$

This magnetic force must be balanced by the electric force from the induced electric field.

$$F_m = F_e \quad \rightarrow \quad q\omega rB = qE \quad \rightarrow \quad E=\omega rB$$

The magnitude of the induced electric field is therefore not constant, but varies linearly with the distance from the axis of rotation[12]. The induced EMF can be found by integrating this electric field with respect to the length of the rod (Equation 6-9).

$$\mathcal{E} = \int_0^L \omega r B dr \quad \rightarrow \quad \mathcal{E} = B\omega \int_0^L r dr$$

$$\mathcal{E} = \frac{B\omega L^2}{2}$$

Once the steady-state electric field in Figure 6.19 is created and the system (*i.e.*, the free electrons in the conductor) are in equilibrium, there will be no net current flowing through the conductor and thus no net magnetic force acting on the conductor. However, if the conductor were to leave the area where the magnetic field is present, this steady-state equilibrium of the free electrons would be perturbed. Indeed, while in the process of leaving the magnetic field, the difference in the strength of the magnetic and electric forces acting on the free electrons in different parts of the conductor would give rise to a net bulk flow of electrons (*i.e.*, a current) and an associated magnetic force acting on the conductor. This magnetic force would act opposite the direction of the conductor's velocity and thus act to pull the conductor back into the magnetic field and maintain the steady-state equilibrium of the system, in agreement with Lenz's law and Le Châtelier's principle.

Consider the physical pendulum shown in Figure 6.22 that consists of a conductor swinging back and forth through a region of space containing a uniform magnetic field. Upon entering or exiting the region of magnetic field, a net magnetic force acts on the conductor, and the direction of this magnetic force is opposite the direction of the velocity of the conductor. This magnetic force thus acts to damp the oscillations of the pendulum.

12 Practice your calculus skills by calculating this electric field using Equation 6-12.

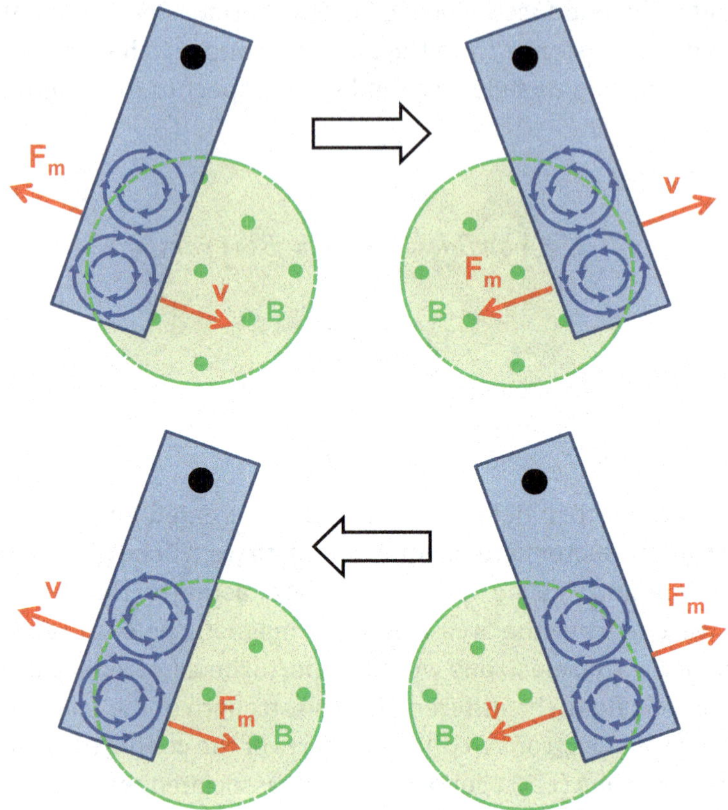

FIGURE 6.22: The magnetic force acting on a conductor swinging as a physical pendulum through a magnetic field will always oppose the motion of the pendulum.

The damping of the oscillations of the pendulum in Figure 6.22 is the basis of electromagnetic braking, such as that used by trains. In this case, however, we can think of the conductor (the rails) as being stationary and the magnetic field (generated by electromagnets on the train) as moving. As this magnetic field is moved past the conducting rails, eddy currents are generated in the rails. These eddy currents then interact with the magnetic field to generate a magnetic force that opposes the motion of the electromagnet. From Newton's third law we know that this force will act on both the conductor (the rails) and the electromagnet (the train), and thus will cause the train to slow. This system of braking has two main benefits. First, it avoids relying upon the friction of two surfaces rubbing against each other (such as the brakes in a car). Second, it transfers the kinetic energy of the train to the kinetic energy of the free electrons in the rail. As we will discuss in Chapter 8, this energy raises the temperature of the rails and is then eventually dissipated into the environment as heat. This is arguably safer than raising the temperature of components of the train itself.

Example 6-8

Problem: In addition to using tethers to change their orbits (Example 4-9), spacecraft and satellites can also use tethers to generate an EMF. What is the magnitude of the EMF induced in a 3 km long tether attached to a spacecraft moving at 4 km/s where the magnitude of the Earth's magnetic field is 20 mT? Assume that the orientation of the tether and the spacecraft's velocity are perpendicular to the Earth's magnetic field.

Solution: The magnitude of the induced electric field is

$$E = vB$$

The induced EMF can then be determined using Equation 6-9. Since this electric field is constant, we have[13]

$$\mathcal{E} = E\int ds \quad \rightarrow \quad \mathcal{E} = E\Delta s \quad \rightarrow \quad \mathcal{E} = EL \quad \rightarrow \quad \mathcal{E} = vBL$$

Substitution gives us

$$\mathcal{E} = \left(4\times10^{3}\frac{m}{s}\right)\left(20\times10^{-6}T\right)\left(3\times10^{3}m\right) \quad \rightarrow \quad \mathcal{E} = 240V$$

Finally, the concept of motional EMF can also be used to explain the induced current in Figure 6.14. As shown in Figure 6.23, as the loop of wire is pulled through the magnetic field, a magnetic force will be exerted on the free electrons in the wire. This magnetic force will cause a bulk movement of the free electrons in the wire; as discussed before, the electrons will move not only because of the magnetic force acting on them but also because of the changing electric field they experience caused by the motion of the other electrons. In some cases, such as in the vertical segment of the loop in Figure 6.23, these magnetic and electric forces will act in concert to move the electrons, whereas in other segments, these forces will be perpendicular to each other. Regardless, the combined effect will be a net flow of electrons through the wire, even in regions where there is no magnetic field present, since in these regions the free electrons will nevertheless

13 We have also chosen our path of integration to be in the same direction as the induced electric field.

move in response to the movement of the free electrons in parts of the wire where the magnetic field is present.

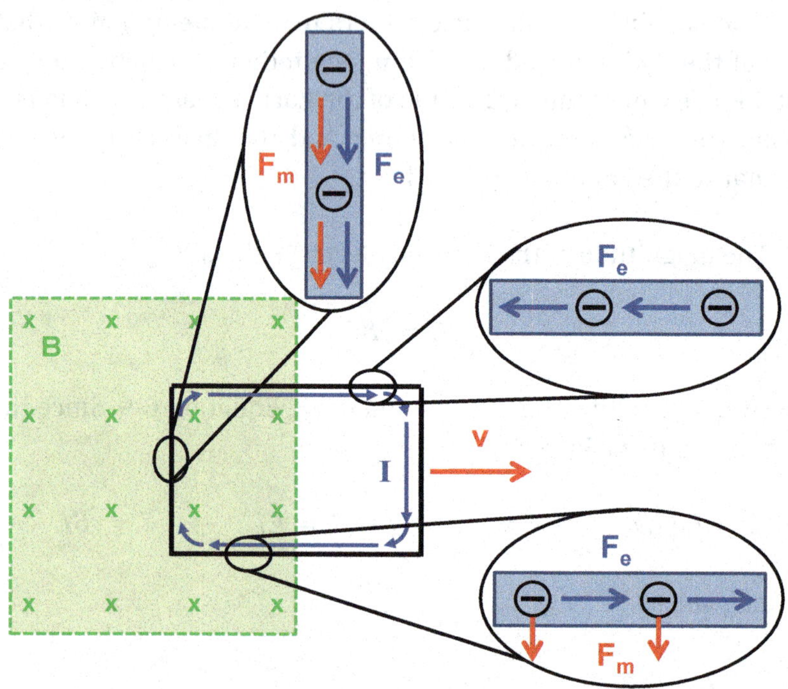

FIGURE 6.23: Motional EMF explanation of the induced current as a loop of wire is pulled out of a magnetic field. Remember that the direction of the current is opposite the direction of the electron's velocity.

Of course, the explanation that electrons obtain the energy required for the induced current from the magnetic field is ultimately unsatisfactory since we know that a magnetic field can do no work on an electric charge (Section 4-3). We can therefore refine our explanation for the current by considering the system from a reference frame moving with the wire. In this reference frame, the free electrons are stationary and the magnetic field is moving. This moving (or changing) magnetic field is observed by the free electrons to be an electric field in their (moving) reference frame (Section 6-2). From Equation 6-1 we see that the magnitude of this electric field is

$$E = vB$$

Not surprisingly, this expression is in agreement with our previous derivation. Thus, although there is no electric field in the stationary reference frame (*i.e.*, the reference frame of the magnetic field), there is an electric field in the reference frame of the moving loop of wire, and it is this electric field that does the necessary amount of work on the free electrons moving as the induced current. Indeed, it is this electric field that is the ultimate origin of the motional EMFs that are generated whenever loops of wire are moved with respect to magnetic fields[14]. Indeed, we could have also used Equation 6-12 to determine the electric field induced in the loop of wire when it moves through the magnetic potential of the magnetic field. It is this electric field that is driving the current in this system.

6-7 A Changing Electric Field Creates a Magnetic Field

Ampère's law (Equation 5-9) can be expanded to include the magnetic field induced by a changing electric field.

$$\oint_{line} \vec{B} \bullet d\vec{s} = \mu_0 \left(I_{enclosed} + \varepsilon_0 \frac{d\Phi_E}{dt} \right) \qquad \textbf{(6-13)}$$

The quantity $\varepsilon_0 \frac{d\Phi_E}{dt}$ is commonly referred to as a displacement current, although this is a poorly chosen name since nothing is actually being displaced and there is no current.

In Section 5-5 we introduced the concept of capacitance and determined the capacitance of a parallel plate capacitor. We can now accurately define a capacitor as two separated *conductors*. These two conductors can be *uncharged* (*i.e.*, each conductor contains no net electric charge) or *charged*, in which case net electric charge of one conductor is equal in magnitude but opposite in (±) sign to the net electric charge of the other conductor.

Let's now consider the magnetic fields induced during the charging of a parallel plate capacitor; the term *charging* refers to the process of creating the separation of charge in the capacitor. The two plates of the capacitor are attached to different wires, both of which are carrying current I, as shown in Figure 6.24.

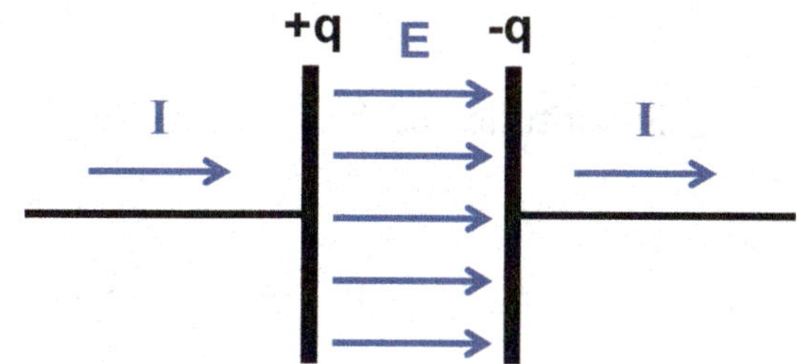

FIGURE 6.24: Charging a parallel plate capacitor with a current.

In Figure 6.24, the current is "depositing" positive charges on the left plate of the capacitor and

14 This is similar to our argument in Section 4-6 that the EMF is the ultimate source of energy for the work done on a current-carrying wire in a magnetic field.

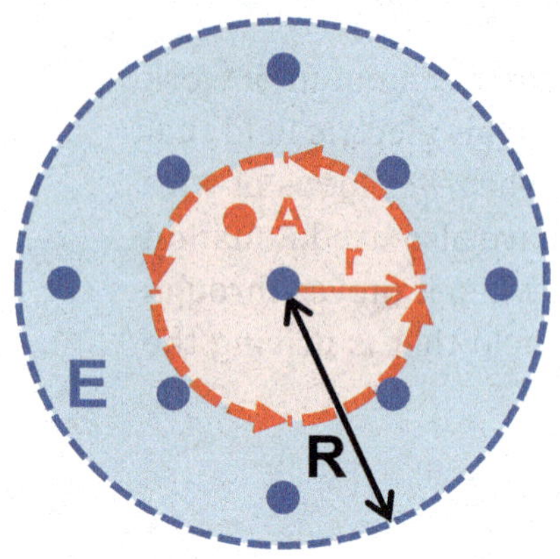

FIGURE 6.25: Determining the induced magnetic field while charging a parallel plate capacitor. The dashed red line indicates the path of integration for determining the electric field associated with a change in magnetic flux. For this path, the surface area vector points out of the page.

"removing" positive charges from the right plate; recall that current is defined as the flow of positive charge (Section 4-4). As the charge separation is created on the plates of the capacitor (*i.e.*, as the capacitor is charged), the magnitude of the electric field between the plates of the capacitor is increasing and, as a result, a magnetic field is induced.

For simplicity, let's assume that the plates of the capacitor are circular disks with a radius of R and that we can ignore any non-uniformity of the electric field between the plates at the edges of the plates. In other words, we will assume that the electric field between the plates is uniform and constant with a magnitude given by Equation 5-5[15], and zero everywhere outside of the plates. We can then define a path for the integration in Equation 6-13 to be a circle of radius r centered on the longitudinal axis of the capacitor, as shown in Figure 6.25.

Let's first determine the induced magnetic *within* the capacitor (*i.e.*, when $r < R$). Since the electric field is uniform we have

$$\frac{d\Phi_E}{dt} = \frac{d}{dt}\left(E\pi r^2\right) \rightarrow \frac{d\Phi_E}{dt} = \pi r^2 \frac{dE}{dt}$$

Furthermore, since the electric field is uniform through this surface, we can also argue from symmetry that the induced magnetic field will have a constant magnitude for all locations along the path of integration. Thus,

$$\oint \vec{B} \bullet d\vec{s} = B(2\pi r)$$

Substitution of these results into Equation 6-13 gives us

$$I_{enclosed} = 0 \rightarrow \oint_{line} \vec{B} \bullet d\vec{s} = \mu_0 \varepsilon_0 \frac{d\Phi_E}{dt} \rightarrow B(2\pi r) = \mu_0 \varepsilon_0 \left(\pi r^2 \frac{dE}{dt}\right)$$

$$B = \frac{\mu_0 \varepsilon_0 r}{2} \frac{dE}{dt}$$

15 This is a reasonable assumption if the radius of the plates is much larger than the separation distance between the plates.

If the magnitude of the electric field is increasing (*i.e.*, if $\frac{dE}{dt} > 0$), then $B > 0$. This means that if the magnitude of the electric field is increasing, the induced magnetic field is in the same direction as the path of integration (*i.e.*, the direction of the induced magnetic field is counterclockwise in Figure 6.26). Similarly, if the magnitude of the electric field is decreasing (*i.e.*, if $\frac{dE}{dt} < 0$), then $B < 0$ and the induced magnetic field is directed clockwise in Figure 6.26.

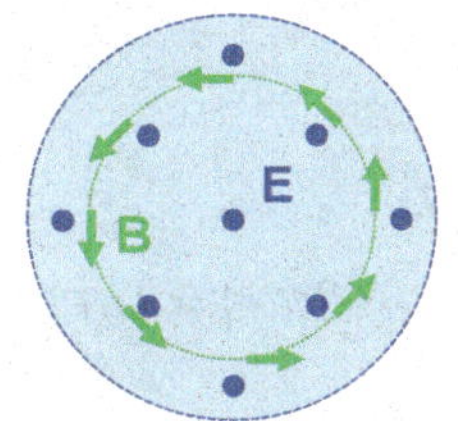

Magnitude of Electric Field Increasing

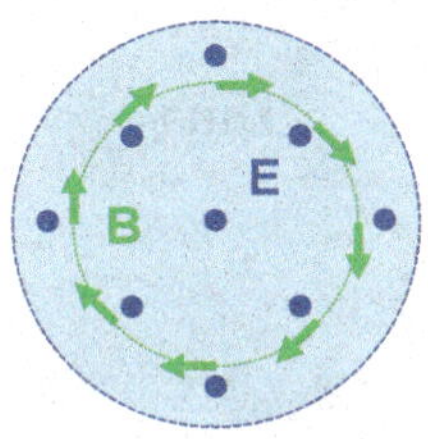

Magnitude of Electric Field Decreasing

FIGURE 6.26: Orientation of the magnetic field induced by a changing electric flux.

The magnitude of the electric field for a parallel plate capacitor is given in Equation 5-5; for a capacitor with plates that are disks of radius R, the area of the plates is πR^2.

$$E = \frac{q}{A\varepsilon_0} \rightarrow E = \frac{q}{\pi R^2 \varepsilon_0} \rightarrow \frac{dE}{dt} = \left(\frac{1}{\pi R^2 \varepsilon_0}\right)\frac{dq}{dt}$$

Thus,

$$B = \frac{\mu_0 \varepsilon_0 r}{2}\left(\frac{1}{\pi R^2 \varepsilon_0}\right)\frac{dq}{dt} \rightarrow B = \left(\frac{\mu_0 r}{2\pi R^2}\right)\frac{dq}{dt}$$

The induced magnetic field also exists outside the capacitor (*i.e.*, when $r > R$). As shown in Figure 6.27, for these locations the magnetic flux is calculated using the area of the cylinder.

$$\frac{d\Phi_E}{dt} = \frac{d}{dt}\left(E\pi R^2\right) \rightarrow \frac{d\Phi_E}{dt} = \pi R^2 \frac{dE}{dt}$$

Since the path for the integration of the electric field is still centered on the longitudinal axis of the cylinder, the electric field will be constant at all points along the path.

$$\oint \vec{B} \cdot d\vec{s} = B(2\pi r)$$

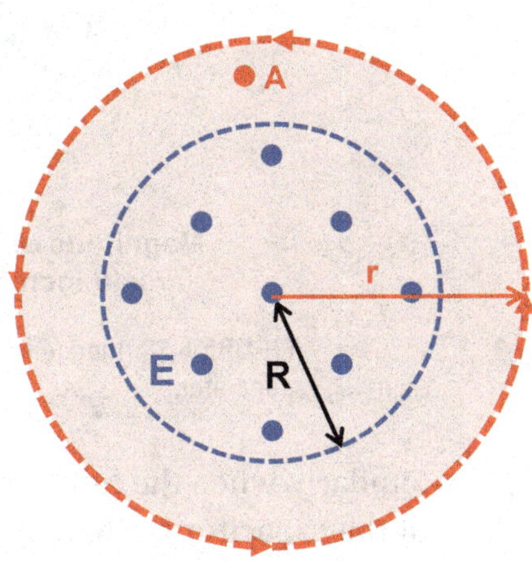

FIGURE 6.27: Determining the induced electric field for locations outside of the capacitor.

Thus,

$$B(2\pi r) = \mu_0 \varepsilon_0 \pi R^2 \frac{dE}{dt} \quad \rightarrow \quad B = \frac{\mu_0 \varepsilon_0 R^2}{2r} \frac{dE}{dt}$$

$$B = \frac{\mu_0 \varepsilon_0 R^2}{2r} \left(\frac{1}{\pi R^2 \varepsilon_0} \right) \frac{dq}{dt} \quad \rightarrow \quad B = \left(\frac{\mu_0}{2\pi r} \right) \frac{dq}{dt}$$

Although not surprising, it is nevertheless interesting to note that the magnitude of the magnetic field outside of the capacitor created by the increasing electric flux inside the capacitor is identical to the magnitude of the magnetic field created by the current in the wires connected to the charging capacitor (Example 5-10 and Equation 4-4), as shown in Figure 6.28.

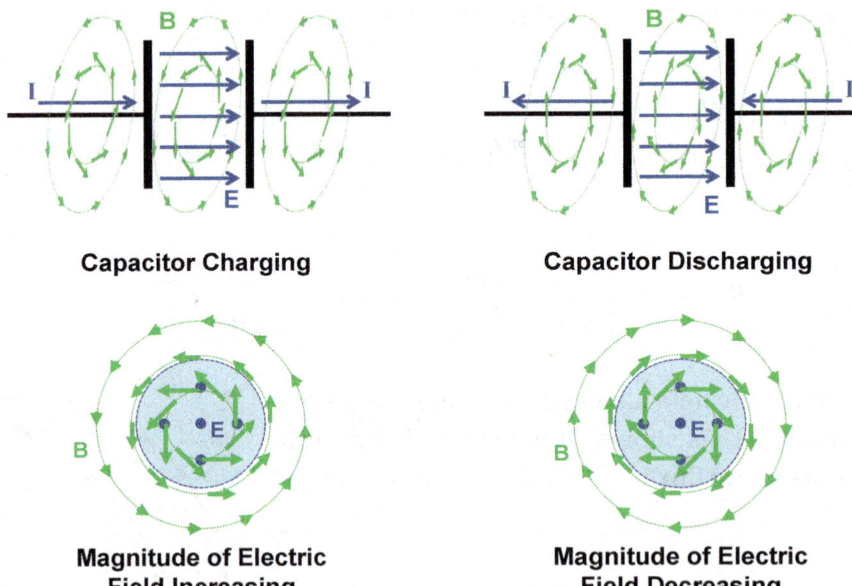

FIGURE 6.28: Induced magnetic fields associated with the charging and discharging of a capacitor.

Similarly, when the capacitor is discharging, the magnitude of the magnetic field outside of the capacitor created by the decreasing electric flux inside the capacitor is identical to the magnitude of the magnetic field created by the current in the wires connected to the capacitor (Figure 6.28).

6-8 Magnetic Energy and Force

The magnetic potential energy of a system of currents can be calculated using Equation 6-14[16].

$$U_m = \frac{1}{2}\sum_i I_i\left(\Phi_B\right)_i \tag{6-14}$$

The variable $\left(\Phi_B\right)_i$ in Equation 6-14 is the total flux through loop i from all sources of magnetic field, including itself.

Example 6-9

Problem: What is the magnetic potential energy of a perfect solenoid with self-inductance L carrying a current I?

Solution: The magnetic potential energy can be determined using Equation 6-14. Since the magnitude of the magnetic field inside the solenoid is constant, the magnetic flux through a single loop is equal to the product of the magnitude of the magnetic field and the cross-sectional area of the solenoid, denoted by A. Furthermore, because the current in each loop is constant we have

$$U_m = \frac{1}{2}\sum_{i=1}^{N} I_i\left(\Phi_B\right)_i \;\rightarrow\; U_m = \frac{1}{2}\sum_{i=1}^{N} IBA \;\rightarrow\; U_m = \frac{1}{2}NIBA$$

Substitution of Equation 5-10 then gives us

$$U_m = \frac{1}{2}NI\left(\mu_0 nI\right)A \;\rightarrow\; U_m = \frac{1}{2}\mu_0 nANI^2$$

The number of loops can be written in terms of the density of loops, n, and the length of the solenoid, l (Equation 5-10).

$$N = nl \;\rightarrow\; U_m = \frac{1}{2}\mu_0 nAnlI^2 \;\rightarrow\; U_m = \frac{1}{2}\mu_0 n^2\left(Al\right)I^2 \;\rightarrow\; U_m = \frac{1}{2}\mu_0 n^2 VI^2$$

16 See Section C-8 for derivation of Equation 6-14.

The variable V denotes the volume of the solenoid. This expression can then be rewritten in terms of the self-inductance of the solenoid (Section 6-3).

$$L = \mu_0 n^2 V \quad \rightarrow \quad U_m = \frac{1}{2} L I^2 \tag{6-15}$$

Now let's consider the situation in which two current-carrying loops of wire, such as those depicted in Figure C.4, are brought together by an external force, denoted by F. The total work done by this force must be equal to the change in the total potential energy of the system. The infinitesimal work (*i.e.*, infinitesimal change in the total potential energy) associated with this force acting over an infinitesimal displacement of the loops would be

$$đW = \vec{F} \cdot d\vec{r} \quad \rightarrow \quad dU_{total} = \vec{F} \cdot d\vec{r}$$

For very small displacements, the system will remain effectively in equilibrium and the net force acting on the system will be zero. Under these conditions, we have

$$\vec{F} = -\vec{F}_m \quad \rightarrow \quad dU_{total} = -\vec{F}_m \cdot d\vec{r} \quad \rightarrow \quad \vec{F}_m = -\nabla U_{total}$$

We could have anticipated this result, of course, from classical mechanics. However, we would like to relate the magnetic force to the magnetic potential energy of the system, rather than simply to the total potential energy of the system. To do this, we first recognize that the total potential energy of the system consists of the magnetic potential energy and the potential energy associated with the original EMFs in the loops (*i.e.*, the sources of the electric potential energy driving the current).

$$U_{total} = U_{EMF} + U_m \quad \rightarrow \quad dU_{total} = dU_{EMF} + dU_m$$

Let's consider two possible scenarios. First, let's assume that the loops are brought together under conditions of constant current in both loops. The magnetic potential energy of the system can be determined from Equation 6-14.

$$U_m = \frac{1}{2}\left(I_1 (\Phi_B)_1 + I_2 (\Phi_B)_2\right)$$

If the currents are constant then

$$dU_m = \frac{1}{2}\left(I_1\left(d\Phi_B\right)_1 + I_2\left(d\Phi_B\right)_2\right)$$

As discussed previously, the changing magnetic fluxes associated with bringing the two loops together will result in induced currents in each loop. In order to maintain a constant current, each original EMF must do extra work to overcome these induced currents. The expression for this work[17] is

$$đW = I_1\left(d\Phi_B\right)_1 + I_2\left(d\Phi_B\right)_2$$

Since the work done by the original EMFs is equal to a decrease in their potential energy, we have

$$dU_{EMF} = -đW \quad \rightarrow \quad dU_{EMF} = -\left(I_1\left(d\Phi_B\right)_1 + I_2\left(d\Phi_B\right)_2\right) \quad \rightarrow \quad dU_{EMF} = -2\left(dU_m\right)$$

Substitution then gives us

$$dU_{total} = -2\left(dU_m\right) + dU_m \quad \rightarrow \quad dU_{total} = -dU_m$$

Therefore,

$$\vec{F}_m = \left(\nabla U_m\right)_I \tag{6-16}$$

The subscript I in Equation 6-16 reminds us that all currents are held constant while calculating the gradient. It follows from Equation 6-16 that *loops of wire with constant currents will always spontaneously arrange themselves so as to maximize their total magnetic potential energy*. In other words, *these loops of wire will arrange themselves so as to maximize the total enclosed magnetic flux*.

We can now return to the discussion at the end of Section 4-11. If the current in the wire and the velocity of the charged particle are both held constant, then Equation 6-16 can be used to determine the force between the particle and the wire.

$$\vec{F}_m = \nabla U_m \quad \rightarrow \quad \vec{F}_m = \left(\frac{\partial}{\partial \rho}\left(\frac{\mu_0 q v I}{2\pi}\ln\left(\frac{2\rho}{\left(4\rho^2 + L^2\right)^{\frac{1}{2}} - L}\right)\right)\right)\hat{\rho}$$

17 See Section C-8.

$$\vec{F}_m = \left(\frac{\mu_0 qvI}{4\pi} \left(-\frac{2L}{\rho\left(4\rho^2 + L\right)^{\frac{1}{2}}} \right) \right) \hat{\rho} \rightarrow \vec{F}_m = -\left(\frac{L\mu_0 qvI}{2\pi\rho\left(4\rho^2 + L\right)^{\frac{1}{2}}} \right) \hat{\rho}$$

As expected, this result is identical to what we obtain using Equation 4-3.

Now let's imagine the scenario in which the loops are brought together under conditions where the magnetic flux through the loops is held constant. If the magnetic flux is constant, there are no induced EMFs, and the original EMFs do no additional work and experience no change in their potential energy.

$$dU_{EMF} = 0 \rightarrow dU_{total} = dU_m$$

$$\vec{F}_m = -\left(\nabla U_m\right)_{\Phi_B} \tag{6-17}$$

The subscript Φ_B in Equation 6-17 reminds us that all magnetic fluxes are held constant while calculating the gradient.

6-9 Maxwell's Equations

Taken together, Equations 5-4, 5-12, 6-3, and 6-13 are referred to as Maxwell's equations, after James Clerk Maxwell who published an early form of them.

$$\oint_{surface} \vec{E} \bullet d\vec{A} = \frac{q_{enclosed}}{\varepsilon_0} \tag{5-4}$$

$$\oint_{surface} \vec{B} \bullet d\vec{A} = 0 \tag{5-12}$$

$$\oint_{line} \vec{E} \bullet d\vec{s} = -\frac{d\Phi_B}{dt} \tag{6-3}$$

$$\oint_{line} \vec{B} \bullet d\vec{s} = \mu_0 \left(I_{enclosed} + \varepsilon_0 \frac{d\Phi_E}{dt} \right) \tag{6-13}$$

Equation 3-8 and Equation 6-12 can also be combined to give a more complete equation for the electric field.

$$\vec{E} = -\nabla V_e - \frac{d\vec{V}_m}{dt} \tag{6-18}$$

These equations, together with the Lorentz force law (Equation 4-3)

$$\vec{F} = q\left(\vec{E} + \vec{v} \times \vec{B}\right) \tag{4-3}$$

provide a good summary of electromagnetic interactions.

Summary

- The electric and magnetic fields in reference frame 1 are related to the electric and magnetic fields in reference frame 2 moving at constant velocity with respect to reference frame 1 by the following equations:

$$\vec{E}_2 = \vec{E}_1 + \vec{v}_{2,1} \times \vec{B}_1$$

$$\vec{B}_2 = \vec{B}_1 - (\mu_0 \varepsilon_0) \vec{v}_{2,1} \times \vec{E}_1$$

- **A changing magnetic flux induces an electric field.**

$$\oint_{line} \vec{E} \bullet d\vec{s} = -\frac{d\Phi_B}{dt}$$

The induced electric field can also be related to a changing magnetic potential.

$$\vec{E} = -\frac{\partial \vec{V}_m}{\partial t}$$

Thus, a general expression for the electric field is

$$\vec{E} = -\nabla V_e - \frac{\partial \vec{V}_m}{\partial t}$$

- **A changing electric flux induces a magnetic field.** Thus, Ampère's law can be written as

$$\oint_{line} \vec{B} \bullet d\vec{s} = \mu_0 \left(I_{enclosed} + \varepsilon_0 \frac{d\Phi_E}{dt} \right)$$

- **Inductance:** the ratio of the magnetic flux through a closed surface to the current generating the magnetic field associated with the flux.

$$L = N \frac{\Phi_B}{I}$$

The inductance can also be calculated from the magnetic potential associated with the magnetic field.

$$L = \oint_{line} \left(\frac{\vec{V}_m}{I} \right) \bullet d\vec{s}$$

The inductance can also be expressed in terms of the induced electric potential associated with a changing current.

$$L = -\frac{\Delta V_e}{\frac{dI}{dt}}$$

The inductance is a scalar quantity defined by the geometry of the magnetic flux. The inductance of a perfect solenoid is

$$L = \mu_0 n^2 V$$

- **Mutual inductance:** the ratio of the magnetic flux through a closed surface to the current generating the magnetic field associated with the flux.

$$M_{21} = \frac{N_2 \left(\Phi_B \right)_{21}}{I_1}$$

Since mutual inductance is reciprocal $M_{21} = M_{12}$.

- **Electromotive force (EMF):** the effective electric potential for a current. The unit of EMF is a volt. Mathematically, we can define an EMF as the work per unit charge that would be done on a point particle with net electric charge as it travels along a path.

$$\mathcal{E} = \frac{W}{q}$$

The EMF can then be expressed in terms of the associated electric field

$$\mathcal{E} = \int \vec{E} \bullet d\vec{s}$$

Thus, for the electric field associated with an electric potential difference

$$\mathcal{E} = -\Delta V_e$$

and for an induced electric field resulting from a changing magnetic field

$$\mathcal{E} = -\frac{d\Phi_B}{dt}$$

- The magnetic potential energy of a system can be calculated using the following equation:

$$U_m = \frac{1}{2}\sum_i I_i \left(\Phi_B\right)_i$$

The magnetic force between two loops of wire carrying constant currents is

$$\vec{F}_m = \left(\nabla U_m\right)_I$$

The magnetic force between two loops of wire maintaining constant magnetic fluxes is

$$\vec{F}_m = -\left(\nabla U_m\right)_{\Phi_B}$$

- The magnetic potential energy of a solenoid is

$$U_m = \frac{1}{2}LI^2$$

Problems

1. The following parallel electric and magnetic fields exist in a laboratory:

$$\vec{E} = \left(5000\frac{\text{V}}{\text{m}}\right)\hat{x} \qquad \vec{B} = \left(0.2\,\text{T}\right)\hat{x}$$

A positively charged point particle is fired through these fields with the following velocity:

$$\vec{v} = \left(2\times10^5\frac{\text{m}}{\text{s}}\right)\hat{y}$$

What are the electric and magnetic fields in the particle's reference frame?

2. A closed loop of wire of radius $R = 0.2$ m is in the presence of a uniform magnetic field, as shown in the figure. The magnitude of the magnetic field is changing according to the following equation:

$$B = B_0\left(1 + e^{-kt}\right)$$

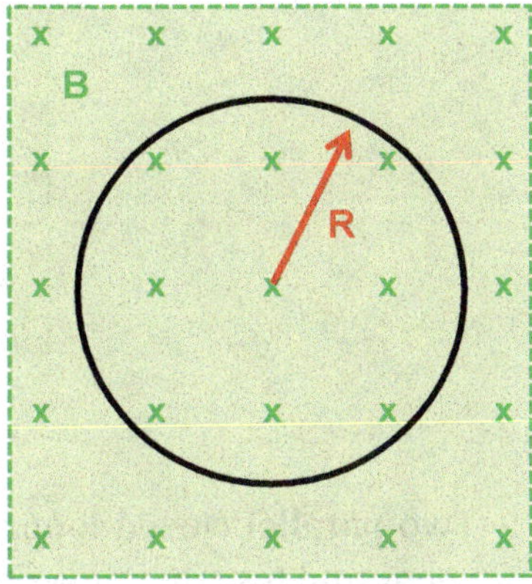

What is the magnitude of the induced electric field? What is its direction?

3. A cylindrical symmetric but non-uniform magnetic field has a magnitude given by the following equation:

$$B = B_0\left(\frac{1 - e^{-\frac{\tau}{t}}}{1 + \rho^2}\right)^2$$

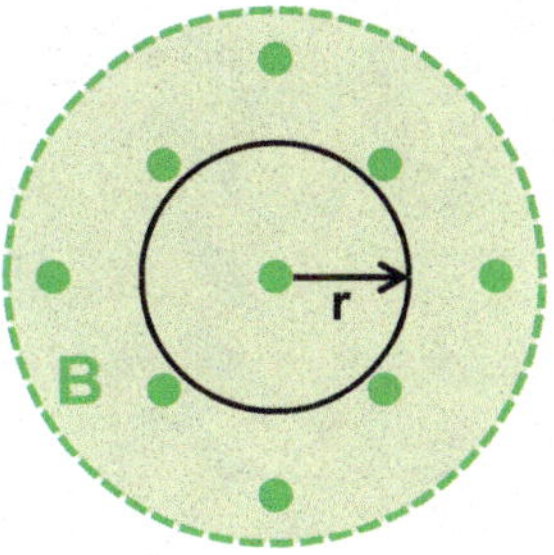

In this equation B_0 and τ are constants. A circular loop of wire is centered on the longitudinal axis of symmetry of the magnetic field as shown in the figure. What is the magnitude of the induced EMF in the loop? What is the direction of the associated current?

4. A rectangular loop of wire is pushed at a constant speed v into a region where a uniform magnetic field exists. What is the magnitude of the induced EMF in the loop? What is the direction of the associated current?

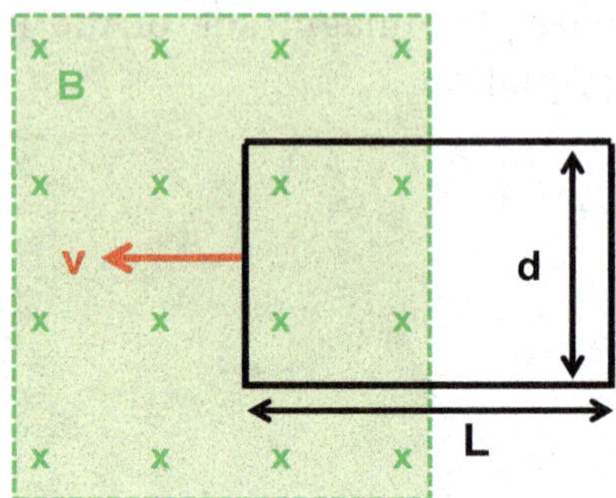

5. Two parallel closed loops of wire are centered on the same longitudinal axis and separated by a distance z as shown in the figure. The bottom loop has a radius R and carries a current I. The radius of the top loop, r, is so small that the magnetic field from the current in the bottom loop can be considered to be constant throughout the top loop (and equal to the magnetic field at the center of the top loop). If the current in the bottom loop is changing, how must the separation distance between the loops change so that no electric field is induced in the top loop? In other words, what is the equation relating I and z that ensures no induced electric field?

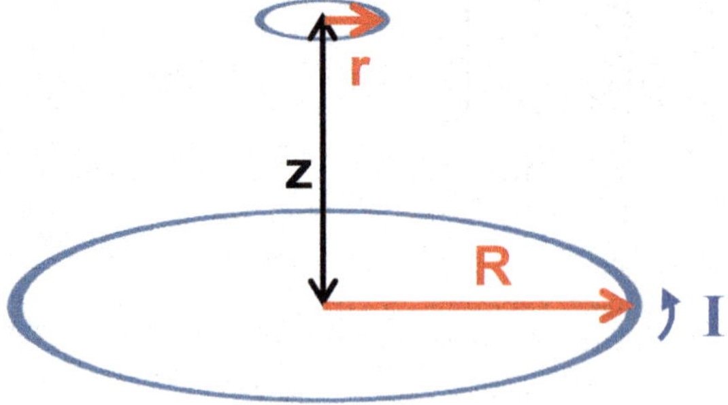

6. A square loop of wire is next to an infinitely long, straight wire, as shown in the figure. The current in the infinitely long wire is changing according to the following equation:

$$I = I_0 \sin(\omega t)$$

What is the magnitude of the induced EMF in the loop?

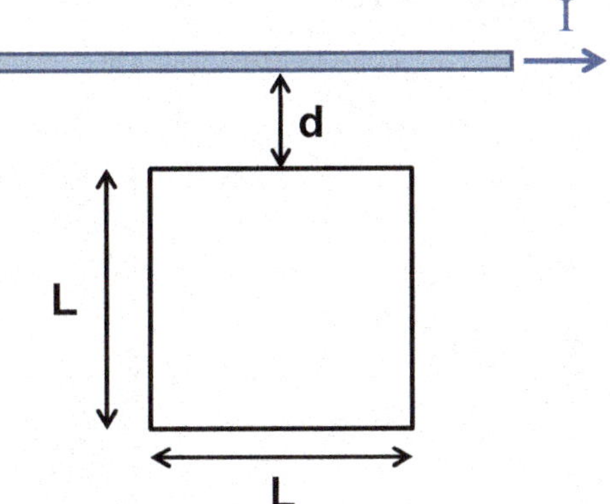

7. A square loop of wire is next to an infinitely long, straight wire, as shown in the figure; the distance between the loop and the wire is initially d_0. The current in the infinitely long wire is constant, but the loop is moving directly away from the wire at speed v. In other words, the distance, d, between the loop and the wire can be written as

$$d = d_0 + vt$$

What is the magnitude of the induced EMF in the loop?

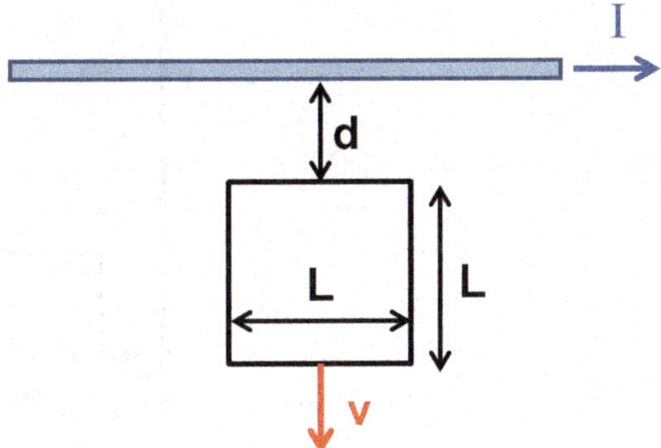

8. A thin, uniformly dense, and rigid rod of length L is rotating around its center of mass at angular velocity ω in the presence of a uniform magnetic field, as shown in the figure. What is the induced motional EMF between the center of the rod and one end of the rod?

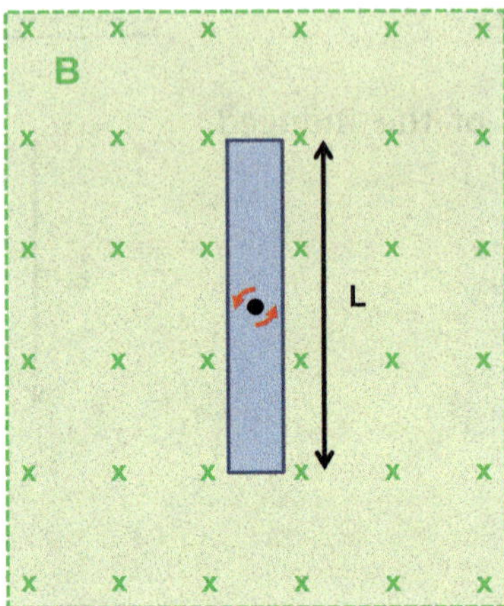

9. What is the mutual inductance of the two loops of wire shown in the figure? Both loops are rectangles, but the lengths of the vertical components of loop 2 are infinite compared to the size of loop 1.

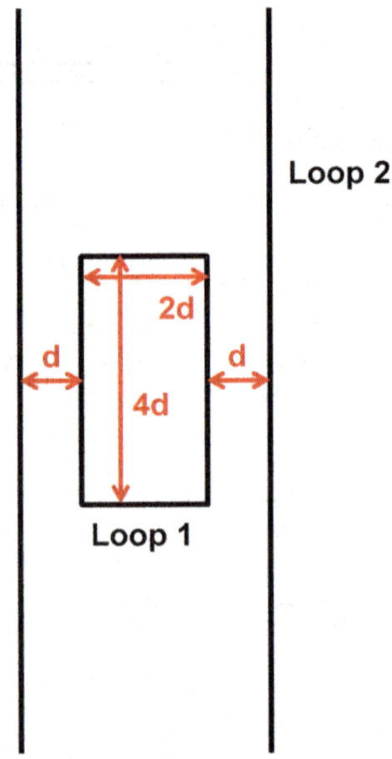

10. The induced electric potential in a solenoid of inductance L is given by the following equation:

$$\Delta V_e = \left(\Delta V_e\right)_0 \left(1 - e^{-kt}\right)$$

In this equation the variable k is a constant. What is the current in the solenoid as a function of time? Assume that the current is zero at $t = 0$.

CHAPTER 7

Electric and Magnetic Properties of Materials

7-1 Introduction

Up to this point, we have focused on magnetic and electric interactions occurring in vacuum. In this chapter, we turn our attention to how these interactions change when they occur in other environments and, by extension, to the electric and magnetic properties of materials. This sets the stage for discussions in later chapters about the application of electric and magnetic interactions to circuits and optics.

7-2 Electric Properties of Conducting Materials

In Section 6-6 we defined a ***conductor*** as a material containing free electrons that could move in response to an external magnetic field.

Conductor: a material containing free electrons that are free to move throughout the material.

If we wait long enough, these free electrons will continue to move until the magnitude of the net electric field they experience is zero (*i.e.*, they will move around until the electric field generated from the free electrons exactly cancels the external electric field at every point inside the conductor)[1]. When this has occurred, we say that the conductor is in *electrostatic equilibrium.*

In electrostatic equilibrium, the electric field is zero everywhere inside the conductor; if it weren't, the free electrons would still be moving and the conductor would not be in a *static* equilibrium. If the electric field is zero inside the conductor, then the electric potential inside the conductor must be constant (Equation 3-8). In addition, we recognize from Gauss's law (Equation 5-4) that if the electric field is zero everywhere inside a conductor, the free electrons must reside entirely on the surface of the conductor. Furthermore, since the conductor is in electrostatic equilibrium (*i.e.*, since the free electrons are not moving), the electric field just outside the surface of the conductor must be perpendicular to the surface of the conductor; if a component of the electric field were parallel to the surface, then the free electrons on the surface would move, and the conductor would not be in a static equilibrium. It follows that the electric potential is constant at all positions on the surface of the conductor (Equation 3-10).

In summary, when a conductor is in electrostatic equilibrium:

- The electric potential is constant everywhere inside the conductor.
- The electric field is zero everywhere inside the conductor.
- The free electrons reside entirely on the surface of the conductor.
- The electric potential is constant at all positions on the surface of the conductor.
- The electric field just outside the surface of the conductor is perpendicular to that surface.

Example 7-1

Problem: Two spherical conductors, A and B, with radii r_A and r_B, respectively, are connected by a conducting wire and separated by a distance much larger than r_A or r_B. The two conductors carry uniformly distributed net electric charges Q_A and Q_B, respectively, and the entire system is in electrostatic equilibrium. What is the ratio of the magnitudes of the electric fields at the surfaces of the two conductors?

1 This is similar to the process causing a motional EMF.

Solution: The electric field near the surface of each conductor can be found using Gauss's law (Equation 5-4). Indeed, from the solution of Example 5-7 we know that the magnitudes of the electric fields near the surface of the two conductors will be

$$E_A = \frac{Q_A}{4\pi r_A^2 \varepsilon_0} \qquad E_B = \frac{Q_B}{4\pi r_B^2 \varepsilon_0}$$

$$\frac{E_A}{E_B} = \frac{\dfrac{Q_A}{4\pi r_A^2 \varepsilon_0}}{\dfrac{Q_B}{4\pi r_B^2 \varepsilon_0}} \rightarrow \frac{E_A}{E_B} = \frac{Q_A r_B^2}{Q_B r_A^2}$$

To proceed further we need an equation that relates the net charges to the radii of the conductors. We can derive this equation from the equations for the electric potentials of the conductors. These electric potentials can be found using Equation 3-10[2].

$$\left(V_e\right)_A = \frac{Q_A}{4\pi r_A \varepsilon_0} \qquad \left(V_e\right)_B = \frac{Q_B}{4\pi r_B \varepsilon_0}$$

Since the entire system is in static equilibrium and the two conductors are connected by a conducting wire, the electric potential must be the same on the surfaces of both conductors.

$$\left(V_e\right)_A = \left(V_e\right)_B \rightarrow \frac{Q_A}{4\pi r_A \varepsilon_0} = \frac{Q_B}{4\pi r_B \varepsilon_0} \rightarrow Q_A = \frac{r_A}{r_B} Q_B$$

Thus,

2 We could have also obtained this equation for the electric potential by modeling each conductor as a point particle with a net charge.

$$\frac{E_A}{E_B} = \frac{\left(\frac{r_A}{r_B} Q_B\right) r_B^2}{Q_B r_A^2} \quad \rightarrow \quad \frac{E_A}{E_B} = \frac{r_B}{r_A}$$

Therefore, the magnitude of the electric field is largest for the sphere with the smallest radius.

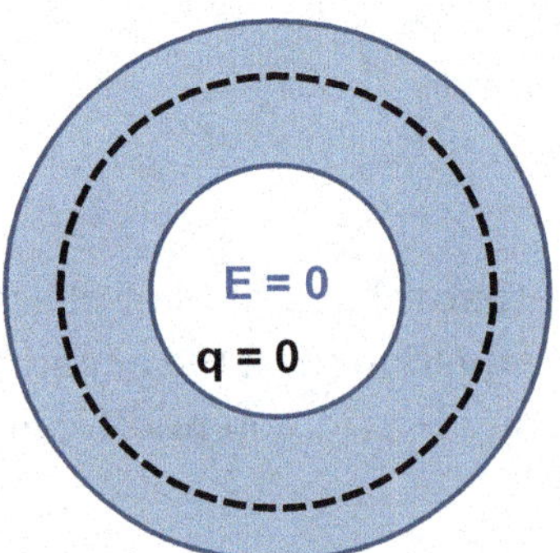

FIGURE 7.1: A conductor that contains a cavity. The net electric field inside the cavity is zero. The dashed line is the boundary of a Gaussian surface inside the conductor.

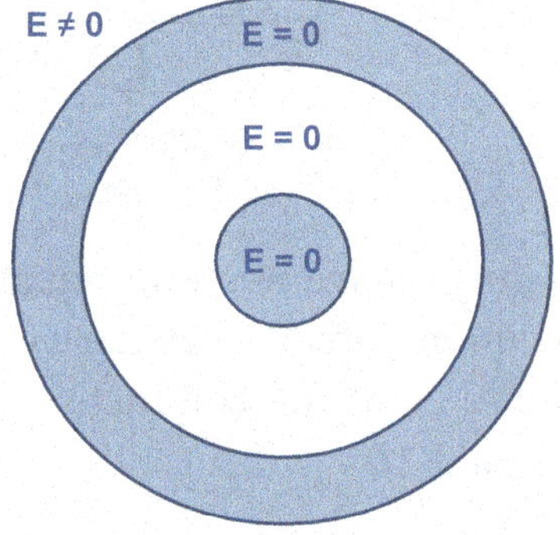

FIGURE 7.2: A conductor inside of another conductor will be shielded from any external electric fields.

It follows from the solution to Example 7-1 that for an irregularly shaped conductor in electrostatic equilibrium, the magnitude of the surface charge density and corresponding electric field is largest at points where the radius of curvature is the smallest (*e.g.*, at corners and sharp points).

Now consider the conductor shown in Figure 7.1 that contains a cavity. We know that under electrostatic equilibrium the electric field inside the conductor must be zero. Therefore, the net electric flux through a Gaussian surface contained inside the conductor must be zero. It follows from Gauss's law (Equation 5-4) that the enclosed net electric charge is zero. Thus, there can be no electric charge on the interior surface of a conductor with a cavity; all of the conductor's net electric charge must reside on the exterior surface of the conductor.

This conclusion is still valid, of course, even if there is another conductor within the cavity, as shown in Figure 7.2. As long as the inner conductor is uncharged, there will be no

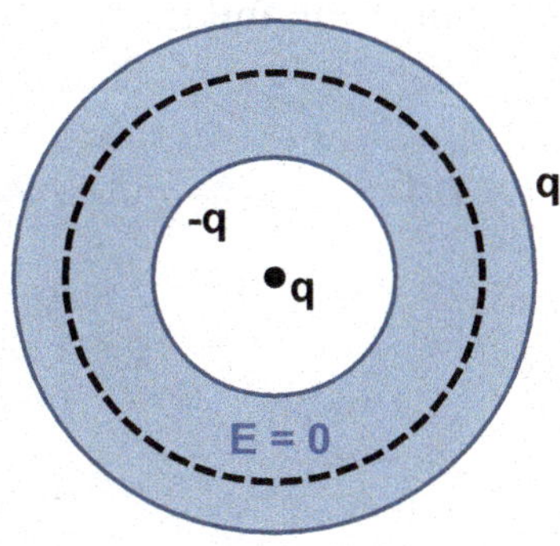

FIGURE 7.3: The distribution of charges on a conductor containing a charge in its cavity. The dashed line is the boundary of a Gaussian surface inside the conductor, where the electric field is zero.

electric field within the inner conductor or the surrounding cavity. We say that the inner conductor is completely *shielded* or *screened* from any external electric fields.

Finally, since the electric field inside a conductor must be zero, if a net electric charge exists inside the cavity of a conductor, the interior surface of the conductor must obtain a net charge of equal magnitude but opposite sign (Figure 7.3). In this way, the net electric charge contained by a Gaussian surface through the conductor will enclosed no net charge, and therefore the electric field inside the conductor will be zero, as required by the conditions of electrostatic equilibrium. (new sentence) To compensate for this net negative charge on the interior surface of the conductor a net positive charge must exist on the exterior surface of the conductor, as shown in Figure 7.3.

Thus a net electric charge inside the cavity of the conductor would make its presence known to someone outside the conductor by means of the induced charge on the exterior surface of the conductor and the electric field associated with it.

Example 7-2

Problem: A spherical conductor with a radius of 12 cm contains a spherical cavity with a radius of 2 cm. A point particle with a net electric charge of -5 nC is located within the cavity. This entire system is in electrostatic equilibrium. What is the magnitude of the electric potential just outside the exterior surface of the conductor?

Solution: Since the conductor is in electrostatic equilibrium, the electric potential will be constant at all locations on the surface of the conductor. Thus, the equipotential surfaces of the electric potential outside the conductor will be spherically symmetric. Because of this, let's choose a spherical shell as the Gaussian surface for determining the electric field using Equation 5-4 (Figure 7.4). This calculation is further simplified by the fact

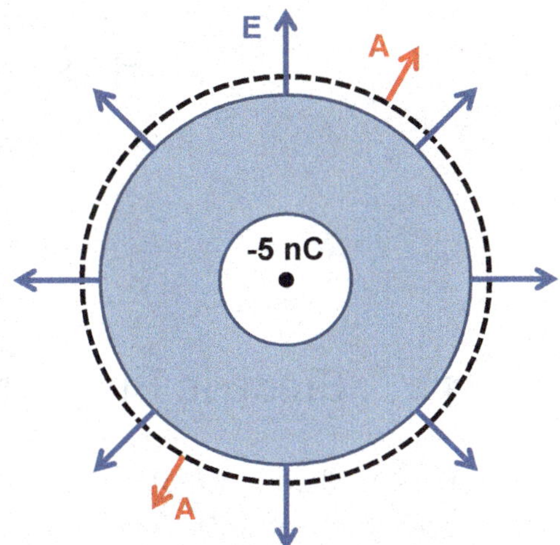

FIGURE 7.4: Determining the magnitude of the electric field outside the conductor in Example 7-2. The Gaussian surface for this problem is a spherical shell, shown in cross-section in black. The direction of the vectors associated with the area of this surface and the gravitational field are indicated by the arrows.

that the electric field is perpendicular to the surface of the conductor and thus to the Gaussian surface as well.

$$\oint_{surface} \vec{E} \bullet d\vec{A} = EA\cos(0°) \rightarrow \oint_{surface} \vec{E} \bullet d\vec{A} = EA$$

$$E\left(4\pi r^2\right) = \frac{q_{enclosed}}{\varepsilon_0} \rightarrow E = \frac{q_{enclosed}}{4\pi\varepsilon_0 r^2}$$

As expected, the electric field outside the conductor is identical to the electric field of a point particle with a net charge located at the center of the conductor. Of course, since the electric field outside the conductor is created by the surface electric charge density on the conductor, it will be the same regardless of where the point particle with net electric charge is located inside the cavity. The electric potential associated with this electric field is found using Equation 3-10[3].

$$V(r) = \left(\frac{q_{enclosed}}{4\pi\varepsilon_0 r}\right)$$

Just outside the surface of the conductor, the electric potential would be

$$V_e = \frac{-5 \times 10^{-9}\,\text{C}}{4\pi\left(8.85 \times 10^{-12}\,\frac{\text{C}^2}{\text{J m}}\right)(0.12\,\text{m})} \rightarrow V_e = -375\,\text{V}$$

7-3 Electric Properties of Insulating Materials

We can define a molecule (or an atom) to be neutral when its net electric charge is zero (Section 3-2). The negative electric charge (from the free and bound electrons) of a neutral molecule is distributed symmetrically around the positively charged nuclei,

3 Naturally, since the electric field is the same as the electric field of a charged point particle, the electric potential is the same as the electric potential of a charged point particle.

and thus the electric dipole moment of a neutral molecule is also zero. When a neutral molecule is placed in an external electric field, there will be "external" electric forces exerted on the electrons and nuclei, which will separate the electrons and nuclei from each other. If the material is a conductor, these forces will cause the free electrons to move to the surface of the conductor (Section 7-2). In an insulator, however, these forces will cause a small physical separation of the bound electrons from their nuclei within each atom of the insulator. Eventually, the "internal" forces of the molecule (such as the electric forces between the electrons and the nuclei) will balance the "external" electric forces, and the molecule will be in an equilibrium state with a net physical separation of positive and negative electric charge, and thus a non-zero electric dipole moment (Section 3-8). We then say that the molecule has an *induced electric dipole moment* and that it (and the entire insulator) has become *polarized*; this process is referred to as ***charge polarization***.

> *Charge polarization*: a small separation of the positive and negative charges in a neutral object.

Furthermore, the direction of the induced electric dipole moment will be the same as the direction of the external electric field.

Some molecules, such as water, have a permanent non-zero electric dipole moment even in the absence of an external electric field; such molecules are frequently referred to as ***polar molecules***.

> *Polar molecule*: a molecule that has a permanent non-zero electric dipole moment.

In the absence of an external electric field these permanent electric dipole moments will generally be randomly oriented so that the total electric dipole moment of a large collection of these molecules will be zero[4]. In the presence of an external electric field, however, there will be a tendency for the molecules to rotate so as to align their permanent electric dipole moments with the external electric field (Section 3-8). Of course, at equilibrium the molecules (and their permanent dipole moments) will retain some random orientation owing to collisions between molecules, thermal fluctuations, *etc.* Nevertheless, there will be a net electric dipole moment (from the sum of the electric dipole moments of all of the molecules) aligned with the external electric field.

A ***dielectric material*** (or simply a ***dielectric***) is an electrical ***insulator*** that can be polarized by an applied electric field.

---

4 This follows from the 2nd law of thermodynamics.

> *Dielectric*: an electrical insulator that can be polarized by an applied electric field.
>
> *Insulator*: a material containing no free electrons. All of the electrons in the material are bound to individual atoms and unable to move throughout the material.

There are also some materials called *electrets* which have a permanent electric dipole moment (*i.e.*, that are polarized) even in the absence of an external electric field. However, these materials are quite rare, and do not have the same technical or commercial applications as materials with permanent magnetic dipole moments (Section 7-4). Of course, it is also possible for material to have higher-order electric multipoles, such as quadrupole moments (Section 3-10), but, as we have seen, the contributions of these moments to the electric potential and electric field decrease in magnitude more rapidly with distance from the molecule than contributions from the electric dipole moment. Thus, we will assume in our macroscopic description of the electrical properties of dielectrics that the dominant features are associated with the electric dipole moments; in other words, *we will assume that neutral matter is equivalent to a collection of electric dipoles.*

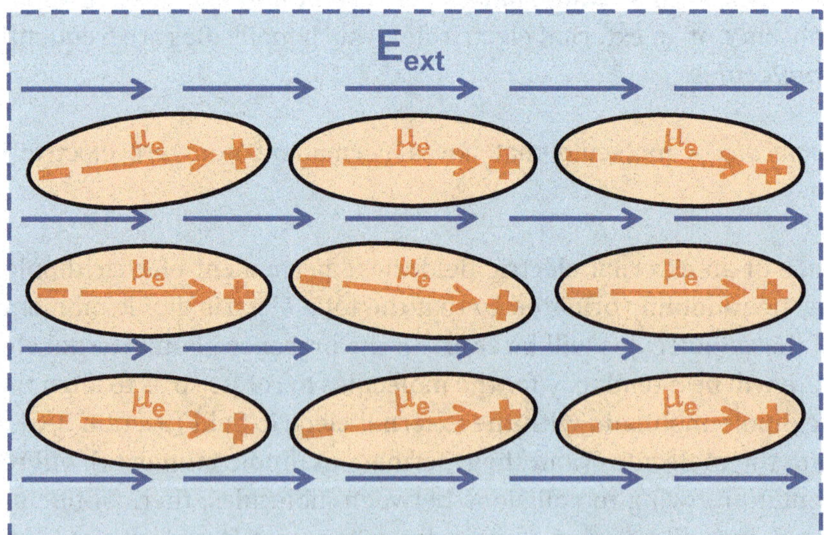

FIGURE 7.5: The induced or permanent electric dipole moments that constitute a dielectric will align themselves with an external electric field.

Let's now consider the situation in which an external electric field, denoted by $\vec{E}_{ext}$ is applied to a dielectric. As shown in Figure 7.5, the induced or permanent electric dipole moments within a dielectric will align themselves with the external electric field.

Consequently, as shown in Figure 7.6, the contributions of the positive and negative electric charges of these dipoles will cancel each other in the interior of the dielectric, leading to effectively zero net electric charge there, but cause net negative and positive electric charges to form on the edges of the dielectric. These electric charges are commonly referred to as polarization charges. This net separation of polarization charges creates an induced electric field (denoted as $\vec{E}_{ind}$ in Figure 7.6) within the dielectric[5]. The direction of this induced electric field is opposite the direction of the external electric field.

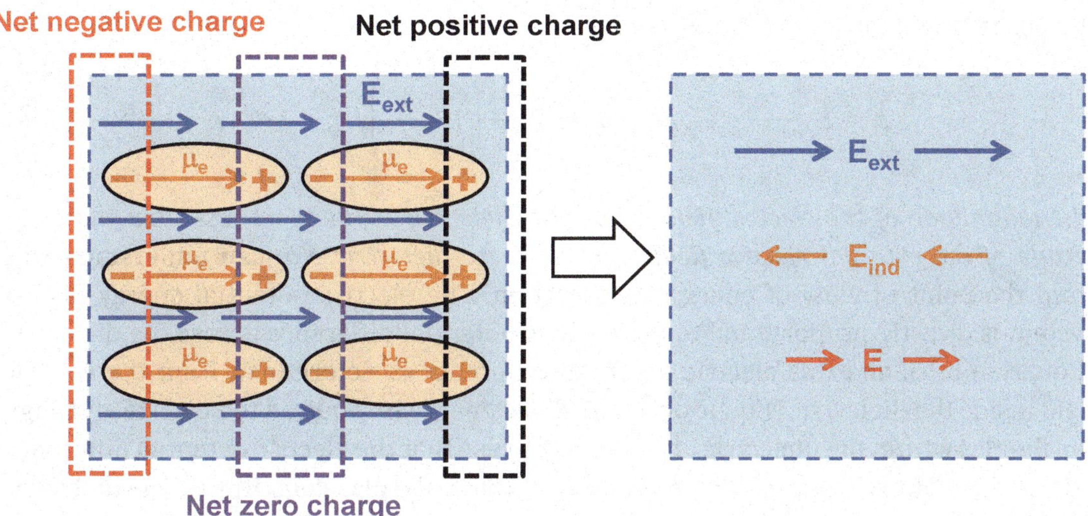

FIGURE 7.6: An external electric field causes a separation of bound electric charges within a dielectric. This separation of charges creates an induced electric field within the dielectric. The net electric field inside the dielectric is the sum of the external and induced electric fields.

Since the induced electric field results from the alignment of the dielectric's induced or permanent dipole moments with the external electric field, it is possible to express the induced electric field in terms of the external electric field. For simplicity, we will limit our consideration of *linear isotropic dielectrics*[6], in which the directions of the induced and external electric fields are opposite[7]. For these systems, the magnitude of the induced electric field is related to the magnitude of the external electric field using Equation 7-1.

$$E_{ind} = \left(1 - \frac{1}{\kappa_e}\right) E_{ext} \tag{7-1}$$

5 This induced electric field is often referred to as the polarization field, albeit with a slightly different definition.

6 The term "linear" denotes a linear relationship between the external and induced electric fields. The term "isotropic" means that the linear relationship is valid for all directions of the external electric field.

7 You should feel inspired to take more advanced physics classes to study non-isotropic linear dielectrics and non-linear dielectrics. These materials have many interesting applications, such as components in lasers.

The scalar quantity κ_e in Equation 7-1 is the *dielectric constant*[8]. For all known substances $\kappa_e \geq 1$; for vacuum[9] $\kappa_e = 1$. It follows from Equation 7-1 that the magnitude of the induced electric field is always less than the magnitude of the external electric field.

The net electric field inside the dielectric is the summation of the external electric field and the induced electric field. In other words, the electric field inside the dielectric results from the electric charges associated with the external electric field (commonly referred to as *free electric charges*) and the bound electric charges of the dielectric itself. Since external and induced electric fields are in opposite directions the magnitude of the electric field inside the conductor is

$$E = E_{ext} - E_{ind} \quad \rightarrow \quad E = E_{ext} - \left(1 - \frac{1}{\kappa_e}\right)E_{ext} \quad \rightarrow \quad E = \frac{E_{ext}}{\kappa_e} \qquad \textbf{(7-2)}$$

The magnitude of the electric field inside the dielectric is therefore less than the magnitude of the external electric field applied to the dielectric. We can understand this from the point of view of energy conservation. The electric potential energy of this system is directly proportional to the electric potential difference across the dielectric (Equation 3-5), and this electric potential difference is determined from the electric field inside the dielectric (Equation 3-10). Since energy is required to polarize and align the dipoles within the dielectric, it must be the case that the electric potential difference across the dielectric is less than it would be if the molecules constituting the dielectric could not be polarized (or if there were no dielectric at all). It follows that the magnitude of the electric field inside the dielectric must also be less than the external electric field applied to the dielectric.

Using Equation 7-2, we can rewrite Equation 5-4 to be true for all electric fields, including those in the presence of dielectrics.

$$\oint_{surface} \vec{E} \bullet d\vec{A} = \frac{q_{enclosed}}{\kappa_e \varepsilon_0} \qquad \textbf{(7-3)}$$

The variable $q_{enclosed}$ in Equation 7-3 is the net *free* electric charge contained within the closed surface of integration. We account for the contributions of the *bound* electric charges through the inclusion of the dielectric constant. Please remember that Equation 7-3 is valid for linear isotropic dielectrics only. Finally, the quantity $\kappa_e \varepsilon_0$ in Equation 7-3 is called the *permittivity* (or absolute permittivity) and is commonly denoted by ε.

8 The dielectric constant is also referred to as the relative permittivity.

9 N.B., people often say "free-space" rather than vacuum.

Example 7-3

Problem: A charged point particle with net electric charge q is embedded in a linear isotropic dielectric with dielectric constant k_e. The dielectric is infinitely large. What is the electric field associated with this point particle?

Solution: Since the dielectric is infinitely large, we can ignore any contributions from bound electric charges on the outer surface of the dielectric, and we can assume that the electric field will all be spherically symmetric. Let's therefore use a spherical shell (with radius r centered on the point particle) as our Gaussian surface for the integration in Equation 7-3; the net free electric charge contained within this surface is the net electric charge of the point particle.

$$E\left(4\pi r^2\right)=\frac{q}{\kappa_e\varepsilon_0} \quad\rightarrow\quad E=\frac{q}{4\pi\kappa_e\varepsilon_0 r^2} \quad\rightarrow\quad \vec{E}=\left(\frac{q}{4\pi\kappa_e\varepsilon_0 r^2}\right)\hat{r}$$

Now let's consider the system shown in Figure 7.7 that consists of a parallel plate capacitor with a linear isotropic dielectric completely filling the space between the plates; the distance between the plates is d. The electric field in-between the plates of the capacitor can be determined using Equation 7-3; the only free electric charges in this system are the charges on the plates of the capacitor. Let's choose a Gaussian surface that contains only the free positive charges, as shown in Figure 7.7.

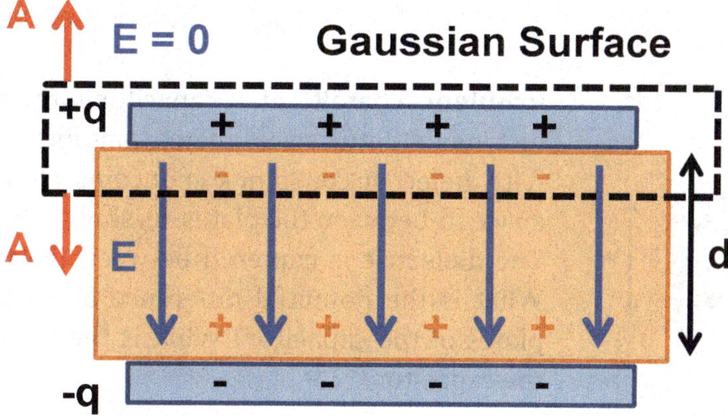

FIGURE 7.7: Determining the electric field for a parallel plate capacitor containing a dielectric.

We will assume that the electric field is uniform everywhere between the plates of the capacitor and zero elsewhere.

$$\oint_{surface} \vec{E} \bullet d\vec{A} = \frac{q_{free}}{\kappa_e \varepsilon_0} \quad \rightarrow \quad \oint_{surface} E\, dA \cos(0°) = \frac{q}{\kappa_e \varepsilon_0}$$

$$E \oint_{surface} dA = \frac{q}{\kappa_e \varepsilon_0} \quad \rightarrow \quad EA = \frac{q}{\kappa_e \varepsilon_0} \quad \rightarrow \quad E = \frac{q}{\kappa_e A \varepsilon_0}$$

As before (Section 5-5), the electric potential difference between the plates of the capacitor can be determined from this electric field using Equation 3-10, and further simplified using the definition of the capacitance of the capacitor (Section 5-5).

$$\Delta V_e = -\frac{q}{A\kappa_e \varepsilon_0} d \quad \rightarrow \quad \Delta V_e = -\frac{q}{\kappa_e C}$$

Since $\kappa_e \geq 1$, the magnitude of the electric potential in-between the plates of a capacitor is always less in the presence of a dielectric. As discussed before, this makes sense because some of the electric potential energy associated with the separation of the free charges on the plates of the capacitor has been used to align the electric dipole moments (and create the associated separation of bound charges) within the dielectric.

Example 7-4

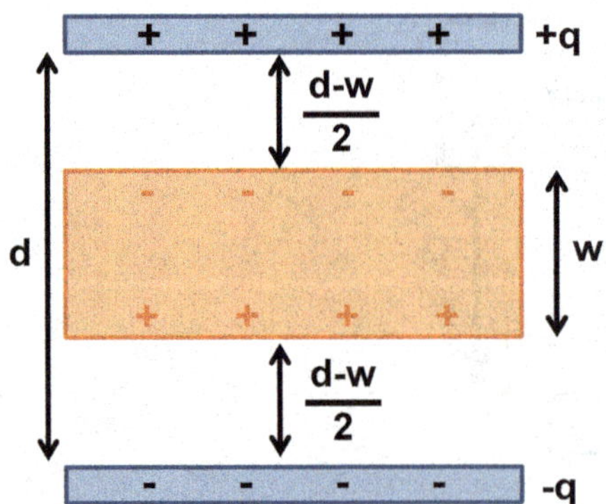

FIGURE 7.8: The parallel plate capacitor in Example 7-4.

Problem: A parallel plate capacitor with a separation distance d between its plates contains a dielectric, with dielectric constant k_e, not completely filling the space in-between the plates, as shown in Figure 7.8. The dielectric is centered between the two plates. What is the potential difference between the two plates of the capacitor? What is the capacitance of the capacitor?

Solution: We can determine the electric field in-between the two plates using Equation 7-3. As shown in Figure 7.9, we will denote the magnitude of the electric field between the plates and the dielectric as E_1, the magnitude of the electric field inside the dielectric as E_2, and the magnitude of the electric field to be zero everywhere else. To determine the values of E_1 and E_2, we will use the two Gaussian surfaces shown in Figure 7.9, and follow the derivation of the electric field for a parallel plate capacitor in Example 5-8.

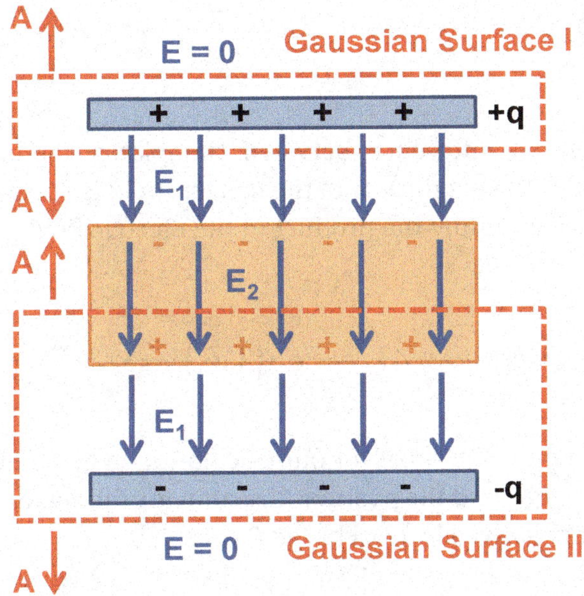

FIGURE 7.9: The Gaussian surfaces for determining the electric field for the capacitor in Example 7-4.

For Gaussian Surface I, we have

$$\kappa_e = 1 \;\rightarrow\; \oint_{surface} \vec{E}_1 \bullet d\vec{A} = \frac{q_{free}}{\varepsilon_0} \;\rightarrow\; \oint_{surface} E_1\, dA \cos(0°) = \frac{q}{\varepsilon_0}$$

Let's assume that the electric field is uniform everywhere between the plates of the capacitor and zero elsewhere.

$$E_1 \oint_{surface} dA = \frac{q}{\varepsilon_0} \;\rightarrow\; E_1 A = \frac{q}{\varepsilon_0} \;\rightarrow\; E_1 = \frac{q}{A\varepsilon_0}$$

In this expression, the variable A is the surface area of each plate of the capacitor.

Applying similar assumptions to Gaussian Surface II gives us

$$\oint_{surface} E_2\, dA \cos(180°) = \frac{-q}{\kappa_e \varepsilon_0}$$

$$-E_2 \oint_{surface} dA = \frac{-q}{\kappa_e \varepsilon_0} \rightarrow E_2 A = \frac{q}{\kappa_e \varepsilon_0} \rightarrow E_2 = \frac{q}{A\kappa_e \varepsilon_0}$$

The electric potential difference between the plates can be found using Equation 3-10. Since the magnitude of the electric field is constant, the electric potential difference is simply the product of the electric field and the distance over which the field acts.

$$\Delta V_e = -\int \vec{E} \bullet d\vec{s} \rightarrow \Delta V_e = -\vec{E} \bullet \left(\int d\vec{s}\right) \rightarrow \Delta V_e = -\vec{E} \bullet \Delta\vec{s}$$

Let's define the origin of the s-axis for our calculation to be at the plate with net positive electric charge, and the positive direction for the s-axis to point toward the plate with net negative electric charge. Then,

$$\Delta V_e = -\left(E_1\left(\frac{d-w}{2} - 0\right) + E_2\left(\frac{d+w}{2} - \frac{d-w}{2}\right) + E_1\left(d - \frac{d+w}{2}\right)\right)$$

$$\Delta V_e = -\left(E_2 w + E_1(d-w)\right) \rightarrow \Delta V_e = -\left(\left(\frac{q}{A\kappa_e \varepsilon_0}\right)w + \left(\frac{q}{A\varepsilon_0}\right)(d-w)\right)$$

$$\Delta V_e = -\frac{q}{A\varepsilon_0}\left(d + w\left(\frac{1-\kappa_e}{\kappa_e}\right)\right)$$

The capacitance of the capacitor can be determined using Equation 5-7.

$$\Delta V_e = -\frac{q}{C} \rightarrow C = -\frac{q}{\Delta V_e}$$

$$C = \frac{A\varepsilon_0}{\left(d + w\left(\frac{1-\kappa_e}{\kappa_e}\right)\right)}$$

Based upon the solution to Example 7-4, we see that the capacitance of a capacitor will increase when a dielectric is placed between the plates. When the dielectric completely fills the space between the plates of the capacitor, the capacitance of the capacitor is

$$C = \kappa_e\left(\frac{A\varepsilon_0}{d}\right) \rightarrow C = \kappa_e C_0 \tag{7-4}$$

The variable C_0 in Equation 7-4 denotes the capacitance of the capacitor in the absence of the dielectric (Equation 5-6). It follows from Equation 7-4 and Equation 5-8 that the electric potential energy stored within a capacitor will be reduced in the presence of the capacitor. The reduction of electric potential energy reflects the fact that some of the electric potential energy associated with the separation of charges on the plates of the capacitor is "used" to polarize the dielectric between the plates.

The electric potential energy of any distribution of particles with net electric charge is affected by the presence of any dielectric in-between the particles. The electric potential energy for a system of two point particles with net electric charges q_1 and q_2 separated by a distance r would be

$$U_e = \frac{q_1 q_2}{4\pi\kappa_e\varepsilon_0 r} \tag{7-5}$$

More generally, as shown in Example 7-3, the magnitude of the electric potential and electric field of a point particle with a net electric charge is also affected by any dielectric surrounding the particle.

$$V_e = \frac{q}{4\pi\kappa_e\varepsilon_0 r} \tag{7-6}$$

$$\vec{E} = \left(\frac{q}{4\pi\kappa_e\varepsilon_0 r^2}\right)\hat{r} \tag{7-7}$$

Example 7-5

Problem: Two point particles, A and B, with net charges +2 nC and −4 nC, are separated by a distance of 10 cm in a polar fluid with a dielectric constant of 40. What is the magnitude of the electric field experienced by particle B? What is the electric potential energy of this system?

Solution: The magnitude of the electric field experienced by particle B can be determined using Equation 7-7. This electric field originates from particle A.

$$E_{A,B} = \frac{q_A}{4\pi\kappa_e\varepsilon_0 r_{A,B}^2} \rightarrow E_{A,B} = \frac{2\times10^{-9}\,\mathrm{C}}{4\pi(40)\left(8.85\times10^{-12}\,\frac{\mathrm{C}^2}{\mathrm{J\,m}}\right)(0.1\,\mathrm{m})^2}$$

$$E_{A,B} = 45\frac{\mathrm{N}}{\mathrm{C}} = 45\frac{\mathrm{V}}{\mathrm{m}}$$

The electric potential energy of this system can be found using Equation 7-5.

$$U_e = \frac{(2\times10^{-9}\,\mathrm{C})(-4\times10^{-9}\,\mathrm{C})}{4\pi(40)\left(8.85\times10^{-12}\,\frac{\mathrm{C}^2}{\mathrm{J\,m}}\right)(0.1\,\mathrm{m})} \rightarrow U_e = -1.8\times10^{-8}\,\mathrm{J}$$

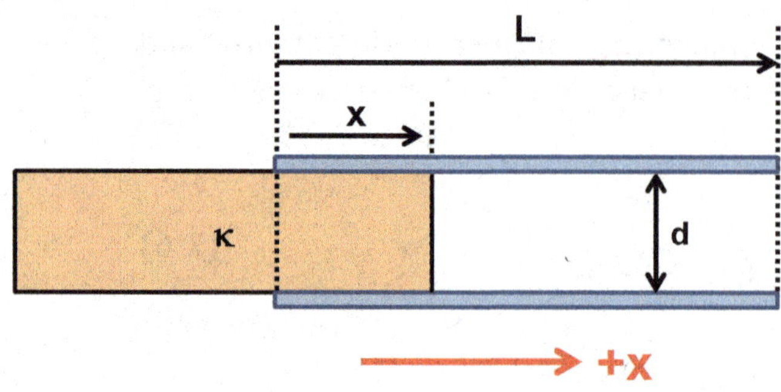

FIGURE 7.10: A dielectric partially inside a capacitor.

When a dielectric is polarized, the resultant net bound electric charges and net free electric charges will all exert electric forces on each other. These forces can be quite difficult to calculate directly using Equation 3-6, but the average force can be determined using an energy-based approach. Let's consider a parallel plate capacitor with square plates; the length of each side of each plate is denoted by *L*. For simplicity, we will assume that the electric field is the same as when the plates are infinitely large. A side view of this system in shown in Figure 7.10.

The electric potential energy of the capacitor is determined using Equation 5-8 and Equation 7-4.

$$U_e = \frac{q^2}{2\kappa_e C_0}$$

The difference in the electric potential energy of the capacitor with and without the dielectric is therefore[10]

$$\Delta U_e = \frac{q^2}{2\kappa_e C_0} - \frac{q^2}{2C_0} \rightarrow \Delta U_e = \frac{q^2}{2C_0}\left(\frac{1}{\kappa_e} - 1\right)$$

Since $\kappa_e \geq 1$, the electric potential energy of the capacitor is smaller in the presence of the dielectric. Thus, in order for the system to minimize its potential energy, the dielectric will be pulled into the capacitor. The average force acting on the dielectric as it is pulled into the capacitor can be determined using Equation 1-1.

$$F_{avg} = -\frac{\Delta U_e}{L} \rightarrow F_{avg} = -\frac{q^2}{2C_0 L}\left(\frac{1}{\kappa_e} - 1\right) \rightarrow F_{avg} = \frac{q^2}{2C_0 L}\left(\frac{\kappa_e - 1}{\kappa_e}\right)$$

Similarly, since the presence of a dielectric reduces the electric potential energy of a capacitor, the presence of a dielectric reduces the net electric force between the positively and negatively charged surfaces of the capacitor (Section 5-5).

If a dielectric is not rigid, the electric forces between the free and bound electric charges will cause the dielectric to deform. This phenomenon is called *electrostriction*. Furthermore, if the magnitude of the external electric field applied to a dielectric is too large, the bound electrons can be stripped from their associated nuclei and flow as a current; in other words, the dielectric becomes a conductor. This process is called ***dielectric breakdown*** (or electrical breakdown).

Dielectric breakdown: The rapid conversion of an insulating dielectric to a conductor in the presence of an electric field.

10 Recall that the dielectric constant is 1 for vacuum.

The magnitude of the electric field required for dielectric breakdown, commonly referred to as the dielectric strength, depends upon the substance. For air, the dielectric strength is around 3×10^6 V/m, and the corresponding dielectric breakdown is visible as lightning. The dielectric strengths of some other common materials are: 7×10^7 V/m (water); 1×10^7 V/m to 1×10^8 V/m (plastic); and 2×10^9 V/m (diamond).

Returning to Section 7-2 we see that dielectric breakdown is most likely to occur at corners and sharp points since the magnitude of the electric field is largest at those locations. Hence, if you want to charge an object to high electric potential and not have it discharge in the air through sparks, you should make sure that the surface of the object is as smooth and regular as possible.

7-4 Magnetic Properties of Materials

In Section 7-3 we assumed that the atoms and molecules in neutral matter were composed of positive and negative electric charges, and were, on the whole, electrically neutral. Let's extend this model by assuming that at least some of these electric charges are not at rest, but are moving around inside the material. It is reasonable to presume that some of these electric charges are moving in closed paths and thus giving rise to *internal* magnetic dipole moments (Section 4-7). These permanently circulating currents are referred to as ***Ampèrian currents***, after André-Marie Ampère who first proposed their existence.

Ampèrian currents: permanently circulating currents in neutral matter.

In the absence of an external magnetic field, it is possible that the internal magnetic dipole moments of these Ampèrian currents—together with the spin and orbital magnetic dipole moments of the electrons themselves (Section 4-9)—combine to give no net magnetic dipole moment for the material. When an external magnetic field is present, however, a torque will be exerted on these internal magnetic dipoles that will cause them to align in the direction of the external magnetic field, thus creating a non-zero internal magnetic dipole moment for the material. We describe this process by saying that a magnetic dipole moment has been *induced* in the material, and that the material has been *magnetized*. More specifically, this process is referred to as paramagnetism (Section 4-9).

Even if the material contains permanent magnetic dipole moments in the absence of an external magnetic field, it is still possible that these magnetic dipole moments are randomly oriented so that the total magnetic dipole moment is zero. In the presence of an external magnetic field, these magnetic dipole moments will feel a torque aligning them with the external magnetic field and thus magnetizing the material. Some materials have the property that, even in the absence of an external magnetic field, the permanent magnetic dipoles are at least partially aligned, and the material is said to be permanently magnetized or to be a permanent magnet. This is an example of ferromagnetism (Section 4-9).

All materials will also exhibit some degree of diamagnetism (Section 4-9), although this effect is typically very small. In the absence of the external magnetic field, the spin and orbital magnetic dipole moments of the electrons combine to produce no net magnetic dipole moment. However, in the presence of an external magnetic field, the electrons experience a Lorentz force (Equation 4-2) that simultaneously changes both their orbital motion and the magnitude of their orbital magnetic dipole moments. The resulting imbalance between the spin and orbital magnetic dipole moments gives rise to a net magnetic dipole moment oriented opposite the direction of the external magnetic field. All conductors also experience an effective diamagnetism through the induction of eddy currents by the external magnetic field (Section 6-6); the induced magnetic dipole moments associated with these currents are oriented opposite the direction of the external magnetic field.

As with the case of the electrical properties of material (Section 7-3), we can expect that molecules can also have higher-order multipole magnetic moments[11]. However, as before, the magnitude of the contributions of these higher-order magnetic moments decrease in magnitude more rapidly with distance from the molecule than contributions from the magnetic dipole moment. Thus, we will assume in our macroscopic description of the magnetic properties of material that the dominate features are associated with the magnetic dipole moments; in other words, *we will assume that matter is equivalent to a collection of magnetic dipoles.*

We will consider only the simplest case—linear isotropic magnetic materials for which the direction of any induced magnetic fields and the external magnetic field are the same[12]. For paramagnetic and ferromagnetic material, the direction of the induced magnetic field is the same as the direction of the external magnetic field, whereas for diamagnetic material, the direction of the induced magnetic field is opposite the direction of the external magnetic field. For linear isotropic magnetic materials, Equation 5-9 can be written as

11 See Section C-4.

12 You should feel inspired to take more advanced physics classes to study non-isotropic and non-linear magnetic materials.

$$\oint_{line} \vec{B} \bullet d\vec{s} = \kappa_m \mu_0 I_{enclosed} \quad \textbf{(7-8)}$$

The variable κ_m in Equation 7-8 is the relative permeability, and is the ratio of the absolute permeability (μ) to the vacuum permeability.

$$\kappa_m = \frac{\mu}{\mu_0} \quad \textbf{(7-9)}$$

For ferromagnetic and paramagnetic materials, $\kappa_m > 1$, and for diamagnetic materials $\kappa_m < 1$. For nearly all linear isotropic magnetic paramagnetic and diamagnetic materials, $|\kappa_m| \approx 1$, but for ferromagnetic materials, κ_m can be much larger than 1 (even larger than 1000). A special type of material, called a superconductor, has a relative permeability of zero. The magnetic dipoles inside the superconductor align themselves exactly antiparallel with the external magnetic field so that $\vec{B} = 0$ everywhere inside a superconductor[13].

Example 7-6

Problem: An electromagnet (*i.e.*, a solenoid) of length *l* is created by wrapping wire around a linear isotropic magnetic material with relative permeability κ_m. What is the magnetic field inside the solenoid? What is the self-inductance of the solenoid? Consider the solenoid to be perfect.

Solution: The magnetic field inside a perfect solenoid containing no magnetic material was determined in Section 5-6 using Equation 5-9. Following a similar derivation, but using Equation 7-8, we see that the magnitude of the magnetic field is

$$B = \kappa_m \mu_0 n I$$

As shown in Figure 7.11, the magnetic field always has the same direction as the external magnetic field. For paramagnetic and ferromagnetic material, the magnitude of the magnetic field is larger than the magnitude of the external

13 This is known as the Meissner effect.

magnetic field. For diamagnetic materials, the magnitude of the magnetic field is smaller than the magnitude of the external magnetic field.

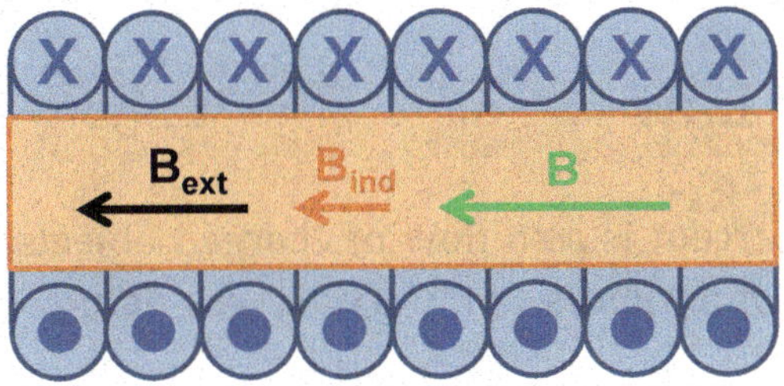

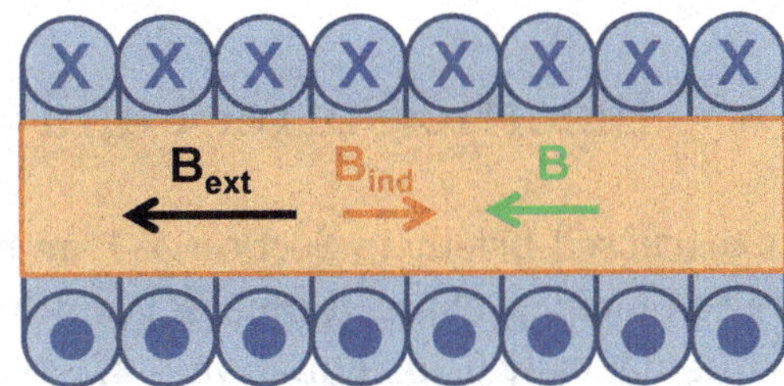

Ferromagnetic and paramagnetic material

Diamagnetic material

FIGURE 7.11: The magnetic field inside the solenoid is the sum of the external magnetic field and the induced magnetic field.

The self-inductance of the solenoid is determined using Equation 6-5. Since the magnetic field inside a perfect solenoid is constant, we have

$$\Phi_B = BA \quad \rightarrow \quad L = N\frac{BA}{I} \quad \rightarrow \quad L = N\frac{\kappa_m \mu_0 nIA}{I}$$

$$N = nl \quad \rightarrow \quad L = \kappa_m \mu_0 n^2 lA \quad \rightarrow \quad L = \kappa_m \mu_0 n^2 V$$

$$L = \kappa_m L_0$$

We should be careful here to note that most ferromagnetic materials are not linear isotropic magnetic materials, and thus the exact relationship between the magnetic field and current is more complicated than Equation 7-8 implies. Nevertheless, it is true that the strength of an electromagnet can be increased many fold by including a ferromagnetic core.

Finally, since the induced magnetic dipole associated with diamagnetism is oriented opposite the external magnetic field, it will experience a magnetic force pushing it away from the source of the external magnetic field (Figure 4.20). Since all objects display some amount of diamagnetism, this phenomenon can be used for levitation. However, because the diamagnetic dipole moment is typically very small, large magnetic fields are required to levitate even small objects.

7-5 Electric Current and Ohm's Law

As discussed briefly in Section 4-4, an electric current is as a flow of charged objects (*i.e.*, as moving charged objects) and has units of amperes (A). Current was defined mathematically using Equation 4-4.

$$I = \frac{dq}{dt}$$

We will now define a new variable, called the current density, which is the current per cross-sectional area in the conductor. The current and the current density are related through Equation 7-10.

$$I = \int \vec{J} \bullet d\vec{A} \qquad \textbf{(7-10)}$$

Example 7-7

Problem: A cylindrical wire with radius R carries a current with non-uniform current density. The magnitude of the current density increases with increasing radial distance from the center of the wire according to the following equation:

$$J(r) = \alpha + \beta r$$

In this equation, α and β are constants. What is the total current in the wire? What is the magnitude of the magnetic field from the current as a function of the radial distance r from the center of wire? Assume that the relative permeability is 1 for this system.

Solution: Let's choose the direction for the differential area vector in Equation 7-10 to be the same as the direction of the current density. Then Equation 7-10 becomes

$$I = \int J\, dA$$

The differential area for the integration is a thin ring of circumference $2\pi r$ and width dr[14].

$$I = \int J(2\pi r)\, dr \quad \rightarrow \quad I = 2\pi \int Jr\, dr \quad \rightarrow \quad I = 2\pi \int (\alpha + \beta r) r\, dr$$

$$I = 2\pi \left(\int_0^R \alpha r + \beta r^2\, dr \right)$$

$$I = 2\pi \left(\alpha r^2 + \beta r^3 \Big|_0^R \right) \quad \rightarrow \quad I = 2\pi \left(\alpha R^2 + \beta R^3 \right)$$

The magnitude of the magnetic field is found using Equation 7-8. The symmetry of the magnetic field will be reflected in the symmetry of the current distribution that generates it. In this case, the magnetic field should exhibit the cylindrical symmetry of the wire carrying the current. We should therefore choose an Ampèrian loop that is a circle centered on the wire/current. By symmetry, we also know that the magnetic field is constant everywhere along the surface of the Ampèrian loop. Thus,

$$\oint_{line} \vec{B} \bullet d\vec{s} = B(2\pi r)$$

From the previous solution, we know that the current passing through this Ampèrian loop when r is less than the radius of the wire is

$$I_{enclosed} = 2\pi \left(\alpha r^2 + \beta r^3 \right)$$

14 See Example 1-2, Example 3-12, Example 5-6, and Example 6-2.

Substitution into Equation 7-8 then gives us

$$B(2\pi r) = \mu_0 2\pi(\alpha r^2 + \beta r^3) \quad \rightarrow \quad B = \mu_0(\alpha r + \beta r^2)$$

When r is greater than the radius of the wire, we have

$$I_{enclosed} = 2\pi(\alpha R^2 + \beta R^3) \quad \rightarrow \quad B(2\pi r) = \mu_0 2\pi(\alpha R^2 + \beta R^3)$$

$$B = \frac{\mu_0(\alpha R^2 + \beta R^3)}{r}$$

As expected, these two equations for the magnetic field are identical when $r = R$.

When an external electric field is applied to an isolated conductor, the free electrons in the conductor will move until a surface charge density is created on the conductor that reduces the total electric field within the conductor to zero (Section 7-2). If, however, the conductor is not isolated and the external electric field (or, equivalently, the external EMF) is maintained, an electric current will continue to flow through the conductor.

We define ***direct current*** (DC) to be the unidirectional flow of electric charge. In ***alternating current*** (AC), the movement of the electric charge periodically reverses direction.

Direct current (DC): the unidirectional flow of electric charge.

Alternating current (AC): a flow of electric charge in which the direction of movement periodically reverses.

As discussed in Section 6-5, creating and maintaining either kind of current requires an input of energy from the source of the EMF that drives the current (*i.e.*, that does the work associated with moving the charges). The instantaneous power associated with this input of energy is equal to the rate of change of the electric potential energy of the electrons, which can be related to the effective electric potential difference of the EMF by using a modified version of Equation 3-3[15].

---

15 Equation 7-11 can also be derived from Equation 6-8.

$$P = \frac{dU_e}{dt} \rightarrow P = \frac{d}{dt}(q\Delta V_e) \rightarrow P = \left(\frac{dq}{dt}\right)\Delta V_e$$

$$P = I\mathcal{E} \quad \textbf{(7-11)}$$

What happens to the energy supplied by the electric potential to the current? The exact description of the flow of current in conductors involves physics that is above the level of this course[16]. Very simply, as the free electrons move through the conductor, they convert their electric potential energy into kinetic energy, while simultaneously reacting with each other and with the atoms (the bound electrons and their associated nuclei). Through these interactions, energy and momentum from the free electrons are transferred to the atoms, where it subsequently results in an increase in the vibrational motion of the atoms around their equilibrium positions and, through the equipartition theorem, is manifest as an increase in the temperature of the conductor. Eventually, the energy is then dissipated from the conductor as heat[17]. For the rest of our consideration of current, however, we will assume a steady-state condition in which the free electrons are, on average, flowing uniformly through the conductor at a constant average speed. We call this speed the *drift speed* and will discuss it further in Section 7-6. Furthermore, we will model the instantaneous power output associated with these processes of energy dissipation[18] as the product of the current and a potential energy change.

$$P = I(\Delta V_e) \quad \textbf{(7-12)}$$

In this model, the energy supplied by the EMF is dissipated from the conductor as heat[19].

Ultimately, it is the electric field associated with the EMF that drives the current. In other words, it is an electric field that supplies the force (Equation 3-11) necessary to move the electrons. The relationship between the electric field applied to a conductor and the current density in the conductor is called Ohm's law.

---

16 You should feel inspired to take advanced courses in quantum mechanics and solid state physics to understand how conductors carry current.

17 As discussed in Section 4-6, the energy and momentum of the electrons can also be exchanged with the walls of a conductor to produce the work required to move the conductor.

18 In Chapter 8, we will discuss the specific power dissipation of different electronic elements interacting with a current.

19 Since the speed of the free electrons is not changing, the kinetic energy of the electrons is constant. Thus, the change in the electric potential energy of the free electrons as they move through the difference in electric potential is transferred into heat, not into a change in kinetic energy.

$$\vec{J} = \sigma \vec{E} \tag{7-13}$$

The variable σ in Equation 7-13 is the conductivity of the conductor and depends upon the physical properties of the conductor. Strictly speaking, Equation 7-13 is valid for only linear isotropic conductors, since for these systems σ is a scalar[20]. The units of conductivity are

$$\frac{\text{C}^2}{\text{Nm}^2\text{s}} = \frac{\text{A}}{\text{Vm}}$$

More typically, the units of conductivity are siemens per meter (S/m); 1 S is equal to 1 A/V. Conductors will have conductivities in the range of 10^6 S/m to 10^7 S/m, whereas insulators will have conductivities smaller than 10^{-16} S/m.

We can now relate the current to the electric field through the integration of Equation 7-13.

$$\int \vec{J} \bullet d\vec{A} = \int \sigma \left(\vec{E} \bullet d\vec{A} \right) \quad \rightarrow \quad I = \int \sigma \left(\vec{E} \bullet d\vec{A} \right)$$

For simplicity let's assume that the electric field is constant across the surface of integration and is directed perfectly perpendicular to the surface (and therefore parallel to the surface area vector).

$$I = \sigma \vec{E} \bullet \left(\int d\vec{A} \right) \quad \rightarrow \quad I = \sigma E A$$

Since the electric field is constant, we can relate its magnitude to an electric potential difference using Equation 3-10. Let's define the positive direction for the integration in this equation to be the same as the direction of the electric field (*i.e.*, the same as the direction of the current).

$$\Delta V_e = -EL$$

The variable L in this equation is the length over which the electric potential difference is measured (*i.e.*, it is the length over which the electric field is acting). Substitution into the equation derived from Ohm's law gives us

20 You really should take upper level physics courses to discover all the fun that happens when non-linear effects are included in physics.

$$E = \frac{-\Delta V_e}{L} \quad \rightarrow \quad I = \sigma\left(\frac{-\Delta V_e}{L}\right)A \quad \rightarrow \quad I = -\left(\frac{\sigma A}{L}\right)\Delta V_e$$

Finally, we can define a new variable, called the resistance, in terms of the variables L, A, and σ.

$$R = \frac{L}{\sigma A} \tag{7-14}$$

The units of resistance are ohms (Ω).

$$1\Omega = 1\frac{1}{\mathrm{S}} = 1\frac{\mathrm{V}}{\mathrm{A}}$$

Thus,

$$I = -\frac{\Delta V_e}{R} \tag{7-15}$$

Equation 7-15 is the macroscopic equivalent of Ohm's law. The minus sign in Equation 7-15 indicates that the direction of the current is always towards lower electric potential, consistent with the definition of current as the flow of positive charge (Section 4-4).

Example 7-8

Problem: A cylindrical wire has a radius of 2 mm, a length of 10 cm, and a conductivity of 10^7 S/m. What is the current flowing through the wire if the electric potential difference across the length of the wire is 24 V?

Solution: The resistance of the wire is determined from Equation 7-14.

$$R = \frac{0.1\,\mathrm{m}}{\left(10^7\,\frac{\mathrm{S}}{\mathrm{m}}\right)\left(\pi(0.002\,\mathrm{m})^2\right)} \quad \rightarrow \quad R = 8\times10^{-4}\,\Omega$$

The current can then be found using Ohm's law (Equation 7-15).

$$I = \frac{24\,\text{V}}{8 \times 10^{-4}\,\Omega} \quad \rightarrow \quad I = \frac{24\,\text{V}}{8 \times 10^{-4}\,\frac{\text{V}}{\text{A}}} \quad \rightarrow \quad I = 3 \times 10^{4}\,\text{A}$$

The equivalent expression for Ohm's law (Equation 7-15) for AC circuits is

$$I = \frac{\mathcal{E}}{Z} \qquad \textbf{(7-16)}$$

The variable Z in Equation 7-16 is called the *impedance*. We will discuss impedance in more detail in Chapter 9, but for now it can be thought of as the AC equivalent of resistance.

7-6 Forces and Current

Microscopically, we can determine the movement of the free electrons from their response to the forces acting on them. Each free electron will experience an electric force because of the electric field at its location. Let's consider only linear isotropic conductors for which Equation 3-11 is valid.

$$F_e = eE$$

We can model the average loss of energy and momentum from the free electron to the atoms in the conductor as an *effective* damping force acting on the free electron. For simplicity, we will assume that the magnitude of this damping force is proportional to the speed of the free electron[21].

$$F_d = \eta v_d$$

21 We can motivate this expression by arguing a faster speed would correspond to a shorter time between interactions with the atoms, and thus larger damping (*i.e.*, more frequent loss of energy and momentum). This simple equation for damping is also what we used for the damping of a simple harmonic oscillator.

In this expression F_d is the magnitude of the effective damping force, η is the associated damping constant, and v_d is the magnitude of the drift velocity of the free electron (*i.e.*, the drift speed). Under a *steady-state approximation*, where the drift velocity is constant, the magnitude of the net force acting on the free electron must be zero. Since the electric and damping forces act in opposite directions on the free electron, we have

$$F_{net} = 0 \quad \rightarrow \quad F_e = F_d \quad \rightarrow \quad eE = \eta v_d$$

We can also define the magnitude of the current density in a conductor macroscopically as the product of the electron's electric charge, the density of free electrons in the conductor (ρ), and the magnitude of the drift velocity of the free electrons.

$$J = e\rho v_d$$

Substitution of this expression into our previous equation gives us

$$J = e\rho v_d \quad \rightarrow \quad v_d = \frac{J}{e\rho} \quad \rightarrow \quad eE = \eta\left(\frac{J}{e\rho}\right)$$

$$J = \left(\frac{e^2\rho}{\eta}\right)E$$

By comparing this result to Equation 7-13, we see that the quantity in the parenthesis is the conductivity of the conductor.

$$\sigma = \frac{e^2\rho}{\eta}$$

An expression relating the electric field to the drift velocity can be found through additional substitution using Equation 7-13.

$$J = e\rho v_d \quad \rightarrow \quad \sigma E = e\rho v_d \quad \rightarrow \quad v_d = \frac{\sigma E}{e\rho}$$

The magnitude of the electric field in this expression can then be related to the corresponding electric potential difference, ΔV_e, driving the movement of the free electrons and the length, L, of the conductor through which the free electrons are moving. If we assume that the electric field is constant in magnitude over this distance, we have[22]

$$E = \frac{\Delta V_e}{L} \quad \rightarrow \quad v_d = \frac{\sigma \Delta V_e}{e \rho L}$$

For simplicity, let's assume that there is one free electron for each atom in the conductor. This corresponds to a density of approximately 10^{29} free electrons/m^3. If we then assume a conductivity of 10^7 S/m, a length of 1 m, and a potential difference of 120 V, the magnitude of the drift velocity (*i.e.*, the drift speed) is

$$v_d = \frac{\left(10^7 \frac{\text{S}}{\text{m}}\right)(120\,\text{V})}{\left(1.6 \times 10^{-19}\,\text{C}\right)\left(10^{29} \frac{1}{\text{m}^3}\right)(1\,\text{m})} \quad \rightarrow \quad v_d = 0.75 \frac{\text{mm}}{\text{s}}$$

This is a very, very small speed. Thus, it is incorrect to associate electric current with a large distance movement of free electrons in the conductor. Of course, a steady-state current does not require that a single free electron traverse the entire length of the conductor. Once the electric potential difference is applied to the conductor, all of the free electrons begin to move, including the free electrons adjacent to the ends of the conductor. This explains why the effects of current, such as a lamp turning on when the corresponding light switch is flipped, occur seemingly instantaneously.

In the presence of a magnetic field, the force acting on a free electron in a conductor is given by Equation 4-3.

$$\vec{F}_e = e\left(\vec{E} + \vec{v}_d \times \vec{B}\right)$$

Under a *steady-state approximation*, where the drift velocity is constant, the magnitude of the net force acting on the free electron must be zero.

22 We ignore the minus sign in Equation 3-10 that indicates the direction of the electric field and the direction of the change in electric potential (*i.e.*, the electric potential difference ΔV_e) since we are concerned with only the magnitudes of these quantities.

$$e\left(\vec{E}+\vec{v}_d\times\vec{B}\right)-\eta\vec{v}_d=0 \quad\rightarrow\quad \vec{v}_d=\left(\frac{e}{\eta}\right)\left(\vec{E}+\vec{v}_d\times\vec{B}\right)$$

The quantity in parenthesis in this equation is called the *electron mobility*[23], which depends upon the conductor and is denoted by μ.

$$\mu=\frac{e}{\eta} \quad\rightarrow\quad \mu=\frac{e}{\frac{e^2\rho}{\sigma}} \quad\rightarrow\quad \mu=\frac{\sigma}{e\rho}$$

The electron mobility is a measure of how fast the free electrons in the current move. Returning to our derivation, we have

$$\vec{v}_d=\mu\left(\vec{E}+\vec{v}_d\times\vec{B}\right)$$

Solving[24] for the drift velocity gives us

$$\vec{v}_d=\frac{\mu}{1+\left(\mu B\right)^2}\left(\vec{E}+\mu\vec{E}\times\vec{B}+\mu^2\left(\vec{E}\bullet\vec{B}\right)\vec{B}\right) \qquad \textbf{(7-17)}$$

Example 7-9

Problem: The electron mobility of the free electrons in a particular material is 0.2 m^2/Vs. The external electric and magnetic fields applied to the material are

$$\vec{E}=\left(5\frac{\mathrm{N}}{\mathrm{C}}\right)\hat{x} \qquad \vec{B}=\left(3\mathrm{T}\right)\hat{x}+\left(4\mathrm{T}\right)\hat{y}$$

What is the drift velocity of these free electrons?

---

23 More generally, this variable is defined to be the carrier mobility. This accounts for the situation where the electric charges in the current are not electrons.

24 See Section C-10 for the derivation of Equation 7-17.

Solution: The drift velocity is determined using Equation 7-17. To simplify our calculations, let's rewire the equation for the magnetic field[25] as

$$1\,\mathrm{T} = 1\frac{\mathrm{Vs}}{\mathrm{m}^2} \quad \rightarrow \quad \vec{B} = \left(3\frac{\mathrm{Vs}}{\mathrm{m}^2}\right)\hat{x} + \left(4\frac{\mathrm{Vs}}{\mathrm{m}^2}\right)\hat{y}$$

Let's examine each term in Equation 7-17 separately,

$$\frac{\mu}{1+(\mu B)^2} = \frac{0.2\frac{\mathrm{m}^2}{\mathrm{Vs}}}{1+\left(\left(0.2\frac{\mathrm{m}^2}{\mathrm{Vs}}\right)\left(5\frac{\mathrm{Vs}}{\mathrm{m}^2}\right)\right)^2} \quad \rightarrow \quad \frac{\mu}{1+(\mu B)^2} = 0.1\frac{\mathrm{m}^2}{\mathrm{Vs}}$$

$$\mu\vec{E}\times\vec{B} = \left(0.2\frac{\mathrm{m}^2}{\mathrm{Vs}}\right)\left(\left(5\frac{\mathrm{N}}{\mathrm{C}}\right)\hat{x}\right)\times\left(\left(3\frac{\mathrm{Vs}}{\mathrm{m}^2}\right)\hat{x} + \left(4\frac{\mathrm{Vs}}{\mathrm{m}^2}\right)\hat{y}\right) \quad \rightarrow \quad \mu\vec{E}\times\vec{B} = \left(4\frac{\mathrm{N}}{\mathrm{C}}\right)\hat{z}$$

$$\mu^2\left(\vec{E}\bullet\vec{B}\right)\vec{B} = \left(0.2\frac{\mathrm{m}^2}{\mathrm{Vs}}\right)^2\left(\left(\left(5\frac{\mathrm{N}}{\mathrm{C}}\right)\hat{x}\right)\bullet\left(\left(3\frac{\mathrm{Vs}}{\mathrm{m}^2}\right)\hat{x} + \left(4\frac{\mathrm{Vs}}{\mathrm{m}^2}\right)\hat{y}\right)\right)\left(\left(3\frac{\mathrm{Vs}}{\mathrm{m}^2}\right)\hat{x} + \left(4\frac{\mathrm{Vs}}{\mathrm{m}^2}\right)\hat{y}\right)$$

$$\mu^2\left(\vec{E}\bullet\vec{B}\right)\vec{B} = \left(9\frac{\mathrm{N}}{\mathrm{C}}\right)\hat{x} + \left(12\frac{\mathrm{N}}{\mathrm{C}}\right)\hat{y}$$

Putting everything together gives us

$$\vec{v}_d = \left(0.1\frac{\mathrm{m}^2}{\mathrm{Vs}}\right)\left(\left(5\frac{\mathrm{N}}{\mathrm{C}}\right)\hat{x} + \left(4\frac{\mathrm{N}}{\mathrm{C}}\right)\hat{z} + \left(9\frac{\mathrm{N}}{\mathrm{C}}\right)\hat{x} + \left(12\frac{\mathrm{N}}{\mathrm{C}}\right)\hat{y}\right)$$

$$\vec{v}_d = \left(0.1\frac{\mathrm{m}^2}{\mathrm{Vs}}\right)\left(\left(14\frac{\mathrm{N}}{\mathrm{C}}\right)\hat{x} + \left(12\frac{\mathrm{N}}{\mathrm{C}}\right)\hat{y} + \left(4\frac{\mathrm{N}}{\mathrm{C}}\right)\hat{z}\right)$$

---

25 See Section 4-2.

$$1\,\text{V} = 1\frac{\text{Nm}}{\text{C}} \quad \rightarrow \quad \vec{v}_d = -\left(0.1\frac{\text{Cm}}{\text{Ns}}\right)\left(\left(14\frac{\text{N}}{\text{C}}\right)\hat{x} + \left(12\frac{\text{N}}{\text{C}}\right)\hat{y} + \left(4\frac{\text{N}}{\text{C}}\right)\hat{z}\right)$$

$$\vec{v}_d = -\left(1.4\frac{\text{m}}{\text{s}}\right)\hat{x} + \left(1.2\frac{\text{m}}{\text{s}}\right)\hat{y} + \left(0.4\frac{\text{m}}{\text{s}}\right)\hat{z}$$

Finally, let's consider the case in which current is flowing through a conductor that is placed in the presence of a magnetic field. As shown in Figure 7.12, the magnetic field will exert a magnetic force on the electric charge in the current, which will lead to a separation of electric charge within the conductor; this is similar to the creation of the motional EMF (Section 6-6). This separation of charge creates an induced electric field (Figure 7.12) that exerts an electric force on the electric charges in the current.

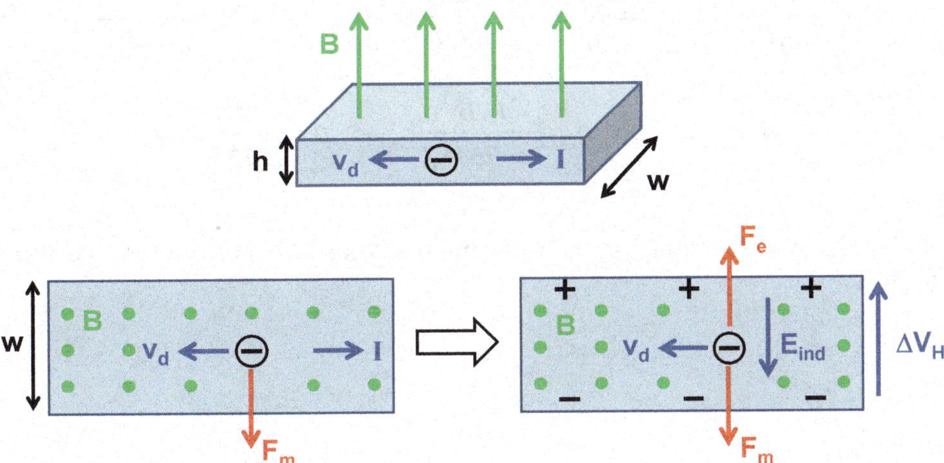

FIGURE 7.12: The deflection of moving charges when a current passes through a magnetic field creates a potential difference called the Hall voltage.

Eventually, a steady-state condition is obtained in which the magnitude of the magnetic force equals the magnitude of the electric force and the electric charges move through the conductor without deflection.

$$F_e = F_m \quad \rightarrow \quad eE_{ind} = ev_d B \quad \rightarrow \quad E_{ind} = v_d B$$

If we assume that the induced electric field is constant throughout the conductor, we can relate the magnitude of induced electric field to the magnitude of the induced change in electric potential (ΔV_H in Figure 7.12) and the width of the conductor (w in Figure 7.12) using Equation 3-10[26].

$$E_{ind} = \frac{\Delta V_H}{w} \rightarrow \frac{\Delta V_H}{w} = v_d B \rightarrow v_d = \frac{\Delta V_H}{wB}$$

The induced electric potential, ΔV_H, is referred to as the Hall voltage, named after the graduate student who first observed it. The process of creating the induced electric potential is also known as the Hall effect. Since measurement of B, ΔV_H, and w are all straightforward, the Hall effect provides a convenient experimental approach for determining the drift speed of charges in a current. Similarly, the Hall effect also allows for an experimental approach to determine the density of free electrons in a conductor.

$$v_d = \frac{J}{e\rho} \rightarrow v_d = \frac{\frac{I}{A}}{e\rho} \rightarrow \frac{\Delta V_H}{wB} = \frac{\frac{I}{A}}{e\rho} \rightarrow \rho = \frac{IwB}{eA\Delta V_H}$$

$$A = wh \rightarrow \rho = \frac{IwB}{ewh\Delta V_H} \rightarrow \rho = \frac{IB}{eh\Delta V_H}$$

The variable h in this equation is the thickness of the conductor, as shown in Figure 7.12.

Example 7-10

Problem: A 2 mm thick conductor carries a current of 2 A. When placed in the presence of a 0.5 T magnetic field, the measured Hall voltage is 2.5 mV. What is the density of free electrons in the conductor?

26 We ignore the minus sign in Equation 3-10 that indicates the direction of the electric field and the direction of the change in electric potential, as shown in Figure 7.12, since we are concerned with only the magnitudes of these quantities.

Solution: Substitution into the previously derived equation gives us

$$\rho = \frac{(2\text{A})(0.5\text{T})}{(1.6 \times 10^{-19}\text{C})(2 \times 10^{-3}\text{m})(2.5 \times 10^{-3}\text{V})}$$

$$\rho = 1.25 \times 10^{24} \frac{1}{\text{m}^3}$$

7-7 Batteries and Generators

The source of the EMF and free electrons required for maintaining direct current in a conductor is often a *battery*. A battery converts chemical potential energy into electric potential energy through chemical (typically redox) reactions. A common type of battery is an electrochemical cell, an example of which is shown in Figure 7.13.

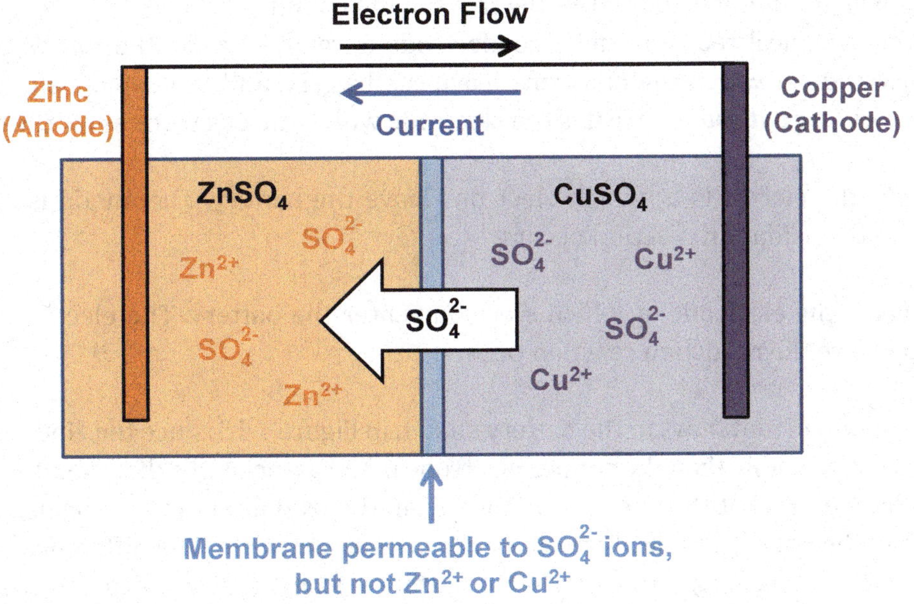

FIGURE 7.13: An electrochemical cell functioning as a battery.

The copper-zinc battery shown in Figure 7.13 consists of two different compartments separated by a semi-permeable membrane and connected by a wire; one compartment contains a zinc ***electrode*** immersed in $ZnSO_4$, and the other compartment contains a copper ***electrode*** immersed in $CuSO_4$.

> *Electrode*: an electrical conductor used to make contact with a nonmetallic part of a battery or a circuit[27].

Both the zinc and copper electrodes will dissolve in their respective solutions, generating zinc, copper, and sulfate ions. The zinc electrode dissolves more readily than the copper electrode, however, which results in more free electrons in the zinc electrode than the copper electrode. This imbalance in the population of free electrons in the two electrodes results in a net flow of free electrons from the zinc electrode to the copper electrode through the connecting wire. Once in the copper electrode, these free electrons will interact with the copper ions in solution to form copper metal on copper electrode (*i.e.*, to solidify the copper from the solution). To maintain the new equilibrium of copper and zinc ions in solution, sulfate ions flow from one compartment to another through a semi-permeable membrane; this membrane is permeable to sulfate ions, but not metal ions. These sulfate ions will then bind with the zinc ions in solution, which decreases the concentration of zinc ions in solution. This, in turn, causes more zinc to dissolve in the solution, which generates more free electrons in the circuit.

For this chemical reaction, the zinc electrode is referred to as the ***anode***, since it is the electrode at which the electrons leave the battery, and the copper electrode is referred to as the ***cathode***, since it is the electrode where the electrons enter the battery.

> *Anode*: the electrode at which electrons leave the battery. The anode is also where the oxidation reaction occurs.
>
> *Cathode*: the electrode at which electrons enter the battery. The electrode is also where the reduction reaction occurs.

As mentioned, current flows in the battery shown in Figure 7.13, since the zinc electrode dissolves more readily than the copper electrode in the associated sulfate solution. More technically, we say that the energy obtained from the oxidation of the zinc electrode is larger than the energy required to reduce the copper electrode. The difference in these energies is the total energy produced by the battery, which is typically expressed in terms of an electric potential difference between the anode and the cathode (Equation 3-5).

---

27 We'll define and discuss circuits in Chapter 8.

Of course, batteries don't maintain a voltage difference between their anodes and cathodes forever (*i.e.*, they can't convert chemical potential energy into electric potential energy forever). Eventually, a uniform chemical potential is established throughout the battery, and there is no more net movement of electrons from the anode to the cathode.

The source of the electromotive force for alternating current is an AC generator (often simply referred to as a generator). A simple AC generator is shown schematically in Figure 7.14.

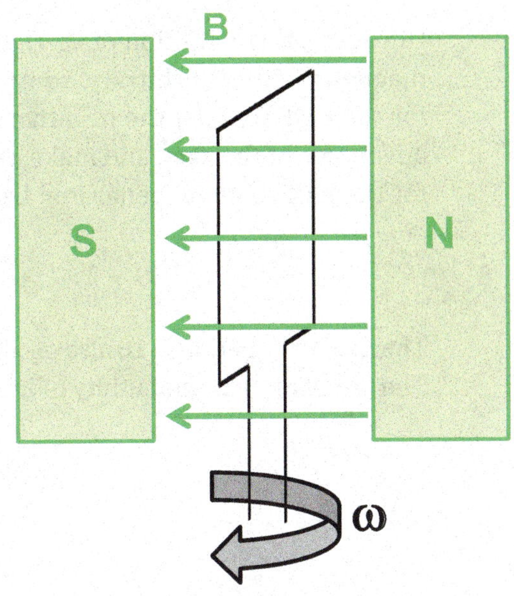

FIGURE 7.14: A simple AC generator.

As the loop of wire rotates, the magnetic flux through the area enclosed by the loop changes, and an electric field is induced in the loop (Equation 6-3). If we assume that the magnetic field is uniform across the enclosed area of the loop, then the magnetic flux through the loop can be determined using Equation 5-11.

$$\Phi_B = \int_{surface} \vec{B} \bullet d\vec{A} \quad \rightarrow \quad \Phi_B = BA\cos\theta$$

In this equation, the angle θ between the vectors $\vec{B}$ and $d\vec{A}$ is related to the angular velocity of the loop's rotation. For simplicity, let's assume that the loop is rotating at a constant angular velocity.

$$\theta = \omega t + \phi_0$$

The variable ϕ_0 denotes the angle at $t = 0$; this is the equivalent of the phase constant for harmonic motion. The rate of change of the magnetic flux is thus

$$\frac{d\Phi_B}{dt} = \frac{d}{dt}\left(BA\cos\left(\omega t + \phi_0\right)\right) \quad \rightarrow \quad \frac{d\Phi_B}{dt} = BA\omega\sin\left(\omega t + \phi_0\right)$$

The induced EMF associated with this change in magnetic flux is found using Equation 6-11.

$$\mathcal{E} = -BA\omega\sin\left(\omega t + \phi_0\right)$$

Of course, the definition of ϕ_0 (*i.e.*, the $t = 0$ orientation of the loop with respect to the magnetic field) is arbitrary, so we could have equally accurately written this equation for the EMF without the negative sign, or with a cosine rather than a sine. We can take advantage of this fact, and make our lives simpler later on, if we use a different notation for the EMF of an AC generator. Let's write this EMF as

$$\mathcal{E} = \mathcal{E}_0 e^{i\omega t} \tag{7-18}$$

The EMF in Equation 7-18 also oscillates at angular frequency ω, but is a complex number[28]. The motivation for and utility of Equation 7-18 will become clear in Chapter 9.

28 See Appendix D.

Summary

- **Conductor:** a material containing free electrons that are able to move throughout the material.

- **Insulator:** a material containing no free electrons. All of the electrons in the material are bound to individual atoms and unable to move throughout the material.

- When a conductor is in electrostatic equilibrium:

    - The electric potential is constant everywhere inside the conductor.
    - The electric field is zero everywhere inside the conductor.
    - The free electrons reside entirely on the surface of the conductor.
    - The electric potential is constant at all positions on the surface of the conductor.
    - The electric field just outside the surface of the conductor is perpendicular to that surface.

- **Charge polarization:** a small separation of the positive and negative charges in a neutral object.

- **Polar molecule:** a molecule that has a permanent non-zero electric dipole moment.

- **Dielectric:** an electrical insulator that can be polarized by an applied electric field.

- **Dielectric breakdown:** the rapid conversion of an insulating dielectric to a conductor in the presence of an electric field.

- Gauss's law for electric fields in linear isotropic dielectrics is

$$\oint_{surface} \vec{E} \bullet d\vec{A} = \frac{q}{\kappa_e \varepsilon_0}$$

  In this expression, κ_e is the dielectric constant.

- The magnitude of the induced electric field in a linear isotropic dielectric is related to the magnitude of the external electric field using the dielectric constant.

$$E_{ind} = \left(1 - \frac{1}{\kappa_e}\right) E_{ext}$$

It follows that the magnitude of the electric field inside the dielectric is therefore less than the magnitude of the external electric field applied to the dielectric.

$$E = \frac{E_{ext}}{\kappa_e}$$

- When a dielectric completely fills the space between the plates of the capacitor, the capacitance of the capacitor is

$$C = \kappa_e \left(\frac{A\varepsilon_0}{d}\right) \quad \rightarrow \quad C = \kappa_e C_0$$

The variable C_0 in this equation denotes the capacitance of the capacitor in the absence of the dielectric.

- **Ampèrian currents:** permanently circulating currents in neutral matter.
- Ampère's law for magnetic fields in linear isotropic magnetic materials is

$$\oint_{line} \vec{B} \cdot d\vec{s} = \kappa_m \mu_0 I_{free}$$

In this expression, κ_m is the relative permeability.

- **Ohm's law:** the current density is the product of the conductivity and the electric field.

$$\vec{J} = \sigma \vec{E}$$

The current is the integral of the current density over a surface.

$$I = \int \vec{J} \cdot d\vec{A}$$

- The drift velocity of an electron in an electric and magnetic field is

$$\vec{v}_d = \frac{\mu}{1+(\mu B)^2}\left(\vec{E} + \mu \vec{E} \times \vec{B} + \mu^2 \left(\vec{E} \bullet \vec{B}\right)\vec{B}\right)$$

The variable μ is the electron mobility and is a measure of how fast free electrons move as part of a current. The electron mobility is a function of the conductivity, the density of free electrons, and the electric charge of the electron.

$$\mu = \frac{\sigma}{e\rho}$$

- The power supplied by an EMF to a current is the product of the EMF and the current.

$$P = I\mathcal{E}$$

- **Ohm's law:** the current flowing through a conductor is proportional to the potential difference driving the current and the resistance in the conductor.

$$I = -\frac{\Delta V_e}{R}$$

For AC currents, the current is proportional to the potential difference and the impedance of the conductor.

$$I = \frac{\Delta V_e}{Z}$$

- An alternating EMF can be written in terms of a complex exponential.

$$\mathcal{E} = \mathcal{E}_0 e^{i\omega t}$$

In this equation, ω is the angular frequency of the oscillation of the current.

Problems

1. A solid insulating sphere carries a net positive electric charge Q distributed uniformly throughout its volume. A conducting spherical shell of inner radius r_a and outer radius r_b is centered on the solid sphere and carries a net negative electric charge $-2Q$. What is the electric field inside the spherical shell (at some distance from the center of the shell between r_a and r_b)? What is the electric field outside of the shell (at some distance greater than r_b)? Assume this system is in electrostatic equilibrium.

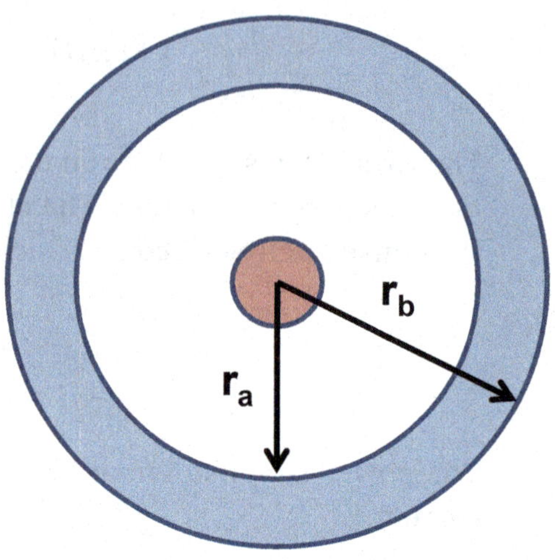

2. A parallel plate capacitor has cross-sectional area A and separation $2d$. One of the plates has a net electric charge of $+Q$ and the other plate has a net electric charge of $-Q$. Two dielectrics, with dielectric constants k_1 and k_2, are placed in series in-between the plates of the capacitor, as shown in the figure. What is the capacitance of this capacitor? What is the potential energy of this capacitor?

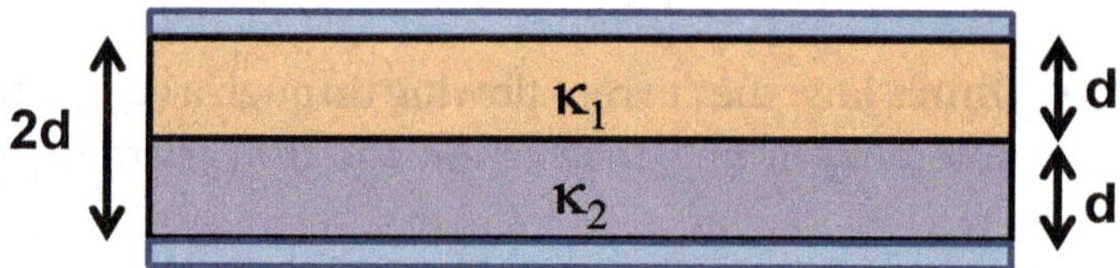

3. Two point particles, A and B, with net charges +2 nC and −4 nC, are separated by a distance of 10 cm in a polar fluid with a dielectric constant of 40. What is the magnitude of the electric force exerted by particle A on particle B?

4. A cylindrical wire with a radius of 2 mm carries a current with non-uniform current density. The current density increases with increasing radial distance from the center of the wire according to the following equation:

$$J(r) = \left(3 \times 10^6 \frac{\text{A}}{\text{m}^3}\right) r$$

What is the total current in the wire? What is the magnitude of the magnetic field from the current at a radial distance of 4 mm from the center of wire? Assume that the relative permeability is 1 for this system.

5. The electron mobility of the free electrons in a particular conductor is 0.2 m^2/Vs. The external electric and magnetic fields applied to the conductor are

$$\vec{E} = \left(5 \frac{\text{N}}{\text{C}}\right) \hat{x} \qquad \vec{B} = \left(-5 \frac{\text{Vs}}{\text{m}^2}\right) \hat{x}$$

What is the drift velocity of these free electrons?

6. A cylindrical wire has a radius of 4 mm, a length of 20 cm, and a conductivity of 10^7 S/m. What is the current flowing through the wire if the electric potential difference across the length of the wire is 120 V?

7. A 0.25 mm thick and 5 mm wide conductor carries a current of 1.8 A. When placed in the presence of a 0.75 T magnetic field, the measured Hall voltage is 3.6 μV. What is the drift velocity of the electrons in the conductor?

Direct Current Circuits

8-1 Introduction

In the preceding chapters of this book, we calculated several electric and magnetic fields, determined how these fields interact with each other, and showed how these interactions are affected by the properties of the material in which they exist. In the remainder of this book, we turn our attention to several applications of these concepts and equations. The first application will be the design and function of electric circuits. In this chapter, we will discuss direct current circuits, and in Chapter 9 we will discuss alternating current circuits.

8-2 Components of Circuits

An ***electric circuit*** is a path through which charged particles can flow.

Electric circuit: a path through which charged particles can flow.

For DC circuits, a battery or equivalent source of energy provides the EMF necessary to drive the current and provides the source (and sink) of the free electrons constituting the current. For AC circuits, an AC generator provides the EMF necessary to drive the current.

The part of the circuit to which the battery or AC generator is connected is referred to as the electric circuit's ***load***.

Electrical load: the portion of an electric circuit that dissipates or stores electrical energy.

We have already introduced three different circuit components that can constitute an electric circuit's load: resistors, capacitors, and inductors. The symbols used for representing these circuit components in circuit diagrams are shown in Table 8.1.

TABLE 8.1: Components of circuits

Name	Symbol
Battery[1]	
AC Generator	
Resistor	
Capacitor	
Inductor	
Light Bulb	
Switch	

A light bulb is included as a separate circuit component in Table 8.1, but will be treated as a resistor which dissipates energy as both light and heat.

1 The longer bar in the symbol for the battery denotes the location of the higher electric potential. In other words, the current gains electric potential energy as it moves from the smaller bar to the longer bar.

A circuit is said to be ***closed*** if it contains a continuous path through which the electrons can flow between the high and low potential terminals of the EMF source. Circuits that are not closed are referred to as ***open*** circuits.

> *Closed electric circuit*: an electric circuit that contains a continuous path through which current flows between the high and low potential terminals of the EMF source.

> *Open electric circuit*: an electric circuit that is not closed.

Although ***transient*** direct currents can flow in either open or closed circuits, ***steady-state*** direct currents can flow only in closed circuits.

> *Transient current*: a current whose magnitude and/or direction varies non-periodically with time. Transient currents typically have an exponential dependence upon time.

> *Steady-state current*: a current whose direction and magnitude are constant or vary periodically.

Alternating currents can flow in open or closed circuits, and can include both transient (*i.e.*, exponentially dependent) and steady-state (*i.e.*, periodically varying) components.

8-3 Kirchhoff's Laws

Consider the simple DC electric circuit shown in Figure 8.1.

This circuit is a closed circuit since there is a continuous path connecting the high and low potential terminals of the battery (*i.e.*, the EMF source). Furthermore, since the source of EMF is a battery, the current flowing in the circuit will be direct current.

Let's evaluate what happens to the electric potential energy of the current flowing in the

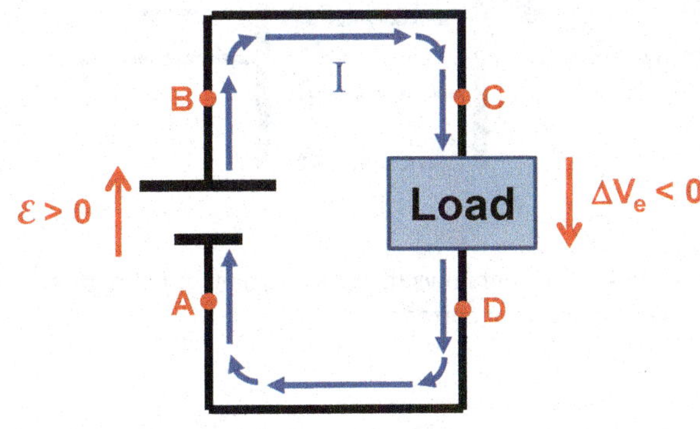

FIGURE 8.1: A simple DC electric circuit.

circuit[2]. Starting at point A, the current gains electric potential energy from the EMF of the battery as it flows to point B; since the current is the flow of particles with positive electric charge, moving from the lower electric potential to the higher electric potential through this path would result in an increase in the electric potential energy of the current (Equation 3-5). As the current moves from point B to point C, some electric potential energy will be loss (dissipated) due to the resistance in the wire (Section 7-5). More electric potential energy is lost as the current moves from point C to point D, since energy is used to power the load (*e.g.*, power a light bulb). Finally, additional electric potential energy is lost as the current flows from point D to point A, due to the resistance in the wire (Section 7-5).

However, since electric potential energy is a state function, the electric potential energy of the current at point A must be the same for all charged particles in the current, whether they have traversed the entire electric circuit or always remained at point A. Thus, the net change in electric potential energy for any closed loop in an electric circuit must be zero. This is the equivalent of stating that the net change in electric potential for any closed loop in an electric circuit must be zero (Equation 3-5). This application of energy conservation to an electric circuit is one of Kirchhoff's rules[3].

Kirchhoff's loop rule: The sum of the electric potential differences across all components around any closed loop in an electric circuit must be zero.

The definition of the change in electric potential across a component in an electric circuit depends upon the direction of the current across/through the component and the direction of the "loop" or path around which Kirchhoff's loop rule is being applied. A summary of the changes in electric potential for the components we will encounter is shown in Table 8.2

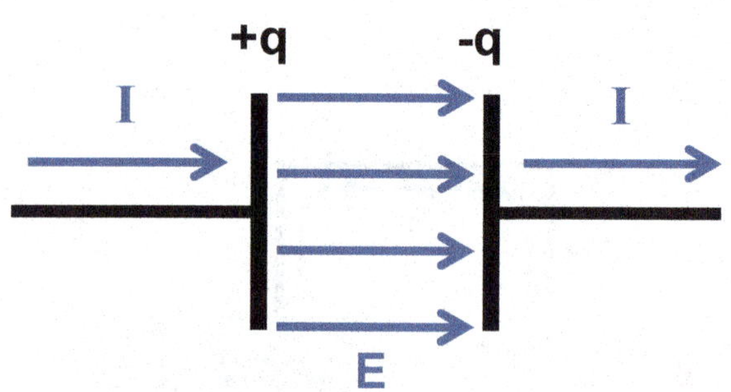

FIGURE 8.2: Current flows through a capacitor in the form of the displacement current.

We will also make the simplifying assumption that there is no change in electric potential across any of the wires in our electric circuits. In other words, all wires have effectively zero resistance.

It is important to note that current does not actually flow through a capacitor; there is no conductor between the two plates of

2 We recognize that the current actually consists of negatively electrically charged electrons moving in the opposite direction, but for simplicity we will discuss the changes in electric potential energy as though positively electrically charged particles constituted the current.

3 Named for Gustav Kirchhoff.

a capacitor. Rather, it is more accurate to say that current flows into one plate of the capacitor and out of the other plate of the capacitor, as shown in Figure 8.2.

TABLE 8.2: Changes in electric potential for Kirchhoff's loop rule

Circuit Component and Current Direction	Direction of Path	
	A to B	B to A
A —||— B (battery)	$\Delta V_e = \mathcal{E}$	$\Delta V_e = -\mathcal{E}$
I →, A —(~)— B	$\Delta V_e = \mathcal{E}_0 e^{i\omega t}$	$\Delta V_e = -\mathcal{E}_0 e^{i\omega t}$
I →, A —||— B (capacitor)	$\Delta V_e = -\frac{q}{C}$	$\Delta V_e = \frac{q}{C}$
I →, A —/\/\/— B (resistor)	$\Delta V_e = -RI$	$\Delta V_e = RI$
I →, A —000— B (inductor)	$\Delta V_e = -L\frac{dI}{dt}$	$\Delta V_e = L\frac{dI}{dt}$
I →, A —(⊗)— B (lamp)	$\Delta V_e = -RI$	$\Delta V_e = RI$

In this way, we can consider the current to *deposit* a net positive electric charge on one plate of the capacitor and *leave behind* a net negative electric charge on the other plate of the capacitor. In-between the two plates, the displacement current flows (Section 6-7) and, as we have seen, the magnetic field associated with the displacement current is identical to the magnetic field of the current; thus the displacement current can be roughly considered to be the *continuation* of the current in the electric circuit in-between the plates of the capacitor. Therefore, for simplicity, we will assume that current flows through capacitors when applying Kirchhoff's rules.

Example 8-1

Problem: Use Kirchhoff's loop rule to determine the magnitude of the current in the electric circuit shown in Figure 8.3.

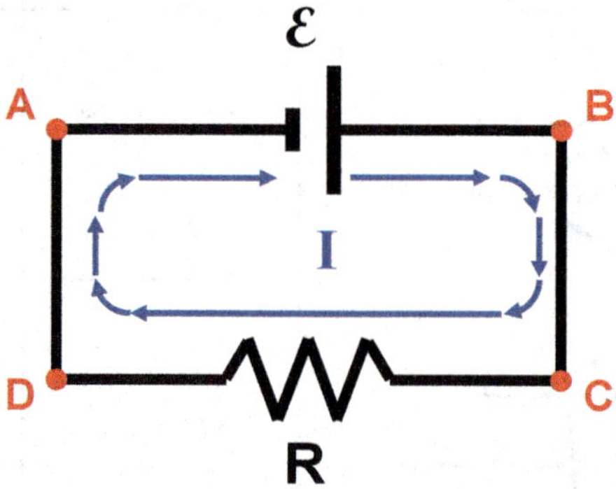

FIGURE 8.3: The electric circuit in Example 8-1.

Solution: The magnitude of the current can be determined using Kirchhoff's loop rule. Let's consider a loop that starts at point A, and then proceeds through points B, C, and D, before returning back to point A; for simplicity, we will refer to this loop as ABCDA. Applying Kirchhoff's loop rule to this loop gives us

$$\mathcal{E} - RI = 0 \quad \rightarrow \quad \mathcal{E} = RI \quad \rightarrow \quad I = \frac{\mathcal{E}}{R}$$

Next, let's consider the loop ADCBA (*i.e.*, the same loop as before, but traversed in the opposite direction). Applying Kirchhoff's loop rule to this loop gives us

$$RI - \mathcal{E} = 0 \quad \rightarrow \quad \mathcal{E} = RI \quad \rightarrow \quad I = \frac{\mathcal{E}}{R}$$

The result is the same regardless of the path chosen for the loop.

What would be the solution to Example 8-1 if a different direction for the current had been specified? Consider the definition of the current shown in Figure 8.4. Applying Kirchhoff's loop rule to the loop ABCDA now gives us

$$\mathcal{E} + RI = 0 \quad \rightarrow \quad \mathcal{E} = -RI \quad \rightarrow \quad I = -\frac{\mathcal{E}}{R}$$

Similarly, applying Kirchhoff's loop rule to the loop ADCBA gives us

$$-\mathcal{E} - RI = 0 \quad \rightarrow \quad \mathcal{E} = -RI \quad \rightarrow \quad I = -\frac{\mathcal{E}}{R}$$

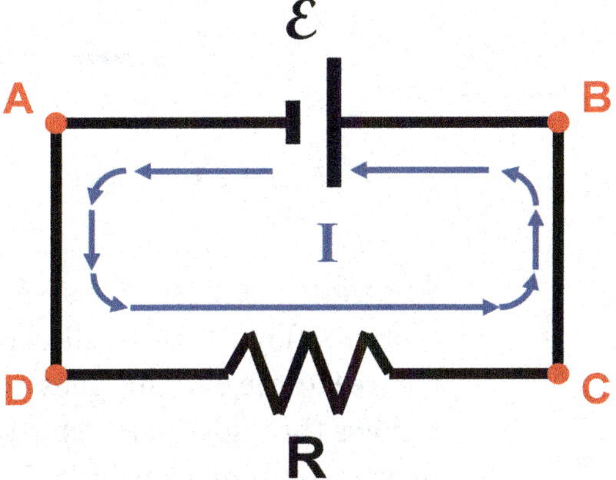

FIGURE 8.4: The electric circuit in Example 8-1 with a different direction of current.

The negative sign in our solution[4] indicates that the true direction of the current is opposite the direction specified in Figure 8.4; the true direction of the current is as indicated in Figure 8.3.

The second of Kirchhoff's rules is derived from the conservation of electric charge (Section 3-2). Since electric charge can neither be created nor destroyed, the total electric charge flowing as current into any given point in an electric circuit must be equal to the total electric charge flowing as current out of that point. Consider the situation shown in Figure 8.5, in which the current entering a *junction* (or *node*) in a circuit is divided into two separate currents.

This conservation of electric current requires that $I_1 = I_2 + I_3$. This application of the conservation of electric charge/current is referred to as ***Kirchhoff's junction rule.***

> **Kirchhoff's junction rule**: The sum of the currents entering any junction must be equal to the sum of the currents leaving that junction.

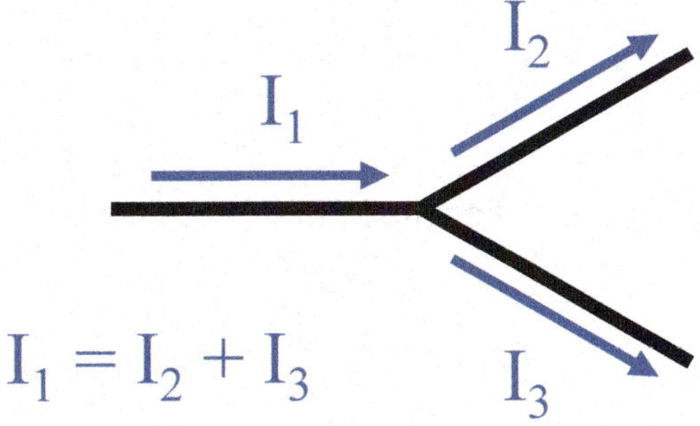

FIGURE 8.5: The conservation of electric charge requires that the total current entering a junction equals the total current leaving that junction.

4 Negative signs have a similar meaning in solutions obtained with Gauss' Law (Section 5-4 and Section 5-5) and Ampère's Law (Section 5-6).

Example 8-2

Problem: Use Kirchhoff's rules to determine the magnitude of the currents in the electric circuit shown in Figure 8.6.

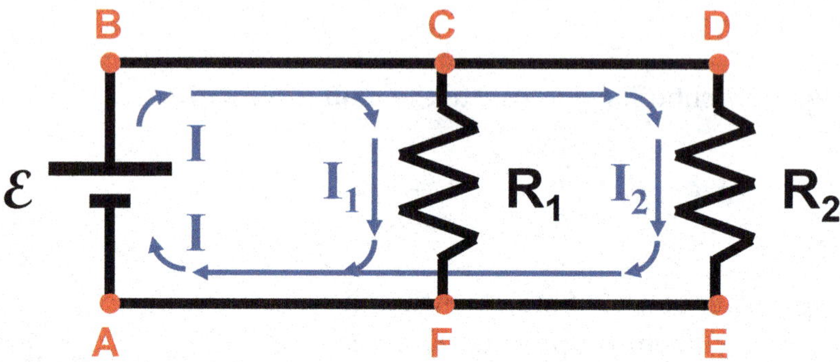

FIGURE 8.6: The electric circuit in Example 8-2.

Solution: The best first step in solving electric circuit problems is to label the currents flowing through each component of the electric circuit. As shown in Figure 8.6, the EMF supplies a current I that divides into I_1 and I_2 at point C. Current I_1 flows through the resistor with resistance R_1 and current I_2 flows through the resistor with resistance R_2. These two currents recombine at point F to reform current I, which flows back to the battery. The next step in the solution is to apply Kirchhoff's loop rule to the electric circuit. For the loop ABCDEFA, we have

$$\mathcal{E} - R_2 I_2 = 0 \rightarrow \mathcal{E} = R_2 I_2 \rightarrow I_2 = \frac{\mathcal{E}}{R_2}$$

For the loop ABCFA, we have

$$\mathcal{E} - R_1 I_1 = 0 \rightarrow \mathcal{E} = R_1 I_1 \rightarrow I_1 = \frac{\mathcal{E}}{R_1}$$

Finally, for the loop CDEFC, we have

$$-R_2 I_2 + R_1 I_1 = 0 \rightarrow R_2 I_2 = R_1 I_1$$

This final equation indicates that the change in electric potential across each resistor is the same. We will discuss this result further in Section 8-5.

Applying Kirchhoff's junction rule to point C gives us

$$I = I_1 + I_2 \quad \rightarrow \quad I = \frac{\mathcal{E}}{R_1} + \frac{\mathcal{E}}{R_2} \quad \rightarrow \quad I = \mathcal{E}\left(\frac{1}{R_1} + \frac{1}{R_2}\right)$$

All of the currents are positive, and thus the directions of the currents shown in Figure 8.6 are correct.

Example 8-3

Problem: What is the magnitude of the current flowing through the 4 Ω resistor in the electric circuit shown in Figure 8.7 when the switch is open? What is the current flowing through the 4 Ω resistor when the switch is closed?

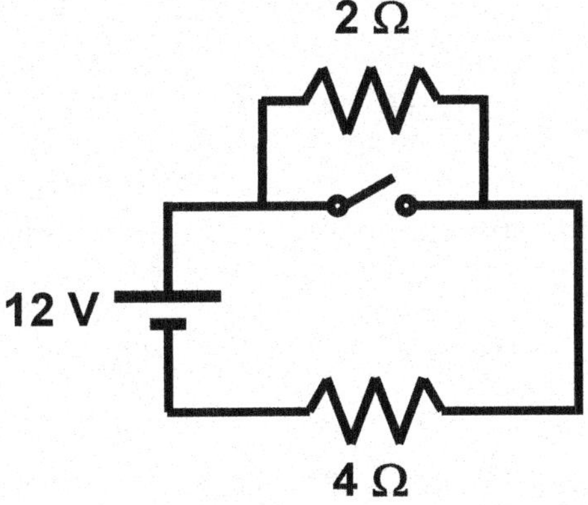

FIGURE 8.7: The electric circuit in Example 8-3.

Solution: As shown in Figure 8.8, the EMF supplies a current I that divides into I_1 and I_2 at point B. Current I_1 flows through the switch and current I_2 flows through the 2 Ω resistor. These two currents recombine at point E to reform current I, which flows through the 4 Ω resistor and then back to the 12 V battery.

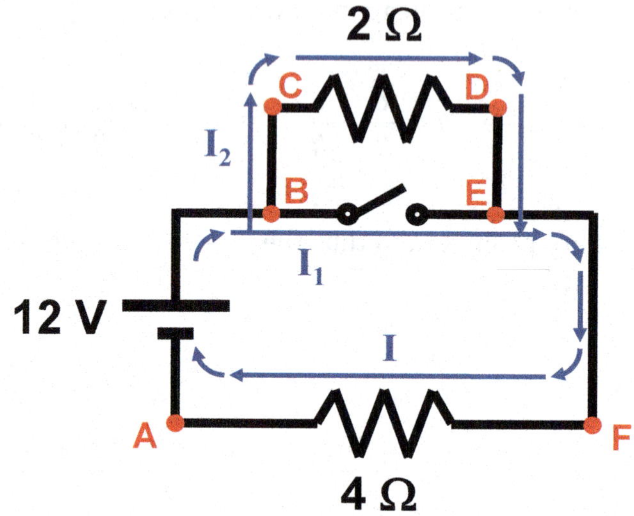

FIGURE 8.8: Designation of the currents flowing in the electric circuit in Example 8-3.

When the switch is open, the current I_1 is equal to zero and from Kirchhoff's junction rule we have

$$I_1 = 0 \rightarrow I_2 = I$$

Applying Kirchhoff's loop rule to the loop ABCDEFA in Figure 8.8 gives us

$$12\,\text{V} - (2\,\Omega)I - (4\,\Omega)I = 0 \rightarrow 12\,\text{V} = (6\,\Omega)I \rightarrow I = 2\,\text{A}$$

When the switch is closed, the current I_1 is not zero. Applying Kirchhoff's loop rule to the loop BCDEB in Figure 8.8 gives us[5]

$$-(2\,\Omega)I_2 + (0\,\Omega)I_1 = 0 \rightarrow (2\,\Omega)I_2 = 0 \rightarrow I_2 = 0$$

5 Recall that we will assume that wires have no resistance.

When the switch is closed, no current flows through the 2 Ω resistor. We can then apply Kirchhoff's loop rule to the loop ABEFA in Figure 8.8 to determine the current flowing through the 4 Ω resistor when the switch is closed.

$$12\,\text{V} - (4\,\Omega)I = 0 \quad \rightarrow \quad 12\,\text{V} = (4\,\Omega)I \quad \rightarrow \quad I = 3\,\text{A}$$

Finally, it is worth noting that all real batteries have an internal resistance, since they are made of material which will have some intrinsic resistance (Figure 8.9).

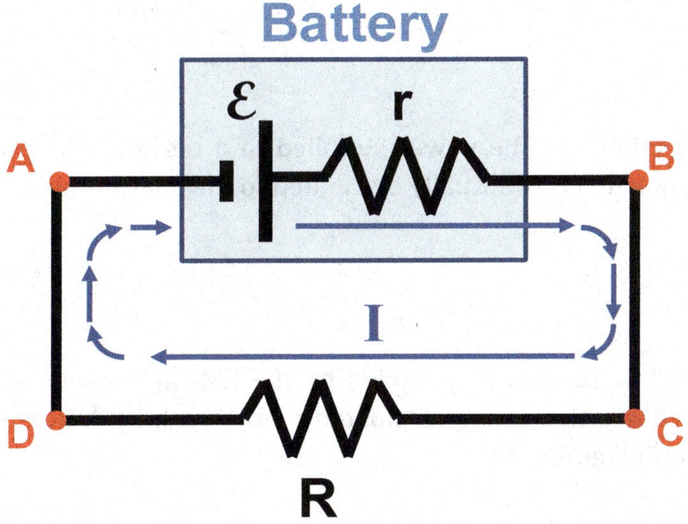

FIGURE 8.9: A real battery contains internal resistance.

Applying Kirchhoff's loop rule to the loop ABCDA of the electric circuit in Figure 8.9 gives us

$$\mathcal{E} - rI - RI = 0 \quad \rightarrow \quad \mathcal{E} = rI + RI \quad \rightarrow \quad \mathcal{E} = (r + R)I \quad \rightarrow \quad I = \frac{\mathcal{E}}{r + R}$$

8-4 Power in Electric Circuits

In Section 7-5, we stated that the power supplied to any component in an electric circuit was equal to the product of the change in electric potential across the component and the electric current flowing through the component (Equation 7-12)

$$P = I(\Delta V_e)$$

Thus, for a resistor

$$P_R = I(RI) \rightarrow P_R = I^2 R = \frac{(\Delta V_e)^2}{R} \tag{8-1}$$

As discussed in Section 7-5, the power supplied to a resistor (or, more generally, to resistance in a conductor) is eventually dissipated to the environment.

Example 8-4

Problem: What is the power supplied by the EMF in the electric circuit in Figure 8.6? What is the power dissipated by the each of the resistors in the electric circuit in Figure 8.6?

Solution: The power supplied by the EMF is the product of the EMF and the current provided by the EMF (Equation 7-11).

$$P_{EMF} = I\mathcal{E} \rightarrow P_{EMF} = \left(\mathcal{E}\left(\frac{1}{R_1} + \frac{1}{R_2}\right)\right)\mathcal{E} \rightarrow P_{EMF} = \mathcal{E}^2\left(\frac{1}{R_1} + \frac{1}{R_2}\right)$$

$$P_{EMF} = \mathcal{E}^2\left(\frac{R_1 + R_2}{R_1 R_2}\right)$$

The power dissipated by the resistors is determined using Equation 8-1. For the resistor with resistance R_1, we have

$$P_{R_1} = I_1^2 R_1 \rightarrow P_{R_1} = \left(\frac{\mathcal{E}}{R_1}\right)^2 R_1 \rightarrow P_{R_1} = \frac{\mathcal{E}^2}{R_1}$$

For the resistor with resistance R_2, we have

$$P_{R_2} = I_2^2 R_2 \rightarrow P_{R_2} = \left(\frac{\mathcal{E}}{R_2}\right)^2 R_2 \rightarrow P_{R_2} = \frac{\mathcal{E}^2}{R_2}$$

As expected, the power supplied to the electric circuit by the EMF equals the power dissipated by the circuit by the resistors.

$$P_{R_1} + P_{R_2} = \frac{\mathcal{E}^2}{R_1} + \frac{\mathcal{E}^2}{R_2} \rightarrow P_{R_1} + P_{R_2} = \mathcal{E}^2\left(\frac{1}{R_1} + \frac{1}{R_2}\right) \rightarrow P_{R_1} + P_{R_2} = \mathcal{E}^2\left(\frac{R_1 + R_2}{R_1 R_2}\right)$$

$$P_{EMF} = P_{R_1} + P_{R_2}$$

In other words, energy is conserved within the electric circuit.

Example 8-5

Problem: The power supplied by the EMF in the electric circuit in Figure 8.9 will be dissipated by the internal resistance of the battery and the load resistance. For which load resistance is the maximum power dissipated by the load resistance?

Solution: The power dissipated by the load resistance is

$$P_R = I^2 R$$

Substitution of the current from our previous solution gives us

$$I = \frac{\mathcal{E}}{r+R} \rightarrow P_R = \left(\frac{\mathcal{E}}{r+R}\right)^2 R \rightarrow P_R = \frac{\mathcal{E}^2 R}{(r+R)^2}$$

The resistance associated with maximum power is therefore

$$\frac{dP_R}{dR} = 0 \rightarrow \frac{d}{dR}\left(\frac{\mathcal{E}^2 R}{(r+R)^2}\right) = 0$$

$$\frac{(r+R)^2 \mathcal{E}^2 - 2\mathcal{E}^2 R(r+R)}{(r+R)^4} = 0 \rightarrow (r+R)^2 = 2R(r+R) \rightarrow r+R = 2R$$

$$R = r$$

The maximum power is supplied to the load resistance when the load resistance equals the internal resistance.

The power supplied to capacitors and inductors is (typically) not dissipated to the environment[6], but rather is stored within these components. The electric potential energy stored in a capacitor is given by Equation 5-8.

$$U_e = \frac{q^2}{2C}$$

The power supplied to an inductor is

$$P = I\left(L\frac{dI}{dt}\right) \rightarrow Pdt = (LI)dI$$

6 We'll discuss this further in Chapter 9.

The integral of this power with respect to time is the total change in the electric potential energy of the inductor.

$$P = \frac{dU_e}{dt} \rightarrow Pdt = dU_e \rightarrow \int P\,dt = \int dU_e \rightarrow \Delta U_e = \int P\,dt$$

If we assume that the current in the circuit is zero at time $t = 0$, we have

$$\int_0^t P\,dt = \int_0^I (LI)\,dI \rightarrow \int_0^t P\,dt = \frac{1}{2}LI^2 \rightarrow \Delta U_e = \frac{1}{2}LI^2$$

Not surprisingly, this is equal to the magnetic potential energy of the solenoid (Equation 6-15). In other words, the energy supplied by the EMF to the inductor is stored as magnetic potential energy in the inductor.

8-5 Series and Parallel Electric Circuits

Electric circuit components that are connected along a single path, so that the same current flows through each of the components, are said to be connected in ***series***.

> *Series electric circuit components*: electric circuit components that have the same current.

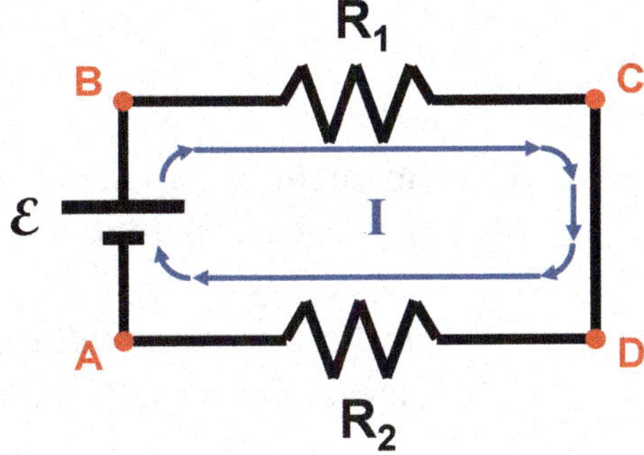

FIGURE 8.10: An electric circuit with two resistors in series.

The resistors in the electric circuit shown in Figure 8.10 are connected in series as the same current flows through each of these resistors.

Applying Kirchhoff's loop rule to the path ABCDA in Figure 8.10 gives us[7]

$$\mathcal{E} - R_1 I - R_2 I = 0 \rightarrow \mathcal{E} = R_1 I + R_2 I \rightarrow \mathcal{E} = (R_1 + R_2)I$$

7 We will assume from now on that all batteries are ideal and have no internal resistance.

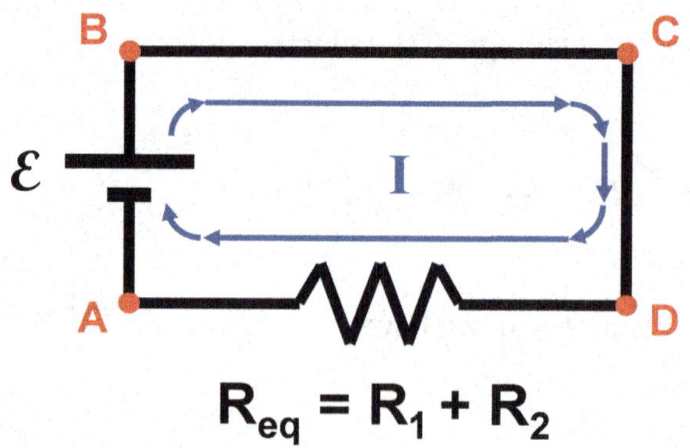

FIGURE 8.11: The equivalent electric circuit to the electric circuit shown in Figure 8.10.

$$I = \frac{\mathcal{E}}{R_1 + R_2}$$

The battery in the electric circuit supplies the EMF and the electrons to drive a current with this magnitude. A similar current would result in the electric circuit shown in Figure 8.11.

Applying Kirchhoff's loop rule to the path ABCDA in Figure 8.11 gives us

$$\mathcal{E} - R_{eq} I = 0 \quad \rightarrow \quad I = \frac{\mathcal{E}}{R_{eq}} \quad \rightarrow \quad I = \frac{\mathcal{E}}{R_1 + R_2}$$

We say that the electric circuit in Figure 8.11 is *equivalent* to the electric circuit in Figure 8.10 because the EMF supplies the same current for both circuits. Therefore, resistors connected in series can be modeled as a single *equivalent* resistor whose resistance is equal to the sum of the resistances of the individual resistors.

$$R_{eq} = R_1 + R_2 + \ldots \qquad \textbf{(8-2)}$$

Electric circuit components connected in ***parallel*** are connected so that the same change in electric potential occurs across each of the components.

> *Parallel electric circuit components*: electric circuit components that have the same difference in electric potential.

The resistors in the electric circuit shown in Figure 8.6 are connected in parallel, so that the change in electric potential across each of these resistors is the same; recall that $I_1 R_1 = I_2 R_2$ for this electric circuit. The current supplied by the EMF in this electric circuit was determined to be

$$I = \mathcal{E}\left(\frac{1}{R_1} + \frac{1}{R_2}\right)$$

The equivalent resistance for this group of resistors is

$$\frac{\mathcal{E}}{R_{eq}} = \mathcal{E}\left(\frac{1}{R_1} + \frac{1}{R_2}\right) \rightarrow \frac{1}{R_{eq}} = \frac{1}{R_1} + \frac{1}{R_2}$$

Hence, resistors connected in parallel can also be modeled as a single equivalent resistor. The inverse of the resistance of this equivalent resistor is equal to the sum of the inverses of the resistances of the individual resistors.

$$\frac{1}{R_{eq}} = \frac{1}{R_1} + \frac{1}{R_2} + \ldots \qquad \textbf{(8-3)}$$

Example 8-6

Problem: What is the equivalent resistance of the resistors in the electric circuit shown in Figure 8.12? What is the current supplied by the EMF in this circuit?

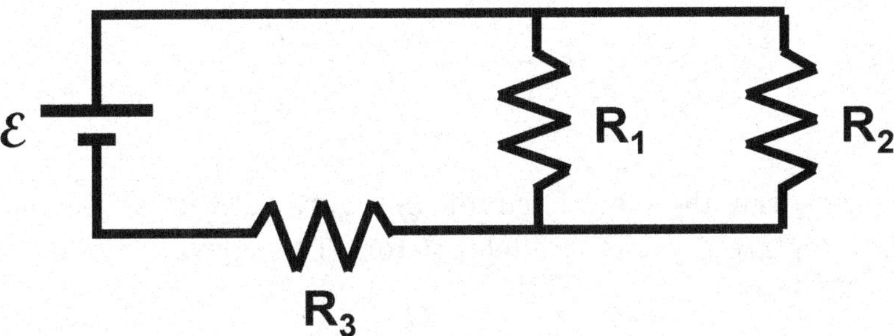

FIGURE 8.12: The electric circuit in Example 8-6.

Solution: We begin by labeling the currents flowing through each component of the electric circuit. As shown in Figure 8.13, the EMF supplies a current I that divides into I_1 and I_2 at point C. Current I_1 flows through the resistor with resistance R_1 and current I_2 flows through the resistor with resistance R_2. These two currents recombine at point F to reform current I, which flows through the resistor with resistance R_3 and then back to the EMF.

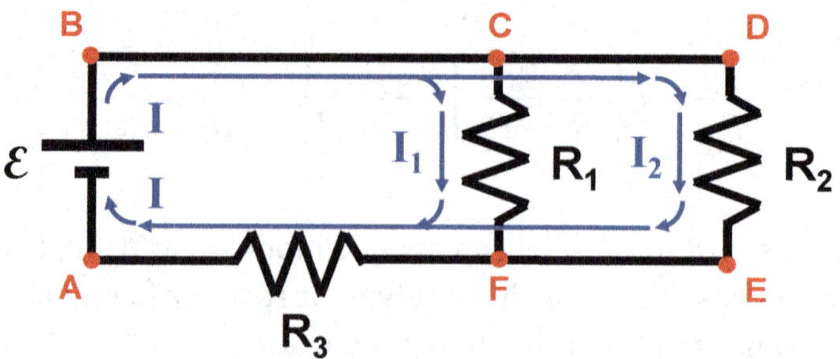

FIGURE 8.13: Designation of the currents flowing in the electric circuit in Example 8-6.

Applying Kirchhoff's loop rule to the loop CDEFC in Figure 8.13 gives us

$$-R_2I_2 + R_1I_1 = 0 \;\rightarrow\; R_2I_2 = R_1I_1 \;\rightarrow\; \left(\Delta V_e\right)_{R_1} = \left(\Delta V_e\right)_{R_2}$$

Since the change in electric potential across these two resistors is the same, these resistors are connected in parallel. The equivalent resistance of these two resistors is thus

$$\frac{1}{R_{eq}} = \frac{1}{R_1} + \frac{1}{R_2} \;\rightarrow\; R_{eq} = \frac{R_1R_2}{R_1 + R_2}$$

We can now redraw the electric circuit in Figure 8.13 using this equivalent resistance to replace the two parallel resistors. This equivalent electric circuit is shown in Figure 8.14.

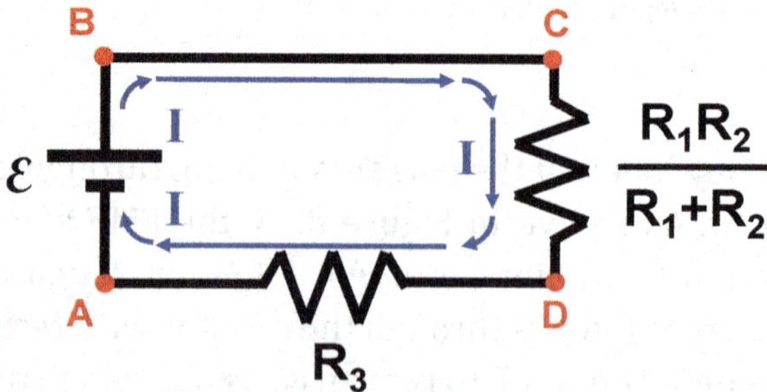

FIGURE 8.14: The equivalent electric circuit of the electric circuit in Figure 8.13.

Applying Kirchhoff's loop rule to the loop ABCDA in Figure 8.14 gives us

$$\mathcal{E}-\left(\frac{R_1R_2}{R_1+R_2}\right)I-R_3I=0 \quad\rightarrow\quad \mathcal{E}=\left(\frac{R_1R_2}{R_1+R_2}+R_3\right)I$$

The equivalent resistance of the electric circuit is therefore

$$\mathcal{E}=R_{eq}I \quad\rightarrow\quad R_{eq}=\frac{R_1R_2}{R_1+R_2}+R_3 \quad\rightarrow\quad R_{eq}=\frac{R_1R_2+R_3\left(R_1+R_2\right)}{R_1+R_2}$$

We also could have arrived at this expression by realizing that since the same current flows through both resistors in the electric circuit in Figure 8.14, those two resistors are in series. The equivalent resistance of these two resistors is thus

$$R_{eq}=\frac{R_1R_2}{R_1+R_2}+R_3 \quad\rightarrow\quad R_{eq}=\frac{R_1R_2+R_3\left(R_1+R_2\right)}{R_1+R_2}$$

The current supplied by the EMF is

$$I=\frac{\mathcal{E}}{\dfrac{R_1R_2+R_3\left(R_1+R_2\right)}{R_1+R_2}} \quad\rightarrow\quad I=\frac{\mathcal{E}\left(R_1+R_2\right)}{R_1R_2+R_3\left(R_1+R_2\right)}$$

These rules for determining the equivalent resistance for resistors connected in series and parallel can also be applied to determine the equivalent capacitance of series and parallel combinations of capacitors and inductors. The only difference is that for capacitors the variable is the inverse of the capacitance, rather than the capacitance itself. This occurs since the difference in electric potential across the plates of a capacitor is proportional to the inverse[8] of the capacitance (Equation 5-7), whereas the difference

8 This is another historical whoops, similar to giving the electron a negative charge. However, this whoops cannot be blamed on Ben Franklin.

in electric potential across a resistor (Equation 7-15) or an inductor (Equation 6-4) is directly proportional to the resistance or inductance, respectively, as shown in Table 8.2.

$$\text{Inductors in series: } L_{eq} = L_1 + L_2 + \ldots \tag{8-4}$$

$$\text{Inductors in parallel: } \frac{1}{L_{eq}} = \frac{1}{L_1} + \frac{1}{L_2} + \ldots \tag{8-5}$$

$$\text{Capacitors in series: } \frac{1}{C_{eq}} = \frac{1}{C_1} + \frac{1}{C_2} + \ldots \tag{8-6}$$

$$\text{Capacitors in parallel: } C_{eq} = C_1 + C_2 + \ldots \tag{8-7}$$

Example 8-7

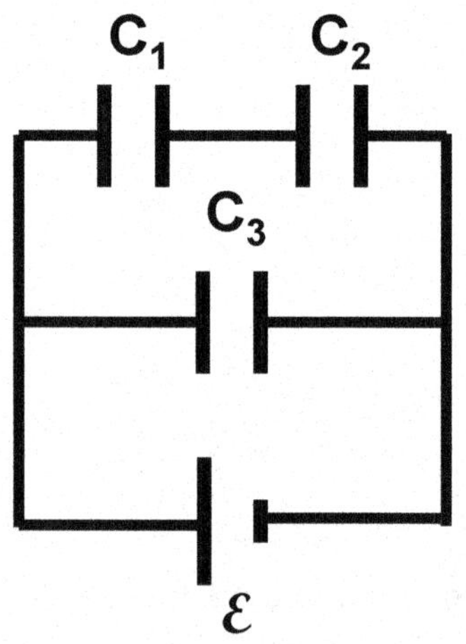

FIGURE 8.15: The electric circuit in Example 8-7.

Problem: What is the equivalent capacitance of the capacitors in the electric circuit shown in Figure 8.15?

Solution: We begin by labeling the currents flowing through each component of the electric circuit. As shown in Figure 8.16, the EMF supplies a current I that divides into I_1 and I_2 at point B. Current I_1 flows through the capacitor with capacitance C_3, and current I_2 flows through the capacitor with capacitance C_1 and the capacitor with capacitance C_2. These two currents recombine at point E to reform current I, which flows back to the EMF.

The capacitor with capacitance C_1 and the capacitor with capacitance C_2 are in series, since the same current flows through both capacitors. The equivalent capacitance of these capacitors is found using Equation 8-6.

$$\frac{1}{C_{eq}} = \frac{1}{C_1} + \frac{1}{C_2} \rightarrow C_{eq} = \frac{C_1C_2}{C_1 + C_2}$$

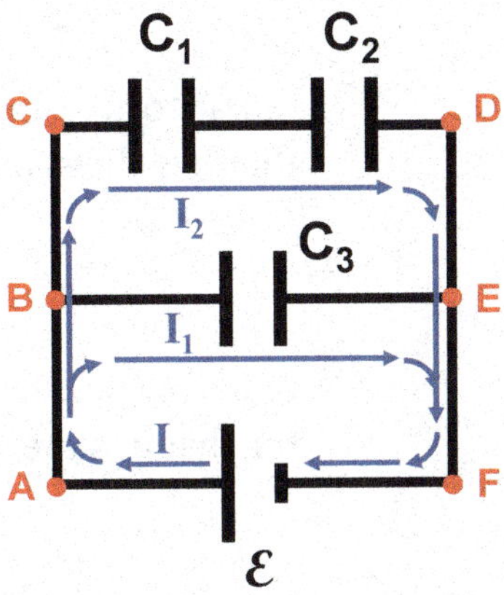

FIGURE 8.16: Designation of the currents flowing in the electric circuit in Example 8-7.

We can now redraw the electric circuit in Figure 8.16 using this equivalent capacitance to replace the two series capacitors. This equivalent electric circuit is shown in Figure 8.17.

Applying Kirchhoff's loop rule to the loop BCDEB in Figure 8.17 gives us

$$-\left(\Delta V_e\right)_{\frac{C_1C_2}{C_1+C_2}} + \left(\Delta V_e\right)_{C_3} = 0 \rightarrow \left(\Delta V_e\right)_{\frac{C_1C_2}{C_1+C_2}} = \left(\Delta V_e\right)_{C_3}$$

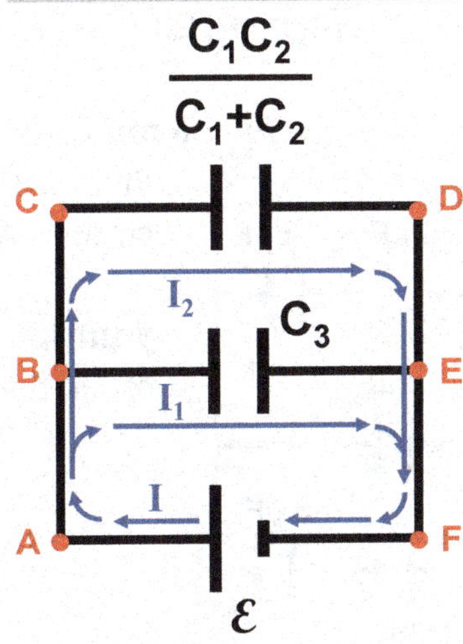

FIGURE 8.17: The equivalent electric circuit of the electric circuit in Figure 8.16.

Since the change in electric potential across these two capacitors is the same, these capacitors are connected in parallel. The equivalent capacitance of these two capacitors can be determined using Equation 8.7.

$$C_{eq} = \frac{C_1C_2}{C_1 + C_2} + C_3 \rightarrow C_{eq} = \frac{C_1C_2 + C_3\left(C_1 + C_2\right)}{C_1 + C_2}$$

This is the equivalent capacitance of the capacitors in the electric circuit.

As shown in Figure 8.18, since the same current flows through capacitors in series, the magnitude of electric charge on these capacitors must be the same.

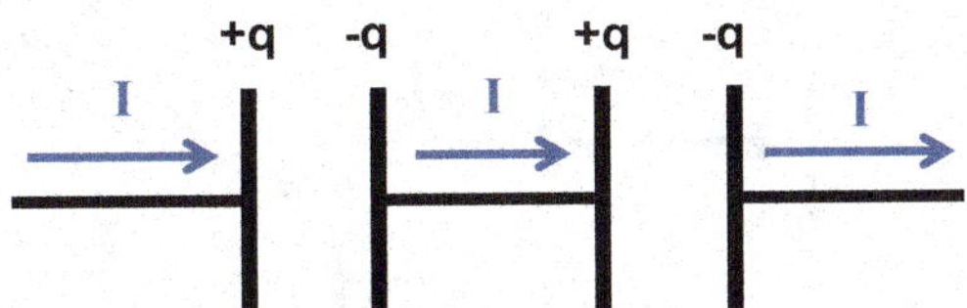

FIGURE 8.18: Capacitors in series contain the same magnitude of electric charge

This leads to two additional statements about equivalent capacitors.

Capacitors connected in series can be replaced by a single equivalent capacitor that has the same electric charge as the actual capacitors.

$$\text{Capacitors in series: } q_1 = q_2 = \ldots = q_{eq}$$

Capacitors connected in parallel can be replaced by a single equivalent capacitor that has the same *total* electric charge as the actual capacitors.

$$\text{Capacitors in parallel: } q_1 + q_2 + \ldots = q_{eq}$$

Example 8-8

Problem: Four capacitors are connected to a 20 V battery, as shown in Figure 8.19. What is the equivalent capacitance of the entire electric circuit? What is the net electric charge on each capacitor?

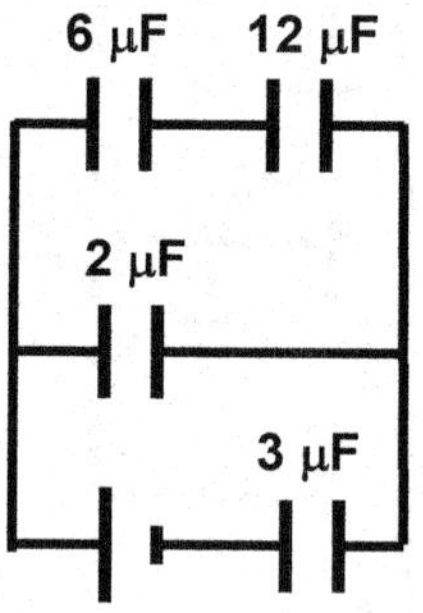

6 μF
12 μF
2 μF
3 μF
20 V

FIGURE 8.19: The electric circuit in Example 8-8.

Solution: Since the 12 μF and 6 μF capacitors are in series with each other, the equivalent capacitance of these two capacitors is therefore

$$\frac{1}{C_{eq}} = \frac{1}{6\,\mu\text{F}} + \frac{1}{12\,\mu\text{F}} \rightarrow \frac{1}{C_{eq}} = \frac{3}{12\,\mu\text{F}} \rightarrow C_{eq} = 4\,\mu\text{F}$$

We can therefore redraw this electric circuit as shown in Figure 8.20.

4 μF
2 μF
3 μF
20 V
6 μF
3 μF
20 V
2 μF
20 V

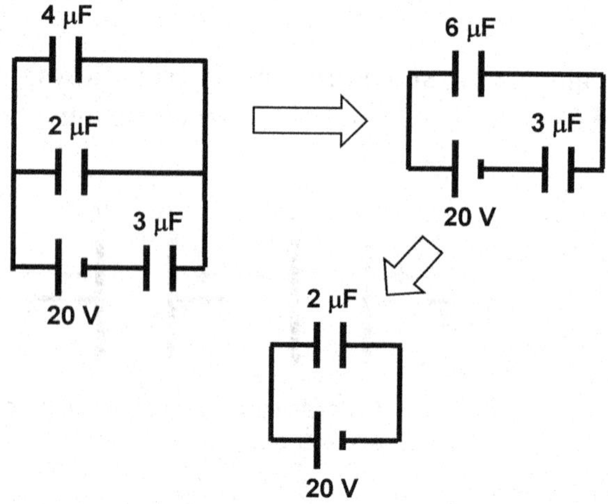

FIGURE 8.20: The equivalent electric circuits for the electric circuit in Figure 8.19.

As shown in Figure 8.20, we can then continue this process of writing equivalent electric circuits. The 2 μF and 4 μF capacitors in Figure 8.20 are in parallel with each other. The equivalent capacitance of these two capacitors is

$$\frac{1}{C_{eq}} = 2\,\mu\text{F} + 4\,\mu\text{F} \quad \rightarrow \quad C_{eq} = 6\,\mu\text{F}$$

Finally, the 3 μF and 6 μF capacitors in Figure 8.20 are in series with each other. The equivalent capacitance of these two capacitors is

$$\frac{1}{C_{eq}} = \frac{1}{3\,\mu\text{F}} + \frac{1}{6\,\mu\text{F}} \quad \rightarrow \quad \frac{1}{C_{eq}} = \frac{3}{6\,\mu\text{F}} \quad \rightarrow \quad C_{eq} = 2\,\mu\text{F}$$

Hence, the final equivalent circuit contains only a single 2 μF capacitor. The electric charge on this single equivalent capacitor is found using Kirchhoff's loop rule.

$$20\,\text{V} - \frac{Q}{2\mu\text{F}} = 0 \quad \rightarrow \quad Q = 40\,\mu\text{C}$$

To determine the electric charge on the capacitors, let's work backwards through the equivalent electric circuits in Figure 8.20. As shown in Figure 8.21, the 3 μF and 6 μF capacitors in Figure 8.20 are in series with each other and therefore must have the same electric charge, and this electric charge must be equal the 40 μC on the equivalent 2 μF capacitor.

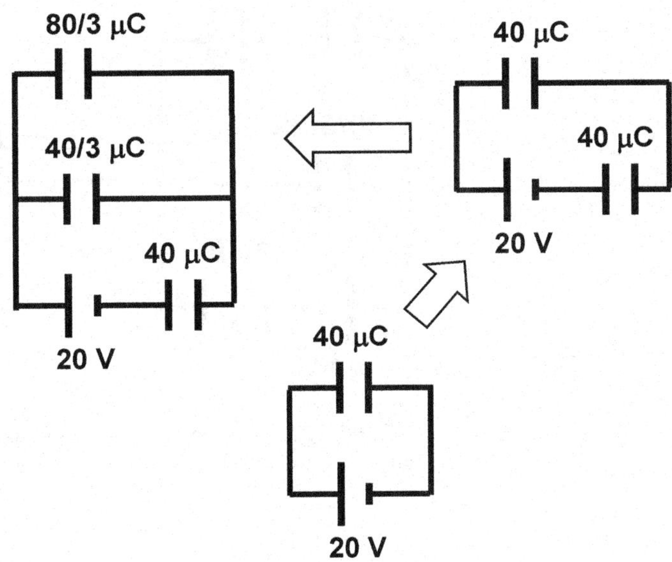

FIGURE 8.21: Determining the electric charge on each capacitor in the electric circuit in Figure 8.19.

The 2 μF and 4 μF capacitors in Figure 8.20 are in parallel with each other, and thus the sum of the electric charge on each capacitor must equal 40 μC. To determine how this electric charge is distributed, we recall that since these two capacitors are in parallel they must have the same electric potential difference.

$$\Delta V_{e,2\mu F} = \Delta V_{e,4\mu F} \quad \to \quad \frac{Q_{e,2\mu F}}{2\,\mu\text{F}} = \frac{Q_{e,4\mu F}}{4\,\mu\text{F}} \quad \to \quad 2Q_{e,2\mu F} = Q_{e,4\mu F}$$

Thus,

$$Q_{e,2\mu F} + Q_{e,4\mu F} = 40\,\mu\text{C} \quad \to \quad Q_{e,2\mu F} + 2Q_{e,2\mu F} = 40\,\mu\text{C} \quad \to \quad 3Q_{e,2\mu F} = 40\,\mu\text{C}$$

$$Q_{e,2\mu F} = \frac{40}{3}\,\mu\text{C} \quad \to \quad Q_{e,4\mu F} = \frac{80}{3}\,\mu\text{C}$$

Finally, since the 12 μF and 6 μF capacitors in the electric circuit in Figure 8.19 are in series with each other, they must have the same electric charge. Specifically, they must have the same 80/3 μC electric charge possessed by the 4 μF equivalent capacitor in Figure 8.21.

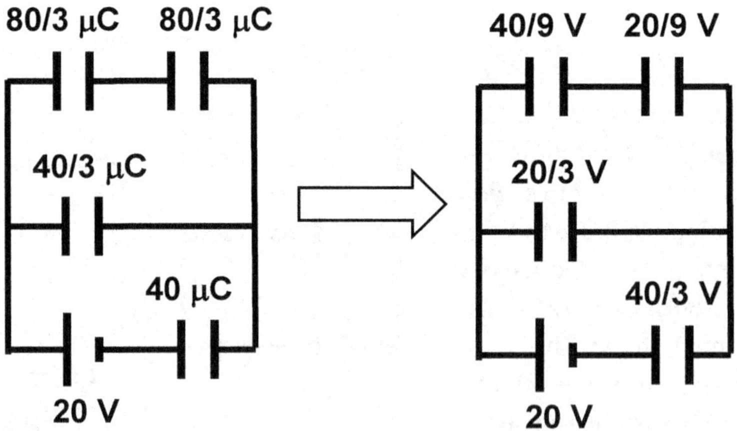

FIGURE 8.22: The electric charge and associated electric potential difference for each of the capacitors in the electric circuit in Figure 8.19.

The electric charges and associated electric potential differences for each capacitor in the electric circuit in Figure 8.19 are shown in Figure 8.22. The sum of the differences in the electric potential across any closed path in this circuit is zero, in agreement with Kirchhoff's loop rule.

We can also apply our thinking about parallel and series circuit components more broadly to solve a wider variety of problems.

Example 8-9

Problem: The parallel plate capacitor in Figure 8.23 contains two different dielectrics, with dielectric constants κ_1 and κ_2, which each cover one-half of the total surface area of the plates.

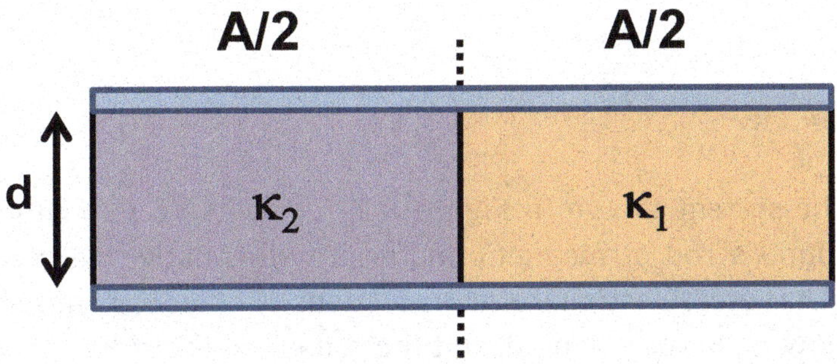

FIGURE 8.23: The capacitor in Example 8-9.

What is the capacitance of this capacitor? Express the answer in terms of the C_0, the capacitor of the capacitor in the absence of any dielectric.

Solution: We can model this capacitor as consisting of two capacitors in parallel, each of which contains a different dielectric. The equation for the equivalent capacitance of these two capacitors is found using Equation 7-4 and Equation 8-7.

$$C_{eq} = \kappa_1 C_1 + \kappa_2 C_2$$

In this equation the variables C_1 and C_2 denote the capacitance of the two capacitors that are in parallel. These capacitances can be determined using Equation 5-6.

$$C_{eq} = \kappa_1\left(\frac{\frac{A}{2}\varepsilon_0}{d}\right) + \kappa_2\left(\frac{\frac{A}{2}\varepsilon_0}{d}\right) \rightarrow C_{eq} = \left(\frac{\frac{A}{2}\varepsilon_0}{d}\right)(\kappa_1 + \kappa_2)$$

$$C_{eq} = \left(\frac{A\varepsilon_0}{d}\right)\left(\frac{\kappa_1 + \kappa_2}{2}\right) \rightarrow C_{eq} = C_0\left(\frac{\kappa_1 + \kappa_2}{2}\right)$$

Example 8-10

Problem: The system shown in Figure 8.24 consists of a solenoid partially wrapped around a rod of magnetic material with relative permeability κ_m. The solenoid has cross-sectional area A and density of loops n. Initially, there is no current flowing in the solenoid, and the rod extends out from the solenoid. What is the magnetic force exerted on the rod when the solenoid carries current I? Assume that the solenoid is perfect.

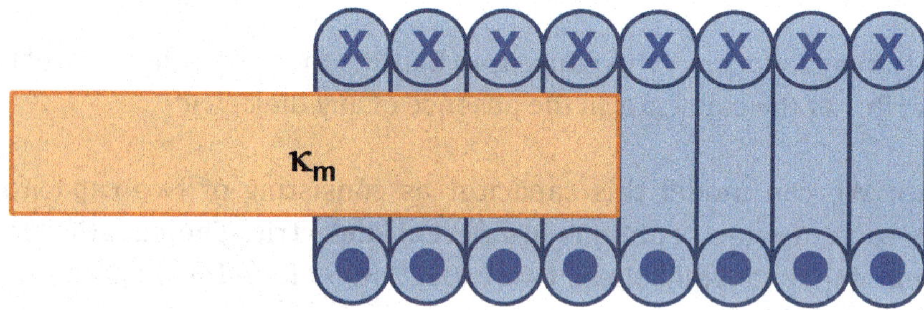

FIGURE 8.24: The solenoid in Example 8-10.

Solution: Let's define an x-axis to be parallel to the longitudinal axis of the solenoid with a positive direction pointing from the rod into the solenoid, as shown in Figure 8.25.

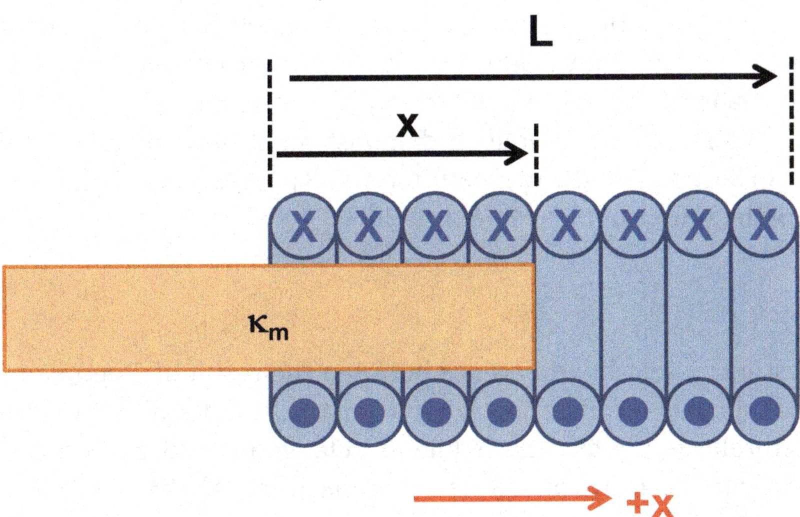

FIGURE 8.25: Defining the reference frame for the system in Figure 8.24.

Let's denote the length of the solenoid as L and the length of the rod that is inside the solenoid as x. We can then model this system as two solenoids connected in series; one solenoid contains the rod and the other does not. The equivalent inductance of these two solenoids is found using Equation 8-4.[9]

$$L = \mu_0 n^2 A \kappa_m x + \mu_0 n^2 A(L - x) \quad \rightarrow \quad L = \mu_0 n^2 A\left(L + (\kappa_m - 1)x\right)$$

The magnetic potential energy of the solenoid is determined using Equation 6-15.

$$U_m = \frac{1}{2}\mu_0 n^2 A\left(L + (\kappa_m - 1)x\right)I^2$$

Since the current in the solenoid is held constant in this example, the magnetic force is determined using Equation 6-16.

$$\vec{F}_m = \left(\nabla U_m\right)_I \quad \rightarrow \quad \vec{F}_m = \left[\frac{1}{2}\mu_0 n^2 A(\kappa_m - 1)I^2\right]\hat{x}$$

9 The inductance of a solenoid with a magnetic core was determined in Example 7-6.

Thus, if $\kappa_m > 1$ (*i.e.*, if the rod is made of paramagnetic or ferromagnetic material), the rod will be drawn into the solenoid. Alternatively, if $\kappa_m < 1$ (*i.e.*, if the rod is made of diamagnetic material), the rod will be expelled from the solenoid. This result is a simplified demonstration of diamagnetic levitation discussed in Section 7-4; the magnetic force will push the diamagnetic material away from the source of the magnetic field.

It is interesting to note the difference between the result of Example 8-10 and the result of the analogous problem involving a capacitor and a dielectric (Figure 7.10). This difference reflects the fact that while all relative permittivities are greater than one (so the capacitor is always pulled into the capacitor), relative permeabilities can be greater than or less than one.

8-6 Transient and Steady-State Behavior in DC Circuits

The electric circuit shown in Figure 8.26 consists of a battery, a resistor, and a switch. When the switch is open (left panel in Figure 8.26), no current will flow in the electric circuit, since there is no conducting path connecting the high and low potential terminals of the battery.

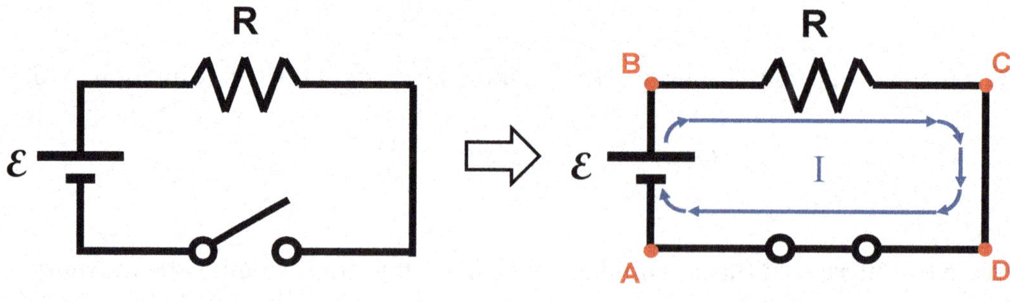

FIGURE 8.26: A DC circuit with only resistance.

When the switch is closed, however, a current can flow in the electric circuit (right panel in Figure 8.26). We can determine the magnitude of this current by applying Kirchhoff's loop rule to the electric circuit. For the loop ABCDA, we have

$$\mathcal{E} - RI = 0 \quad \rightarrow \quad \mathcal{E} = RI \quad \rightarrow \quad I = \frac{\mathcal{E}}{R}$$

Of course, since the electric circuit forms a closed "loop" of wire, there must be some self-inductance in the electric circuit (Section 6-3). We can model this self-inductance by including an inductor in series with the resistor in the electric circuit, as shown in Figure 8.27.

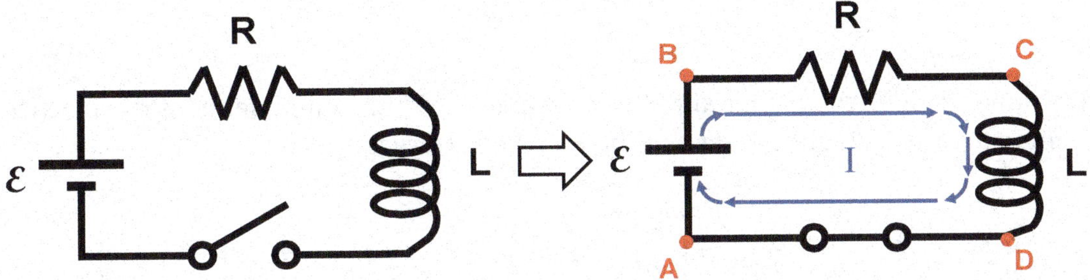

FIGURE 8.27: A DC circuit with resistance and inductance.

As before, no current will flow in the electric circuit shown in Figure 8.27 when the switch is open (left panel), but current will flow when the switch is closed (right panel). Applying Kirchhoff's loop rule to loop ABCDA in Figure 8.27 gives us

$$\mathcal{E} - RI - L\frac{dI}{dt} = 0 \quad \rightarrow \quad \mathcal{E} = L\frac{dI}{dt} + RI$$

You can verify that a solution to this equation, assuming that the current is zero immediately after the switch is closed, is given by

$$I(t) = \frac{\mathcal{E}}{R}\left(1 - e^{-\frac{R}{L}t}\right)$$

When the switch in this electric circuit is closed, a current will start to flow in the electric circuit. This sudden change in the magnitude of the current in the electric circuit will induce an EMF in the inductor that will oppose the direction of this current (Section 6-3). The magnitude of this induced EMF will be directly proportional to the magnitude of the rate of current change (Table 8-2) and will asymptotically approach zero as the current in the electric circuit asymptotically approaches its *steady-state*[10] value. Once the current in the electric circuit is no longer changing (*i.e.*, when the current is at its steady-state value), there is no induced EMF in the inductor, and the electric circuit in Figure 8.27 becomes equivalent to the electric circuit in Figure 8.26. Indeed, you can verify that the limit of the current for the electric circuit in Figure 8.27 as time approaches infinity is the same as the current in electric circuit in Figure 8.26.

$$\lim_{t\to\infty} I(t) = \frac{\mathcal{E}}{R}$$

Let's now consider the power supplied/dissipated by the components of the electric circuit in Figure 8.27. The power supplied by the battery is

$$P_{battery} = I\mathcal{E} \;\rightarrow\; P_{battery} = \left(\frac{\mathcal{E}}{R}\left(1 - e^{-\frac{R}{L}t}\right)\right)\mathcal{E} \;\rightarrow\; P_{battery} = \frac{\mathcal{E}^2}{R}\left(1 - e^{-\frac{R}{L}t}\right)$$

The power supplied to the resistor is

$$P_{resistor} = I^2 R \;\rightarrow\; P_{resistor} = \frac{\mathcal{E}^2}{R^2}\left(1 - e^{-\frac{R}{L}t}\right)^2 R$$

$$P_{resistor} = \frac{\mathcal{E}^2}{R}\left(1 - 2e^{-\frac{R}{L}t} + e^{-2\frac{R}{L}t}\right)$$

The power supplied to the inductor is

$$P_{inductor} = IL\frac{dI}{dt} \;\rightarrow\; P_{inductor} = \frac{\mathcal{E}}{R}\left(1 - e^{-\frac{R}{L}t}\right)L\left(\frac{\mathcal{E}}{R}\frac{R}{L}e^{-\frac{R}{L}t}\right)$$

$$P_{inductor} = \frac{\mathcal{E}^2}{R}\left(e^{-\frac{R}{L}t} - e^{-2\frac{R}{L}t}\right)$$

10 The term "steady-state" denotes the situation when the parameters are no longer changing with time (Section 8-2). In this case, it refers to the situation when the current flowing in the electric circuit is constant.

As expected, the power supplied by the battery is equal to the sum of the power supplied to the resistor and the power supplied to the inductor.

$$P_{resistor} + P_{inductor} = \frac{\mathcal{E}^2}{R}\left(1 - 2e^{-\frac{R}{L}t} + e^{-2\frac{R}{L}t}\right) + \frac{\mathcal{E}^2}{R}\left(e^{-\frac{R}{L}t} - e^{-2\frac{R}{L}t}\right)$$

$$P_{resistor} + P_{inductor} = \frac{\mathcal{E}^2}{R}\left(1 - e^{-\frac{R}{L}t}\right) \rightarrow P_{resistor} + P_{inductor} = P_{battery}$$

The power (*i.e.*, the energy) supplied to the resistor is eventually dissipated from the system as heat (Section 7-5). The power supplied to the inductor is stored in the magnetic potential energy of the circuit's self-inductance (Equation 6-15).

$$U_m = \frac{1}{2}LI^2 \rightarrow U_m = \frac{\mathcal{E}^2 L}{2R^2}\left(1 - e^{-\frac{R}{L}t}\right)^2$$

Once the current in the electric circuit has achieved its steady-state value, the total magnetic potential energy of the electric circuit is

$$\lim_{t\to\infty} U_m = \frac{\mathcal{E}^2 L}{2R^2}$$

The time courses for the current in the circuit, the power supplied by the battery, and the power supplied to the resistor and the inductor are shown in Figure 8.28.

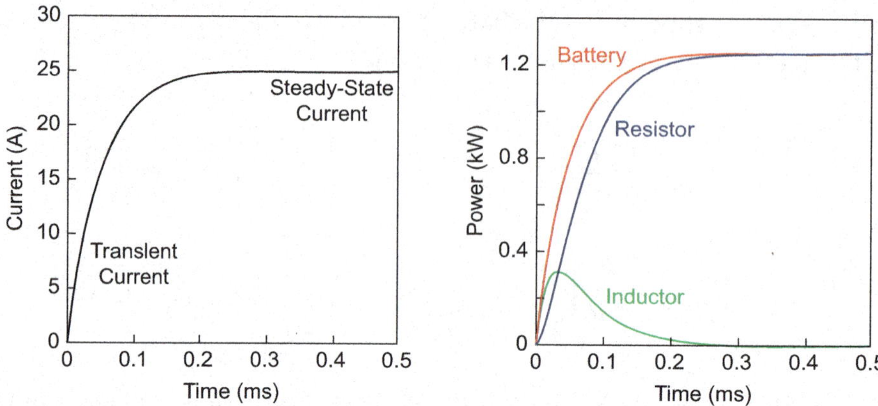

FIGURE 8.28: (Left panel) The current in the electric circuit in Figure 8.27. The initial changing (*i.e.*, transient current) exponentially approaches a constant steady-state value. (Right panel) The power supplied by the battery to the resistor and inductor in the electric circuit in Figure 8.27. The area under the curve of power supplied to the inductor versus time is the total energy stored in the inductor. For both panels, $R = 2\,\Omega$, $L = 100$ H, and $\mathcal{E} = 50$ V.

Now let's look at the electric circuit shown in Figure 8.29. When the switch is in the orientation shown in the left panel, the battery is connected to the rest of the electric circuit, and the electric circuit is identical to the electric circuit in Figure 8.27.

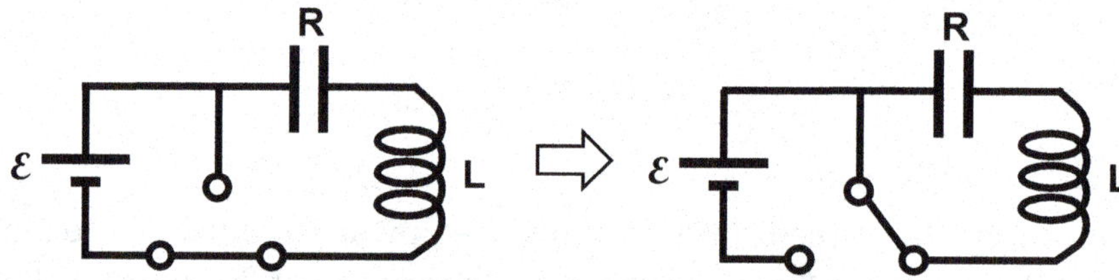

FIGURE 8.29: The energy store in an inductor can power a circuit.

After the current in this circuit has achieved its steady-state value, the switch is then moved to the orientation shown in the right panel of Figure 8.29, with the battery now disconnected from the rest of electric circuit. Applying Kirchhoff's loop rule to the portion of the electric circuit containing only the resistor and the inductor gives us

$$L\frac{dI}{dt} + RI = 0$$

You can verify that the solution for this equation with an initial current equal to the steady-state current from before is

$$I(t) = \frac{\mathcal{E}}{R} e^{-\frac{R}{L}t}$$

In this electric circuit, the energy stored in the inductor supplies the EMF for the circuit. Indeed, we see that the total energy dissipated by the resistor is equal to the magnetic potential energy initially stored in the inductor. Recall from classical mechanics that a change in energy is related to power through integration.

$$P = \frac{dE}{dt} \quad \rightarrow \quad \int dE = \int P\,dt \quad \rightarrow \quad \Delta E = \int P\,dt$$

Substitution gives us

$$P_{resistor} = \frac{\mathcal{E}^2}{R} e^{-\frac{2R}{L}t} \quad \rightarrow \quad \Delta E = \int_0^\infty \frac{\mathcal{E}^2}{R} e^{-\frac{2R}{L}t}\,dt \quad \rightarrow \quad \Delta E = \frac{\mathcal{E}^2}{R}\left(-\frac{L}{2R}\right)\left(e^{-\frac{2R}{L}t}\Big|_0^\infty\right)$$

$$\Delta E = \frac{\mathcal{E}^2}{R}\left(-\frac{L}{2R}\right)(-1) \quad \rightarrow \quad \Delta E = \frac{\mathcal{E}^2 L}{2R^2} = U_m$$

Next let's consider the electric circuit shown in Figure 8.30 that consists of a battery and a capacitor. When the switch is open (left panel in Figure 8.30), no current will flow in the electric circuit, since there is no conducting path connecting the high and low potential terminals of the battery.

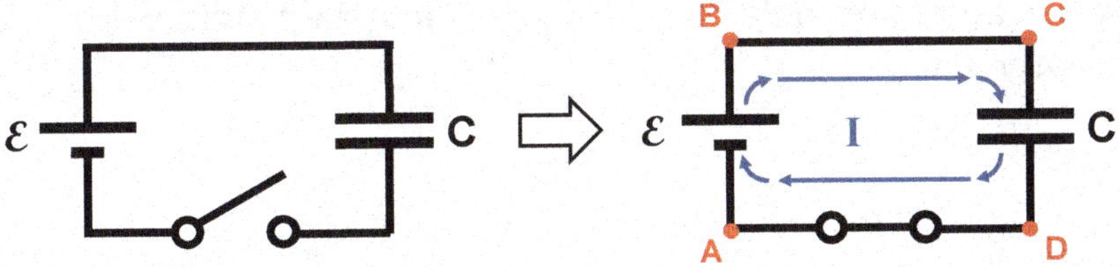

FIGURE 8.30: A DC circuit with only capacitance.

When the switch is closed, however, a current can flow in the electric circuit (right panel in Figure 8.30). We can determine the magnitude of this current by applying Kirchhoff's loop rule to the electric circuit. For the loop ABCDA in Figure 8.30 we have

$$\mathcal{E} - \frac{q}{C} = 0 \quad \rightarrow \quad q = C\mathcal{E}$$

It follows from this result that no current flows in the electric circuit.

$$I = \frac{dq}{dt} \quad \rightarrow \quad I = 0$$

Of course, the idealized electric circuit in Figure 8.30 can never be created since the wires or the battery itself will always have some resistance. We can model this resistance by including a resistor in series with the capacitor and battery, as shown in Figure 8.31.

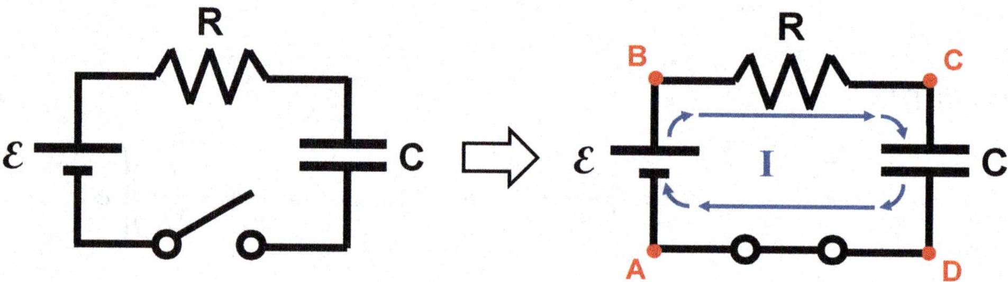

FIGURE 8.31: A DC circuit with resistance and capacitance.

Applying Kirchhoff's loop rule to the loop ABCDA in Figure 8.31 gives us

$$\mathcal{E} - RI - \frac{q}{C} = 0 \quad \rightarrow \quad \mathcal{E} - R\left(\frac{dq}{dt}\right) - \frac{q}{C} = 0$$

You can verify that a solution to this equation (including the condition that the capacitor is initially uncharged) is

$$q(t) = \mathcal{E}C\left(1 - e^{-\frac{t}{RC}}\right) \quad \rightarrow \quad I(t) = \frac{\mathcal{E}}{R} e^{-\frac{t}{RC}}$$

When the switch in this electric circuit is closed, a current will start to flow in the electric circuit. As the current flows, electric charge will build up on the plates of the capacitor, and consequently the electric potential difference across the plates of the capacitor will increase. The capacitor will operate in the circuit as an *effective battery* whose EMF is opposite in orientation to the actual battery. The net EMF in the electric circuit will therefore decrease over time, and, as a result, the current in the circuit will also decrease over time. Eventually, the potential difference across the plates of the capacitor exactly equals the EMF of the battery, at which point there is no longer current in the electric circuit. In other words, the steady-state current in the electric circuit is zero as predicted from the analysis of the electric circuit in Figure 8.30. Indeed, the equations for the current and capacitor charge in the electric circuit in Figure 8.31 recapitulate the results obtained from the electric circuit in Figure 8.30 in the limit as time approaches infinity.

Let's now consider the power supplied/dissipated by the components of the electric circuit in Figure 8.31. The power supplied by the battery is

$$P_{battery} = I\mathcal{E} \rightarrow P_{battery} = \left(\frac{\mathcal{E}}{R}e^{-\frac{t}{RC}}\right)\mathcal{E} \rightarrow P_{battery} = \frac{\mathcal{E}^2}{R}e^{-\frac{t}{RC}}$$

The power supplied to the resistor is

$$P_{resistor} = I^2R \rightarrow P_{resistor} = \left(\frac{\mathcal{E}}{R}e^{-\frac{t}{RC}}\right)^2 R \rightarrow P_{resistor} = \frac{\mathcal{E}^2}{R}e^{-2\frac{t}{RC}}$$

The power supplied to the capacitor is

$$P_{capacitor} = I\left(\frac{q}{C}\right) \rightarrow P_{capacitor} = \frac{\mathcal{E}}{R}e^{-\frac{t}{RC}}\left(\frac{\mathcal{E}C\left(1-e^{-\frac{t}{RC}}\right)}{C}\right)$$

$$P_{capacitor} = \frac{\mathcal{E}^2}{R}\left(e^{-\frac{t}{RC}} - e^{-2\frac{t}{RC}}\right)$$

As expected, the power supplied by the battery is equal to the sum of the power supplied to the resistor and the power supplied to the capacitor.

$$P_{resistor} + P_{capacitor} = \frac{\mathcal{E}^2}{R}e^{-2\frac{t}{RC}} + \frac{\mathcal{E}^2}{R}\left(e^{-\frac{t}{RC}} - e^{-2\frac{t}{RC}}\right)$$

$$P_{resistor} + P_{capacitor} = \frac{\mathcal{E}^2}{R}e^{-\frac{t}{RC}} \rightarrow P_{resistor} + P_{capacitor} = P_{battery}$$

The power (*i.e.*, the energy) supplied to the resistor is eventually dissipated from the system as heat (Section 7-5). The power supplied to the capacitor is stored in the electric potential energy of the capacitor (Equation 5-8).

$$U_e = \frac{q^2}{2C} \rightarrow U_e = \frac{1}{2C}\left(\mathcal{E}C\left(1-e^{-\frac{t}{RC}}\right)\right)^2 \rightarrow U_e = \frac{\mathcal{E}^2C}{2}\left(1-e^{-\frac{t}{RC}}\right)^2$$

The time courses for the current in the circuit, the power supplied by the battery, and the power supplied to the resistor and the capacitor are shown in Figure 8.32.

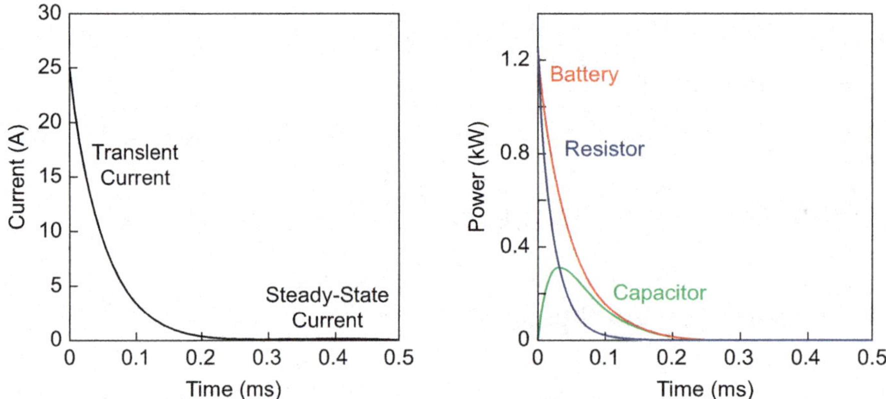

FIGURE 8.32: (Left panel) The current in the electric circuit in Figure 8.31. The initial changing (*i.e.*, transient current) exponentially approaches a constant steady-state value of zero. (Right panel) The power supplied by the battery to the resistor and capacitor in the electric circuit in Figure 8.31. The area under the curve of power supplied to the capacitor versus time is the total energy stored in the capacitor. For both panels, $R = 2\ \Omega$, $C = 10\ \mu F$, and $\mathcal{E} = 50$ V.

Of course, as discussed previously, there is always some self-inductance in an electric circuit. We can model this self-inductance by including an inductor in series with the resistor, capacitor, and battery, as shown in Figure 8.33. This configuration is referred to as a series RLC DC circuit.

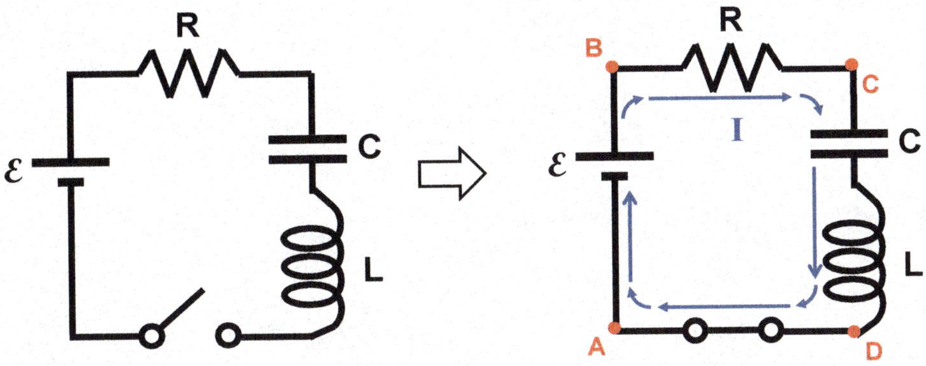

FIGURE 8.33: A series RLC DC circuit.

Applying Kirchhoff's loop rule to the loop ABCDA in Figure 8.33 gives us

$$\mathcal{E} - RI - \frac{q}{C} - L\frac{dI}{dt} = 0 \quad \rightarrow \quad \mathcal{E} - R\left(\frac{dq}{dt}\right) - \frac{q}{C} - L\left(\frac{d^2q}{dt^2}\right) = 0$$

This looks exactly like the differential equation for a damped harmonic oscillator. You can verify that a solution of this equation (including the condition that the capacitor is initially uncharged) is

$$q(t) = \mathcal{E}C\left(1 + \left(\frac{\beta_2}{\beta_1 - \beta_2}\right)e^{-\beta_1 t} - \left(\frac{\beta_1}{\beta_1 - \beta_2}\right)e^{-\beta_2 t}\right)$$

where the constants β_1 and β_2 are

$$\beta_1 = \frac{CR + \sqrt{C^2R^2 - 4LC}}{2CL} \qquad \beta_2 = \frac{CR - \sqrt{C^2R^2 - 4LC}}{2CL}$$

We can also confirm that this solution recapitulates our previous solution in the appropriate limits. For example, if there were no capacitor in the circuit (*i.e.*, if C = ∞[11]) then

11 Recall from Table 8.2 that $\Delta V_e = \frac{q}{C}$. Therefore, C must be infinite for there to be no change in electric potential across the capacitor (*i.e.*, for there to be no capacitor).

$$\lim_{C\to\infty}\beta_1 = \frac{R}{L} \qquad \lim_{C\to\infty} e^{-\beta_1 t} = e^{-\frac{R}{L}t}$$

$$\lim_{C\to\infty}\beta_2 = 0 \qquad \lim_{C\to\infty} e^{-\beta_2 t} = 1$$

Therefore, when no capacitor is present

$$\lim_{C\to\infty} Q(t) = 0 \qquad \lim_{C\to\infty} I(t) = \frac{\mathcal{E}}{R}\left(1 - e^{-\frac{R}{L}t}\right)$$

This expression for the current is identical to that derived for the electric circuit in Figure 8.27.

Example 8-11

Problem: The electric circuit in Figure 8.34 is in steady-state with the switch in the position shown in the left panel. The switch is then moved to the position shown in the right panel of Figure 8.34. What is the equation for the electric charge on the capacitor in this new configuration? What happens to the energy in the circuit?

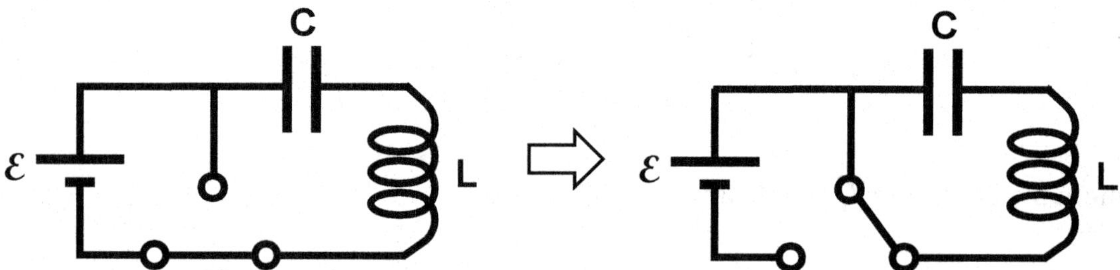

FIGURE 8.34: The electric circuit in Example 8-11. Initially the circuit is a series LC circuit (left panel), but after the switch is closed, the EMF is removed from the circuit (right panel).

Solution: Applying Kirchhoff's loop rule to the electric circuit in the right panel of Figure 8.34 gives us

$$-\frac{q}{C} - L\frac{dI}{dt} = 0 \quad \rightarrow \quad -\frac{q}{C} - L\left(\frac{d^2q}{dt^2}\right) = 0 \quad \rightarrow \quad \left(\frac{d^2q}{dt^2}\right) = -\left(\frac{1}{LC}\right)q$$

This is an equation for simple harmonic oscillation. The electric charge in the system oscillates between being stored on the plates of the capacitor and flowing in the circuit as current. The angular frequency of this oscillation is

$$\omega = \left(\frac{1}{LC}\right)^{\frac{1}{2}}$$

We can determine the equation for the electric charge using the initial conditions of the problem. Since the system was initially in a steady-state with no current flowing in the electric circuit, the capacitor would have been fully charged with a potential difference equal to $\mathcal{E}$. The equations for the electric charge stored on the capacitor and the current in the electric circuit are

$$q = C\mathcal{E}\cos(\omega t)$$

$$I = \frac{dq}{dt} \rightarrow I = -C\mathcal{E}\omega\sin(\omega t)$$

In these equations, we have defined $t = 0$ to be the time when the switch was moved. The positive and negative signs in these equations simply denote the orientation of electric charge on the plates of the capacitor and the direction of the current, as shown in Figure 8.35.

The oscillations of the location of the electric charge in this electric circuit will result in oscillations of the location of the potential energy in the electric circuit. The potential energy in the electric circuit oscillates between being stored as electric potential energy in the capacitor and as magnetic potential energy in the inductor. The electric potential energy of the capacitor is found using Equation 5-8.

$$q = C\mathcal{E}\cos(\omega t) \rightarrow U_e = \frac{(C\mathcal{E}\cos(\omega t))^2}{2C} \rightarrow U_e = \frac{\mathcal{E}^2 C}{2}\cos^2(\omega t)$$

The equation for the magnetic potential energy of the inductor is found using Equation 6-15.

$$I = \frac{dq}{dt} \rightarrow I = -C\mathcal{E}\omega \sin(\omega t) \rightarrow U_m = \frac{\mathcal{E}^2 C^2 \omega^2 L}{2} \sin^2(\omega t)$$

$$\omega = \left(\frac{1}{LC}\right)^{\frac{1}{2}} \rightarrow U_m = \frac{\mathcal{E}^2 C^2 L}{2}\left(\frac{1}{LC}\right)\sin^2(\omega t) \rightarrow U_m = \frac{\mathcal{E}^2 C}{2}\sin^2(\omega t)$$

As expected, the sum of the electric potential energy of the capacitor and the magnetic potential energy of the inductor is a constant.

$$U_e + U_m = \frac{\mathcal{E}^2 C}{2}\cos^2(\omega t) + \frac{\mathcal{E}^2 C}{2}\sin^2(\omega t) \rightarrow U_e + U_m = \frac{\mathcal{E}^2 C}{2}$$

The oscillations of the electric charge and the potential energy of this system are shown in Figure 8.35.

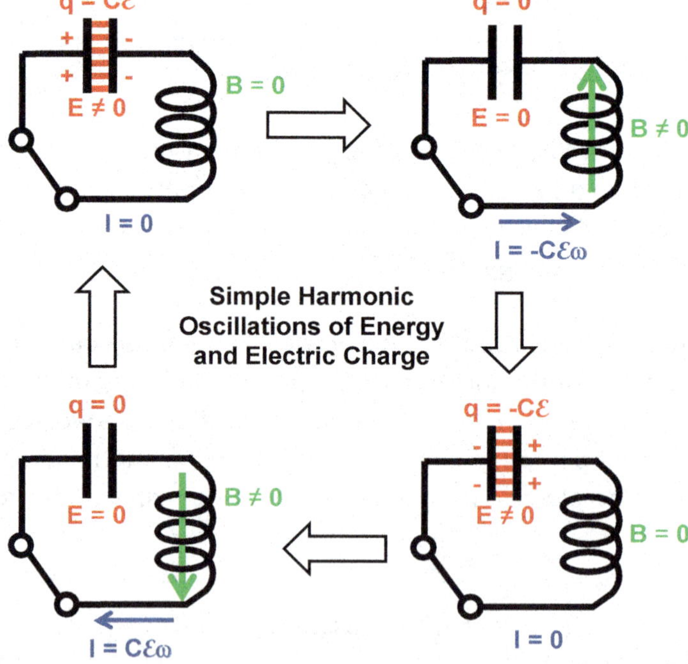

FIGURE 8.35: The cycle of simple harmonic oscillations in a series LC circuit.

8-7 DC Circuits with Alternative EMFs

As we learned in Chapter 6, a DC EMF can also be created without a battery. Consider the electric circuit shown in Figure 8.36, which is in the presence of a uniform magnetic field.

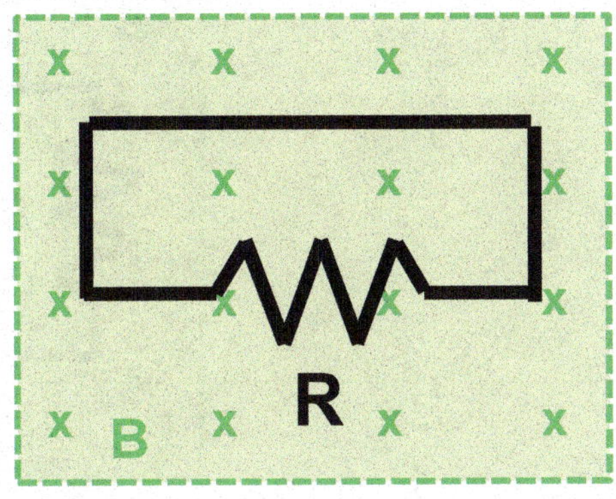

FIGURE 8.36: A DC circuit whose EMF results from a changing magnetic flux.

If the magnetic flux through the electric circuit changes, an EMF will be induced in the electric circuit (Equation 6-11). If the cross-sectional area of the electric circuit in Figure 8.36 is constant and equal to A, the magnitude of this EMF will be

$$\mathcal{E} = -\frac{d}{dt}(BA) \rightarrow |\mathcal{E}| = A\frac{dB}{dt}$$

We can determine the magnitude of the electric current flowing in this circuit by examining what happens to the energy in this system. The energy supplied to the circuit by the EMF is dissipated by the resistor. Thus, the power supplied by the EMF must equal the power supplied to the resistor.

$$P_{EMF} = P_{resistor} \rightarrow I|\mathcal{E}| = I^2R \rightarrow |\mathcal{E}| = IR \rightarrow A\left(\frac{dB}{dt}\right) = IR$$

$$I = \frac{A}{R}\left(\frac{dB}{dt}\right)$$

The direction of the current in the electric circuit can be determined using Lenz's law (Section 6-3).

Example 8-12

Problem: The system in Figure 8.37 consists of a U-shaped wire upon which a conducting rod can freely move. The rod maintains electrical contact with the wire so that current can flow through this system as an electric circuit. This electric circuit contains a single resistor, and the entire system is in the presence of a uniform magnetic field.

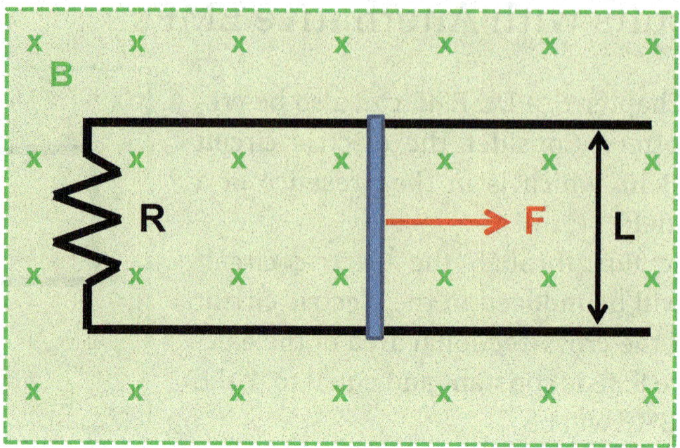

FIGURE 8.37: The system in Example 8-12.

What is the power dissipated by the resistor as the conducting rod is moved to the right in Figure 8.37?

Solution: As the rod is moved to the right, the magnetic flux through the electric circuit changes. This will induce an electric field, an associated EMF, and an associated current in the electric circuit. To determine the EMF let's define the direction of the surface area vector to be opposite the direction of the magnetic field, as shown in Figure 8.38.

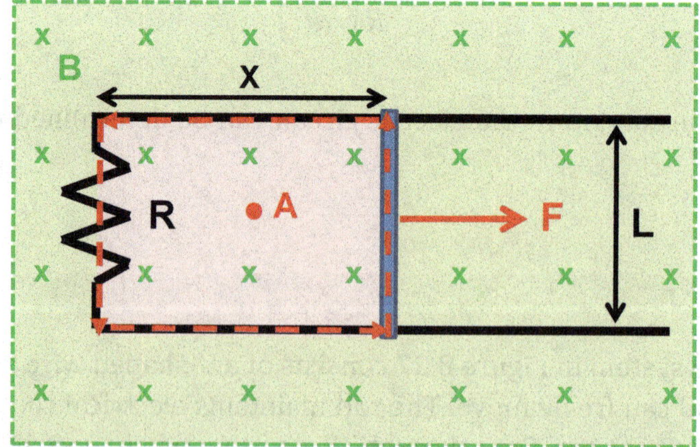

FIGURE 8.38: The area used for calculating the magnetic flux through the electric circuit. The corresponding "path of integration" is shown as dashed arrows.

Let's define the variable x to denote the distance between the left end of the electric circuit and the conducting bar, as shown in Figure 8.38. Then, since the magnitude of the magnetic field is constant, we have

$$\Phi_B = BA\cos\left(180^\circ\right) \rightarrow \Phi_B = -BA \rightarrow \Phi_B = -BLx$$

The induced EMF can be determined using Equation 6-11.

$$\mathcal{E} = -\frac{d\Phi_B}{dt} \rightarrow \mathcal{E} = -\frac{d}{dt}\left(-BLx\right) \rightarrow \mathcal{E} = BL\frac{dx}{dt}$$

The derivative in this expression is the speed of the conducting rod.

$$\frac{dx}{dt} = v \rightarrow \mathcal{E} = BLv$$

The positive sign in this solution indicates that the direction of the EMF (*i.e.*, the direction of the induced current) is the same as the direction of the corresponding "path of integration" that defines the area (*i.e.*, counter-clockwise, as shown in Figure 8.39).

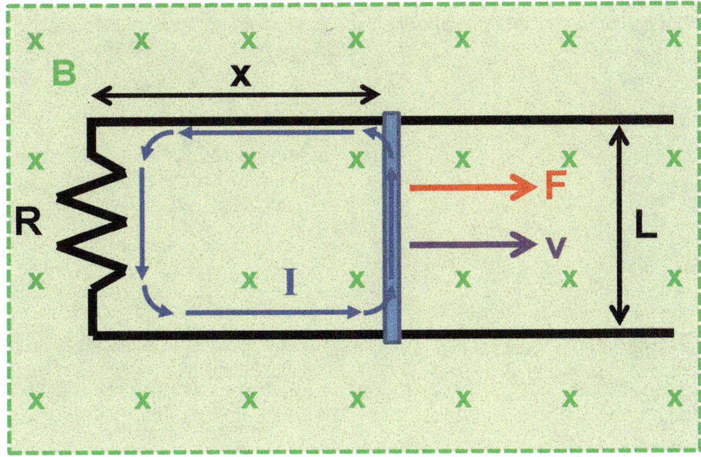

FIGURE 8.39: The direction of the induced current in the electric circuit in Figure 8.37.

Hence, the power dissipated by the resistor is

$$P_{resistor} = I^2 R \;\rightarrow\; P_{resistor} = \left(\frac{BLv}{R}\right)^2 R \;\rightarrow\; P_{resistor} = \frac{B^2 L^2 v^2}{R}$$

This power comes from the EMF, which in turn comes from the work done by the force pulling the rod. For simplicity, let's assume that the rod is moving at constant speed. In that case, the magnitude of the external force F in Figure 8.39 must equal the magnitude of the magnetic force acting on the rod (Section 4-6).

$$F = F_m \;\rightarrow\; F = ILB \;\rightarrow\; F = \left(\frac{BLv}{R}\right) LB \;\rightarrow\; F = \frac{B^2 L^2 v}{R}$$

The power supplied by this force is thus

$$P_F = \vec{F} \bullet \vec{v} \;\rightarrow\; P_F = Fv \;\rightarrow\; P_F = \frac{B^2 L^2 v^2}{R}$$

As expected, the power supplied by the external force is equal to the power dissipated by the resistor.

Summary

- **Electric circuit:** a path through which charge can flow. A closed electric circuit is a continuous path between the high and low potential terminals of an EMF. An open electric circuit is a discontinuous path between the high and low potential terminals of an EMF.

- **Electrical load:** the portion of an electric circuit that dissipates or stores electrical energy.

- **Transient current:** a current whose magnitude and/or direction varies non-periodically with time. Transient currents typically have an exponential dependence upon time.

- **Steady-state current:** a current whose direction and magnitude are constant or vary periodically.

- **Kirchhoff's loop rule:** The sum of the electric potential differences across all components around any closed loop in an electric circuit must be zero.

- **Kirchhoff's junction rule:** The sum of the currents entering any junction must be equal to the sum of the currents leaving that junction.

- Circuit components connected in series are connected so that the same current flows through each of the components.

- Circuit components connected in parallel are connected so that the same change in electric potential occurs across each of the components.

- The equivalent resistance of resistors connected in series or in parallel are

$$\text{Resistors in series: } R_{eq} = R_1 + R_2 + \ldots$$

$$\text{Resistors in parallel: } \frac{1}{R_{eq}} = \frac{1}{R_1} + \frac{1}{R_2} + \ldots$$

- The equivalent inductance of inductors connected in series or in parallel are

$$\text{Inductors in series: } L_{eq} = L_1 + L_2 + \ldots$$

$$\text{Inductors in parallel: } \frac{1}{L_{eq}} = \frac{1}{L_1} + \frac{1}{L_2} + \ldots$$

- The equivalent capacitance of capacitors connected in series or in parallel are

$$\text{Capacitors in series: } \frac{1}{C_{eq}} = \frac{1}{C_1} + \frac{1}{C_2} + \ldots$$

$$\text{Capacitors in parallel: } C_{eq} = C_1 + C_2 + \ldots$$

- Capacitors connected in series can be replaced by a single equivalent capacitor that has the same electric charge as the actual capacitors.

$$\text{Capacitors in series: } q_1 = q_2 = \ldots = q_{eq}$$

Capacitors connected in parallel can be replaced by a single equivalent capacitor that has the same *total* electric charge as the actual capacitors.

$$\text{Capacitors in parallel: } q_1 + q_2 + \ldots = q_{eq}$$

Problems

1. What is the current in each resistor in the electric circuit shown in the figure?

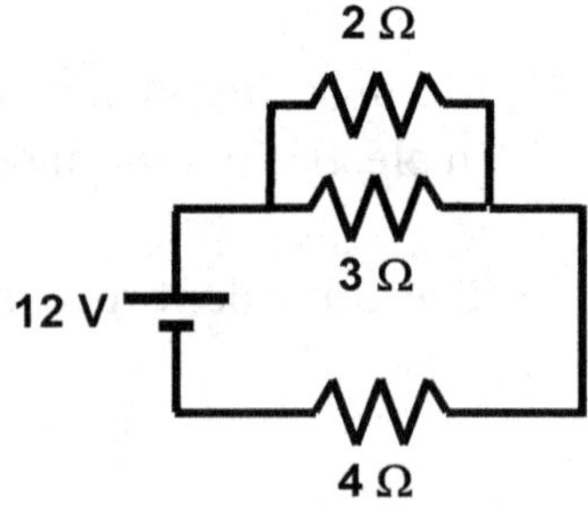

2. What is the current in the electric circuit shown in the figure?

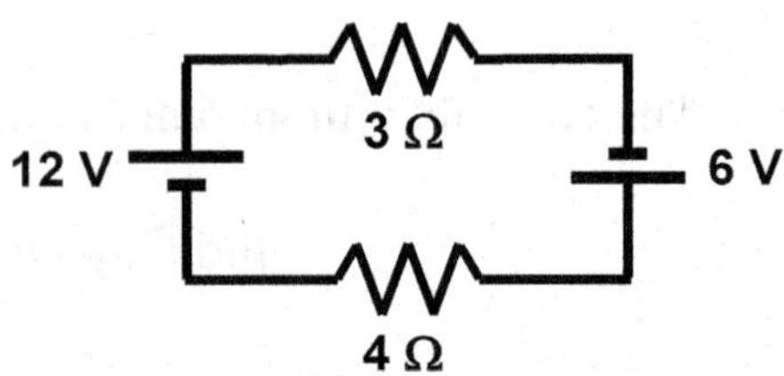

3. What is the current in each resistor in the electric circuit shown in the figure?

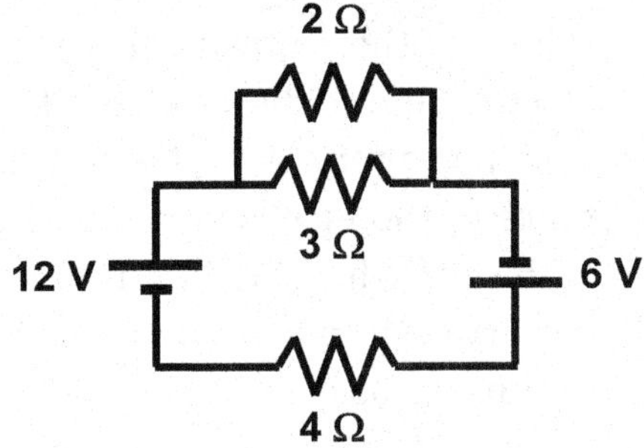

4. What is the equivalent resistance of the electric circuit shown in the figure? What is the current supplied by the battery to the electric circuit?

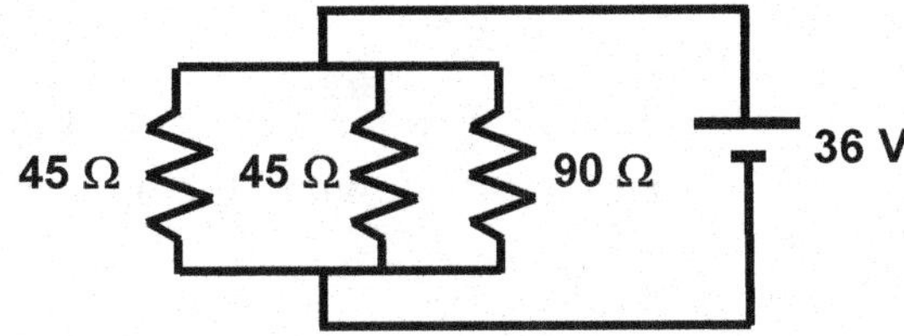

5. What is the equivalent resistance of the electric circuit shown in the figure? What is the current supplied by the battery to the electric circuit?

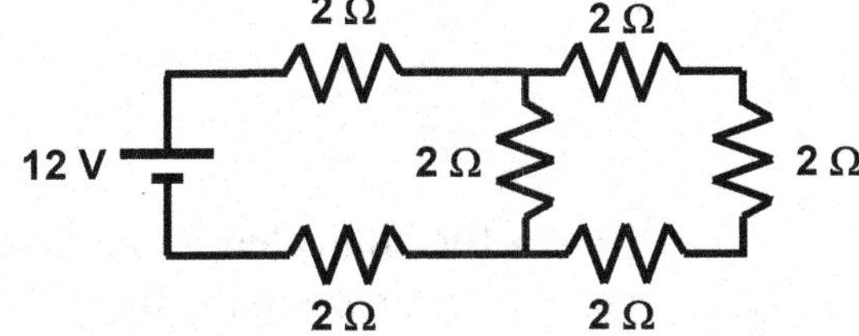

6. What is the equivalent resistance of the electric circuit shown in the figure? What is the current supplied by the battery to the electric circuit?

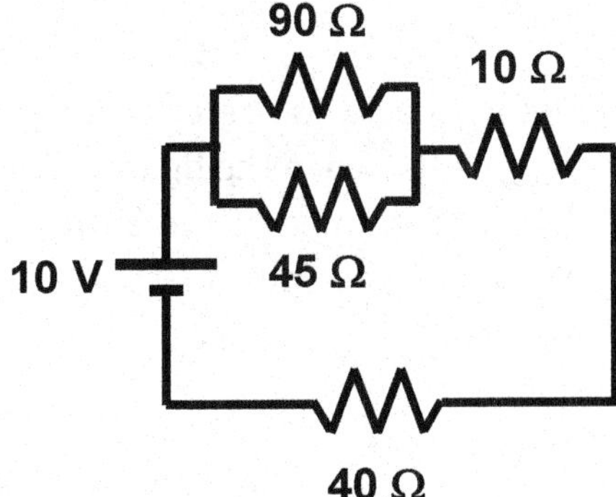

7. What is the equivalent resistance of the electric circuit shown in the figure when the switch is open? What is the equivalent resistance of the electric circuit when the switch is closed? What is the current supplied by the battery to the electric circuit in these configurations of the electric circuit?

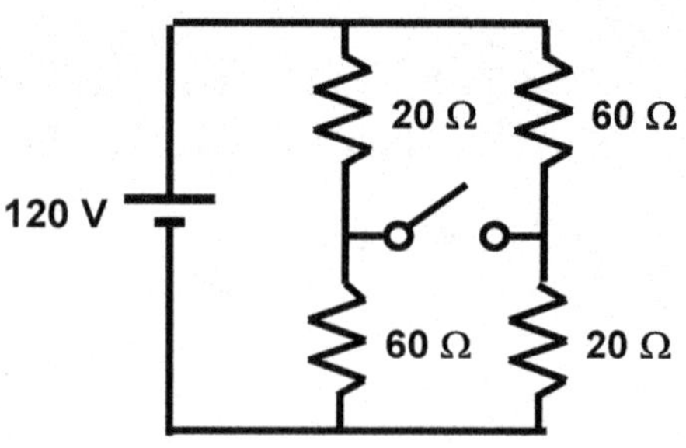

8. What is the equivalent inductance of the electric circuit shown in the figure?

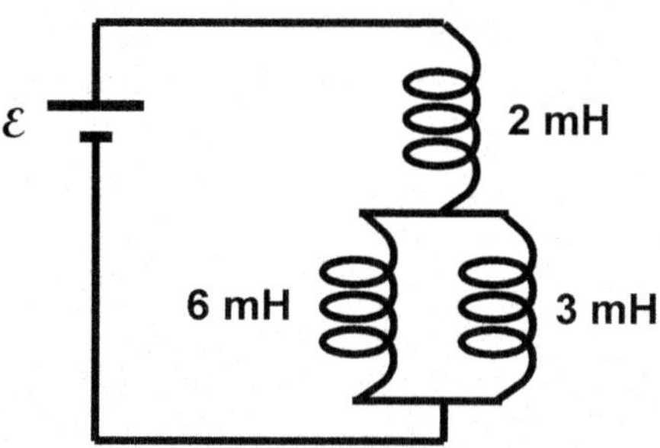

9. What is the equivalent inductance of the electric circuit shown in the figure?

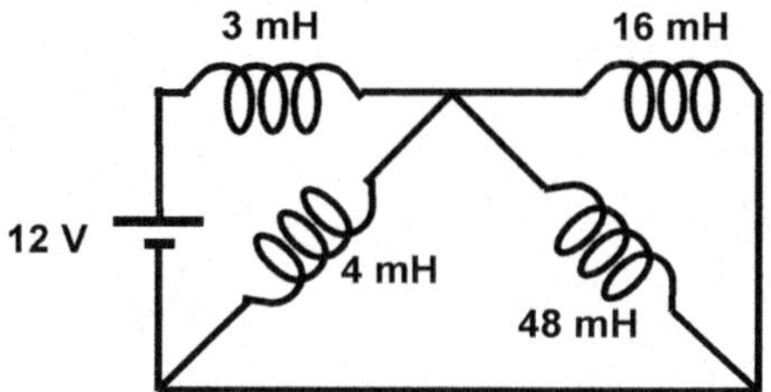

10. Four capacitors are connected to a 12 V battery, as shown in the figure. What is the equivalent capacitance of the entire electric circuit? What is the net electric charge on each capacitor?

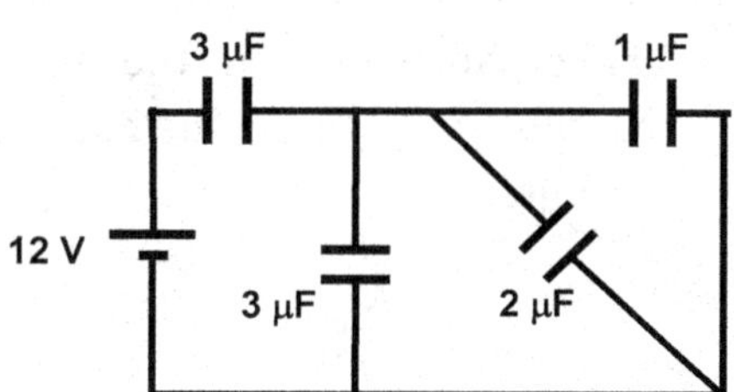

11. After the switch is closed in the electric circuit shown in the figure, current will start flow and the capacitor will charge. How long will it take for the electric charge on the capacitor to reach 50% of its final value? How much energy will be dissipated by each resistor during this time interval?

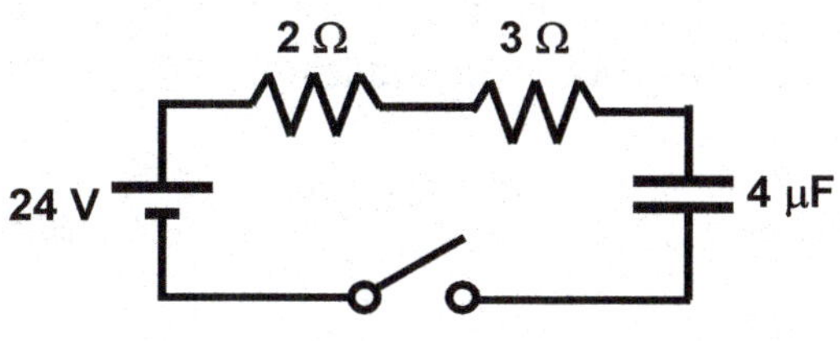

12. After the switch is closed in the electric circuit shown in the figure, current will start flow and the capacitor will charge. How long will it take for the current to reach 75% of its steady-state value? How much energy will be dissipated by each resistor during this time interval?

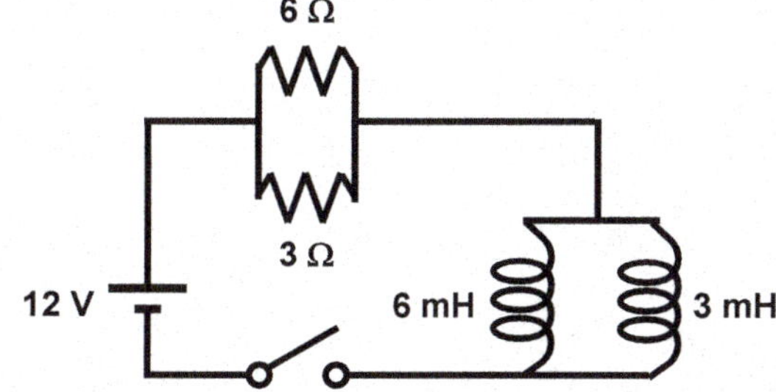

13. The electric circuit shown in the figure has a cross-sectional area of 100 cm^2. The resistance of the resistor in the electric circuit is 5 Ω. The capacitor in the electric circuit has a capacitance of 2 μF and is initially uncharged. The magnitude of the magnetic field is increasing at a constant rate of 0.2 T/s. How long will it take for the capacitor to obtain a charge of 2 nC?

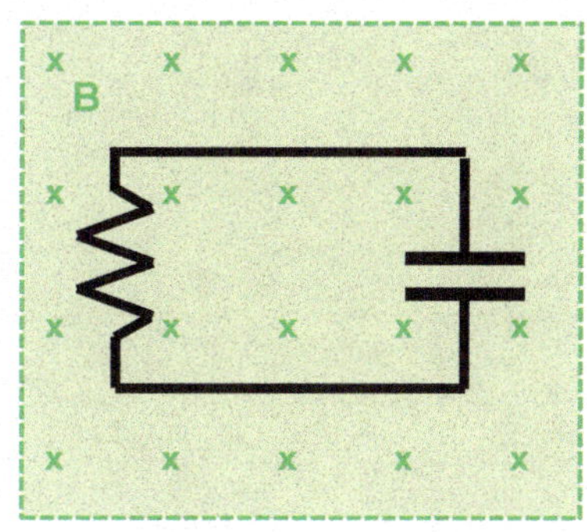

14. A square electric circuit contains a single resistor with resistance R. The length of each side of the square is w. The electric circuit is initially a distance d away from an infinitely long wire carrying current I, as shown in the

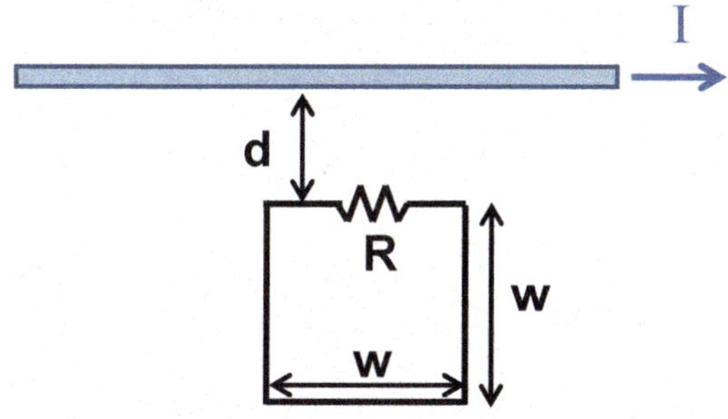

figure. The current in the wire then begins to increase in magnitude according to the following equation:

$$I(t) = I_0\left(2 - e^{-kt}\right)$$

What is the current in the circuit as a function of time? What is the power dissipated by the resistor as a function of time?

CHAPTER 9

Alternating Current Circuits

9-1 Introduction

In this chapter, we turn our attention to AC circuits. We will largely ignore any transient currents in these circuits, such as those discussed in Section 8-6, and instead focus on steady-state AC currents.

In Equation 7-18, we introduced the following complex notation for an alternating EMF:

$$\mathcal{E} = \mathcal{E}_0 e^{i\omega t} \tag{9-1}$$

The motivation for using this notation is that it simplifies our calculations for the current in an AC circuit and for the differences in electric potential across the components of an AC circuit. We see from Equation D-5 that an equivalent expression for the alternating EMF would be

$$\mathcal{E} = \mathcal{E}_0 \left[\cos(\omega t) + i \sin(\omega t) \right]$$

We can therefore consider the EMF to be the real part of Equation 9-1.

$$\mathcal{E} = \mathcal{E}_0 \cos(\omega t)$$

Similarly, the current in an AC circuit is the real part of the complex current that we determine. The motivation for keeping the complex notation is that it will allow us to determine more easily not only the currents in AC circuits but also the differences in the phases of oscillations of these currents and the AC EMF. If you are inexperienced with this use of complex numbers, it is strongly recommended that you read Appendix D before continuing with this chapter.

9-2 Impedance and Ohm's Law

As discussed in Section 7-5, the equivalent expression for Ohm's law in AC circuits is

$$I = \frac{\mathcal{E}}{Z} \tag{9-2}$$

The variable Z in Equation 9-2 is called the ***impedance***.

> *Impedance*: the effective resistance of an electric circuit or component to alternating current. Impedance is a complex number and is denoted by the variable Z. The unit of impedance is the ohm (Ω).

The impedance of a resistor is

$$Z_R = R \tag{9-3}$$

The impedance of an inductor is

$$Z_L = i\omega L \tag{9-4}$$

The impedance of a capacitor is

$$Z_C = -\frac{i}{\omega C} \tag{9-5}$$

The variable ω in Equation 9-4 and Equation 9-5 is the angular frequency of oscillation of the AC EMF in the circuit. Using Equation D-6, we can write the impedance in terms of a complex exponential.

$$Z = |Z| e^{i\phi} \quad \rightarrow \quad Z = \left[\left(\mathrm{Re}(Z) \right)^2 + \left(\mathrm{Im}(Z) \right)^2 \right]^{\frac{1}{2}} e^{i\phi} \tag{9-6}$$

In Equation 9-6, Re(Z) denotes the real part of the impedance and Im(Z) denotes the imaginary part of the impedance. The angle ϕ in Equation 9-6 is determined from the imaginary and real parts of the impedance using Equation 9-7[1].

$$\tan\phi = \frac{\mathrm{Im}(Z)}{\mathrm{Re}(Z)} \tag{9-7}$$

Ohm's law in AC circuits (Equation 9-2) can therefore be written as

$$I = \frac{\mathcal{E}}{Z} \quad \rightarrow \quad I = \frac{\mathcal{E}_0 e^{i\omega t}}{|Z| e^{i\phi}} \quad \rightarrow \quad I = I_0 e^{i(\omega t - \phi)} \tag{9-8}$$

It follows from Equation 9-8 that the oscillations of the current and the oscillations of the EMF will be out of phase if the corresponding impedance has an imaginary component; $\phi \neq 0$ if the impedance has an imaginary component. If the impedance has only a real component, then the oscillations of the current are in-phase with the oscillations of the EMF; $\phi = 0$ if the impedance has only a real component. The variable I_0 in Equation 9-8 is the *amplitude* of the oscillations of the current.

$$I_0 = \frac{\mathcal{E}_0}{|Z|} \tag{9-9}$$

Let's use Ohm's law to determine the current in the electric circuit shown in Figure 9.1

The impedance of this circuit is simply the impedance of the resistor. Hence, from Equation 9-3 and Equation 9-7 we have

$$Z = R \quad \rightarrow \quad \tan\phi = \frac{0}{R} \quad \rightarrow \quad \phi = 0$$

1 Equation 9-7 is the equivalent of Equation D-3.

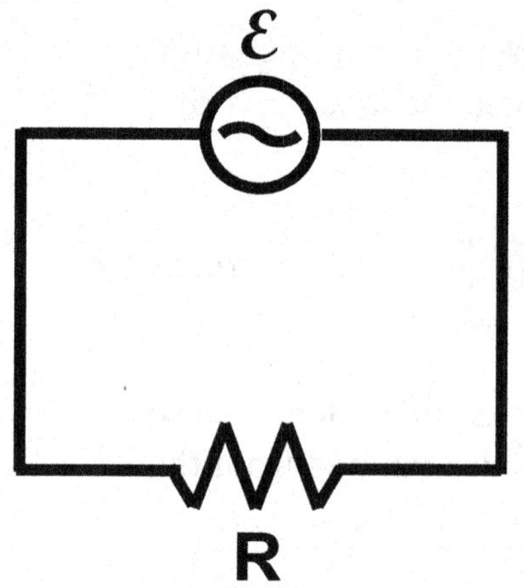

FIGURE 9.1: An AC circuit with only resistance.

Substitution into Equation 9-8 and Equation 9-9 gives us gives us

$$Z = R \;\rightarrow\; |Z| = R \;\rightarrow\; I_0 = \frac{\mathcal{E}_0}{R} \;\rightarrow\; I = \frac{\mathcal{E}_0}{R} e^{i\omega t}$$

The current is thus oscillating in phase with the EMF.

$$\mathcal{E} \sim e^{i\omega t} \qquad I \sim e^{i\omega t}$$

The real parts of the EMF and the current are

$$\mathcal{E} = \mathcal{E}_0 \cos(\omega t) \qquad I = \frac{\mathcal{E}_0}{R}\cos(\omega t)$$

As expected, when the angular frequency of the oscillation of the EMF is zero, the current in this electric circuit matches the current of the DC electric circuit in Example 8-1.

Example 9-1

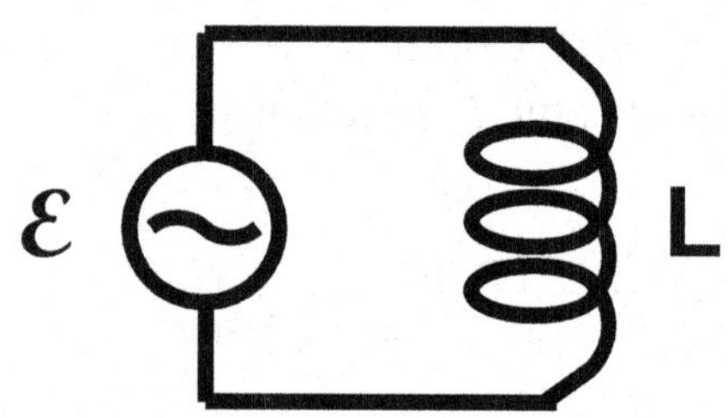

FIGURE 9.2: The AC circuit in Example 9-1.

Problem: What is the difference in phase between the oscillations of the EMF and the oscillations of the current in Figure 9.2? What is the equation for the current in this electric circuit? What is the equation for the change in electric potential across the inductor in this electric circuit?

Solution: The impedance of this circuit is simply the impedance of the inductor. Thus, from Equation 9-4 and Equation 9-7, we have

$$Z = i\omega L \;\rightarrow\; \tan\phi = \frac{\omega L}{0} \;\rightarrow\; \tan\phi = \infty \;\rightarrow\; \phi = \frac{\pi}{2}$$

Substitution into Equation 9-8 and Equation 9-9 gives us gives us

$$Z = i\omega L \quad \rightarrow \quad |Z| = \omega L \quad \rightarrow \quad I_0 = \frac{\mathcal{E}_0}{\omega L} \quad \rightarrow \quad I = \frac{\mathcal{E}_0}{\omega L} e^{i\left(\omega t - \frac{\pi}{2}\right)}$$

The oscillations of the current are thus π/2 radians out of phase with the oscillations of the EMF.

$$\mathcal{E} \sim e^{i\omega t} \qquad\qquad I \sim e^{i\left(\omega t - \frac{\pi}{2}\right)}$$

The change in the electric potential across the inductor is determined using Equation 6-4.

$$\Delta V_e = -L\frac{dI}{dt} \quad \rightarrow \quad \Delta V_e = -L(i\omega)\left(\frac{\mathcal{E}_0}{\omega L}\right)e^{i\left(\omega t - \frac{\pi}{2}\right)} \quad \rightarrow \quad \Delta V_e = -i\mathcal{E}_0 e^{i\left(\omega t - \frac{\pi}{2}\right)}$$

Applying Equation D-5 then gives us

$$\Delta V_e = -i\mathcal{E}_0\left[\cos\left(\omega t - \frac{\pi}{2}\right) + i\sin\left(\omega t - \frac{\pi}{2}\right)\right] \quad \rightarrow \quad \Delta V_e = -i\mathcal{E}_0\left[\sin(\omega t) - i\cos(\omega t)\right]$$

$$\Delta V_e = -\mathcal{E}_0\left[\cos(\omega t) + i\sin(\omega t)\right]$$

The potential difference across the inductor is the real part of this expression.

$$\Delta V_e = -\mathcal{E}_0 \cos(\omega t)$$

Thus, the oscillations of the electric potential across the inductor are in phase with the oscillations of the EMF.

$$\mathcal{E} \sim \cos(\omega t) \qquad\qquad \Delta V_e \sim \cos(\omega t)$$

This is consistent with Kirchhoff's loop rule.

Example 9-2

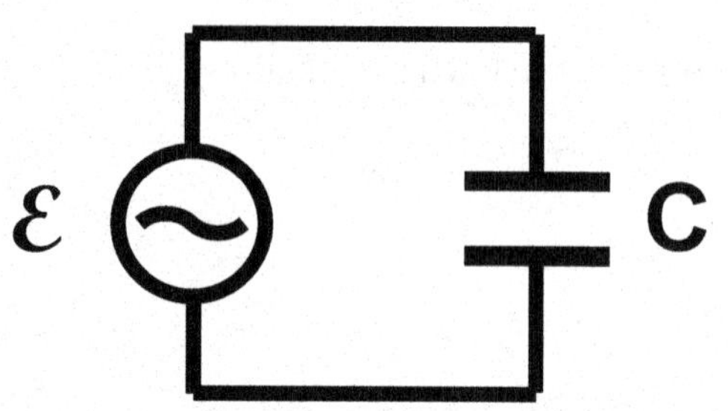

FIGURE 9.3: The AC circuit in Example 9-2.

Problem: What is the difference in the phase of the oscillations of the EMF and the oscillations of the AC current in Figure 9.3? What is the equation for the current in this electric circuit? What is the equation for the change in electric potential across the capacitor in this electric circuit?

Solution: The impedance of this circuit is simply the impedance of the capacitor. Thus, from Equation 9-5 and Equation 9-7, we have

$$Z = -\frac{i}{\omega C} \quad \rightarrow \quad \tan\phi = \frac{-\frac{1}{\omega C}}{0} \quad \rightarrow \quad \tan\phi = -\infty \quad \rightarrow \quad \phi = -\frac{\pi}{2}$$

The current in the electric circuit is found using Equation 9-8.

$$Z = -\frac{i}{\omega C} \quad \rightarrow \quad |Z| = \frac{1}{\omega C} \quad \rightarrow \quad I = \left(\frac{\mathcal{E}_0}{\frac{1}{\omega C}}\right) e^{i\left(\omega t + \frac{\pi}{2}\right)} \quad \rightarrow \quad I = \left(\mathcal{E}_0 \omega C\right) e^{i\left(\omega t + \frac{\pi}{2}\right)}$$

The oscillations of the current are π/2 radians out of phase with the oscillations of the EMF.

$$\mathcal{E} \sim e^{i\omega t} \qquad I \sim e^{i\left(\omega t + \frac{\pi}{2}\right)}$$

The electric charge on the capacitor is then found through integration using Equation 4-4.

$$I = \frac{dq}{dt} \quad \rightarrow \quad \int dq = \int I\,dt \quad \rightarrow \quad q = \int I\,dt$$

$$q = \int \left(\mathcal{E}_0 \omega C\right) e^{i\left(\omega t + \frac{\pi}{2}\right)} dt \quad \rightarrow \quad q = \left(\frac{\mathcal{E}_0 \omega C}{i\omega}\right) e^{i\left(\omega t + \frac{\pi}{2}\right)} \quad \rightarrow \quad q = \left(\frac{\mathcal{E}_0 C}{i}\right) e^{i\left(\omega t + \frac{\pi}{2}\right)}$$

We can rewrite this expression in terms of trigonometric functions using Equation D-5.

$$q = \left(\frac{\mathcal{E}_0 C}{i}\right)\left[\cos\left(\omega t + \frac{\pi}{2}\right) + i\sin\left(\omega t + \frac{\pi}{2}\right)\right] \rightarrow q = \left(\frac{\mathcal{E}_0 C}{i}\right)\left[-\sin(\omega t) + i\cos(\omega t\right.$$

$$q = \mathcal{E}_0 C\left[i\sin(\omega t) + \cos(\omega t)\right]$$

The actual electric charge on the capacitor is the real part of this expression.

$$q = C\mathcal{E}_0 \cos(\omega t)$$

The change in the electric potential across the capacitor is determined using Equation 5-7.

$$\Delta V_e = \frac{q}{C} \rightarrow \Delta V_e = \frac{C\mathcal{E}_0 \cos(\omega t)}{C} \rightarrow \Delta V_e = \mathcal{E}_0 \cos(\omega t)$$

Thus, the oscillations of the electric potential across the inductor are in phase with the oscillations of the EMF.

$$\mathcal{E} \sim \cos(\omega t) \qquad \Delta V_e \sim \cos(\omega t)$$

This is consistent with Kirchhoff's loop rule.

9-3 Series and Parallel AC Circuits

The equivalent impedance of a group of circuit components connected in series is

$$Z_{eq} = Z_1 + Z_2 + \ldots \tag{9-10}$$

The equivalent impedance of a group of circuit components connected in parallel is

$$\frac{1}{Z_{eq}} = \frac{1}{Z_1} + \frac{1}{Z_2} + \ldots \tag{9-11}$$

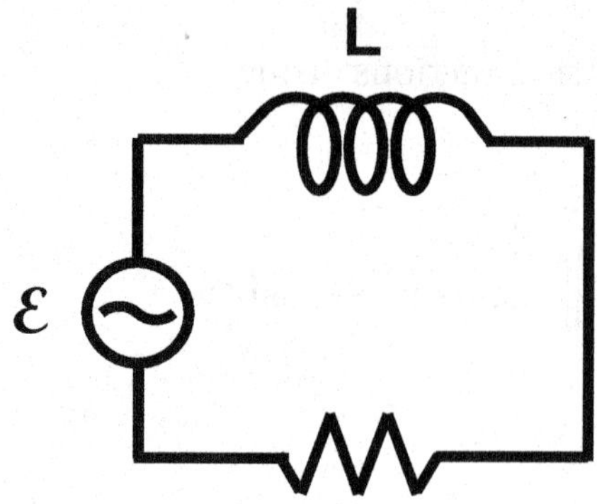

FIGURE 9.4: An AC circuit with resistance and inductance.

Let's consider the electric circuit shown in Figure 9.4 that contains an inductor and a resistor in series with each other.

The impedance of this circuit is found using Equation 9-3, Equation 9-4, and Equation 9-10.

$$Z_{eq} = R + i\omega l$$

Thus, from Equation 9-7, we have

$$Z = R + i\omega L \quad \rightarrow \quad \tan\phi = \frac{\omega L}{R} \quad \rightarrow \quad \phi = \tan^{-1}\left(\frac{\omega L}{R}\right)$$

We see that for this electric circuit, the oscillations of the current are not in-phase with the oscillations of the EMF. This occurs because the impedance of the circuit has an imaginary component. The current in the electric circuit is determined using Equation 9-8.

$$Z = R + i\omega L \quad \rightarrow \quad |Z| = \left(R^2 + \omega^2 L^2\right)^{\frac{1}{2}} \quad \rightarrow \quad I = \left(\frac{\mathcal{E}_0}{\left(R^2 + \omega^2 L^2\right)^{\frac{1}{2}}}\right) e^{i(\omega t - \phi)}$$

$$I_0 = \frac{\mathcal{E}_0}{\left(R^2 + \omega^2 L^2\right)^{\frac{1}{2}}}$$

We see that the amplitude of the current in this electric circuit decreases as the angular frequency of oscillation of the EMF increases. Thus, the electric circuit shown in Figure 9.4 is commonly referred to as a *high-frequency filter*[2], since it attenuates the magnitude of currents oscillating at high angular frequency.

2 This circuit is also called a *low-pass filter* since it has a small impedance for low frequency oscillations.

Example 9-3

Problem: What is the electric current in the circuit shown in Figure 9.5?

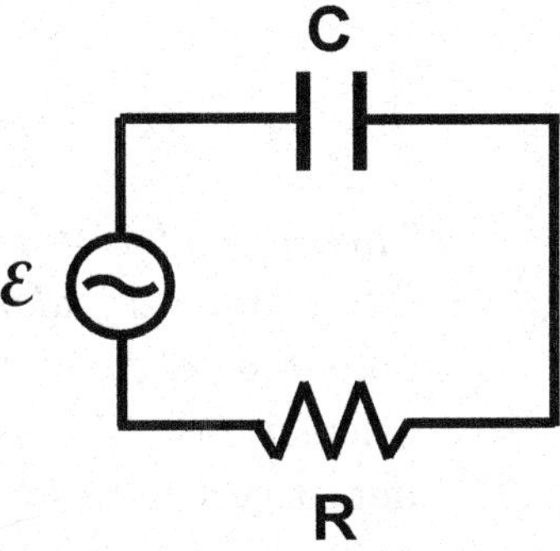

FIGURE 9.5: The electric circuit in Example 9-3.

Solution: The impedance of this circuit is found using Equation 9-3, Equation 9-5, and Equation 9-10.

$$Z_{eq} = R + \left(-\frac{i}{\omega C}\right) \rightarrow Z_{eq} = R - \frac{i}{\omega C}$$

Thus, from Equation 9-7, we have

$$Z = R - \frac{i}{\omega C} \rightarrow \tan\phi = \frac{-\frac{1}{\omega C}}{R} \rightarrow \phi = \tan^{-1}\left(\frac{-\frac{1}{\omega C}}{R}\right) \rightarrow \phi = -\tan^{-1}\left(\frac{1}{\omega CR}\right)$$

We see that for this electric circuit, the oscillations of the current are not in-phase with the oscillations of the EMF. This occurs because the impedance of the circuit has an imaginary component. The current in the electric circuit is determined using Equation 9-8.

$$Z = R - \frac{i}{\omega C} \rightarrow |Z| = \left(R^2 + \frac{1}{\omega^2 C^2}\right)^{\frac{1}{2}} \rightarrow I = \left(\frac{\mathcal{E}_0}{\left(R^2 + \frac{1}{\omega^2 C^2}\right)^{\frac{1}{2}}}\right) e^{i(\omega t - \phi)}$$

$$I_0 = \frac{\mathcal{E}_0}{\left(R^2 + \frac{1}{\omega^2 C^2}\right)^{\frac{1}{2}}} \rightarrow I_0 = \frac{\mathcal{E}_0}{\left(\frac{R^2\omega^2 C^2 + 1}{\omega^2 C^2}\right)^{\frac{1}{2}}} \rightarrow I_0 = \frac{\mathcal{E}_0 \omega C}{\left(R^2\omega^2 C^2 + 1\right)^{\frac{1}{2}}}$$

We see that the amplitude of the current in this circuit decreases as the angular frequency of oscillation of the EMF decreases. Thus, the electric circuit shown in Figure 9.5 is commonly referred to as a *low-frequency filter*[3], since it attenuates the magnitude of currents oscillating at low angular frequency.

Therefore, *placing a capacitor into an AC circuit attenuates the amplitude of currents oscillating at low angular frequency, and placing an inductor into an AC circuit attenuates the amplitude of currents oscillating at high angular frequency*. This makes sense from our calculations of DC circuits in Chapter 8. A capacitor is an open circuit and therefore will not carry steady-state DC current (*i.e.*, a capacitor has an infinite "effective resistance" to current with $\omega = 0$). In contrast, there will be no induced EMF in an inductor carrying steady current (*i.e.*, an inductor has zero "effective resistance" to current with $\omega = 0$).

Example 9-4

Problem: What is the impedance of the electric circuit shown in Figure 9.6? What is the amplitude of the current supplied by the EMF in this circuit?

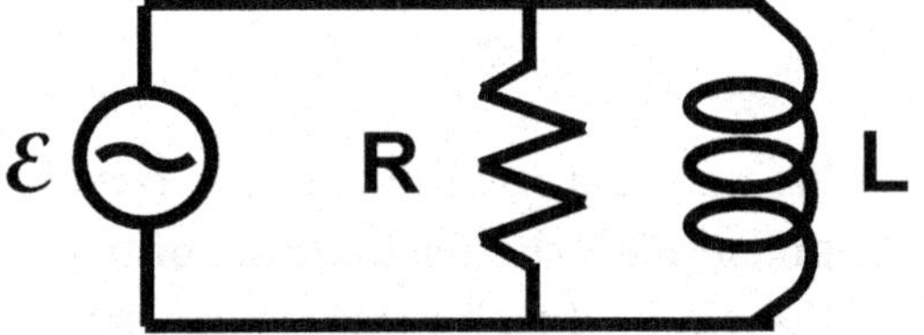

FIGURE 9.6: The electric circuit in Example 9-4.

Solution: The resistor and inductor are in parallel with each other. The equivalent impedance of these two components is determined using Equation 9-11.

$$\frac{1}{Z_{eq}} = \frac{1}{R} + \frac{1}{i\omega L} \rightarrow \frac{1}{Z_{eq}} = \frac{R + i\omega L}{i\omega LR} \rightarrow Z_{eq} = \frac{i\omega LR}{R + i\omega L}$$

We can rewrite this impedance as

$$Z_{eq} = \left(\frac{i\omega LR}{R + i\omega L}\right)\left(\frac{R - i\omega L}{R - i\omega L}\right) \rightarrow Z_{eq} = \left(\frac{1}{R^2 + \omega^2 L^2}\right)\left[\omega^2 L^2 R + i\left(\omega LR^2\right)\right]$$

$$Z_{eq} = \frac{\left(\left(\omega^2 L^2 R\right)^2 + \left(\omega LR^2\right)^2\right)^{\frac{1}{2}}}{R^2 + \omega^2 L^2} e^{i \tan^{-1}\left(\frac{R}{\omega L}\right)}$$

3 This circuit is also called a *high-pass filter* since it has a small impedance for high frequency oscillations.

The amplitude of the current supplied by the EMF is determined using Equation 9-9.

$$I_0 = \frac{\mathcal{E}_0}{\frac{\left(\left(\omega^2 L^2 R\right)^2 + \left(\omega L R^2\right)^2\right)^{\frac{1}{2}}}{R^2 + \omega^2 L^2}} \rightarrow I_0 = \frac{\mathcal{E}_0\left(R^2 + \omega^2 L^2\right)}{\left(\left(\omega^2 L^2 R\right)^2 + \left(\omega L R^2\right)^2\right)^{\frac{1}{2}}}$$

The electric circuit shown in Figure 9.7 is typically referred to as a parallel LC circuit.

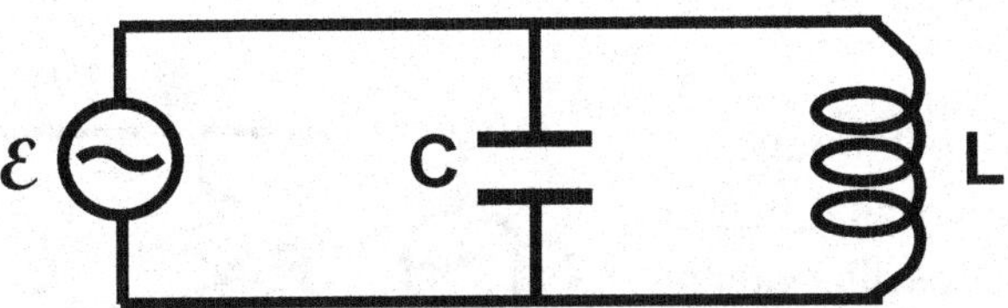

FIGURE 9.7: A parallel LC circuit.

The impedance of this electric circuit is

$$\frac{1}{Z_{eq}} = \frac{1}{-\frac{i}{\omega C}} + \frac{1}{i\omega L} \rightarrow \frac{1}{Z_{eq}} = i\omega C - \frac{i}{\omega L} \rightarrow \frac{1}{Z_{eq}} = \frac{i\omega^2 LC - i}{\omega L}$$

$$Z_{eq} = \frac{\omega L}{i\left(\omega^2 LC - 1\right)} \rightarrow Z_{eq} = \frac{-i\omega L}{\omega^2 LC - 1}$$

$$Z_{eq} = \left(\frac{\omega L}{\omega^2 LC - 1}\right) e^{i\tan^{-1}\left(\frac{\frac{\omega L}{\omega^2 LC-1}}{0}\right)} \rightarrow Z_{eq} = \left(\frac{\omega L}{\omega^2 LC - 1}\right) e^{i\frac{\pi}{2}}$$

The amplitude of the current supplied by the EMF in this circuit is determined using Equation 9-9.

$$I_0 = \frac{\mathcal{E}_0}{\left(\frac{\omega L}{\omega^2 LC - 1}\right)} \rightarrow I_0 = \frac{\mathcal{E}_0\left(\omega^2 LC - 1\right)}{\omega L}$$

This amplitude depends upon the angular frequency of oscillation of the EMF and is exactly zero when $\omega = \left(\frac{1}{LC}\right)^{\frac{1}{2}}$. Thus, the parallel LC circuit will "filter out" an exact angular frequency of oscillation of electric current.

Example 9-5

Problem: What is the difference in phase between the oscillations of the current and the oscillations of the EMF for the parallel RLC circuit shown in Figure 9.8? What is the amplitude of the current supplied by the EMF in this circuit?

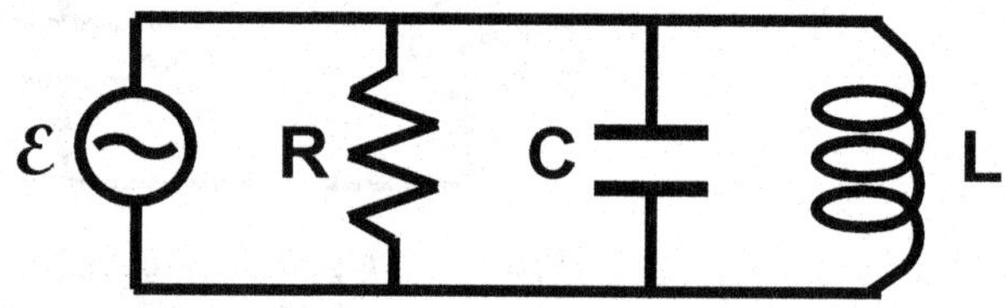

FIGURE 9.8: A parallel RLC circuit.

Solution: The resistor, capacitor, and inductor are all in parallel with each other. The equivalent impedance of these components is

$$\frac{1}{Z_{eq}} = \frac{1}{R} + \frac{1}{i\omega L} + \frac{1}{\frac{-i}{\omega C}} \rightarrow \frac{1}{Z_{eq}} = \frac{1}{R} - \frac{i}{\omega L} + i\omega C \rightarrow \frac{1}{Z_{eq}} = \frac{\omega L - iR + i\omega^2 CRL}{R\omega L}$$

$$Z_{eq} = \frac{R\omega L}{\omega L + i\left(\omega^2 CRL - R\right)} \rightarrow Z_{eq} = \frac{R\omega L}{\omega^2 L^2 + \left(\omega^2 CRL - R\right)^2}\left(\omega L - i\left(\omega^2 CRL - R\right)\right)$$

The difference in phase between the oscillations of the current and the oscillations of the EMF is determined using Equation 9-7.

$$\tan\phi = \frac{-\left(\omega^2 CRL - R\right)}{\omega L} \rightarrow \phi = -\tan^{-1}\left(\frac{\omega^2 CRL - R}{\omega L}\right)$$

The amplitude of the current supplied by the EMF is found using Equation 9-9.

$$|Z_{eq}| = \frac{R\omega L}{\omega^2 L^2 + (\omega^2 CRL - R)^2}\left((\omega L)^2 + (\omega^2 CRL - R)^2\right)^{\frac{1}{2}}$$

$$|Z_{eq}| = \frac{R\omega L}{\left(\omega^2 L^2 + (\omega^2 CRL - R)^2\right)^{\frac{1}{2}}} \rightarrow I_0 = \frac{\mathcal{E}_0}{\frac{R\omega L}{\left(\omega^2 L^2 + (\omega^2 CRL - R)^2\right)}}$$

$$I_0 = \frac{\mathcal{E}_0\left((\omega L)^2 + (\omega^2 CRL - R)^2\right)^{\frac{1}{2}}}{R\omega L}$$

The current supplied by the EMF in this electric circuit is a minimum when

$$\omega^2 CRL - R = 0 \rightarrow \omega = \frac{1}{(LC)^{\frac{1}{2}}}$$

So far, we have determined the impedance and current in an electric circuit for any angular frequency of oscillation of the EMF; in other words, our algebraic solutions for these quantities have included the angular frequency (ω) as a variable. Carrying along the variable ω through all of the algebra of these calculations can be tedious, as we have seen. Therefore, when the angular frequency of the EMF is known, we should simplify our calculations by expressing all impedances in terms of their values at that specific angular frequency.

Example 9-6

Problem: The EMF in the electric circuit shown in Figure 9.9 is described by the following equation:

$$\mathcal{E} = (60\,\mathrm{V})e^{i\left(20\frac{rad}{s}\right)t}$$

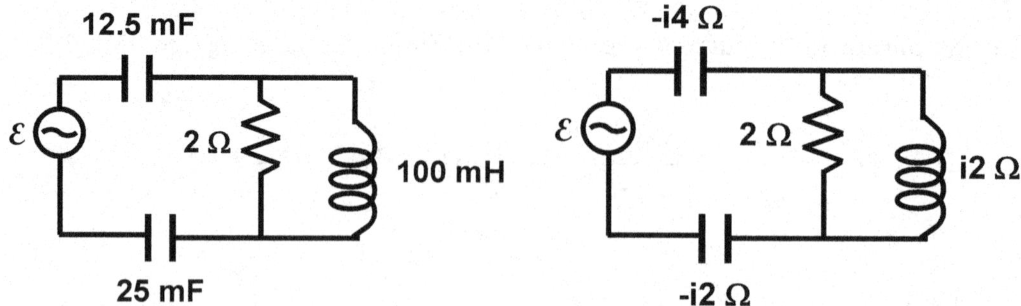

FIGURE 9.9: The electric circuit in Example 9-6.

FIGURE 9.10: The impedances of the components of the electric circuit in Example 9-6 when $\omega = 20$ rad/s.

What is the difference in phase between the oscillations of the current and the oscillations of the EMF for this circuit? What is the amplitude of the current supplied by the EMF in this circuit?

Solution: We can redraw the circuit in terms of the specific impedances of the components at $\omega = 20$ rad/s using Equation 9-3, Equation 9-4, and Equation 9-5. This redrawn electric circuit is shown in Figure 9.10.

The inductor and the resistor are in parallel with each other. The equivalent impedance of these two components is

$$\frac{1}{Z_{eq}} = \frac{1}{2\Omega} + \frac{1}{i2\Omega} \rightarrow \frac{1}{Z_{eq}} = \frac{i+1}{i2\Omega} \rightarrow Z_{eq} = \frac{i2\Omega}{i+1}$$

This equivalent impedance is in series with the two capacitors. Thus, the equivalent impedance of the entire electric circuit is

$$Z_{eq} = \frac{i2\Omega}{i+1} + (-i2\Omega) + (-i4\Omega) \rightarrow Z_{eq} = \frac{i2\Omega}{i+1} - i6\Omega$$

$$Z_{eq} = \frac{i2\Omega - i6\Omega + 6\Omega}{i+1} \rightarrow Z_{eq} = \frac{6\Omega - i4\Omega}{i+1}$$

$$Z_{eq} = \left(\frac{6\Omega - i4\Omega}{i+1}\right)\left(\frac{-i+1}{-i+1}\right) \rightarrow Z_{eq} = \frac{2\Omega - i10\Omega}{2} \rightarrow Z_{eq} = 1\Omega - i5\Omega$$

The difference in phase between the oscillations of the current and the oscillations of the EMF is determined using Equation 9-7.

$$Z_{eq} = 1\Omega - i5\Omega \rightarrow \tan\phi = \frac{-5\Omega}{1\Omega} \rightarrow \phi = -\tan^{-1}(5) \rightarrow \phi = -1.37$$

The amplitude of the current supplied by the EMF is found using Equation 9-9.

$$Z_{eq} = 1\Omega - i5\Omega \rightarrow |Z_{eq}| = \left((1\Omega)^2 + (5\Omega)^2\right)^{\frac{1}{2}} \rightarrow |Z_{eq}| = 5.1\Omega$$

$$I_0 = \frac{\mathcal{E}_0}{|Z_{eq}|} \rightarrow I_0 = \frac{60\,\text{V}}{5.1\Omega} \rightarrow I_0 = 11.8\,\text{A}$$

9-4 Power in AC Circuits

We can substitute Equation 9-8 into Equation 7-11 to determine the instantaneous power supplied by the EMF in an AC circuit. The real part of the instantaneous power supplied is

$$P_{EMF} = I\mathcal{E} \rightarrow P_{EMF} = \left(I_0\cos(\omega t + \phi)\right)\left(\mathcal{E}_0\cos(\omega t)\right)$$

We can simplify the first cosine term using a trigonometric identity (Equation A-1).

$$P_{EMF} = I_0\mathcal{E}_0\left(\cos(\omega t)\cos(\phi) - \sin(\omega t)\sin(\phi)\right)\cos(\omega t)$$

$$P_{EMF} = I_0\mathcal{E}_0\left(\cos^2(\omega t)\cos(\phi) - \sin(\omega t)\cos(\omega t)\sin(\phi)\right)$$

Rather than the instantaneous power, which varies in time, the average power is the more interesting/meaningful quantity for an AC circuit. The average power supplied by the EMF is given in Equation 9-12[4].

$$\left(P_{EMF}\right)_{avg} = \frac{1}{2}I_0\mathcal{E}_0\cos(\phi) \qquad \textbf{(9-12)}$$

4 See Section C-11 for the derivation of Equation 9-12.

The variable I_0 in Equation 9-12 is the amplitude of the current supplied by the EMF (Equation 9-9). The cosine function in Equation 9-12 can also be written in terms of the impedance using Equation D-4.

$$\left(P_{EMF}\right)_{avg} = \frac{1}{2} I_0 \mathcal{E}_0 \frac{\text{Re}(Z)}{|Z|} \tag{9-13}$$

Similarly, the instantaneous power dissipated by a resistor in an AC circuit can be determined using Equation 8-1.

$$P_R = I^2 R \;\rightarrow\; P_R = \left(I_0 \cos(\omega t + \phi)\right)^2 R \;\rightarrow\; P_R = I_0^2 R \cos^2(\omega t + \phi)$$

The average power dissipated by the resistor is

$$\left(P_R\right)_{avg} = \frac{1}{2} I_0^2 R \tag{9-14}$$

The variable I_0 in Equation 9-14 is the amplitude of the current flowing through the resistor.

Example 9-7

Problem: What is the average power supplied by the EMF in the electric circuit shown in Figure 9.2?

Solution: The average power supplied by the EMF is determined using Equation 9-12; the phase angle ϕ for this circuit was determined previously to be $\pi/2$.

$$\phi = \frac{\pi}{2} \;\rightarrow\; P_{avg} = \frac{1}{2} I_0 \mathcal{E}_0 \cos\left(\frac{\pi}{2}\right) \;\rightarrow\; P_{avg} = 0$$

So what happens to the energy in the electric circuit shown in Figure 9.2? The energy supplied by the EMF to the electric circuit is "stored" as magnetic potential energy in the inductor. This stored energy is then "given back" to the EMF when the direction of the current reverses. Therefore, the energy in the electric circuit is oscillating between being stored in the EMF and being stored in the inductor, and no energy is dissipated from the electric circuit. Since the energy of the system is not changing, the average power supplied by the EMF must be zero. Similarly, the average power supplied to the electric circuit shown in Figure 9.3 is also zero. In this case, the energy oscillates between being stored in the EMF and being stored in the electric potential energy of the capacitor.

Example 9-8

Problem: The EMF in the electric circuit shown in Figure 9.11 is described by the following equation:

$$\mathcal{E} = (60V)e^{i\left(20\frac{rad}{s}\right)t}$$

What is the average power supplied by the EMF in this electric circuit?

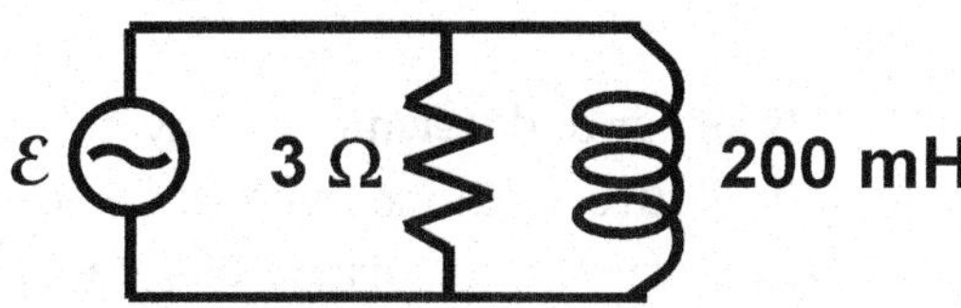

FIGURE 9.11: The electric circuit in Example 9-8.

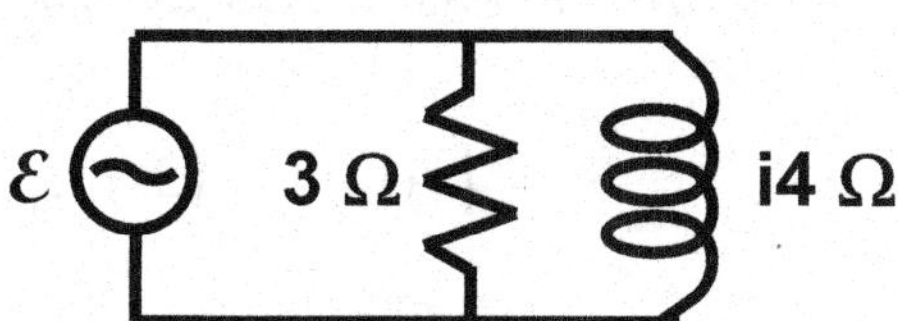

FIGURE 9.12: The electric circuit in Example 9-8. when $\omega = 20$ rad/s.

Solution: The equivalent electric circuit at the angular frequency of the EMF oscillation is shown in Figure 9.12.

Since the inductor and resistor are in parallel with each other, the equivalent impedance of the electric circuit is

$$\frac{1}{Z_{eq}} = \frac{1}{3\Omega} + \frac{1}{i4\Omega} \rightarrow \frac{1}{Z_{eq}} = \frac{4i+3}{i12\Omega} \rightarrow Z_{eq} = \frac{i12\Omega}{3+4i}$$

$$Z_{eq} = \left(\frac{i12\Omega}{3+4i}\right)\left(\frac{3-4i}{3-4i}\right) \rightarrow Z_{eq} = \left(\frac{48\Omega + i36\Omega}{3^2+4^2}\right) \rightarrow Z_{eq} = \left(\frac{48}{25}\Omega\right) + i\left(\frac{36}{25}\Omega\right)$$

The amplitude of the current supplied by the EMF is

$$|Z_{eq}| = \left(\left(\frac{48}{25}\Omega\right)^2 + \left(\frac{36}{25}\Omega\right)^2\right)^{\frac{1}{2}} \rightarrow |Z_{eq}| = 2.4\Omega$$

$$I_0 = \frac{60\,\text{V}}{2.4\Omega} \rightarrow I_0 = 25\,\text{A}$$

The average power supplied by the EMF is determined using Equation 9-13.

$$\left(P_{EMF}\right)_{avg}=\frac{1}{2}I_0\mathcal{E}_0\frac{\text{Re}(Z)}{|Z|}\ \rightarrow\ \left(P_{EMF}\right)_{avg}=\frac{1}{2}(25\,\text{A})(60\,\text{V})\left(\frac{\frac{48}{25}\,\Omega}{2.4\,\Omega}\right)$$

$$\left(P_{EMF}\right)_{avg}=600\,\text{W}$$

In order to determine the average power dissipated by the resistor in the electric circuit in Figure 9.12, we need to determine the current flowing through the resistor. We can determine this current using Kirchhoff's rules[5]. As shown in Figure 9.13, the EMF supplies a current I that divides into I_1 and I_2 at point C. Current I_1 flows through the resistor and current I_2 flows through the inductor. These two currents recombine at point F to reform current I, which flows back to the battery. Applying Kirchhoff's loop rule to the loop CDEF in Figure 9.13 gives us

$$(i4\Omega)I_2-(3\Omega)I_1=0\ \rightarrow\ 4iI_2=3I_1\ \rightarrow\ I_2=-i\frac{3}{4}I_1$$

Then, by applying Kirchhoff's junction rule to point C or point F we obtain

$$I_1+I_2=I$$

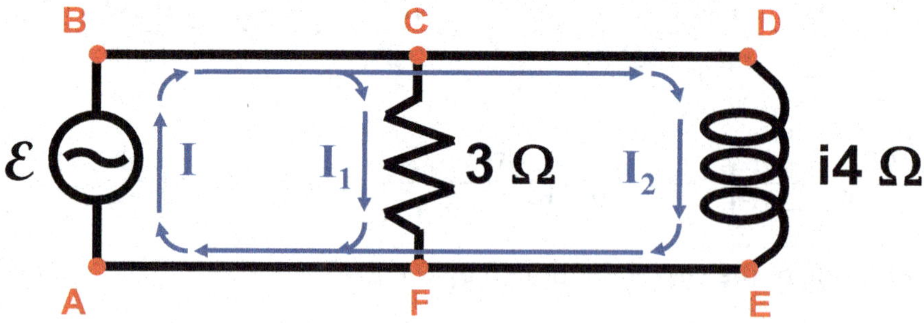

FIGURE 9.13: The electric circuit in Example 9-8 when $\omega = 20$ rad/s. The instantaneous currents at one instant of time are also shown.

5 You should feel inspired to take more physics and math courses to learn the limitations of applying Kirchhoff's rules to AC circuits. For now, we will just ignore them.

Hence,

$$I_1 + \left(-i\frac{3}{4}I_1\right) = I \rightarrow (4-3i)I_1 = 4I \rightarrow I_1 = \frac{4}{4-3i}I$$

The presence of a complex number in this expression tells us that the current flowing through the resistor is not in phase with the current supplied by the EMF[6]. We can rewrite the equation relating I_1 and I using Equation D-6.

$$I_1 = \frac{4}{\left(4^2+3^2\right)^{\frac{1}{2}} e^{i\tan^{-1}\left(\frac{-3}{4}\right)}} I \rightarrow I_1 = \frac{4}{5} e^{i\tan^{-1}\left(\frac{3}{4}\right)} I$$

The current I supplied by the EMF is determined from the equivalent impedance of the circuit.

$$Z_{eq} = \left(\frac{48}{25}\Omega\right) + i\left(\frac{36}{25}\Omega\right) \rightarrow \tan\phi = \frac{\frac{36}{25}\Omega}{\frac{48}{25}\Omega} \rightarrow \phi = \tan^{-1}\left(\frac{3}{4}\right)$$

$$I = (25\,A)e^{i\left(\omega t - \tan^{-1}\left(\frac{3}{4}\right)\right)}$$

Thus,

$$I_1 = \frac{4}{5} e^{i\tan^{-1}\left(\frac{3}{4}\right)} \left((25\,A)e^{i\left(\omega t - \tan^{-1}\left(\frac{3}{4}\right)\right)}\right) \rightarrow I_1 = (20\,\text{A})e^{i\omega t}$$

It is interesting to note that the oscillation of the current flowing through the resistor is in phase with the oscillation of the EMF (but out of phase with the current supplied by the EMF); this is consistent with energy being conserved in the electric circuit. The average power dissipated by the resistor is

$$\left(P_R\right)_{avg} = \frac{1}{2}(20\,\text{A})^2(3\Omega) \rightarrow \left(P_R\right)_{avg} = 600\,\text{W}$$

6 This is similar to how the presence of an imaginary component in the impedance tells us that the oscillations of the current are out of phase with the oscillations of the EMF.

As expected, the average power dissipated by the resistor is exactly equal to the average power supplied by the EMF.

All of the power supplied by the EMFs in an AC circuit is dissipated by the resistors in the circuit.

In other words,

The real part of the impedance is associated with energy dissipation. An AC circuit will dissipate energy only when the real part of its impedance is not zero.

The imaginary part of the impedance is associated with energy storage.

For a sinusoidally oscillating EMF, such as those we have discussed so far, it is common to describe the EMF and current in terms of their root-mean-square[7] (or rms) values rather than the amplitudes of their oscillations.

$$\mathcal{E}_{rms}^2 = \frac{\mathcal{E}_0^2}{2} \qquad I_{rms}^2 = \frac{I_0^2}{2}$$

$$\left(P_{EMF}\right)_{avg} = I_{rms}\,\mathcal{E}_{rms}\frac{\text{Re}(Z)}{|Z|} \tag{9-15}$$

$$\left(P_R\right)_{avg} = I_{rms}^2 R \tag{9-16}$$

There is no real benefit to using Equations 9-15 and 9-16 rather than Equation 9-13 and 9-14 for a sinusoidally oscillating EMF. However, Equation 9-15 and Equation 9-16 are universally valid for any EMF (and current), regardless of whether it is sinusoidal.

Example 9-9

Problem: The EMF in the electric circuit shown in Figure 9.14 is described by the following equation:

$$\mathcal{E} = \left(40\,\text{V}\right)e^{i\left(100\frac{rad}{s}\right)t}$$

7 We previously saw the root mean square operation in thermodynamics when we applied the equipartition theorem to determine the speed of molecules of an ideal gas.

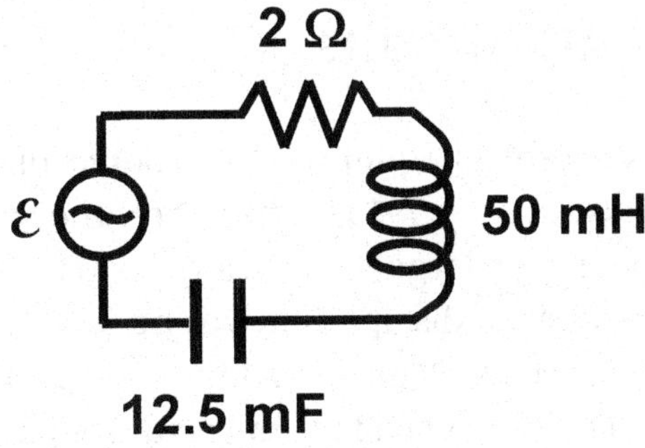

FIGURE 9.14: The electric circuit in Example 9-9.

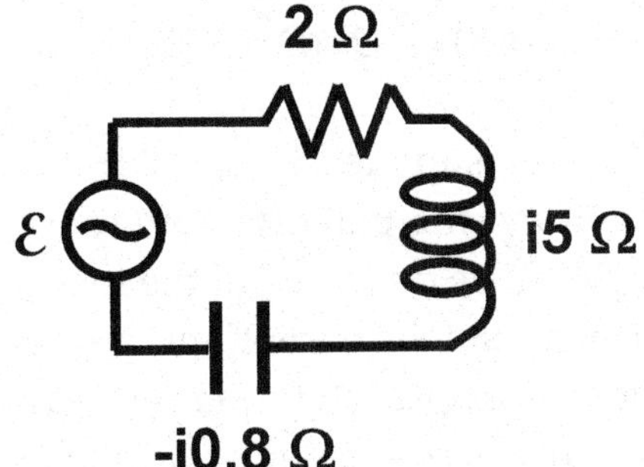

FIGURE 9.15: The electric circuit in Example 9-9 when $\omega = 100$ rad/s.

What is the rms current in the circuit? What is the average power dissipated by the resistor in the circuit?

Solution: The equivalent electric circuit at $\omega = 100$ rad/s is shown in Figure 9.15.

The impedance of the electric circuit is

$$\frac{1}{Z_{eq}} = 2\Omega + i5\Omega + (-i0.8\Omega) \quad \rightarrow \quad Z_{eq} = 2\Omega + i4.2\Omega$$

$$|Z_{eq}| = \left((2\Omega)^2 + (4.2\Omega)^2\right)^{\frac{1}{2}} \quad \rightarrow \quad |Z_{eq}| = 4.7\Omega$$

The amplitude of the current supplied by the EMF is

$$I_0 = \frac{\mathcal{E}_0}{Z_{eq}} \quad \rightarrow \quad I_0 = \frac{40\,\text{V}}{4.7\Omega} \quad \rightarrow \quad I_0 = 8.5\,\text{A}$$

The rms current supplied by the EMF is

$$I_{rms}^2 = \frac{I_0^2}{2} \quad \rightarrow \quad I_{rms}^2 = \frac{(8.5\,\text{A})^2}{2} \quad \rightarrow \quad I_{rms}^2 = 36.1\,\text{A}^2 \quad \rightarrow \quad I_{rms} = 6.0\,\text{A}$$

The average power dissipated by the resistor[8] is

$$(P_R)_{avg} = I_{rms}^2 R \quad \rightarrow \quad (P_R)_{avg} = (36.1\,\text{A}^2)(2\Omega) \quad \rightarrow \quad (P_R)_{avg} = 72.2\,\text{W}$$

8 The current supplied by the EMF is the same as the current flowing through the resistor since all components of this electric circuit are in series.

9-5 Additional Applications of AC Circuits

The power supplied by an EMF to generate a current is equal to the product of the current and the potential difference of the EMF (Equation 7-11). Therefore, the same energy can be delivered in the form of electricity using a large current and a small EMF or using a small current and a large EMF. Since energy dissipation will be larger for the former approach (Equation 9-16), the latter approach is preferred for transporting energy over long distances through power lines. The EMF used for long distance power transmission is typically greater than 100 kV; we refer to the wires carrying this power as *high-voltage transmission lines.*

Of course, it is not necessarily useful or safe to have such a large EMF in your house. Indeed, we would prefer to have no more than a few hundred volts of EMF there. A device called a transformer (Figure 9.16) is used to change the EMF between two different electric circuits.

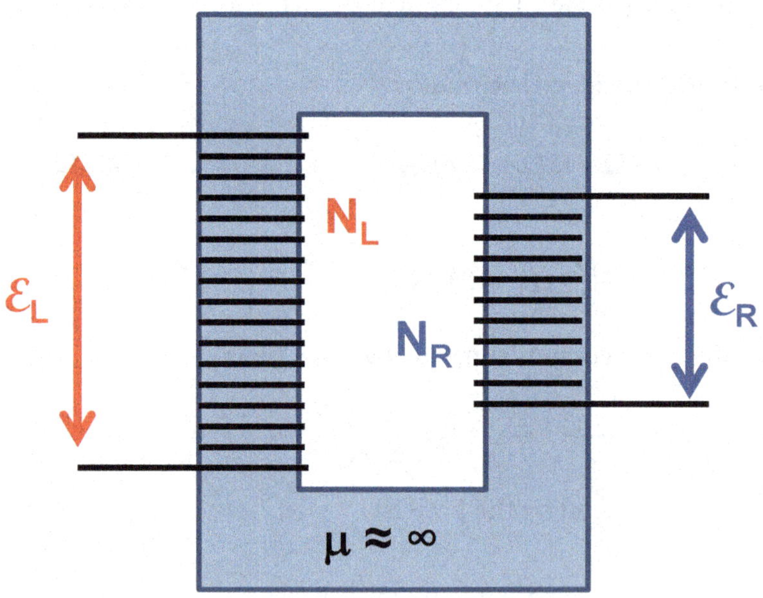

FIGURE 9.16: An ideal transformer.

As shown in Figure 9.16, a transformer consists of two coils of wire wrapped around a magnetic material (in this case, we have assumed that the relative permeability[9] of the magnetic material is infinite). The coil on the left consists of N_L turns of wire, and the coil on the right consists of N_R turns of wire. When a current passes through the coils on the left, a magnetic field is created. This magnetic field magnetizes the magnetic

9 See Section 7-4.

material, and since the relative permeability is infinite, the same magnetic field in the coils on the left passes through the coils on the right. In other words, the magnetic flux through each coil on the left is equal to the magnetic flux through each coil on the right. If the total magnetic flux through the right and left coils are denoted by $(\Phi_B)_R$ and $(\Phi_B)_L$, respectively, we have

$$\frac{(\Phi_B)_R}{N_R} = \frac{(\Phi_B)_L}{N_L} \rightarrow (\Phi_B)_R = \frac{N_R}{N_L}(\Phi_B)_L$$

Taking the derivative of both sides of this equation with respect to time and subsequently applying Equation 6-11 yields

$$\frac{d(\Phi_B)_R}{dt} = \left(\frac{N_R}{N_L}\right)\frac{d(\Phi_B)_L}{dt} \rightarrow \mathcal{E}_R = \left(\frac{N_R}{N_L}\right)\mathcal{E}_L$$

Therefore, we can use a transformer to convert a high EMF into a low EMF and a low EMF into a high EMF. However, although transformers can change the electric potential in transient direct currents, they do not work for steady-state direct currents; there is no change in the magnetic flux for steady-state direct currents. Thus, the only steady-state currents applicable for transformers are alternating currents. It is for this reason that long distance power transmission typically uses alternating current.

Example 9-10

Problem: What ratio of turns of wire in a transformer would be required to transform 120 kV to 120 V?

Solution: Substitution into the previously derived equation gives us

$$120\,\text{kV} = \left(\frac{N_{120\,\text{kV}}}{N_{120\,\text{V}}}\right)(120\,\text{V}) \rightarrow \frac{N_{120\,\text{kV}}}{N_{120\,\text{V}}} = \frac{120\,\text{kV}}{120\,\text{V}}$$

$$\frac{N_{120\,\text{kV}}}{N_{120\,\text{V}}} = 1000$$

Of course, the EMF is usually changed through a series of transformers, rather than a single transformer.

The basic model for high voltage AC (HVAC) power transmission begins with an AC EMF between 10 kV and 40 kV generated at the power plant. This EMF is then "stepped-up" using a transformer to more than 100 kV for transmission, before being "stepped-down" using a transformer to a few hundred volts for use in a home. Unfortunately, when alternating current moves through a conductor, the current density[10] becomes non-uniform and is largest near the surface of the conductor. This phenomenon, known as the *skin effect*, causes the effective resistance of the conductor to be higher than expected from its conductivity and physical dimensions (Equation 7-14). The skin effect increases in magnitude with increasing frequency of oscillation of the current. In other words, as the oscillation frequency increases, the current becomes more concentrated near the surface, and the effective resistance of the conductor increases. This problem can be mitigated using specially woven wires, but often it is avoided completely by transmitting power through high voltage direct current (HVDC).

Another useful application of an AC circuit is a metal detector. A metal detector consists of two coils: a transmitter coil and a receiver coil. When an alternating current is passed through the transmitter coil, an oscillating magnetic field is generated. As shown in Figure 9.17, this oscillating magnetic field will induce oscillating eddy currents in metal[11].

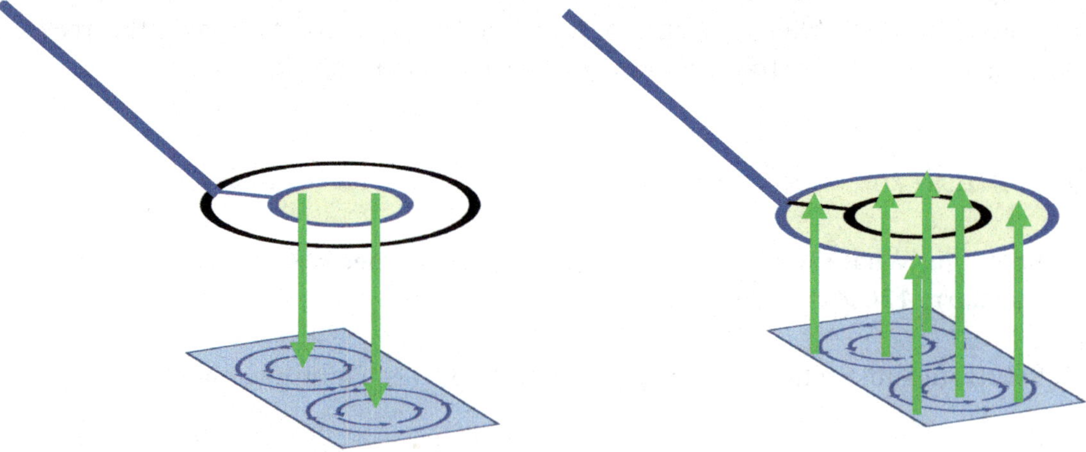

FIGURE 9.17: A metal detector detects the magnetic field associated with induced eddy currents. The inner coil is the transmitter coil, and the outer coil is the receiver coil.

These eddy currents will give rise to an additional magnetic field that can be detected by the receiver coil. Furthermore, since different metals will have different magnetic

10 Section 7-5.

11 Section 6-6.

properties, the phase difference between the oscillations of the EMF in the transmitting coil and the oscillations of the magnetic field associated with the eddy currents in the metal will be vary among metals. Careful measurements of this phase difference can thus enable metal detectors to differentiate different metals.

A similar phenomenon is exploited by magnetic resonance imaging (MRI). In this case, different tissue types and structures respond differently to an oscillating magnetic field applied to them. The differences in these responses enable doctors to differentiate the tissue types and detect abnormalities in the tissue, such as cancer.

A more commonplace, and perhaps more practical, application of AC circuits is induction cooking. In induction cooking, an alternating AC current (flowing in an electric circuit in the cooktop) creates an alternating magnetic field and associated alternating magnetic flux in a metal pot/pan on the cooktop. This alternating magnetic flux induces alternating eddy currents in the pot/pan, which then dissipate energy (because of the resistance of the pot/pan) in the form of heat[12].

A more entertaining application of a simple AC circuit is found in an electric guitar. A permanent magnet placed near each metal string will magnetize a portion of the string. When the string vibrates, the flux from this magnetized portion through a pickup coil will change. The pickup coil is usually wrapped around the permanent magnet as shown in Figure 9.18. The oscillating EMF in the pickup coil associated with the vibrations of the string will create an AC current that can be amplified and sent to loudspeakers.

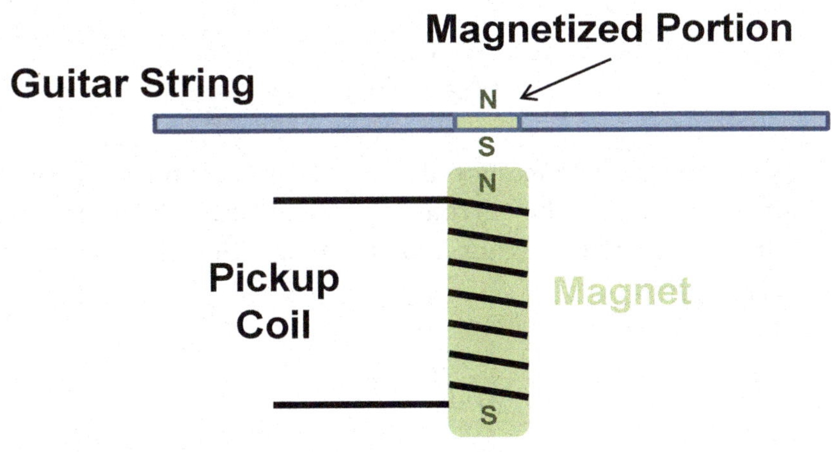

FIGURE 9.18: The pickup coil in an electric guitar.

12 This is similar to how the use of magnetic braking systems on trains heats the rails, rather than the train (Section 6-6).

Finally, wireless battery chargers also rely upon alternating current and the associated oscillating magnetic fields to transmit energy. The mutual induction between the AC circuit in the charger and the circuit containing the battery allows for power to be transferred from the charger to the battery "wirelessly" through a magnetic field. Specifically, the magnetic field generated by the charger induces current in the circuit containing the battery, which then charges the battery (a simple example of this process was discussed in Section 8-7).

Example 9-11

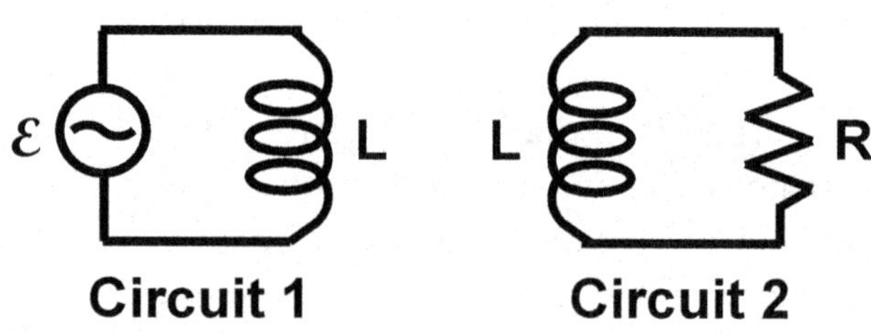

FIGURE 9.19: The two electric circuits in Example 9-11.

Problem: The two electric circuits in Figure 9.19 have a mutual inductance *M*. What is the power dissipated by the resistor in electric circuit 2?

Solution: As shown in Equation 6-6, the mutual inductance relates the current in electric circuit 1 to the magnetic flux through electric circuit 2 (caused by the magnetic field of the current electric circuit 1).

$$M = \frac{(\Phi_B)_{21}}{I_1}$$

Similarly, since mutual inductance is reciprocal (Section 6-3), the same mutual inductance relates the current in electric circuit 2 to the magnetic flux through electric circuit 1 (caused by the magnetic field of current electric circuit 2).

$$M = \frac{(\Phi_B)_{12}}{I_2}$$

The induced EMF in electric circuit 2 can then be calculated using Equation 6-11.

$$(\Phi_B)_{21} = MI_1 \quad \rightarrow \quad \frac{d(\Phi_B)_{21}}{dt} = M\frac{dI_1}{dt}$$

$$|\mathcal{E}| = \frac{d\Phi_B}{dt} \rightarrow |\mathcal{E}_2| = \frac{d(\Phi_B)_{21}}{dt} \rightarrow |\mathcal{E}_2| = M\frac{dI_1}{dt}$$

Similarly, the induced EMF in electric circuit 1 is

$$|\mathcal{E}_1| = M\frac{dI_2}{dt}$$

Applying Kirchhoff's loop rule to electric circuit 1 then gives us[13]

$$\mathcal{E} - L\frac{dI_1}{dt} - M\frac{dI_2}{dt} = 0$$

Applying Kirchhoff's loop rule to circuit 2 then gives us

$$-I_2R - L\frac{dI_2}{dt} + M\frac{dI_1}{dt} = 0$$

The induced EMF through the mutual inductance is positive in this equation, since it drives the current in this circuit. Let's solve these two equations by guessing the following solutions for the two currents:

$$I_1 = (I_0)_1 e^{i\omega t} \qquad I_2 = (I_0)_2 e^{i\omega t}$$

Substitution gives us

$$\mathcal{E}_0 e^{i\omega t} - L(I_0)_1 e^{i\omega t} i\omega - M(I_0)_2 e^{i\omega t} i\omega = ($$

$$\left(\mathcal{E}_0 - i\omega L(I_0)_1 - i\omega M(I_0)_2\right)e^{i\omega t} = 0 \rightarrow \mathcal{E}_0 - i\omega L(I_0)_1 - i\omega M(I_0)_2 = 0$$

$$-(I_0)_2 e^{i\omega t} R - L(I_0)_2 e^{i\omega t} i\omega + M(I_0)_1 e^{i\omega t} i\omega = 0$$

$$\left(-(I_0)_2 R - i\omega L(I_0)_2 + i\omega M(I_0)_1\right)e^{i\omega t} = 0 \rightarrow -(I_0)_2 R - i\omega L(I_0)_2 + i\omega M(I_0)_1 = 0$$

13 Note that according to Lenz's law the direction of induced EMF in electric circuit 1, resulting from the mutual inductance of the two electric circuits, must be opposite the direction of the original EMF in electric circuit 1.

Solving for the current in electric circuit 1 gives us

$$\left(I_0\right)_1 = \left(\frac{R + i\omega L}{i\omega M}\right)\left(I_0\right)_2$$

Substituting this expression then yields

$$\mathcal{E}_0 - i\omega L\left(\frac{R + i\omega L}{i\omega M}\right)\left(I_0\right)_2 - i\omega M\left(I_0\right)_2 = 0$$

$$\mathcal{E}_0 = \left(i\omega L\left(\frac{R + i\omega L}{i\omega M}\right) + i\omega M\right)\left(I_0\right)_2 \quad \rightarrow \quad \left(I_0\right)_2 = \frac{M\mathcal{E}_0}{RL + i\left(\omega L^2 + \omega M^2\right)}$$

The current flowing through the resistor is

$$\left(I_0\right)_2 = \frac{M\mathcal{E}_0}{RL + i\left(\omega L^2 + \omega M^2\right)} e^{i\omega t}$$

$$\left(I_0\right)_2 = \frac{M\mathcal{E}_0}{\left(\left(RL\right)^2 + \left(\omega L^2 + \omega M^2\right)^2\right)^{\frac{1}{2}} e^{i\tan^{-1}\left(\frac{\omega L^2 + \omega M^2}{RL}\right)}} e^{i\omega t}$$

$$\left(I_0\right)_2 = \frac{M\mathcal{E}_0}{\left(\left(RL\right)^2 + \left(\omega L^2 + \omega M^2\right)^2\right)^{\frac{1}{2}}} e^{i\left(\omega t - \tan^{-1}\left(\frac{\omega L^2 + \omega M^2}{RL}\right)\right)}$$

Thus, the average power dissipated by the resistor is

$$\left(P_R\right)_{avg} = \frac{1}{2}\left(\frac{M\mathcal{E}_0}{\left(\left(RL\right)^2 + \left(\omega L^2 + \omega M^2\right)^2\right)^{\frac{1}{2}}}\right)^2 R$$

$$\left(P_R\right)_{avg} = \frac{M^2\mathcal{E}_0^2 R}{2\left(\left(RL\right)^2 + \left(\omega L^2 + \omega M^2\right)^2\right)}$$

Let's check this solution in a few limits. If there is no resistor in circuit 2 or no mutual inductance between the two circuits, then no power is dissipated, as expected.

$$\lim_{R\to 0}\left(P_R\right)_{avg} = 0 \qquad \lim_{M\to 0}\left(P_R\right)_{avg} = 0$$

9-6 Retarded Time and Non-Resistive Energy Dissipation

We introduced the concept of fields and potentials to explain how one object can "sense" another through a long-range interaction, such as the gravitational interaction. We claimed that the presence of an object affects the space around it and that this alteration of space can then be detected by other objects. We never stated how that alteration of space propagated from the object, however. In other words, how long does it take for the field/potential to be established? Or how long does it take the field/potential to adjust if an object changes its position?

For example, the alternating current flowing in the wire in Figure 9.20 will create an alternating magnetic field. Let's use the variable c to denote the speed at which the magnetic field (*i.e.*, the alternation of space associated with the magnetic field) propagates from the wire. The information about current will then require a time r/c to propagate from the wire to the position P in Figure 9.20. We could therefore include this time or propagation into our equation for the current to determine the "effective" current seen at the positon P.

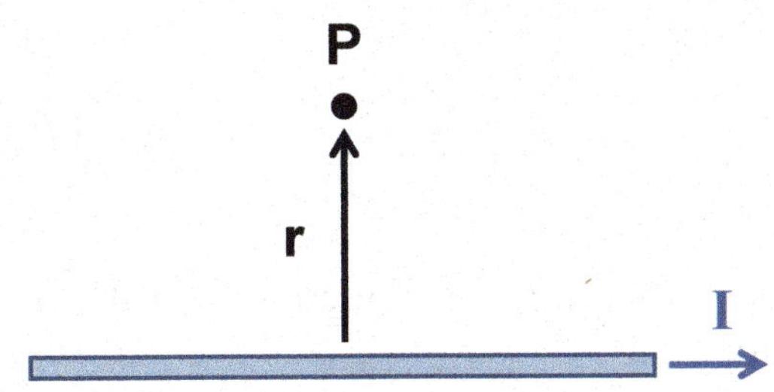

FIGURE 9.20: A current-carrying wire.

$$I = I_0 e^{i\omega\left(t-\frac{r}{c}\right)} \tag{9-17}$$

The term in parenthesis in Equation 9-17 is referred to as the *retarded time* and the current in Equation 9-17 as the *retarded current*, since they are delayed (or retarded) by the propagation time of the magnetic field. The magnetic potential corresponding to this retarded current is found using Equation 4-12.

$$\vec{V}_m = \left(\frac{\mu_0}{4\pi}\right)\int \frac{I_0 e^{i\omega\left(t-\frac{r}{c}\right)} d\vec{s}}{r} \quad (9\text{-}18)$$

Equation 9-18 and Equation 6-7 can be used to determine the self-inductance of a circular loop of wire. This self-inductance and therefore the circular loop's impedance has both real and imaginary components[14]. Thus, even if there is no resistance in the loop of wire, energy will nevertheless be dissipated[15]. The real part of the impedance of a circular loop of cross-sectional area A is

$$\mathrm{Re}(Z) \approx \frac{\mu_0 A^2 \omega^4}{6\pi c^3}$$

This is often also called the *radiative resistance*, and is denoted by R_{rad}.

$$R_{rad} \approx \frac{\mu_0 A^2 \omega^4}{6\pi c^3} \quad (9\text{-}19)$$

The average power radiated by a circular loop of wire is then found using Equation 8-1 and Equation 9-19.

$$\left(P_{rad}\right)_{avg} = I_{rms}^2 R_{rad} \quad (9\text{-}20)$$

The variable I_{rms} in Equation 9-20 is the oscillating rms current responsible for the radiation.

Example 9-12

Problem: What is average power radiated by a circular loop of wire with a radius of 0.5 m carrying an rms current of 2 A oscillating at ω = 60 rad/s?

14 See derivation in Section C-12.
15 Section 9-4.

What is the average power when $\omega = 6 \times 10^5$ rad/s? What is the average power when $\omega = 6 \times 10^8$ rad/s? Assume that the speed of propagation is 3×10^8 m/s.

Solution: Substitution of the variables into Equation 9-19 yields

$$R_{rad} \approx \frac{\left(4\pi\times10^{-7}\,\frac{\text{kgm}}{\text{C}^2}\right)\left(\pi(0.5\,\text{m})^2\right)^2\omega^4}{6\pi\left(3\times10^8\,\frac{\text{m}}{\text{s}}\right)^3} \rightarrow R_{rad} \approx \left(1.52\times10^{-33}\,\Omega\text{s}^4\right)\omega^4$$

$$\omega = 60\,\frac{\text{rad}}{\text{s}} \rightarrow R_{rad} \approx 1.97\times10^{-26}\,\Omega$$

$$\omega = 6\times10^5\,\frac{\text{rad}}{\text{s}} \rightarrow R_{rad} \approx 1.97\times10^{-10}\,\Omega$$

$$\omega = 6\times10^8\,\frac{\text{rad}}{\text{s}} \rightarrow R_{rad} \approx 197\,\Omega$$

The average power associated with each of these angular frequencies of oscillation is determined using Equation 9-20.

$$\omega = 60\,\frac{\text{rad}}{\text{s}} \rightarrow \left(P_{rad}\right)_{avg} = (2\text{A})^2\left(1.97\times10^{-26}\,\Omega\right) \rightarrow \left(P_{rad}\right)_{avg} = 7.88\times10^{-26}\,\text{W}$$

$$\omega = 6\times10^5\,\frac{\text{rad}}{\text{s}} \rightarrow \left(P_{rad}\right)_{avg} = (2\text{A})^2\left(1.97\times10^{-10}\,\Omega\right) \rightarrow \left(P_{rad}\right)_{avg} = 7.88\times10^{-10}\,\text{W}$$

$$\omega = 6\times10^8\,\frac{\text{rad}}{\text{s}} \rightarrow \left(P_{rad}\right)_{avg} = (2\text{A})^2\left(197\,\Omega\right) \rightarrow \left(P_{rad}\right)_{avg} = 788\,\text{W}$$

We can understand the origin of this dissipation of energy by considering the behavior of a charged particle that constitutes the current. A stationary charged particle creates an electric field, but no magnetic field. A charged particle moving at constant velocity creates an electric field and a constant magnetic field. An accelerating charged particle, however, creates a changing magnetic field, which will then induce a changing electric field, which will then induce a changing magnetic field, and so on. The energy dissipated by the accelerating charged particle creates these changing fields, and these changing fields then *transmit* this energy to other charged particles. This process will be the subject of Chapter 10.

The acceleration responsible for the energy dissipation from a circular loop of current is the acceleration associated with the oscillations of the alternating current (*i.e.*, the charged particles are accelerating in an AC circuit since the direction of their motion is continually changing). It is also generally true that an accelerating charged particle will dissipate energy. Let's denote the speed of propagation of the information about the electric and magnetic fields of the charged particle by the variable *c*[16]. If a charged particle is moving at a speed much less than *c* the power associated with the dissipation of energy for its acceleration is given in Equation 9-21.

$$P = \frac{q^2 a^2}{6\pi\varepsilon_0 c^3} \tag{9-21}$$

The variables q and a in Equation 9-21 denote the net electric charge and acceleration of the charged particle.

In our calculation of the energy dissipation of a circular loop of wire, we included only the acceleration associated with the change in the direction of the alternating current. According to Equation 9-21, however, any acceleration should cause a moving charged particle to dissipate energy, thus we should include a correction term for the centripetal acceleration associated with the movement of the electrons (which constitute the current) around the circular loop. However, since the magnitude of the speed of the electrons is so small (Section 7-6), we can ignore this contribution. Indeed, the magnitude of the centripetal acceleration associated with the drift velocity of the electrons for circular loop of wire with a radius of 0.5 m is

$$v_d = 0.75\times10^{-3}\,\frac{\text{m}}{\text{s}} \;\rightarrow\; a_c = \frac{\left(0.75\times10^{-3}\,\frac{\text{m}}{\text{s}}\right)^2}{0.5\text{m}} \;\rightarrow\; a_c = 1.13\times10^{-6}\,\frac{\text{m}^2}{\text{s}}$$

Substitution into Equation 9-21 then gives us[17]

16 We will return to this speed of propagation in Section 10-3.

17 We'll assume in this calculation that $c = 3 \times 10^8$ m/s. Justification for this assumption will be provided in Section 10-3.

$$P = \frac{\left(1.6 \times 10^{-19}\,\text{C}\right)^2 \left(1.13 \times 10^{-6}\,\frac{\text{m}^2}{\text{s}}\right)^2}{6\pi\left(8.85 \times 10^{-12}\,\frac{\text{C}^2}{\text{J m}}\right)\left(3 \times 10^8\,\frac{\text{m}}{\text{s}}\right)^3} \rightarrow P = 7.26 \times 10^{-66}\,\text{W}$$

This is an insignificant amount of power dissipation.

9-7 The Bohr Model for the Hydrogen Atom, Part IV

In the Bohr model for the atom, the electron is in a circular orbit around the proton. Thus, the electron is accelerating (*i.e.*, it is experiencing centripetal acceleration) and must therefore be dissipating energy. We can determine the power associated with this dissipation of energy using Equation 9-21.

$$a = \frac{v^2}{R_{orbit}} \rightarrow P = \frac{q^2 v^4}{6\pi\varepsilon_0 c^3 R_{orbit}^2}$$

Substitution of our previous derived parameters (Section 3-9) gives us[18]

$$P = \frac{\left(1.6 \times 10^{-19}\,\text{C}\right)^2 \left(2.25 \times 10^{6}\,\frac{\text{m}}{\text{s}}\right)^4}{6\pi\left(8.85 \times 10^{-12}\,\frac{\text{C}^2}{\text{J m}}\right)\left(3 \times 10^8\,\frac{\text{m}}{\text{s}}\right)^3 \left(5.3 \times 10^{-11}\,\text{m}\right)^2} \rightarrow P = 5.19 \times 10^{-8}\,\text{W}$$

As the electron dissipates energy, it will move closer to the proton, and eventually it will collide with the proton and the atom will collapse. The total time required to dissipate all of the electric potential energy and kinetic energy (*i.e.*, the ionization energy) of the electron is

$$\Delta t = \frac{\Delta E}{P} \rightarrow \Delta t = \frac{2.17 \times 10^{-18}\,\text{J}}{5.19 \times 10^{-8}\,\text{W}} \rightarrow \Delta t = 41.8\,\text{ps}$$

Of course, the power will not be constant as the electron spirals in towards the proton. Nevertheless, this calculation demonstrates that the Bohr model of the atom is unstable.

18 As before, we'll assume in this calculation that $c = 3 \times 10^8$ m/s. Justification for this assumption will be provided in Section 10-3.

Summary

- **Impedance:** the effective resistance of an electric circuit or component to alternating current. Impedance is a complex number, has the unit ohm, and is denoted by the variable *Z*. The impedance can be written as

$$Z = |Z| e^{i\phi} \rightarrow Z = \left[\left(\mathrm{Re}(Z)\right)^2 + \left(\mathrm{Im}(Z)\right)^2 \right]^{\frac{1}{2}} e^{i\phi}$$

The angle ϕ in this equation is determined using the following equation.

$$\tan\phi = \frac{\mathrm{Im}(Z)}{\mathrm{Re}(Z)}$$

This angle is the difference in phase between the oscillations of the EMF and the oscillations of the current supplied by the EMF.

- The impedance of a resistor is

$$Z_R = R$$

- The impedance of an inductor is

$$Z_L = i\omega L$$

- The impedance of a capacitor is

$$Z_C = -\frac{i}{\omega C}$$

- The equivalent impedance of a group of circuit components connected in series is

$$Z_{eq} = Z_1 + Z_2 + \ldots$$

The equivalent impedance of a group of circuit components connected in parallel is

$$\frac{1}{Z_{eq}} = \frac{1}{Z_1} + \frac{1}{Z_2} + \ldots$$

- Ohm's law for steady-state AC circuits is written as

$$I = \frac{\mathcal{E}}{Z} \quad \rightarrow \quad I = \frac{\mathcal{E}_0 e^{i\omega t}}{|Z| e^{i\phi}} \quad \rightarrow \quad I = I_0 e^{i(\omega t - \phi)}$$

  The amplitude of the current is

$$I_0 = \frac{\mathcal{E}_0}{|Z|}$$

- The root-mean-square (rms) current is proportional to the amplitude of the oscillating current.

$$I_{rms}^2 = \frac{I_0^2}{2}$$

- The root-mean-square (rms) EMF is proportional to the amplitude of the oscillating EMF.

$$\mathcal{E}_{rms}^2 = \frac{\mathcal{E}_0^2}{2}$$

- The average power supplied by the EMF is

$$\left(P_{EMF}\right)_{avg} = \frac{1}{2} I_0 \mathcal{E}_0 \cos(\phi) \quad \rightarrow \quad \left(P_{EMF}\right)_{avg} = \frac{I_0 \mathcal{E}_0 \operatorname{Re}(Z)}{2|Z|}$$

  The variable I_0 in this equation is the amplitude of the oscillating current supplied by the EMF to the electric circuit. This equation is frequently written in terms of the rms EMF and current.

$$\left(P_{EMF}\right)_{avg} = \frac{I_{rms} \mathcal{E}_{rms} \operatorname{Re}(Z)}{|Z|}$$

  All of the power supplied by an AC EMF is dissipated by resistors in the electric circuit. The average power dissipated by a resistor is

$$\left(P_R\right)_{avg} = \frac{1}{2} I_0^2 R$$

The variable I_0 in this equation is the amplitude of the oscillating current flowing in the resistor. This equation is frequently written in terms of the rms current.

$$\left(P_R\right)_{avg} = I_{rms}^2 R$$

- The real part of the impedance of an electric circuit is associated with energy dissipation. If the real part of the impedance is zero, no energy will be dissipated by the circuit.

- The imaginary part of the impedance of an electric circuit is associated with energy storage. The average power supplied to a circuit will be zero if the impedance of the circuit is purely imaginary. An imaginary impedance further indicates that the oscillations of the EMF are not in phase with the oscillations of the current supplied by the EMF.

- The average power radiated by a current is proportional to the radiative resistance of the current.

$$\left(P_{rad}\right)_{avg} = I_{rms}^2 R_{rad}^2$$

- The power radiated by an accelerating charged particle is

$$P = \frac{q^2 a^2}{6\pi\varepsilon_0 c^3}$$

The variables q, a, and c in this equation are the net electric charge of the particle, the acceleration of the particle, and the speed of propagation of the information about the particle's electric and magnetic fields.

Problems

1. A differential equation for a damped harmonic oscillator is

$$m\frac{d^2x}{dt^2} + b\frac{dx}{dt} + kx = 0$$

Verify that the equation $x = e^{iwt}$ is a solution for this equation and determine the value(s) of ω that satisfy the equation. In other words, determine the solution for the differential equation for a damped harmonic oscillator listed above.

2. What is the impedance of the electric circuit shown in the figure? What is the average power supplied by the EMF when $\omega = 20$ rad/s and $\mathcal{E}_0 = 20$V? What is the rms current in the inductor when $\omega = 100$ rad/s and $\mathcal{E}_0 = 20V$? What is the average power dissipated by the resistor when $\omega = 60$ rad/s and $\mathcal{E}_0 = 20$V? What are the average energies stored in the capacitor and inductor when $\omega = 60$ rad/s and $\mathcal{E}_0 = 20$V?

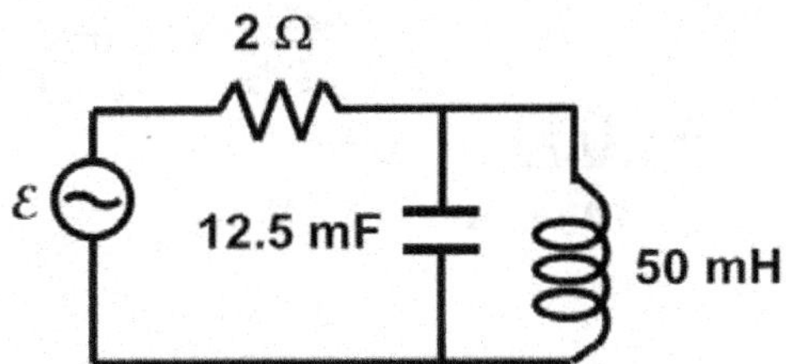

3. What is the impedance of the electric circuit shown in the figure? What is the average power supplied by the EMF when $\omega = 200$ rad/s and $\mathcal{E}_0 = 40$V? What is the rms current in the capacitor when $\omega = 20$ rad/s and $\mathcal{E}_0 = 40$V? What is the average power dissipated by the resistor and the average energy stored in the capacitor when $\omega = 60$ rad/s and $\mathcal{E}_0 = 40$V?

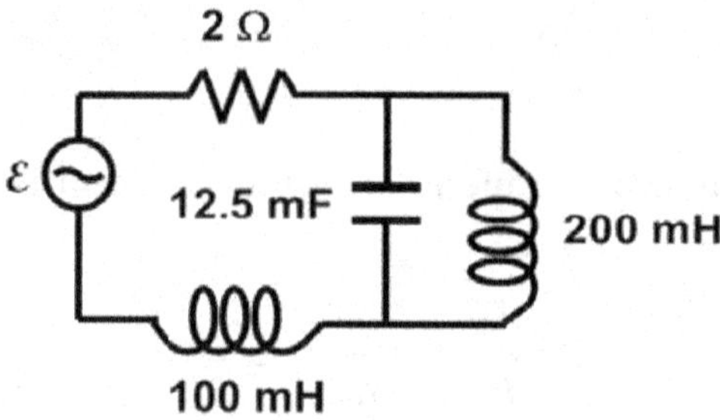

4. What is the impedance of the electric circuit shown in the figure? What is the average power dissipated by the resistor when $\omega = 20$ rad/s and $\mathcal{E}_0 = 40$V? What is the average power dissipated by the resistor when $\omega = 200$ rad/s and $\mathcal{E}_0 = 40$V? What is the average power dissipated by the resistor and the average power stored in the inductors when $\omega = 60$ rad/s and $\mathcal{E}_0 = 40$V?

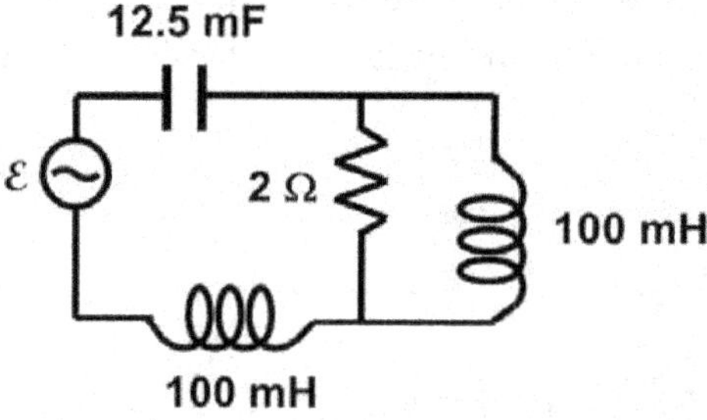

5. The two resistors in the electric circuit shown in the figure have the same resistance, 2 Ω. What is the average power dissipated by each resistor when the EMF oscillates at 10 rad/s? What is the average power dissipated by each resistor when the EMF oscillates at 1000 rad/s?

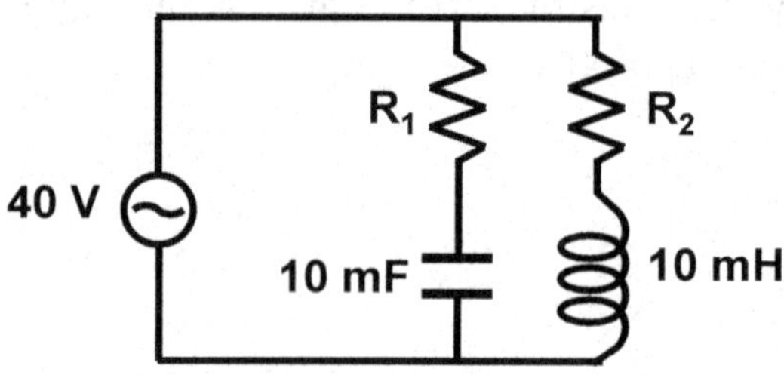

6. The two electric circuits shown in the figure have a mutual inductance *M*.

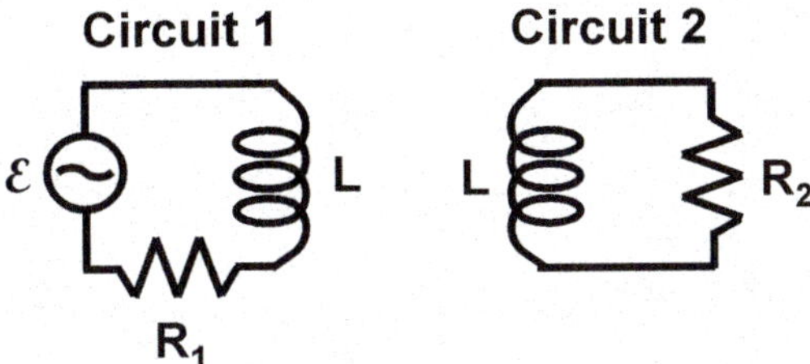

At what angular frequency of oscillation is the maximum power dissipated by the resistor R_2?

7. A thin and uniformly positively charged ring of radius *R* and net electric charge *Q* is lying in the *x-y* plane. The *z*-axis is perpendicular to the plane of the ring and passes through the center of the ring. The ring is rotating around the *z*-axis with angular frequency *ω*. What is the average power dissipated by this system because of the centripetal acceleration of the electric charges that constitute of the ring? What is the radiative resistance of the ring?

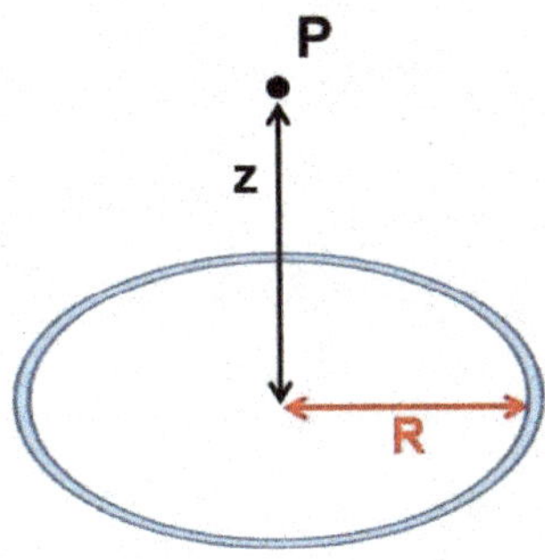

CHAPTER 10

Electromagnetic Waves

10-1 Introduction

In Section 9-6 we introduced the concept of retarded time to explain how the disturbances of space associated with mass and electric charge are propagated away from these sources. In this chapter, we begin our discussion about the details of how these disturbances are propagated and what form these disturbances take.

10-2 Derivation of Wave Equation

An alternating current will create an alternating magnetic field. This alternating magnetic field will then induce an alternating electric field, which can then create an alternating current in a neighboring wire (*i.e.*, an induced current), as shown in Figure 10.1. Based upon the discussion in Section 9-6, we could describe this process by saying that the energy dissipated by the alternating current (*i.e.*, by the accelerating electrons) is used to create the induced current. In this way, energy is transferred from the alternating current to the induced current.

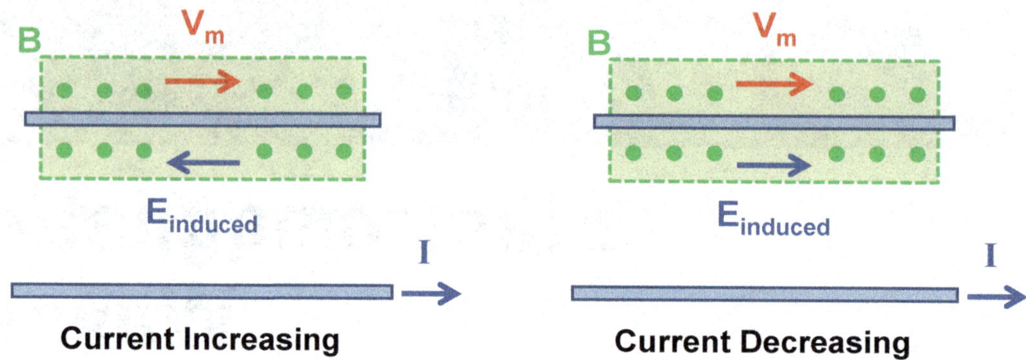

FIGURE 10.1: A current produces a magnetic potential and magnetic field. When this current changes, an electric field is induced.

We argued in Section 9-6 that this transfer of energy occurred through alternating magnetic and electric fields. To further quantify this process, let's define the positive z-axis to point from the lower wire to the upper wire in Figure 10.1. The z-axis is therefore also the direction of propagation of energy between the two wires in this system. From the right-hand rule we conclude that the direction of the magnetic field from the current in the lower wire in Figure 10.1 is out of the page (Section 4-5); let's define this direction to be the positive y-axis, as shown in Figure 10.2. As this magnetic field changes, an electric field will be induced along the x-axis, which is parallel to the lower wire. The magnitude of this induced electric field will change as the magnitude of the magnetic field changes. This will then induce a new magnetic field along the y-axis, which itself will induce a new electric field along the x-axis. This process will repeat itself with the electric field oscillating along the x-axis and the magnetic field oscillating on the y-axis until the energy is transferred to (*i.e.*, the electric field reaches) the second wire in Figure 10.1 and causes an induced current in that wire.

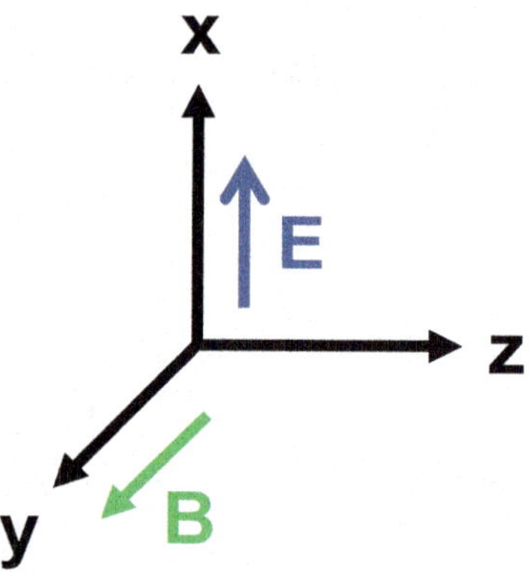

FIGURE 10.2: Orientation of the magnetic field and electric field for energy propagation in the system in Figure 10.1 at one instant of time. The positive z-axis is the direction of energy transfer.

From Faraday's law of induction (Equation C-3[1]) and Equation B-9, we have

$$\nabla \times \vec{E} = -\frac{\partial \vec{B}}{\partial t}$$

$$\left(\frac{\partial E_z}{\partial y} - \frac{\partial E_y}{\partial z}\right)\hat{x} + \left(\frac{\partial E_x}{\partial z} - \frac{\partial E_z}{\partial x}\right)\hat{y} + \left(\frac{\partial E_y}{\partial x} - \frac{\partial E_x}{\partial y}\right)\hat{z} = -\frac{\partial}{\partial t}\left(B_y \hat{y}\right)$$

Since the electric field oscillates along the x-axis, it will have only an x-axis component.

$$E_y = E_z = 0 \quad \rightarrow \quad \left(\frac{\partial E_x}{\partial z}\right)\hat{y} + \left(-\frac{\partial E_x}{\partial y}\right)\hat{z} = -\frac{\partial}{\partial t}\left(B_y \hat{y}\right)$$

In order for this equation to be valid

$$\frac{\partial E_z}{\partial y} = 0$$

In other words, E_x is a function of the variable z only. Hence,

$$\left(\frac{\partial E_x}{\partial z}\right)\hat{y} = \left(-\frac{\partial B_y}{\partial t}\right)\hat{y} \quad \rightarrow \quad \frac{\partial E_x}{\partial z} = -\frac{\partial B_y}{\partial t} \tag{10-1}$$

Similarly, using Ampère's law (Equation C-5)[2] we have[3]

$$\nabla \times \vec{B} = \mu_0 \varepsilon_0 \frac{\partial \vec{E}}{\partial t} \quad \rightarrow \quad \nabla \times \left(B_y \hat{y}\right) = \mu_0 \varepsilon_0 \frac{\partial}{\partial t}\left(E_x \hat{x}\right)$$

$$\left(-\frac{\partial B_y}{\partial z}\right)\hat{x} + \left(\frac{\partial B_y}{\partial x}\right)\hat{z} = \mu_0 \varepsilon_0 \frac{\partial}{\partial t}\left(E_x \hat{x}\right)$$

In order for this equation to be valid

1 See Section C-13 for the derivation of this form of Faraday's law of induction.

2 See Section C-14 for the derivation of this form of Ampère's law.

3 We assume here that the relative permittivity and relative permeability are both 1 for the material in-between the two wires in Figure 10.1.

$$\frac{\partial B_y}{\partial x} = 0$$

In other words, B_y must be a function of the variable z only. Hence,

$$-\frac{\partial B_y}{\partial z} = \mu_0 \varepsilon_0 \frac{\partial E_x}{\partial t} \tag{10-2}$$

Taking the partial derivative of Equation 10-1 with respect to z yields

$$\frac{\partial}{\partial z}\left(\frac{\partial E_x}{\partial z}\right) = \frac{\partial}{\partial z}\left(-\frac{\partial B_y}{\partial t}\right) \quad \rightarrow \quad \frac{\partial^2 E_x}{\partial z^2} = -\frac{\partial^2 B_y}{\partial t \, \partial z}$$

Taking the partial derivative of Equation 10-2 with respect to t yields

$$\frac{\partial}{\partial t}\left(-\frac{\partial B_y}{\partial z}\right) = \frac{\partial}{\partial t}\left(\mu_0 \varepsilon_0 \frac{\partial E_x}{\partial t}\right) \quad \rightarrow \quad -\frac{\partial^2 B_y}{\partial t \, \partial z} = \mu_0 \varepsilon_0 \frac{\partial^2 E_x}{\partial t^2}$$

Combining these two expressions gives us

$$\frac{\partial^2 E_x}{\partial z^2} = \mu_0 \varepsilon_0 \frac{\partial^2 E_x}{\partial t^2}$$

We can make this equation more general by including the possibility that the material in-between the two wires in Figure 10.1 has a relative permittivity and/or a relative permeability different from one.

$$\frac{\partial^2 E_x}{\partial z^2} = \kappa_m \mu_0 \kappa_e \varepsilon_0 \frac{\partial^2 E_x}{\partial t^2} \tag{10-3}$$

Similarly, Equation 10-1 and Equation 10-2 can be combined to yield

$$\frac{\partial^2 B_y}{\partial z^2} = \kappa_m \mu_0 \kappa_e \varepsilon_0 \frac{\partial^2 B_y}{\partial t^2} \tag{10-4}$$

Equation 10-3 and Equation 10-4 are referred to as *one-dimensional wave equations.* In other words, the electric and magnetic fields oscillate and propagate as *waves.* Furthermore, since these oscillations are coupled to each other (*i.e.*, the changing electric field generates the magnetic field, and the changing magnetic field generates the electric field), we refer to the coupled propagating electric and magnetic waves as a single *electromagnetic wave.*

10-3 Waves

A ***wave*** is a disturbance that propagates and carries energy.

> *Wave*: a disturbance that propagates and transmits energy. This transfer of energy occurs without an accompanying transfer of matter.

A familiar example of a wave is a water wave. If a rock is dropped into a still body of water, waves ripple out from the location where the rock entered the water. During the collision with the water, some of the kinetic energy of the rock is transferred to the motion of the water molecules (*i.e.,* some of the energy is used to displace the water and create the wave). When the water waves encounter an object in the water (a cork, *e.g.*), that object will oscillate up and down as it "rides" the wave. In this way, the kinetic energy of the wave is transferred to the kinetic and potential energy of the oscillating object. This is an example of how a wave transmits energy.

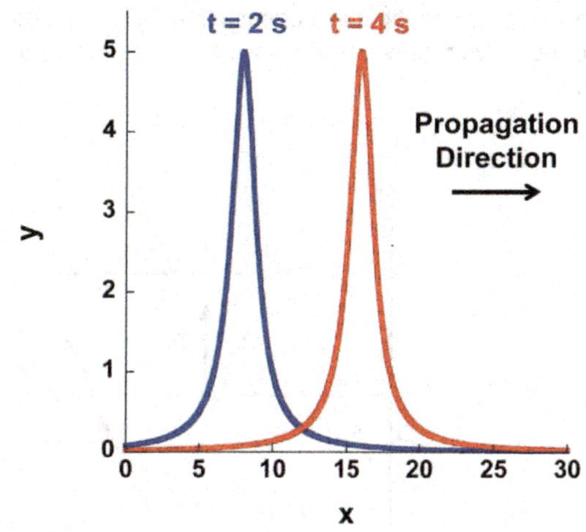

FIGURE 10.3: A transverse pulse propagating in the positive x-axis direction.

A mathematically and visually simple example of a wave is a pulse, as shown in Figure 10.3; the equation for this pulse is

$$y(x,t) = \frac{5}{(x-4t)^2+1}$$

The pulse in Figure 10.3 is travelling in the positive x-axis direction; as time progresses, the center of the pulse moves in the positive x-axis direction. The magnitude of the pulse along the y-axis is the magnitude of the disturbance (*e.g.,* the magnitude of the displacement) associated with the pulse. A rock dropped into water will create a cylindrically symmetric pulse (or pulses).

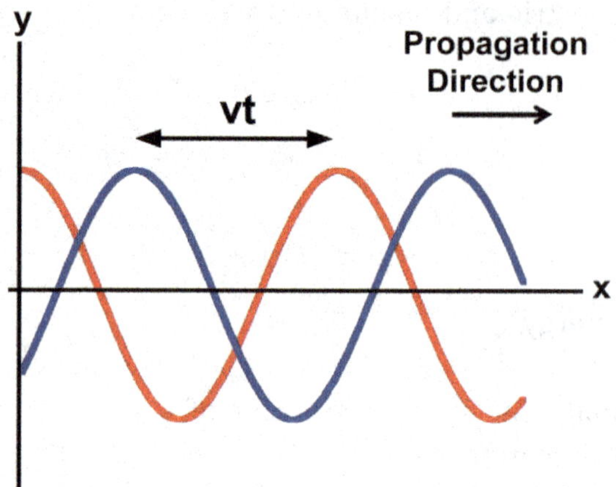

FIGURE 10.4: A sinusoidal travelling wave. The blue curve represents the wave at one instant of time, and the red curve represents the wave at a later instant of time.

The waves created by a single perturbation, such as a rock colliding with water, will be pulses that propagate away from the perturbation. In contrast, a continuous perturbation, such as someone continually splashing the water, will create a ***travelling wave***.

Travelling wave: a periodic continuous wave that moves at constant speed.

A simple example of a travelling wave is a sinusoidal wave, shown in Figure 10.4

The travelling wave is moving in the positive x-axis direction at speed v. A snapshot of a sinusoidal travelling wave at one instant of time is shown in Figure 10.5

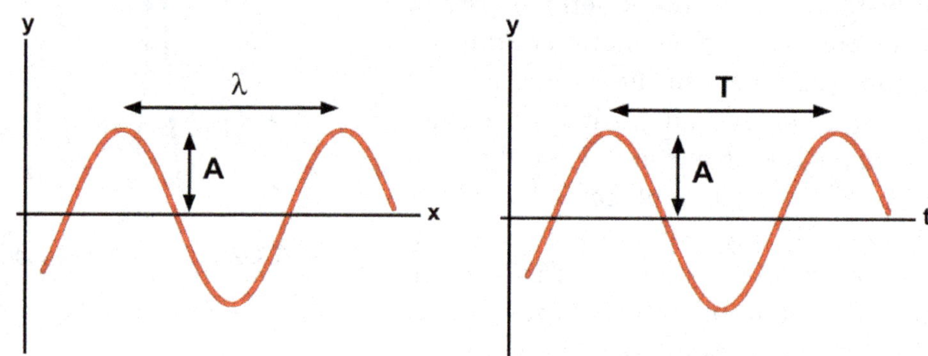

FIGURE 10.5: Definition of the amplitude (A), wavelength (λ), and period (T) of a sinusoidal travelling wave.

The wavelength of the wave, denoted by λ, is the minimum distance between any two identical points on adjacent oscillations of the wave. Similarly, the period of the wave, denoted by T, is the minimum time between any two identical points on adjacent oscillations of the wave. Finally, the amplitude of the wave, denoted by A, is a measure of the magnitude of the disturbance.

The speed of a sinusoidal travelling wave is the product of the wave's wavelength and frequency.

$$\lambda f = v \tag{10-5}$$

The frequency of the wave is the inverse of the period of the wave[4].

$$f = \frac{1}{T} \tag{10-6}$$

The wavenumber of the wave is the defined in Equation 10-7.

$$k = \frac{2\pi}{\lambda} \tag{10-7}$$

The speed of the wave can thus be written in terms of the wavenumber and the angular frequency of oscillation of the wave as

$$v = \frac{\omega}{k} \tag{10-8}$$

Example 10-1

Problem: A wave has a wavelength of 5 m and a speed of 10 m/s. What is the period of this wave? What is the wavenumber of this wave?

Solution: The frequency of the wave is calculated using Equation 10-5.

$$f = \frac{v}{\lambda} \quad \rightarrow \quad f = \frac{10\frac{\text{m}}{\text{s}}}{5\text{m}} \quad \rightarrow \quad f = 2\,\text{Hz}$$

The wavenumber of the wave is determined using Equation 10-7.

$$k = \frac{2\pi}{5\text{m}} \quad \rightarrow \quad 1.3\text{m}^{-1}$$

4 Recall from classical mechanics that the unit of frequency is the hertz (Hz).

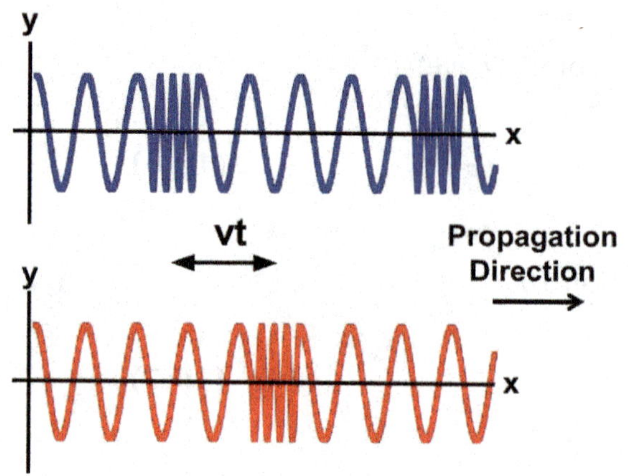

FIGURE 10.6: A longitudinal travelling wave. The blue curve represents the wave at one instant of time, and the red curve represents the wave at a later instant of time.

In a ***transverse wave*** the disturbance is oriented perpendicular to the direction of propagation of the wave. In a ***longitudinal wave*** the disturbance is oriented parallel to the direction of propagation of the wave.

Transverse wave: a wave in which the disturbance is perpendicular to the direction of propagation of the wave.

Longitudinal wave: a wave in which the disturbance is parallel to the direction of propagation of the wave.

Ripples on water and the vibrations of a string are both examples of transverse waves. Sound waves are the most common examples of longitudinal waves; the perturbations of a sound wave are differences in air pressure. A longitudinal travelling wave is shown in Figure 10.6. Earthquakes produce both transverse and longitudinal waves called seismic waves.

10-4 Electromagnetic Waves

As shown in Figure 10.3, Figure 10.4, and Figure 10.6 the perturbation associated with a wave is a function of position and time. The relationship between the position and time dependence of a wave is described by the linear wave equation (Equation 10-9).

$$\frac{\partial^2 y}{\partial x^2} = \frac{1}{v^2}\frac{\partial^2 y}{\partial t^2} \tag{10-9}$$

In Equation 10-9, the variable y represents the magnitude (*i.e.*, size) of the perturbation associated with the wave as a function of the variables x (position) and t (time). A solution[5] to Equation 10-9 is

$$y = y_0 e^{i(kx-\omega t)} \tag{10-10}$$

5 See Section C-15.

Therefore, solutions to the wave equation for electric and magnetic waves (Equation 10-3 and Equation 10-4) are

$$E_x = E_0 e^{i(kz-\omega t+\phi_0)} \qquad B_y = B_0 e^{i(kz-\omega t+\phi_0)} \tag{10-11}$$

The variables E_0 and B_0 in Equation 10-11 are the maximum magnitudes (*i.e.*, the amplitudes) of the electric field and magnetic field, respectively. The term within the parentheses is called the *phase* of the wave, and is usually denoted by ϕ.

$$\phi = kz - \omega t + \phi_0$$

The variable ϕ_0 in this equation is the phase constant[6] and depends upon the initial conditions (position, *e.g.*) of the wave. By comparing Equation 10-4 with Equation 10-11, we see that the speed of an electromagnetic wave is

$$v = \frac{1}{\left(\kappa_m \mu_0 \kappa_e \varepsilon_0\right)^{\frac{1}{2}}}$$

For electromagnetic waves propagating through vacuum, $\kappa_e = 1$ and $\kappa_m = 1$. The speed of propagation for these waves is

$$v = \frac{1}{\left((1)\left(4\pi \times 10^{-7}\,\frac{\text{kgm}}{\text{C}^2}\right)(1)\left(8.85 \times 10^{-12}\,\frac{\text{C}^2}{\text{J m}}\right)\right)^{\frac{1}{2}}} \quad \rightarrow \quad v = 3 \times 10^8\,\frac{\text{m}}{\text{s}}$$

This speed for the propagation of an electromagnetic wave in vacuum is denoted by the variable c.

$$c = 3 \times 10^8\,\frac{\text{m}}{\text{s}}$$

Furthermore, substitution of Equation 10-11 into Equation 10-1 yields

$$\frac{\partial}{\partial z}\left(E_0 e^{i(kz-\omega t+\phi_0)}\right) = -\frac{\partial}{\partial t}\left(B_0 e^{i(kz-\omega t+\phi_0)}\right) \quad \rightarrow \quad E_0 ike^{i(kz-\omega t+\phi_0)} = B_0 i\omega e^{i(kz-\omega t+\phi_0)}$$

6 Similar definitions of phase and phase constant were used in classical mechanics for describing simple harmonic motion.

$$E_0 k = B_0 \omega \quad \rightarrow \quad \frac{E_0}{B_0} = \frac{\omega}{k} \quad \rightarrow \quad \frac{E_0}{B_0} = v \tag{10-12}$$

For electromagnetic waves propagating through vacuum we have

$$\frac{E_0}{B_0} = c \tag{10-13}$$

The ratio of the magnitude of the electric field to the magnitude of the magnetic field is constant, at every instant, and is equal to the speed of the electromagnetic wave.

Example 10-2

Problem: An electromagnetic wave has a wavelength of 1 km and is propagating through vacuum. What are the wavenumber, frequency, and period of this wave?

Solution: The wavenumber of the wave is determined using Equation 10-7.

$$k = \frac{2\pi}{1000\,\mathrm{m}} \quad \rightarrow \quad 6.3 \times 10^{-3}\,\mathrm{m}^{-1}$$

The frequency of the wave is calculated using Equation 10-5.

$$f = \frac{v}{\lambda} \quad \rightarrow \quad f = \frac{3 \times 10^{8}\,\frac{\mathrm{m}}{\mathrm{s}}}{1000\,\mathrm{m}} \quad \rightarrow \quad f = 3 \times 10^{5}\,\mathrm{Hz} \quad \rightarrow \quad f = 300\,\mathrm{kHz}$$

Finally, we can use Equation 10-6 to determine the period of the wave.

$$T = \frac{1}{f} \quad \rightarrow \quad T = \frac{1}{3 \times 10^{5}\,\mathrm{Hz}} \quad \rightarrow \quad T = 3.3 \times 10^{-6}\,\mathrm{s} \quad \rightarrow \quad T = 3.3\,\mu\mathrm{s}$$

The electromagnetic wave is a *transverse wave* because the electric and magnetic fields are oscillating perpendicular to the direction of propagation of the wave (See Figure 10.7). Since the electric and magnetic fields of the electromagnetic wave are oscillating in a plane (for the wave in Figure 10.7, this plane is the *x-y* plane) the electromagnetic wave is also referred to as a *plane wave*.

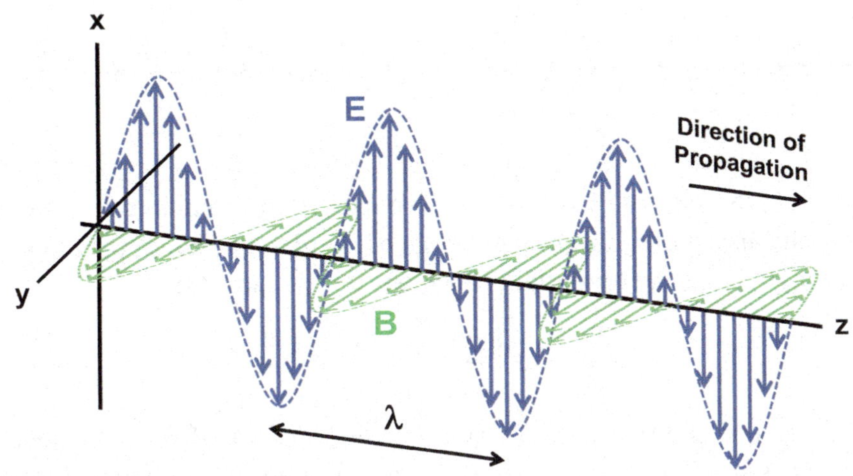

FIGURE 10.7: An electromagnetic wave propagating along the *z*-axis. The electric field is oscillating along the *x*-axis and the magnetic is oscillating along the *y*-axis.

The electric and magnetic field components of the electromagnetic wave in Figure 10.7 can be written as

$$\vec{E} = \left[E_0 e^{i(kz-\omega t+\phi_0)} \right] \hat{x} \qquad \vec{B} = \left[B_0 e^{i(kz-\omega t+\phi_0)} \right] \hat{y}$$

In Section 10-3, we described a wave as a disturbance that propagates and transmits energy. But how exactly does an electromagnetic wave transmit energy between the two wires in Figure 10.1? One way to think about is to assume that some of the energy supplied to the lower wire to alternate the current[7] is used to generate an electromagnetic wave that then propagates from the lower wire to the upper wire. When this electromagnetic wave encounters the upper wire, the electric field of the electromagnetic wave will generate a current in the upper wire. Since the electric field in the electromagnetic wave is oscillating in time, the generated current will be an alternating current.

Another model for how the electromagnetic wave transmits energy is to say that the electromagnetic wave itself *stores energy*. In Section 5-5, we argued that the separation of net electric charge on the two plates of the capacitor resulted in the electric potential energy of the capacitor. In other words, the work required to physically separate the

7 There must be an external EMF driving the AC current in this wire.

electric charges (*i.e.*, overcome the attractive force between them) resulted in the system having electric potential energy[8]. We can equivalently express this energy in terms of the electric field between the plates of the capacitor by combining Equation 5-5, Equation 5-8, and Equation 7-4.

$$U_e = \frac{q^2}{2C} \rightarrow U_e = \frac{\left(EA\kappa_e\varepsilon_0\right)^2}{2\left(\frac{A\kappa_e\varepsilon_0}{d}\right)} \rightarrow U_e = \frac{\kappa_e\varepsilon_0 E^2\left(Ad\right)}{2} \rightarrow U_e = \frac{\kappa_e\varepsilon_0 E^2 V}{2}$$

In this expression, the variable V is the volume enclosed by the plates (*i.e.*, the volume is the product of the area and separation distance of the plates). Hence, the energy density of the capacitor (the electric potential energy per unit volume) is

$$\frac{U_e}{V} = \frac{\kappa_e\varepsilon_0 E^2}{2} \rightarrow u_e = \frac{\kappa_e\varepsilon_0 E^2}{2} \qquad \textbf{(10-14)}$$

We could therefore argue that the energy required to create the charge separation of the capacitor is "stored" in the electric field that now exists between the two plates of the capacitor. Indeed, we know that an electric field has the ability to do work on objects with net electric charge. If such an object was put in between the plates of the capacitor, it would feel an electric force (Equation 3-11) and experience an associated acceleration. Hence, the object would gain kinetic energy from its interaction with the electric field. This increase in the kinetic energy of the object would accompany a decrease in the electric potential energy of the capacitor, since it would change the net electric charge distribution of the system[9]. This decrease in the electric potential energy of the system would result in a decrease in the magnitude of the electric field between the plates of the capacitor. In this way, we could say that the electric field transferred energy to the object.

Similarly, we can define a magnetic energy density for a solenoid/inductor using Equation 5-10, Equation 6-15, and Equation 7-9.

$$U_m = \frac{1}{2}LI^2 \rightarrow U_m = \frac{1}{2}\left(\kappa_m\mu_0 n^2 V\right)\left(\frac{B}{\kappa_m\mu_0 n}\right)^2 \rightarrow U_m = \frac{1}{2\kappa_m\mu_0}B^2 V$$

$$\frac{U_m}{V} = \frac{1}{2\kappa_m\mu_0}B^2 \rightarrow u_m = \frac{1}{2\kappa_m\mu_0}B^2 \qquad \textbf{(10-15)}$$

8 Recall that we can directly relate work to potential energy for a conservative force.

9 The moving charge itself is part of the system, and by moving in-between the plates of the capacitor, the distribution of the net electric charge for the entire system would change.

The magnetic field can store magnetic potential energy through its associated magnetic flux (Equation 6-14). This potential energy can be used to drive a current through the creation of an induced EMF (Equation 6-4) if the associated magnetic flux changes.

Although Equation 10-14 and Equation 10-15 were derived for specific systems (a capacitor and a solenoid/inductor), these equations are nevertheless universally valid for all systems[10]. Thus, we can describe the energy of an electromagnetic wave as being stored in the oscillating electric and magnetic fields constituting the wave. The total instantaneous energy density of the electromagnetic field is therefore

$$u_{total} = u_e + u_m \quad \rightarrow \quad u_{total} = \frac{\kappa_e \varepsilon_0 E^2}{2} + \frac{1}{2\kappa_m \mu_0} B^2$$

Substituting Equation 10-12 into this expression gives us

$$B = \frac{E}{v} \quad \rightarrow \quad u_{total} = \frac{\kappa_e \varepsilon_0 E^2}{2} + \frac{1}{2\kappa_m \mu_0}\left(\frac{E}{v}\right)^2 \quad \rightarrow \quad u_{total} = \frac{1}{2}\left(\kappa_e \varepsilon_0 + \frac{1}{\kappa_m \mu_0 v^2}\right)E^2$$

$$v = \frac{1}{(\kappa_m \mu_0 \kappa_e \varepsilon_0)^{\frac{1}{2}}} \quad \rightarrow \quad u_{total} = \frac{1}{2}\left(\kappa_e \varepsilon_0 + \frac{\kappa_m \mu_0 \kappa_e \varepsilon_0}{\kappa_m \mu_0}\right)E^2 \quad \rightarrow \quad u_{total} = \kappa_e \varepsilon_0 E^2$$

The instantaneous energy density of the electromagnetic wave is thus proportional to the square of the electric field. We recall from classical mechanics that the energy of a harmonic oscillator was proportional to the square of the amplitude of the oscillation. It is therefore not surprising that the instantaneous energy density of an electromagnetic wave, which consists of oscillating electric and magnetic fields, is proportional to the square of the amplitude of these fields.

Since the electric field oscillates periodically, however, the average energy density is a more useful quantity that the instantaneous energy density. Since the electric field is oscillating sinusoidally, the average total energy density is[11]

$$(u_{total})_{avg} = \frac{1}{2}\kappa_e \varepsilon_0 E_0^2 \tag{10-16}$$

10 You should feel inspired to take advanced physics courses to learn why this is true.

11 See the derivation of Equation 9-12 in Section C-11 to understand the origin of the one-half in Equation 10-16.

The variable E_0 in Equation 10-16 is the maximum magnitude of the electric field (see Equation 10-11). In other words, E_0 is the amplitude of the oscillation of the electric field.

Example 10-3

Problem: The maximum magnitude of the electric field of an electromagnetic wave propagating through vacuum is 600 N/C. What is the maximum magnitude of the magnetic field of this electromagnetic wave? What is the average total energy density of this electromagnetic wave?

Solution: The maximum magnitude of the magnetic field is determined using Equation 10-13.

$$B_0 = \frac{E_0}{c} \rightarrow B_0 = \frac{600\frac{\text{N}}{\text{C}}}{3\times10^8\frac{\text{m}}{\text{s}}} \rightarrow B_0 = 2\times10^{-6}\,\text{T} \rightarrow B_0 = 2\,\mu\text{T}$$

The average total energy density is calculated using Equation 10-16.

$$\left(u_{total}\right)_{avg} = \frac{1}{2}(1)\left(8.85\times10^{-12}\frac{\text{C}^2}{\text{J m}}\right)\left(600\frac{\text{N}}{\text{C}}\right)^2 \rightarrow \left(u_{total}\right)_{avg} = 1.6\times10^{-6}\frac{\text{J}}{\text{m}^3}$$

10-5 The Poynting Vector

The Poynting vector (Equation 10-17), named after John Henry Poynting, is the rate of flow of electromagnetic energy per unit area for an electromagnetic wave[12]. In other words, it is a power flux and has the units W/m².

$$\vec{S} = \frac{1}{\kappa_m \mu_0}\left(\vec{E}\times\vec{B}\right) \qquad \textbf{(10-17)}$$

12 See Section C-16 for the derivation of Equation 10-17.

The direction of the Poynting vector is also the direction of propagation of the electromagnetic wave. It is important to note, however, that the Poynting vector is a valid expression for energy flow for any distribution of electric and magnetic fields, not just electromagnetic waves.

Example 10-4

Problem: An infinitely long wire of radius R carries a current I distributed uniformly throughout the wire. What is the Poynting vector at the surface of the wire?

Solution: Let's define the z-axis to point in the direction of the current in the wire. The magnetic field from such a wire was calculated in Section 4-5 and Example 5-10. At the surface of the wire, this magnetic field is

$$\vec{B} = \left(\frac{I\mu_0}{2\pi R}\right)\hat{\varphi} \rightarrow \vec{B} = \left(\frac{I\mu_0 R}{2\left(\pi R^2\right)}\right)\hat{\varphi} \rightarrow \vec{B} = \left(\frac{J\mu_0 R}{2}\right)\hat{\varphi}$$

The electric field is determined from Ohm's law (Equation 7-13).

$$\vec{E} = \left(\frac{J}{\sigma}\right)\hat{z}$$

The Poynting vector can now be determined using Equation 10-17.

$$\vec{S} = \frac{1}{\mu_0}\left(\left(\frac{J}{\sigma}\right)\hat{z} \times \left(\frac{J\mu_0 R}{2}\right)\hat{\varphi}\right) \rightarrow \vec{S} = \left(\frac{J^2 R}{2\sigma}\right)\left(\hat{z} \times \hat{\varphi}\right)$$

$$\vec{S} = \left(\frac{J^2 R}{2\sigma}\right)\left(-\hat{\rho}\right)$$

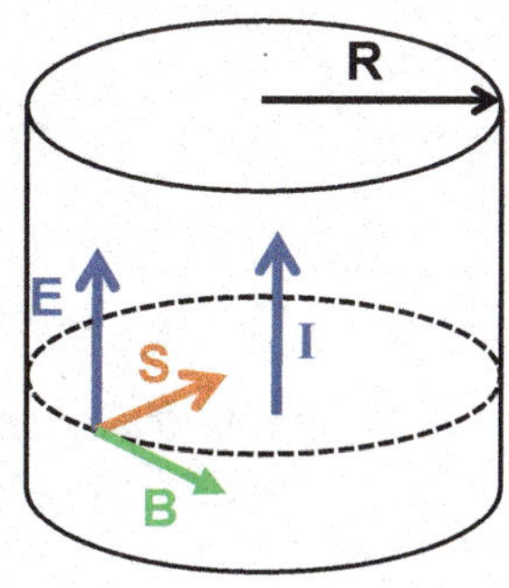

FIGURE 10.8: Direction of the electric field, magnetic field, and Poynting vector for an infinitely long current-carrying wire.

The Poynting vector is thus perpendicular to the surface of the wire and directed radially inward, as shown in Figure 10.8.

Hence, in order for the current to remain constant, there must be a steady flow of energy into the wire. The ultimate source of this energy is an EMF, such as a battery.

The average magnitude of the Poynting vector is called the *wave intensity*, denoted by *I*. For a sinusoidally oscillating electromagnetic wave, we have

$$I = S_{avg} \rightarrow I = \frac{1}{2}\left(\frac{E_0 B_0}{\kappa_m \mu_0}\right) \rightarrow I = \frac{1}{2}\left(\frac{E_0^2}{\kappa_m \mu_0 v}\right) \rightarrow I = \frac{1}{2} v \kappa_e \varepsilon_0 E_0^2 \quad \textbf{(10-18)}$$

Substituting Equation 10-14 into Equation 10-18 allows us to relate the wave intensity to the average total energy density of the wave.

$$I = S_{avg} = v\left(u_{total}\right)_{avg} \quad \textbf{(10-19)}$$

The wave intensity is the product of the average total energy density and the speed of the wave.

Example 10-5

Problem: The maximum magnitude of the electric field of an electromagnetic wave propagating through vacuum is 600 N/C. What is the average intensity of this electromagnetic wave?

Solution: The average intensity is determined using Equation 10-18.

$$I = \frac{1}{2} v \kappa_e \varepsilon_0 E_0^2 \rightarrow I = \frac{1}{2}\left(3 \times 10^8 \,\frac{\text{m}}{\text{s}}\right)(1)\left(8.85 \times 10^{-12} \,\frac{\text{C}^2}{\text{J m}}\right)\left(600 \,\frac{\text{N}}{\text{C}}\right)^2$$

$$I = 370 \,\frac{\text{W}}{\text{m}^2}$$

The integral of the Poynting vector over a surface is the power transmitted through that surface by the electromagnetic wave.

$$P = \int \vec{S} \bullet d\vec{A} \quad \textbf{(10-20)}$$

Example 10-6

Problem: The average intensity of the electromagnetic radiation from the Sun on the surface of the Earth is 10^3 W/m^2. This radiation is incident upon a 7 m × 14 m roof that is covered in solar panels. If the solar panels can convert 30% of incident radiation into electric power, what is the average electric power supplied by the solar cells while exposed to the sunlight?

Solution: The average power absorbed by the solar panels can be determined using Equation 10-19 and Equation 10-20.

$$P = \int \vec{S} \bullet d\vec{A} \rightarrow P_{avg} = S_{avg} \int dA \rightarrow P_{avg} = S_{avg} A \rightarrow P_{avg} = IA$$

$$P_{avg} = \left(10^3 \frac{\text{W}}{\text{m}^2}\right)(7\,\text{m})(14\,\text{m}) \rightarrow P_{avg} = 9.8 \times 10^4\,\text{W}$$

The average electric power supplied by the solar cells is thus

$$\left(P_{electric}\right)_{avg} = \left(9.8 \times 10^4\,\text{W}\right)(0.3) \rightarrow \left(P_{electric}\right)_{avg} = 2.9 \times 10^4\,\text{W}$$

10-6 Polarization

The ***polarization*** of an electromagnetic wave is defined to be the direction of the oscillation of the electric field of the wave.

Polarization: the direction of oscillations of a transverse wave. The polarization of an electromagnetic wave is defined to be the direction of the oscillation of the electric field of the wave.

For example, the polarization of the electromagnetic wave in Figure 10.7 is along the x-axis. Furthermore, the electromagnetic wave in Figure 10.7 is said to be ***linearly polarized*** (or *plane polarized*) since the polarization (*i.e.*, the oscillations of the electric field) are always along the same axis.

> *Linearly polarized*: an electromagnetic wave is linearly polarized if the oscillations of the electric field are always along the same axis.

Most sources of electromagnetic waves found in nature are not polarized. This means that the directions of electric field oscillations of the individual electromagnetic waves produced by these sources are different from one another (but always perpendicular to the direction of propagation of the associated wave). Such collections of electromagnetic waves are commonly referred to as unpolarized electromagnetic waves[13].

A ***polarizer*** is a material that will absorb the energy of an electromagnetic wave if the electric field has a specific polarization.

> *Polarizer*: a material that absorbs the energy from an electromagnetic wave with a specific polarization.

Polarizers are often constructed of long, linear chains of molecules aligned symmetrically along a particular direction. If the electric field of the electromagnetic wave is polarized (*i.e.,* is oscillating) with the same orientation as these chains, then the electric field can accelerate the electrons in these molecules to move along the chains. In this way, some of the energy of the electromagnetic wave is transferred to the kinetic and potential energies of the electrons in the polarizer. If, however, the electric field of the electromagnetic wave is not oriented in the direction of these chains, little energy will be absorbed by the polarizer.

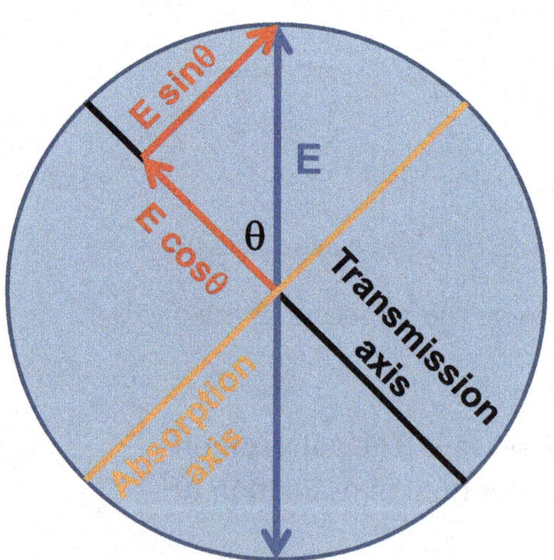

FIGURE 10.9: An electromagnetic wave with a polarization at an angle θ with the transmission axis of the polarizer. Only a component of the incident electric field will pass through the polarizer.

Since the polarizer works by absorbing energy from the electromagnetic wave, the intensity of an electromagnetic wave must be reduced by passing through a polarizer (Equation 10-19). The absorption and transmission axes of a polarizer are the axes along which the polarizer will have maximum and minimum, respectively, absorption of incident electric fields (see Figure 10.9); the polarization and transmission axes are perpendicular to each other.

As shown in Figure 10.9, the component of the incident electric field parallel to the absorption axis of the polarizer will be absorbed. The component of the incident electric field parallel to the transmission axis will pass through the polarizer

13 We'll learn in Chapter 11 that this is also called unpolarized light.

unabsorbed. Thus, if the incident polarization of the electromagnetic wave is at an angle θ with respect to the transmission axis of the polarizer, the component of the electric field that passes through the polarizer unabsorbed (*i.e.*, the component of the electric field transmitted through the polarizer) is

$$E_{unabsorbed} = E_0 \cos\theta \quad \rightarrow \quad E_{transmitted} = E_0 \cos\theta$$

Therefore, using Equation 10-18, we find that the intensity of the electromagnetic wave passing through the polarizer (which is also referred to as the transmitted intensity) is

$$I_{transmitted} = I_0 \cos^2\theta \qquad \textbf{(10-21)}$$

The variable I_0 in Equation 10-21 denotes the incident intensity of the electromagnetic wave. Equation 10-21 is known as the law of Malus, named after Étienne-Louis Malus.

Example 10-7

Problem: A linearly polarized electromagnetic wave is incident on a polarizer whose transmission axis is at 45° with respect to its polarization. The transmitted light then passes through a second polarizer whose transmission axis is at 45° with respect to the transmission axis of the first polarizer, as shown in Figure 10.10.

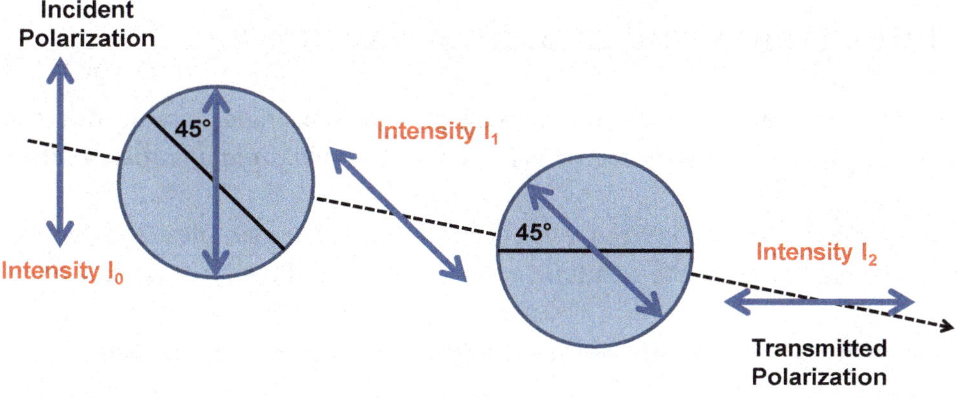

FIGURE 10.10: The combination of polarizers in Example 10-6.

What is the intensity of the electromagnetic wave transmitted through the second polarizer?

Solution: The intensity of the transmitted electromagnetic wave through the first polarizer, which we can denote as I_1, is determined using Equation 10-21.

$$I_1 = I_0\left[\cos\left(45°\right)\right]^2 \rightarrow I_1 = \frac{1}{2}I_0$$

This intensity is then incident on the second polarizer. The polarization of this electromagnetic wave is the same as the transmission axis of the first polarizer, and thus is at 45° with respect to the transmission axis of the second polarizer. The intensity of the transmitted electromagnetic wave through the second polarizer, which we can denote as I_2, is determined using Equation 10-21.

$$I_2 = I_1\left[\cos\left(45°\right)\right]^2 \rightarrow I_2 = \frac{1}{2}I_1 \rightarrow I_2 = \frac{1}{4}I_0$$

25% of the intensity incident on the first polarizer. The polarization of this transmitted light is rotated 90° from the polarization of the incident light.

10-7 Interference and Standing Waves

When two (or more) waves are simultaneously present at a single location, the resultant wave is the sum of the two waves. This is known as the ***principle of superposition***.

Principle of superposition: When two (or more) waves are simultaneously present at a single location, the resultant wave is the sum of the two waves.

We call this process of combining waves ***interference*** and the waves are said to *interfere*.

Interference: the process of combining two or more waves to create a resultant wave.

Constructive interference: the interference of two or more waves which differ in phase by an integer multiple of 2π. The amplitude of the resultant wave is the sum of the amplitudes of the individual waves (Figure 10.11).

Destructive interference: the interference of two or more waves which differ in phase by an integer multiple of π. The amplitude of the resultant wave is zero (Figure 10.11).

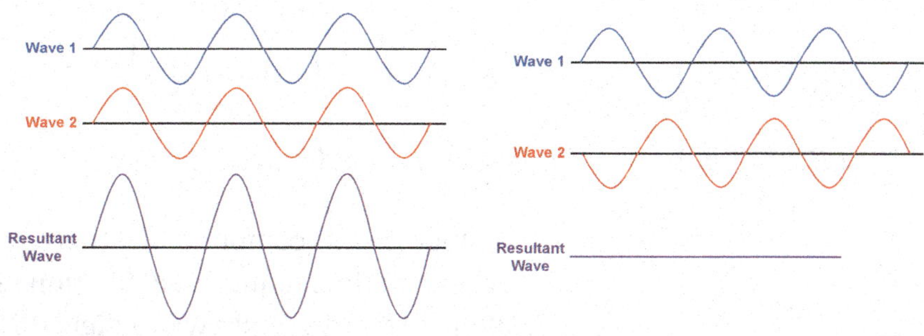

FIGURE 10.11: Constructive and destructive interference of two waves.

Let's consider the superposition of two electromagnetic waves with the same angular frequency travelling in opposite directions. Let's denote the electric and magnetic fields of these waves using the subscripts 1 and 2. For further simplicity, let's assume that the phase constant is zero.

$$\vec{E}_1 = \left[(E_0)_1 e^{i(kz-\omega t)}\right]\hat{x} \qquad \vec{B}_1 = \left[(B_0)_1 e^{i(kz-\omega t)}\right]\hat{y}$$

$$\vec{E}_2 = \left[(E_0)_2 e^{i(kz+\omega t)}\right]\hat{x} \qquad \vec{B}_2 = \left[(B_0)_2 e^{i(kz+\omega t)}\right]\hat{y}$$

The resultant wave, which we will denote using the subscript 3, is thus

$$\vec{E}_3 = \left[(E_0)_1 e^{i(kz-\omega t)} + (E_0)_2 e^{i(kz+\omega t)}\right]\hat{x} \qquad \vec{B}_3 = \left[(B_0)_1 e^{i(kz-\omega t)} + (B_0)_2 e^{i(kz+\omega t)}\right]\hat{y}$$

Let's now just consider the real part of these fields and use a trigonometric identity (Equation A-1) to simplify our expression. After combining similar terms, we have

$$\vec{E}_3 = \left[\cos(kz)\cos(\omega t)\left((E_0)_1 + (E_0)_2\right) + \sin(kz)\sin(\omega t)\left((E_0)_2 - (E_0)_1\right)\right]\hat{x}$$

$$\vec{B}_3 = \left[\cos(kz)\cos(\omega t)\left((B_0)_1 + (B_0)_2\right) + \sin(kz)\sin(\omega t)\left((B_0)_2 - (B_0)_1\right)\right]\hat{y}$$

The resultant wave is no longer a travelling wave, since the *kz* and *wt* terms have now been separated from each other. This wave is referred to as a ***standing wave***.

Standing wave: a periodic continuous wave that is stationary. Each portion of a standing wave oscillates with constant amplitude.

Let's consider the specific case when the amplitudes of the electric and magnetic fields of the two original waves are equal.

$$(E_0)_1 = (E_0)_2 \quad \rightarrow \quad \vec{E}_3 = \left[2(E_0)_1 \cos(kz)\cos(\omega t)\right]\hat{x}$$

$$(B_0)_1 = (B_0)_2 \quad \rightarrow \quad \vec{B}_3 = \left[2(B_0)_1 \cos(kz)\cos(\omega t)\right]\hat{y}$$

A plot of *x*-axis component of this electric field, at several time points, as a function of position along the *z*-axis is shown in Figure 10.12.

As shown in Figure 10.12, there are positions along the *z*-axis for which the magnitude of the electric field is always zero. These points are called ***nodes***.

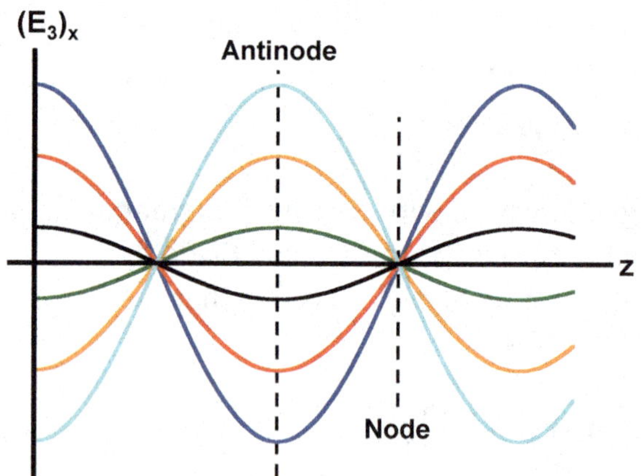

FIGURE 10.12: The electric field of a standing electromagnetic wave as a function of position along the *z*-axis. The different color traces refer to the magnitude of the *x*-axis component of the electric field at different times.

Node: a position along a standing wave where the wave has minimum amplitude. This corresponds to a position of destructive interference.

There are also positions the *z*-axis for which the magnitude of the electric field experiences the largest fluctuations. These points are called ***antinodes***.

Antinode: a position along a standing wave where the wave has maximum amplitude. This corresponds to a position of constructive interference.

The separation distance between adjacent nodes (or antinodes) is determined from the wavenumber of the standing wave.

$$\cos(kz) = 0 \quad \rightarrow \quad kz = \pi \quad \rightarrow \quad \frac{2\pi z}{\lambda} = \pi \quad \rightarrow \quad z = \frac{\lambda}{2}$$

Each node (or antinode) is thus separated by a distance equal to one-half the wavelength of the standing wave.

Example 10-8

Problem: An electromagnetic standing wave in vacuum has a frequency of 100 MHz. What is the distance between adjacent antinodes of this standing wave?

Solution: The spacing between antinodes is determined by the wavelength of the standing wave.

$$\Delta z = \frac{\lambda}{2} \rightarrow \Delta z = \frac{1}{2}\left(\frac{c}{f}\right) \rightarrow \Delta z = \frac{1}{2}\left(\frac{3\times 10^{8}\,\frac{\text{m}}{\text{s}}}{100\times 10^{6}\,\text{Hz}}\right) \rightarrow \Delta z = 1.5\,\text{m}$$

Now let's calculate the superposition of two electromagnetic waves with the different angular frequencies travelling in the same direction. Let's denote the electric and magnetic fields of these waves using the subscripts 1 and 2. As before, for the sake of simplicity let's assume that the phase constant is zero. We will further assume that the magnitudes of the electric and magnetic fields of the two waves are the same.

$$\vec{E}_1 = \left[(E_0)e^{i(k_1 z-\omega_1 t)}\right]\hat{x} \qquad \vec{B}_1 = \left[(B_0)e^{i(k_1 z-\omega_1 t)}\right]\hat{y}$$

$$\vec{E}_2 = \left[(E_0)e^{i(k_2 z-\omega_2 t)}\right]\hat{x} \qquad \vec{B}_2 = \left[(B_0)e^{i(k_2 z-\omega_2 t)}\right]\hat{y}$$

The resultant wave, which we will denote using the subscript 3, is thus

$$\vec{E}_3 = (E_0)\left[e^{i(k_1 z-\omega_1 t)} + e^{i(k_2 z-\omega_2 t)}\right]\hat{x} \qquad \vec{B}_3 = (B_0)\left[e^{i(k_1 z-\omega_1 t)} + e^{i(k_2 z-\omega_2 t)}\right]\hat{y}$$

Let's now just consider the real part of these fields and use a trigonometric identity (Equation A-5) to simplify our expression.

$$\vec{E}_3 = \left[2(E_0)\cos\left(\frac{(k_1 - k_2)z - (\omega_1 - \omega_2)t}{2}\right)\cos\left(\frac{(k_1 + k_2)z - (\omega_1 + \omega_2)t}{2}\right)\right]\hat{x}$$

We can define the average wavenumber and average angular frequency to be

$$k_{avg} = \frac{k_1 + k_2}{2} \qquad \omega_{avg} = \frac{\omega_1 + \omega_2}{2}$$

and the modulation wavenumber and modulation angular frequency to be

$$k_m = \frac{k_1 - k_2}{2} \qquad \omega_m = \frac{\omega_1 - \omega_2}{2}$$

With these definitions, the equation for the resultant electric field is

$$\vec{E}_3 = \left[2(E_0)\cos(k_m z - \omega_m t)\cos(k_{avg} z - \omega_{avg} t)\right]\hat{x}$$

This resultant wave can be regarded as a travelling wave of angular frequency ω_{avg} that has a time-varying (or modulated) amplitude.

$$\vec{E}_3 = \left[A\cos(k_{avg} z - \omega_{avg} t)\right]\hat{x} \qquad A = 2(E_0)\cos(k_m z - \omega_m t)$$

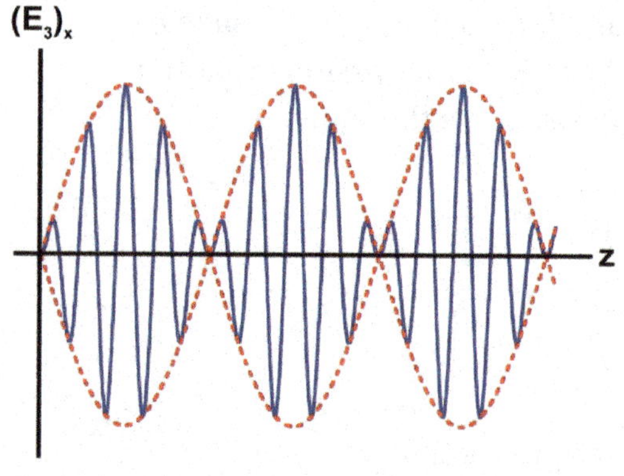

FIGURE 10.13: The electric field of the resultant electromagnetic wave from the addition of two electromagnetic waves with nearly equal angular frequencies. The dashed red line indicates the modulated amplitude of the oscillations of the resultant electric field.

When the angular frequencies are very close to each other, $\omega_{avg} >> \omega_m$ and the modulated amplitude changes slowly, as shown in Figure 10.13.

The intensity of the resultant electromagnetic wave is proportional to the amplitude of the wave (Equation 10-18)[14].

$$I \approx \cos^2(k_m z - \omega_m t) \rightarrow I \approx \frac{1 + \cos(2k_m z - 2\omega_m t)}{2}$$

The intensity thus oscillates at an angular frequency $2\omega_m$ or $\omega_1 - \omega_2$, which is known as the *beat angular frequency*; the beat frequency is the difference in the frequencies of oscillation of the two waves.

14 This expression is simplified using Equation A-4.

Example 10-9

Problem: Two electromagnetic waves with angular frequencies of 1220 kHz and 1200 kHz interfere with each other to form a modulated electromagnetic wave. What is the beat frequency of this modulated wave?

Solution: The beat frequency is simply the difference in the two frequencies.

$$f_{beat} = 1220\,\text{kHz} - 1200\,\text{kHz} \quad \rightarrow \quad f_{beat} = 20\,\text{kHz}$$

The rate at which the beats move is known as the group velocity, and is denoted by v_g.

$$v_g = \frac{\omega_m}{k_m} \quad \rightarrow \quad v_g = \frac{\omega_1 - \omega_2}{k_1 - k_2} \quad \rightarrow \quad v_g = \frac{\Delta\omega}{\Delta k}$$

When the difference in the angular frequencies is small, the group velocity is approximately equal to

$$v_g = \frac{d\omega}{dk}$$

Therefore, from Equation 10-8, we have

$$v_g = v + k\frac{dv}{dk} \tag{10-22}$$

If the speed of the wave is independent of the wavenumber of the wave, then the group velocity is equal to the wave speed.

Finally, let's calculate the superposition of two linearly polarized electromagnetic waves travelling in the same direction with the same angular frequency. The first electromagnetic wave is polarized along the x-axis, and the second electromagnetic wave is polarized along the y-axis.

$$\vec{E}_1 = \left[\left(E_0\right)_1 e^{i\left(kz-\omega t+\left(\phi_0\right)_1\right)}\right]\hat{x} \qquad \vec{E}_2 = \left[\left(E_0\right)_2 e^{i\left(kz-\omega t+\left(\phi_0\right)_2\right)}\right]\hat{y}$$

The electric field of the resultant wave, which we will denote using the subscript 3, is

$$\vec{E}_3 = \left[\left(E_0\right)_1 \left(e^{i\left(kz - \omega t + (\phi_0)_1\right)} \right) \hat{x} + \left(E_0\right)_2 \left(e^{i\left(kz - \omega t + (\phi_0)_2\right)} \right) \hat{y} \right]$$

$$\vec{E}_3 = \left[\left(\left(E_0\right)_1 e^{i(\phi_0)_1} \right) \hat{x} + \left(\left(E_0\right)_2 e^{i(\phi_0)_2} \right) \hat{y} \right] e^{i(kz - \omega t)}$$

Let's rewrite this expression in terms of the difference in the phase constants of the two waves.

$$\Delta\phi = (\phi_0)_2 - (\phi_0)_1 \quad \rightarrow \quad \vec{E}_3 = \left[\left(E_0\right)_1 \hat{x} + \left(\left(E_0\right)_2 e^{i\Delta\phi} \right) \hat{y} \right] e^{i\left(kz - \omega t + (\phi_0)_1\right)}$$

The resultant electromagnetic wave is thus linearly polarized as long as $\Delta\phi$ is an integer multiple of π. However, if the two electromagnetic waves are out of phase by $\pi/2$ (*i.e.*, when $\Delta\phi = \pi/2$) the electric field of the resultant electromagnetic wave is

$$\vec{E}_3 = \left[\left(E_0\right)_1 \hat{x} + \left(\left(E_0\right)_2 e^{\mp i\frac{\pi}{2}} \right) \hat{y} \right] e^{i\left(kz - \omega t + (\phi_0)_1\right)}$$

The real part of this electric field is thus

$$\vec{E}_3 = \left[\left(\left(E_0\right)_1 \cos\left(kz - \omega t + (\phi_0)_1\right) \right) \hat{x} \pm \left(\left(E_0\right)_2 \sin\left(kz - \omega t + (\phi_0)_1\right) \right) \hat{y} \right]$$

The resultant wave is no longer linearly polarized since the electric field is rotating in the *x-y* plane with an angular frequency ω. We describe this electromagnetic wave as being ***elliptically polarized***.

Elliptically polarized: an electromagnetic wave is elliptically polarized if the electric field vector rotates and changes in magnitude as the electromagnetic wave propagates.

A special type of elliptical polarization called ***circular polarization*** occurs when the magnitudes of the electric fields of the two combined electromagnetic waves are equal.

Circularly polarized: an electromagnetic wave is circularly polarized if the electric field vector rotates, but maintains a constant magnitude as the electromagnetic wave propagates.

In this case, the electric field of the resultant electromagnetic wave is

$$\left(E_0\right)_1 = \left(E_0\right)_2 = E_0 \quad \rightarrow \quad \vec{E}_3 = E_0\left[\left(\cos\left(kz - \omega t + \left(\phi_0\right)_1\right)\right)\hat{x} \pm \left(\sin\left(kz - \omega t + \left(\phi_0\right)_1\right)\right)\hat{y}\right]$$

The electric field corresponding to the positive sign in this equation is said to be right-circularly polarized. Similarly, the electric field corresponding to the negative sign in this equation is said to be left-circularly polarized. The rotation of the electric field in a right-circularly polarized electromagnetic wave is shown in Figure 10.14.

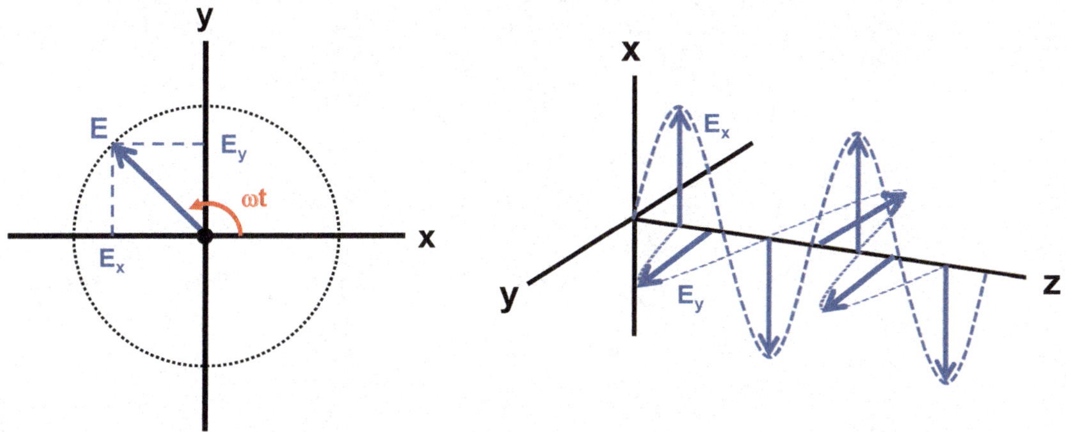

FIGURE 10.14: A right-circularly polarized electromagnetic wave.

Several 3D movies rely upon circularly polarized light to produce stereoscopic images. The 3D glasses contain a right-circularly polarized filter in one lens and a left-circularly polarized filter in the other lens. An advantage of using circularly polarized light over linearly polarized light is that audience members can tilt and turn their heads without affecting the 3D effect.

10-8 Antennae and Radiation

In Section 9-6, we showed that a small loop of current will dissipate energy through a non-resistive mechanism. Rather than dissipating energy as heat, as a resistor does, the loop of current dissipates energy in the form of electromagnetic radiation. We refer to the loop of wire carrying the current as an ***antenna***.

Antenna: an electrical device which can convert electric power into electromagnetic waves, and can convert electromagnetic waves into electric power.

In Example 9-12, we showed that the power dissipated was a function of the frequency of the oscillations of the current. Thus, the ability of the loop antenna to generate an electromagnetic wave (or convert an electromagnetic wave into electrical power) depends upon the frequency or wavelength of electromagnetic wave. Indeed, substitution of Equation 10-7 and Equation 10-8 into Equation 9-19 gives us

$$\omega = \frac{2\pi c}{\lambda} \quad \rightarrow \quad R_{rad} \approx \frac{\mu_0 A^2}{6\pi c^3}\left(\frac{2\pi c}{\lambda}\right)^4 \quad \rightarrow \quad R_{rad} \approx \frac{\mu_0 8\pi^3 c}{3}\left(\frac{A}{\lambda^2}\right)^2$$

$$R_{rad} \approx \frac{\left(4\pi \times 10^{-7}\,\frac{\text{kgm}}{\text{C}^2}\right)8\pi^3\left(3\times 10^8\,\frac{\text{m}}{\text{s}}\right)}{3}\left(\frac{A}{\lambda^2}\right)^2$$

$$R_{rad} \approx \left(3.1\times 10^4\,\Omega\right)\left(\frac{A}{\lambda^2}\right)^2 \tag{10-23}$$

Example 10-10

Problem: What is the value of R_{rad} for a circular loop antenna with a circumference equal to one-fourth the wavelength of the electromagnetic wave?

Solution: We can determine the area from the circumference. If we denote the radius of the loop as R, we have

$$2\pi R = \frac{\lambda}{4} \quad \rightarrow \quad R = \frac{\lambda}{8\pi} \quad \rightarrow \quad A = \pi\left(\frac{\lambda}{8\pi}\right)^2 \quad \rightarrow \quad A = \frac{\lambda^2}{64\pi}$$

Substitution into Equation 10-23 gives us

$$R_{rad} \approx \left(3.1\times 10^4\,\Omega\right)\left(\frac{\frac{\lambda^2}{64\pi}}{\lambda^2}\right)^2 \quad \rightarrow \quad R_{rad} \approx 0.77\,\Omega$$

Another common antenna is the short dipole antenna, shown in Figure 10.15.

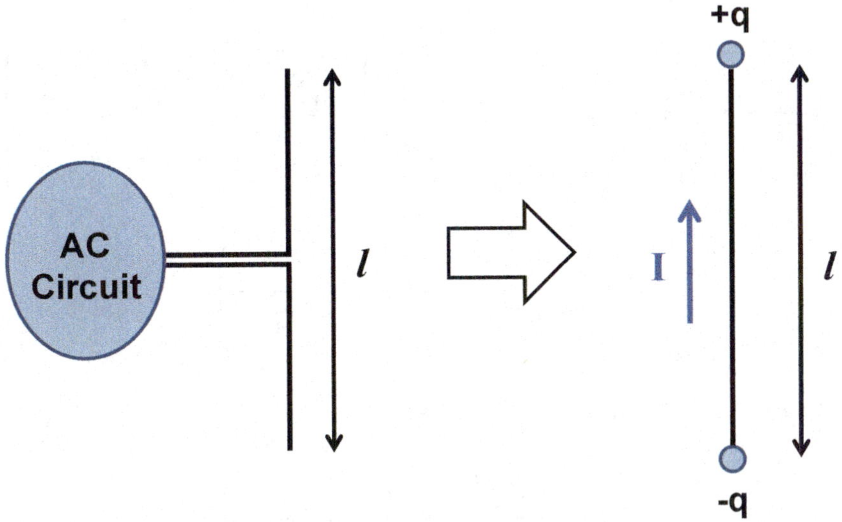

FIGURE 10.15: A linear dipole antenna can be modeled as an electric dipole and an alternating current.

As shown in Figure 10.15, we can model the short dipole antenna as an electric dipole with an alternating current flowing between the separated electric charges; the magnitude (and sign) of the electric charges will also vary in time. Let's determine the magnitude of the electric and magnetic fields at a position P, as shown in Figure 10.16

We begin by recognizing that we must use retarded time (Section 9-6) for all of our calculations. Thus, we can express the electric charges in the electric dipole in Figure 10.16 as

$$q = q_0 e^{i\omega\left(t - \frac{r_A}{c}\right)} \qquad -q = -q_0 e^{i\omega\left(t - \frac{r_B}{c}\right)}$$

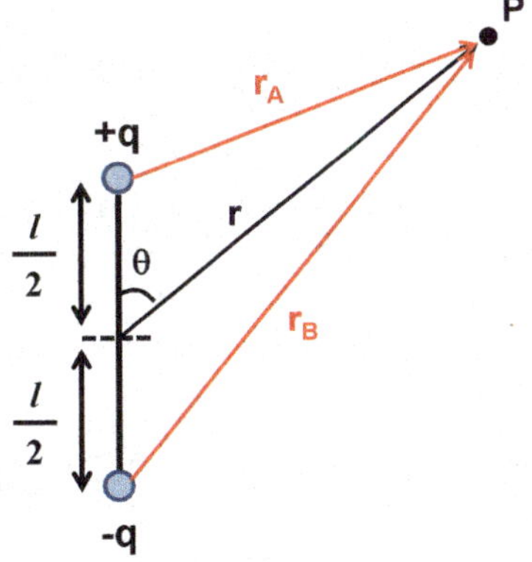

FIGURE 10.16: Determining the electric and magnetic fields from a linear dipole antenna.

We can also relate these electric charges to the electric current in Figure 10.15 using Equation 4-4.

$$I = I_0 e^{i\omega\left(t - \frac{r}{c}\right)} \rightarrow q = \int I_0 e^{i\omega\left(t - \frac{r}{c}\right)} dt \rightarrow q = \frac{I_0}{i\omega} e^{i\omega\left(t - \frac{r}{c}\right)}$$

The electric potential for the electric dipole is thus[15]

$$V_e = \frac{1}{4\pi\varepsilon_0}\left(\frac{q}{r_A} + \frac{-q}{r_B}\right) \rightarrow V_e = \frac{1}{4\pi\varepsilon_0}\left(\frac{\frac{I_0}{i\omega}e^{i\omega\left(t-\frac{r_A}{c}\right)}}{r_A} + \frac{\frac{-I_0}{i\omega}e^{i\omega\left(t-\frac{r_B}{c}\right)}}{r_B}\right)$$

$$V_e = \frac{I_0}{i\omega 4\pi\varepsilon_0}\left(\frac{e^{i(\omega t - kr_A)}}{r_A} - \frac{e^{i(\omega t - kr_B)}}{r_B}\right)$$

Under conditions that $r >> l$ and $\lambda >> l$, we have[16]

$$V_e = \frac{I_0 l e^{i(\omega t - kr)}\cos\theta}{4\pi\varepsilon_0 c}\left(\frac{1}{r} + \frac{c}{i\omega r^2}\right)$$

$$\vec{V}_m = \left[\frac{\mu_0 I_0 l e^{i(\omega t - kr)}}{4\pi r}\right]\left[(\cos\theta)\hat{r} - (\sin\theta)\hat{\theta}\right]$$

The real parts of the corresponding electric and magnetic fields are[17]

$$E_r = \frac{I_0 l\cos\theta}{2\pi\varepsilon_0}\left[\frac{\cos(\omega t - kr)}{cr^2} + \frac{\sin(\omega t - kr)}{\omega r^3}\right]$$

$$E_\theta = \frac{I_0 l\sin\theta}{4\pi\varepsilon_0}\left[-\frac{\omega\sin(\omega t - kr)}{c^2 r} + \frac{\cos(\omega t - kr)}{cr^2} + \frac{\sin(\omega t - kr)}{\omega r^3}\right]$$

$$B_\varphi = \frac{\mu_0 I_0 l\sin\theta}{4\pi}\left(\frac{\cos(\omega t - kr)}{r^2} - \frac{\omega\sin(\omega t - kr)}{cr}\right)$$

15 We will assume that the dipole is radiating into vacuum.
16 See Section C-17.
17 See Section C-17.

Let's now examine these fields until two different limiting cases. First, when the distance r is small (but still greater than l), these fields are approximately

$$E_r = \frac{I_0 l \cos\theta}{2\pi\varepsilon_0}\left[\frac{\sin(\omega t - kr)}{\omega r^3}\right]$$

$$E_\theta = \frac{I_0 l \sin\theta}{4\pi\varepsilon_0}\left[\frac{\sin(\omega t - kr)}{\omega r^3}\right]$$

$$B_\varphi = \frac{\mu_0 I_0 l \sin\theta}{4\pi}\left(\frac{\cos(\omega t - kr)}{r^2}\right)$$

The Poynting vector for these fields is found using Equation 10-17.

$$\vec{S} = \left[\frac{I_0 l}{2\pi}\right]^2 \sin(\omega t - kr)\cos(\omega t - kr)\left\{\left(\frac{\sin\theta\cos\theta}{3\varepsilon_0\omega r^5}\right)(\hat{r}\times\hat{\varphi}) + \frac{\sin^2\theta}{4\omega\varepsilon_0 r^5}(\hat{\theta}\times\hat{\varphi})\right\}$$

$$\vec{S} = \left[\frac{I_0 l}{2\pi}\right]^2 \sin(\omega t - kr)\cos(\omega t - kr)\left\{\left(\frac{\sin\theta\cos\theta}{3\varepsilon_0\omega r^5}\right)(-\hat{\theta}) + \frac{\sin^2\theta}{4\omega\varepsilon_0 r^5}(\hat{r})\right\}$$

We can now simplify this expression using trigonometric identities (Equation A-1, Equation A-2, Equation A-3, and Equation A-5). After grouping the similar terms together, we have

$$\vec{S} = \left[\frac{I_0 l}{2\pi}\right]^2 \sin(2\omega t)\cos(2kr)\left\{\left(\frac{\sin\theta\cos\theta}{3\varepsilon_0\omega r^5}\right)(-\hat{\theta}) + \frac{\sin^2\theta}{4\omega\varepsilon_0 r^5}(\hat{r})\right\}$$

The Poynting vector thus forms a standing wave (Section 10-7). As expected, the average Poynting vector is zero.

$$\vec{S}_{avg} = 0$$

No energy is radiated away from the dipole in the region close to the linear dipole antenna. This region is referred to as the *near field* (Figure 10.17), and energy flow is described as being *reactive*. In the near field, any energy placed into the electromagnetic fields (*i.e.,* into the electromagnetic waves) by the current of the linear dipole antenna is taken back by the antenna during the subsequent cycle of oscillation of the current. Thus, when averaged over many cycles of oscillation, no energy is radiated in this region. This is similar to the behavior of an AC circuit containing no resistors. The energy placed into a capacitor or an inductor in an AC circuit is always returned to the EMF, resulting in no average power being supplied by the EMF to the circuit.

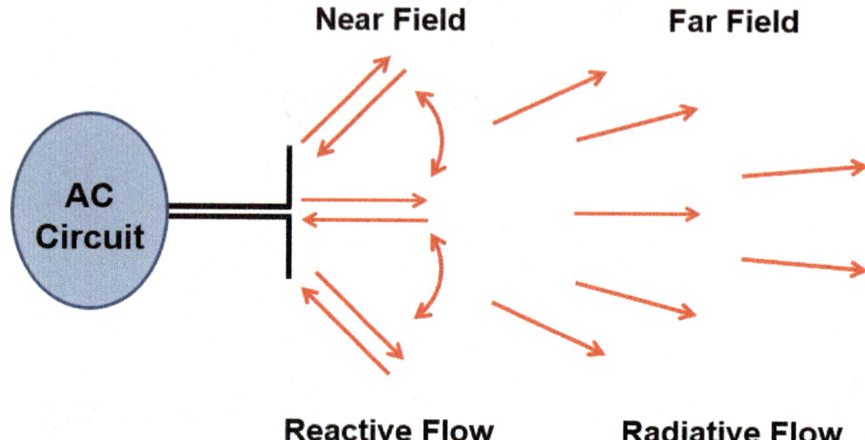

FIGURE 10.17: The flow of energy (red arrows) in the near-field and far-field of a linear dipole antenna.

When the distance r is large, the real parts of the electric and magnetic fields of the linear dipole antenna are

$$E_r \approx 0$$

$$E_\theta = \frac{I_0 l \sin\theta}{4\pi\varepsilon_0}\left[-\frac{\omega\sin(\omega t - kr)}{c^2 r}\right]$$

$$B_\varphi = \frac{\mu_0 I_0 l \sin\theta}{4\pi}\left(-\frac{\omega\sin(\omega t - kr)}{cr}\right)$$

Thus, in this region, referred to as the *far field*, the Poynting vector is

$$\vec{S} = \left[\frac{I_0 l \sin\theta}{4\pi\varepsilon_0}\left[-\frac{\omega\sin(\omega t - kr)}{c^2 r}\right]\right]\left[\frac{I_0 l \sin\theta}{4\pi}\left(-\frac{\omega\sin(\omega t - kr)}{cr}\right)\right]\left[\hat{\theta}\times\hat{\varphi}\right]$$

$$\vec{S} = \left[\frac{I_0 l \sin\theta\omega\sin(\omega t - kr)}{4\pi rc}\right]^2\left[\frac{\hat{r}}{\varepsilon_0 c}\right]$$

Since the Poynting vector in the far field is a travelling wave, energy will be dissipated by the linear dipole antenna in this region. The average Poynting vector in the far field is

$$\vec{S}_{avg} = \left[\frac{I_0^2 l^2 \sin^2 \theta \omega^2}{32\varepsilon_0 \pi^2 r^2 c^3} \right] \hat{r} \qquad \textbf{(10-24)}$$

In the far field, the pattern of radiated energy is symmetric with respect to the ϕ-axis. The intensity of this radiation is maximum when $\theta = 90°$ (perpendicular to the longitudinal axis of the antenna), minimum when $\theta = 0°$ or $180°$ (parallel to the longitudinal axis of the antenna).

Example 10-11

Problem: What is the radiative resistance of a linear dipole antenna?

Solution: The average radiative power is determined using Equation 10-20. The integral in this equation is over a spherical shell with radius *r*. In other words, the radiative power is the integral of the Poynting vector over the surface area of a spherical shell surrounding the antenna.

$$P_{avg} = \int \vec{S}_{avg} \bullet d\vec{A} \quad \rightarrow \quad P_{avg} = \int_0^{2\pi} \int_0^{\pi} \left[\frac{I_0^2 l^2 \sin^2 \theta \omega^2}{32\varepsilon_0 \pi^2 r^2 c^3} \right] r^2 \sin\theta \, d\theta \, d\phi$$

$$P_{avg} = \frac{I_0^2 l^2 \omega^2}{32\varepsilon_0 \pi^2 c^3} \int_0^{2\pi} \int_0^{\pi} \sin^3 \theta \, d\theta \, d\phi \quad \rightarrow \quad P_{avg} = \frac{I_0^2 l^2 \omega^2}{16\varepsilon_0 \pi c^3} \int_0^{\pi} \sin^3 \theta \, d\theta$$

$$\int_0^{\pi} \sin^3 \theta \, d\theta = \frac{4}{3} \quad \rightarrow \quad P_{avg} = \frac{I_0^2 l^2 \omega^2}{12\varepsilon_0 \pi c^3} \quad \rightarrow \quad P_{avg} = \frac{I_{rms}^2 l^2 \omega^2}{6\varepsilon_0 \pi c^3}$$

The radiative resistance can then be calculated using Equation 9-20.

$$R_{rad} = \frac{\left(P_{rad}\right)_{avg}}{I_{rms}^2} \quad \rightarrow \quad R_{rad} = \frac{l^2 \omega^2}{6\varepsilon_0 \pi c^3}$$

The substitution of Equation 10-7 and Equation 10-8 into the solution for Example 10-10 gives us

$$\omega = \frac{2\pi c}{\lambda} \quad \rightarrow \quad R_{rad} \approx \frac{l^2}{6\varepsilon_0 \pi c^3}\left(\frac{2\pi c}{\lambda}\right)^2 \quad \rightarrow \quad R_{rad} = \left(\frac{2\pi}{3\varepsilon_0 c}\right)\left(\frac{l}{\lambda}\right)^2$$

$$R_{rad} = \left(\frac{2\pi}{3\left(8.85\times 10^{-12}\,\frac{\mathrm{C}^2}{\mathrm{J\,m}}\right)\left(3\times 10^{8}\,\frac{\mathrm{m}}{\mathrm{s}}\right)}\right)\left(\frac{l}{\lambda}\right)^2$$

$$R_{rad} = (789\,\Omega)\left(\frac{l}{\lambda}\right)^2 \qquad \textbf{(10-25)}$$

Example 10-12

Problem: What is the radiative resistance of a linear dipole antenna with a length equal to 1/10 the wavelength of the emitted electromagnetic radiation? What rms current would be required to radiate an average power of 10^3 W?

Solution: The radiative resistance is determined using Equation 10-25.

$$R_{rad} = (789\,\Omega)\left(\frac{\frac{1}{10}\lambda}{\lambda}\right)^2 \quad \rightarrow \quad R_{rad} = 7.89\,\Omega$$

The amplitude of the oscillating current required for this average power can be determined using Equation 9-20.

$$I_{rms} = \left(\frac{(P_{rad})_{avg}}{R_{rad}}\right)^{\frac{1}{2}} \quad \rightarrow \quad I_{rms} = \left(\frac{1000\,\mathrm{W}}{7.89\,\Omega}\right)^{\frac{1}{2}} \quad \rightarrow \quad I_{rms} = 11.3\,\mathrm{A}$$

An antenna can receive as well as transmit radiation. It is simplest to model a receiving antenna as an equivalent surface area over which it extracts energy from a passing electromagnetic wave. This absorbed energy *creates* an EMF in the antenna that drives a current in the rest of the electric circuit attached to the antenna. For a linear dipole antenna, this EMF results from the oscillations of the electric field of the incident electromagnetic along the length of the antenna (Equation 6-9). For a circular loop antenna, the EMF results from the change in magnetic flux through the loop associated with the oscillating magnetic field of the incident electromagnetic wave (Equation 6-11).

The effective surface area for the absorption of energy by an antenna can be expressed using Equation 9-20 and Equation 10-20.

$$A_{eff} = \frac{I_{rms}^2 R_{rad}}{S_{avg}} \tag{10-26}$$

The rms current in Equation 20-26 is generated from the energy absorbed by the antenna. This current can also be expressed in terms of the effective EMF created in the antenna by absorbing energy from the electromagnetic wave. The impedance in this case would be the sum of the impedance of the antenna and the impedance of the load of the rest of the electric circuit that is in series with the antenna.

$$A_{eff} = \frac{\mathcal{E}_{eff}^2 R_{rad}}{\left(Z_{antenna} + Z_{load}\right)^2 S_{avg}}$$

By analogy to the solution to Example 8-5, we know that the maximum power will be transferred to the load when the impedance of the load is equal to the impedance of the antenna. In other words, the maximum power is transferred to the load when the resistance of the load is equal to the radiative resistance. In this situation, the equation for the maximum effective area becomes

$$\left(A_{eff}\right)_{\max} = \frac{\mathcal{E}_{eff}^2 R_{rad}}{\left(R_{rad} + R_{rad}\right)^2 S_{avg}} \rightarrow \left(A_{eff}\right)_{\max} = \frac{\mathcal{E}_{eff}^2}{4R_{rad} S_{avg}} \tag{10-27}$$

Example 10-13

Problem: What is the maximum effective area for a short dipole antenna? Assume that the dipole is in a vacuum.

Solution: The effective EMF is a maximum when the short dipole is aligned parallel to the electric field of the incident electromagnetic wave. We can determine the magnitude of this EMF using Equation 3-10 and Equation 6-10.

$$\varepsilon_{eff} = E_0 l$$

In this equation, the variable l is the length of the dipole antenna and the variable E_0 is the magnitude of the electric field of the incident electromagnetic wave. The average Poynting vector is calculated using Equation 10-13 and Equation 10-17[18].

$$S_{avg} = \frac{1}{\mu_0}\left(E_0 B_0\right) \rightarrow S_{avg} = \frac{1}{\mu_0}\left(E_0\left(\frac{E_0}{c}\right)\right) \rightarrow S_{avg} = \frac{E_0^2}{\mu_0 c}$$

Substitution into Equation 10-27 gives us

$$\left(A_{eff}\right)_{\max} = \frac{\left(E_0 l\right)^2}{4\left(\frac{l^2\omega^2}{6\varepsilon_0\pi c^3}\right)\left(\frac{E_0^2}{\mu_0 c}\right)} \rightarrow \left(A_{eff}\right)_{\max} = \frac{6\varepsilon_0\mu_0\pi c^4}{4\omega^2}$$

Simplifying this expression using Equation 10-7 and Equation 10-8 then yields

$$\left(A_{eff}\right)_{\max} = \frac{6\varepsilon_0\mu_0\pi c^4}{4\left(\frac{2\pi c}{\lambda}\right)^2} \rightarrow \left(A_{eff}\right)_{\max} = \frac{3}{8\pi}\lambda^2$$

We see from the solution to Example 10-13 that a linear dipole antenna can absorb power over an effective surface area proportional to λ^2, regardless of the length of the antenna.

18 Recall from Section 10-4 that $\vec{E}$ is perpendicular to $\vec{B}$ for an electromagnetic wave.

Summary

- **Wave:** a disturbance that propagates and transmits energy. This transfer of energy occurs without an accompanying transfer of matter.
- **Travelling wave:** a periodic continuous wave that moves at constant speed.
- **Standing wave:** a periodic continuous wave that is stationary. Each portion of a standing wave oscillates with constant amplitude.
- **Node:** a position along a standing wave where the wave has minimum amplitude.
- **Antinode:** a position along a standing wave where the wave has maximum amplitude.
- **Transverse wave:** a wave in which the disturbance is perpendicular to the direction of propagation of the wave.
- **Longitudinal wave:** a wave in which the disturbance is parallel to the direction of propagation of the wave.
- **Polarization:** the direction of oscillations of a transverse wave. The polarization of an electromagnetic wave is defined to be the direction of the oscillation of the electric field of the wave.
- **Linearly polarized:** an electromagnetic wave is linearly polarized if the oscillations of the electric field are always along the same axis.
- **Elliptically polarized:** an electromagnetic wave is elliptically polarized if the electric field vector rotates and changes in magnitude as the electromagnetic wave propagates.
- **Circularly polarized:** an electromagnetic wave is circularly polarized if the electric field vector rotates, but maintains a constant magnitude, as the electromagnetic wave propagates.
- **Polarizer:** a material that absorbs the energy from an electromagnetic wave with a specific polarization.

- The product of the wavelength and frequency of a wave is the speed of the wave.

$$\lambda f = v$$

The frequency of the wave is the inverse of the period of the wave

$$f = \frac{1}{T}$$

The wavenumber of a wave is defined to be

$$k = \frac{2\pi}{\lambda}$$

Thus, the speed of a wave can be written as

$$v = \frac{\omega}{k}$$

- The one-dimensional wave equation is

$$\frac{\partial^2 y}{\partial x^2} = \frac{1}{v^2}\frac{\partial^2 y}{\partial t^2}$$

A common solution to this equation is

$$y = y_0 e^{i(kx - \omega t)}$$

- The one-dimensional wave equations for the electric and magnetic fields of an electromagnetic wave are

$$\frac{\partial^2 E_x}{\partial z^2} = \kappa_m \mu_0 \kappa_e \varepsilon_0 \frac{\partial^2 E_x}{\partial t^2}$$

$$\frac{\partial^2 B_y}{\partial z^2} = \kappa_m \mu_0 \kappa_e \varepsilon_0 \frac{\partial^2 B_y}{\partial t^2}$$

Common solutions to these equations are

$$E_x = E_0 e^{i(kz - \omega t + \phi_0)} \qquad B_y = B_0 e^{i(kz - \omega t + \phi_0)}$$

The quantity in parentheses in these equations is the phase of the wave, denoted by ϕ, and the constant ϕ_0 is the phase constant. The speed of an electromagnetic wave is

$$v = \frac{1}{\left(\kappa_m \mu_0 \kappa_e \varepsilon_0\right)^{\frac{1}{2}}}$$

The speed of an electromagnetic wave in vacuum is denoted by the variable c.

$$c = \frac{1}{\left(\mu_0 \varepsilon_0\right)^{\frac{1}{2}}}$$

- The average total energy density of an electromagnetic wave is

$$\left(u_{total}\right)_{avg} = \frac{1}{2}\kappa_e \varepsilon_0 E_0^2$$

- The Poynting vector is the rate of flow of electromagnetic energy per unit area for an electromagnetic wave; it has the units W/m^2.

$$\vec{S} = \frac{1}{\kappa_m \mu_0}\left(\vec{E} \times \vec{B}\right)$$

The direction of the Poynting vector is the direction of propagation of the electromagnetic wave. The integral of the Poynting vector over a surface area is the power transmitted through that surface by the electromagnetic wave.

$$P = \int \vec{S} \cdot d\vec{A}$$

- **Wave intensity:** the average magnitude of the Poynting vector. The wave intensity is denoted by the variable I, and for a sinusoidally oscillating electromagnetic is

$$I = S_{avg} \quad \rightarrow \quad I = \frac{1}{2} v \kappa_e \varepsilon_0 E_0^2$$

The wave intensity is also the product of the average total energy density and the speed of the wave.

$$I = S_{avg} = v\left(u_{total}\right)_{avg}$$

- The intensity of the electromagnetic wave passing through a linear polarizer is

$$I_{transmitted} = I_0 \cos^2\theta$$

This equation is also known as law of Malus.

- **Principle of superposition:** when two (or more) waves are simultaneously present at a single location, the resultant wave is the sum of the two waves.

- **Interference:** the process of combining two or more waves to create a resultant wave.

- **Constructive interference:** the interference of two or more waves which differ in phase by an integer multiple of 2π. The amplitude of the resultant wave is the sum of the amplitudes of the individual waves.

- **Destructive interference:** the interference of two or more waves which differ in phase by an integer multiple of π. The amplitude of the resultant wave is zero.

- **Antenna:** an electrical device which can convert electric power into electromagnetic waves, and can convert electromagnetic waves into electric power.

  The radiative resistance of a circular loop antenna with cross-sectional area A is

$$R_{rad} \approx \left(3.1\times10^4\,\Omega\right)\left(\frac{A}{\lambda^2}\right)^2$$

  The radiative resistance of a linear dipole antenna with length l is

$$R_{rad} = \left(789\,\Omega\right)\left(\frac{l}{\lambda}\right)^2$$

The effective area of an antenna for absorbing electromagnetic radiation is

$$A_{eff} = \frac{I_{rms}^2 R_{rad}}{S_{avg}}$$

The maximum effective area of an antenna for absorbing electromagnetic radiation is

$$\left(A_{eff}\right)_{\max} = \frac{\mathcal{E}_{eff}^2}{4R_{rad}S_{avg}}$$

Problems

1. Show that the following equation is a solution to the linear wave equation:

$$y(x,t) = \frac{5}{(x-4t)^2+1}$$

   What is the speed of this wave?

2. A wave has a wavelength of 5 m and a speed of 10 m/s. What is the period of this wave? What is the wavenumber of this wave? What is the angular frequency of this wave?

3. A wave has a frequency of 60 Hz and a speed of 240 m/s. What is the period of this wave? What is the wavenumber of this wave? What is the wavelength of this wave?

4. What are the wavelength and wavenumber of an electromagnetic wave whose frequency is 60 Hz?

5. The maximum magnitude of the magnetic field of an electromagnetic wave is 9 μT. What is the maximum magnitude of the electric field of this electromagnetic wave?

What is the average total energy density of this electromagnetic wave? What is the wave intensity of this electromagnetic wave?

6. A linearly polarized electromagnetic wave is incident on a polarizer whose transmission axis is at 30° with respect to its polarization. The transmitted light then passes through a second polarizer whose transmission axis is at 30° with respect to the transmission axis of the first polarizer. Finally, the transmitted light passes through a third polarizer whose transmission axis is at 30° with respect to the transmission axis of the second polarizer. What is the intensity of the electromagnetic wave transmitted through the third polarizer?

7. A linearly polarized electromagnetic wave is incident on a polarizer whose transmission axis is at 18° with respect to its polarization. The transmitted light then passes through 4 additional polarizers, each of which has a transmission axis at 18° with respect to the preceding polarizer. What is the intensity of the light passing through the final polarizer?

8. An electromagnetic standing wave has a frequency of 1200 kHz. What is the distance between adjacent antinodes of this standing wave?

9. A 5 cm linear dipole antenna is absorbing power from electromagnetic radiation with a frequency of 2 GHz. What incident radiation is required for an rms current of 1 mA to be generated in the electric circuit connected to the antenna? Assume that the antenna is in vacuum and absorbing power with its maximum effective area.

10. What is the maximum effective area for a circular loop antenna? Assume that that the antenna is in a vacuum. What is maximum effective area for an antenna consisting of N circular loops of wire?

CHAPTER 11

Geometric Optics

11-1 Introduction

In Chapter 10, we learned how to describe electromagnetic waves, including how fast they move and how they transmit energy. We now turn our attention to interactions of electromagnetic waves with material, especially the simple interactions that constitute geometrical optics. In geometrical optics, the wave nature of electromagnetic waves is ignored for the sake of mathematical simplicity. In Chapter 12, we will discuss how the wave nature of electromagnetic waves manifests itself in other interactions.

11-2 The Electromagnetic Spectrum

The range of all possible wavelengths/frequencies for electromagnetic waves is called the ***electromagnetic spectrum*** (Figure 11.1).

> *Electromagnetic spectrum*: the range of all possible wavelengths/frequencies for electromagnetic waves.

Some electromagnetic waves with relatively large wavelengths are used for communication (radio waves, TV waves, microwaves), whereas electromagnetic waves with short wavelengths, such as x-rays, have applications in medicine.

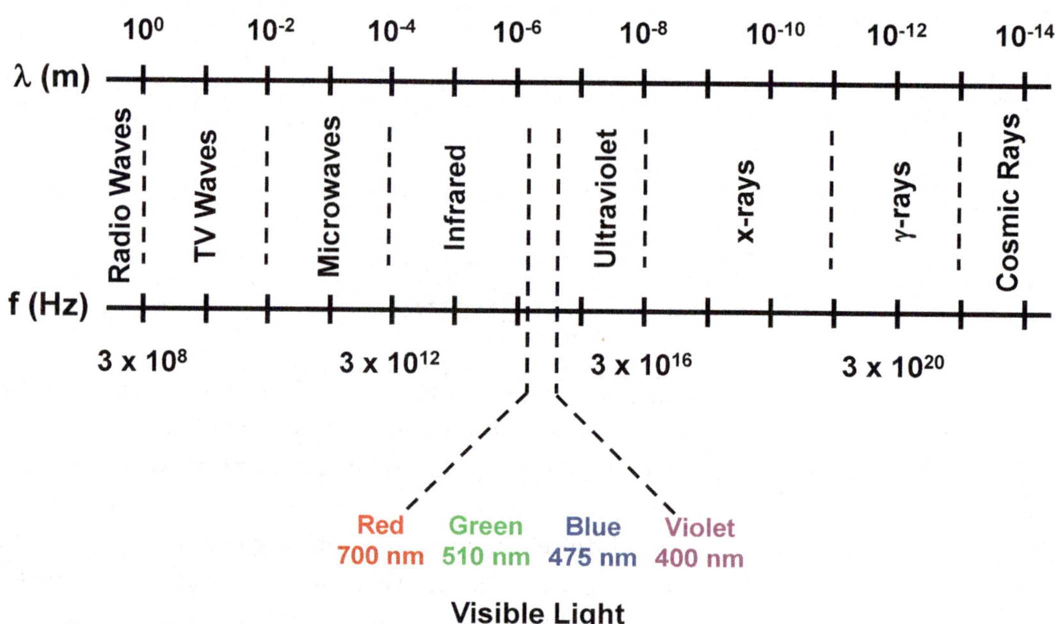

FIGURE 11.1: The electromagnetic spectrum.

The electromagnetic spectrum also includes visible light (wavelengths between 400 nm and 700 nm), which will be the focus of much of the discussion in Chapter 11 and Chapter 12. While we are very familiar and comfortable with the concept of color, we should remember that color itself is not an intrinsic property of visible light, but rather a perception of the wavelength of light according to our eyes and brain. For example, what we perceive as white light is a mixture of all colors of visible light together.

In our discussion of geometric optics, we will use a ***ray model*** of electromagnetic waves, in which we define a *ray* as a line in the direction of the Poynting vector of the electromagnetic wave. The ray is thus perpendicular to the plane of polarization[1] of the electromagnetic wave.

> *Ray model:* an idealized model of electromagnetic waves in which each wave is represented by a ray that points in the direction of energy flow for that wave.

Electromagnetic rays travel in straight lines within the same material (vacuum, glass, *etc.*). A ***ray diagram*** is a pictorial representation of the propagation of electromagnetic waves.

> *Ray diagram:* a pictorial representation of the propagation of electromagnetic waves. The process of making a ray diagram is called *ray tracing.*

For systems containing many electromagnetic waves, however, we will represent only a subset of the waves as rays in the ray diagram.

11-3 Index of Refraction and Dispersion

The speed of an electromagnetic wave is determined by the permittivity and permeability of the medium through which the wave travels as shown in Section 10-4.

$$v = \frac{1}{\left(\kappa_m \mu_0 \kappa_e \varepsilon_0\right)^{\frac{1}{2}}}$$

This equation can be rewritten as

$$v = \frac{1}{\left(\kappa_m \kappa_e\right)^{\frac{1}{2}} \left(\mu_0 \varepsilon_0\right)^{\frac{1}{2}}} \quad \rightarrow \quad v = \frac{c}{\left(\kappa_m \kappa_e\right)^{\frac{1}{2}}} \quad \rightarrow \quad v = \frac{c}{n} \qquad \textbf{(11-1)}$$

The variable n in equation 11-1 is called the *index of refraction* of the medium and depends upon the material properties of the medium.

$$n = \left(\kappa_m \kappa_e\right)^{\frac{1}{2}}$$

1 See Section 10-6

Since $\kappa_e \geq 1$ and $|\kappa_m| \approx 1$, the index of refraction is always greater than or equal to one[2].

$$n \geq 1 \quad \rightarrow \quad v \leq c$$

Furthermore, since $|\kappa_m| \approx 1$, the index of refraction can be approximately written exclusively in terms of the dielectric constant.

$$n^2 = \kappa_e \tag{11-2}$$

The index of refraction also depends upon the wavelength/frequency of the electromagnetic wave. Specifically, the index of refraction increases as the frequency of the electromagnetic wave decreases. Thus, *the index of refraction increases as the wavelength of the electromagnetic wave increases.* This behavior is called ***dispersion.***

> *Dispersion:* the dependence of the index of refraction on the wavelength of the electromagnetic wave.

When an electromagnetic wave moves between media with different indices of refraction, the speed of the wave changes (Equation 11-1). The frequency of the wave remains the same due to continuity requirements of the electric field at the boundary between the two media[3]. Therefore, we see from Equation 10-5 and Equation 10-7 that the wavelength and wavenumber of an electromagnetic wave change as the wave moves between media with different indices of refraction (Figure 11.2).

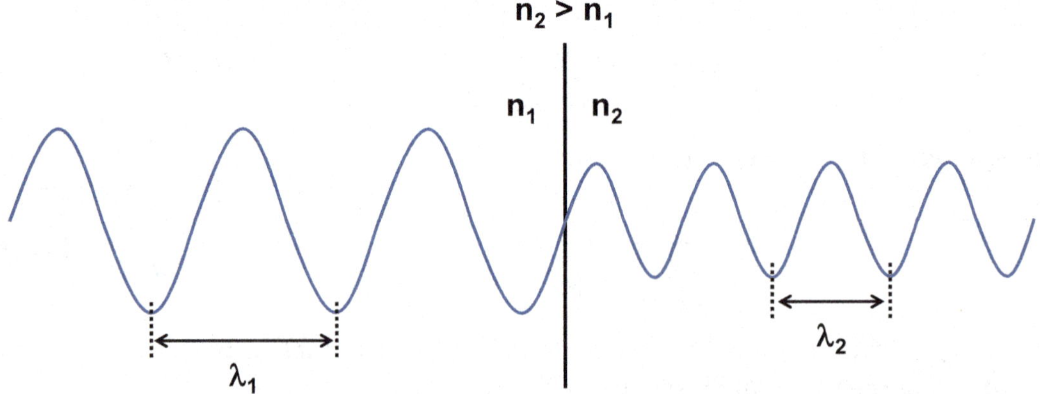

FIGURE 11.2: The wavelength of an electromagnetic wave depends upon the index of refraction of the medium in which the wave propagates.

2 The index of refraction of vacuum is equal to 1.

3 You should feel inspired to take an advanced course in electricity and magnetism to learn more about these requirements.

Combining Equation 10-5 and Equation 11-1 gives us

$$\lambda f = \frac{c}{n}$$

For the electromagnetic wave in Figure 11.3, we have

$$\lambda_1 f = \frac{c}{n_1} \quad \rightarrow \quad \lambda_1 n_1 = \frac{c}{f}$$

$$\lambda_2 f = \frac{c}{n_2} \quad \rightarrow \quad \lambda_2 n_2 = \frac{c}{f}$$

Because the frequency of the electromagnetic wave is the same in both media, we have

$$\lambda_1 n_1 = \lambda_2 n_2$$

The wavelength (λ_n) and wavenumber (k_n) of an electromagnetic wave in a medium with index of refraction n are therefore

$$\lambda_n = \frac{\lambda}{n} \quad \rightarrow \quad k_n = \frac{k}{n} \qquad \textbf{(11-3)}$$

The variables λ and k in Equation 11-3 are the wavelength and wavenumber, respectively, of the electromagnetic wave in vacuum ($n = 1$). Since the index of refraction of any medium is greater than or equal to 1, the wavelength of an electromagnetic wave is always less than or equal to its value in vacuum.

11-4 Reflection and Refraction

When an electromagnetic wave encounters a change in index of refraction (when light passes between vacuum and air, *e.g.*) some of the energy of the incident electromagnetic wave will pass through the boundary (*i.e.*, the wave will be transmitted through the boundary, but at a lower intensity) and some of the energy of the incident electromagnetic wave will *reflect* off the boundary (as a separate electromagnetic wave). Since the index of refraction is different on the two sides of the boundary, the speeds of the incident and transmitted wave will not be the same. This difference in speeds causes the electromagnetic wave to change direction as it crosses the boundary[4]. The changing of the direction of an electromagnetic wave moving from one medium to another is called ***refraction***.

> *Refraction*: the change in the direction of an electromagnetic wave as it moves from a medium with one index of refraction to a medium with a different index of refraction.

We can use the ray diagram in Figure 11.3 to demonstrate the change in the direction of the electromagnetic wave following refraction. The dashed line in Figure 11.3 is perpendicular to the boundary and is commonly referred to as the *normal axis* for the boundary. The angle of incidence (θ_1 in Figure 11.3) and the angle of refraction (θ_2 in Figure 11.3) are both measured with respect to the normal axis.

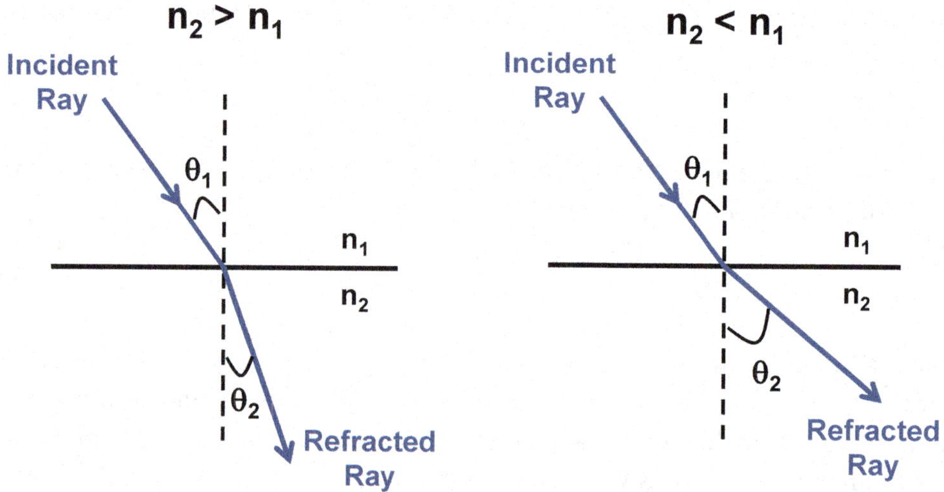

FIGURE 11.3: The angle of refraction (θ_2) depends upon the angle of incidence (θ_1), as well as the indices of refraction on both sides of the boundary. The dashed line is perpendicular to the boundary, and is therefore called the normal axis.

These two angles are related by the indices of refraction n_1 and n_2, according to Snell's law (Equation 11-4).

$$n_1 \sin\theta_1 = n_2 \sin\theta_2 \tag{11-4}$$

4 We'll discuss this in Section 12-2.

When a ray is transmitted into a material with a higher index of refraction, it bends toward the normal. In contrast, when a ray is transmitted into a material with a lower index of refraction, it bends away from the normal.

Example 11-1

Problem: An electromagnetic wave travelling through vacuum ($n = 1$) is incident upon a boundary with glass ($n = 1.5$), as shown in Figure 11.4. The wave is refracted upon passing into the glass, and then refracted again leaving the glass. What is the angle θ in Figure 11.4?

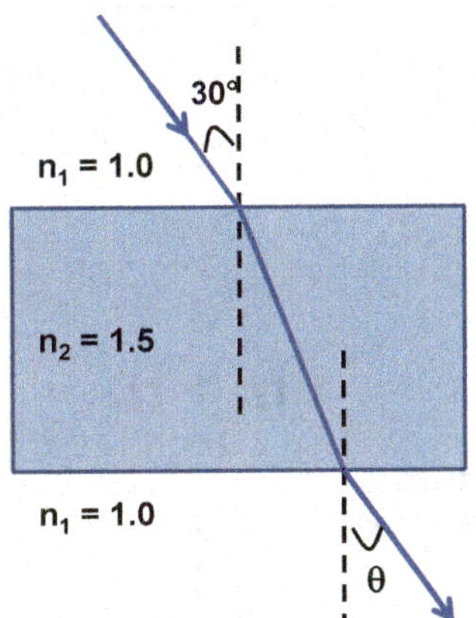

FIGURE 11.4: The ray diagram for Example 11-1.

Solution: The refracted angle when crossing over the first boundary is found using Equation 11-4. Let's denote this angle as θ_2.

$$\sin\theta_2 = \frac{n_1 \sin\theta_1}{n_2} \quad \rightarrow \quad \sin\theta_2 = \frac{(1.0)\sin(30°)}{(1.5)} \quad \rightarrow \quad \sin\theta_2 = \frac{1}{3}$$

The angle θ_2 is the incident angle for the second refraction and thus can be related to the angle θ in Figure 11.4 using Equation 11-4.

$$\sin\theta = \frac{n_2 \sin\theta_2}{n_1} \quad \rightarrow \quad \sin\theta = \frac{(1.5)\left(\frac{1}{3}\right)}{(1.0)} \quad \rightarrow \quad \sin\theta = \frac{1}{2} \quad \rightarrow \quad \theta = 30°$$

The final refracted angle is identical to the initial incident angle since the indices of refraction are identical on either side of the glass.

Since the index of refraction depends upon the frequency/wavelength of the electromagnetic wave (Section 11-3), the angle of refraction will also depend upon the frequency/wavelength of the electromagnetic wave. This is the underlying principle explaining how prisms work. As shown in Figure 11.5, white light incident on a prism will be refracted on the other side of the prism into many different rays of light, each at a separate color (*i.e.*, a separate wavelength).

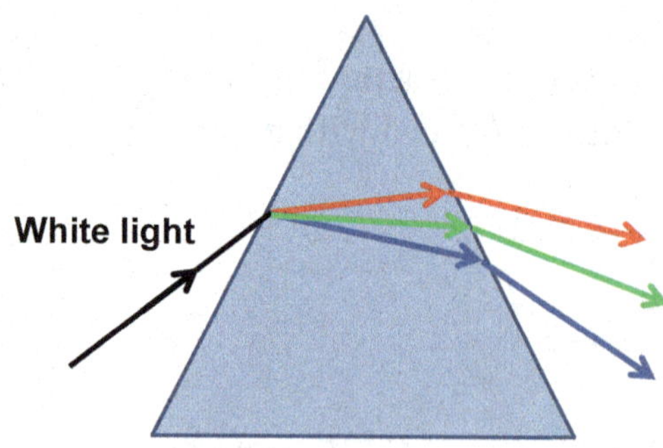

FIGURE 11.5: A prism separates white light into its color components.

The separation angles between the different colors of light depend upon how the index of refraction of the prism varies with the wavelengths of the light (Section 11-3) and therefore depends upon the material used to make the prism. It is through a similar process that water droplets in the air refract sunlight into a rainbow.

As mentioned previously, not all of the energy of an electromagnetic wave will be refracted when the wave encounters a change in the index of refraction. Some of the energy is reflected from the boundary as a separate electromagnetic wave. According to the *law of reflection*, the angle of reflection is equal to the angle of incidence, as shown in Figure 11.6.

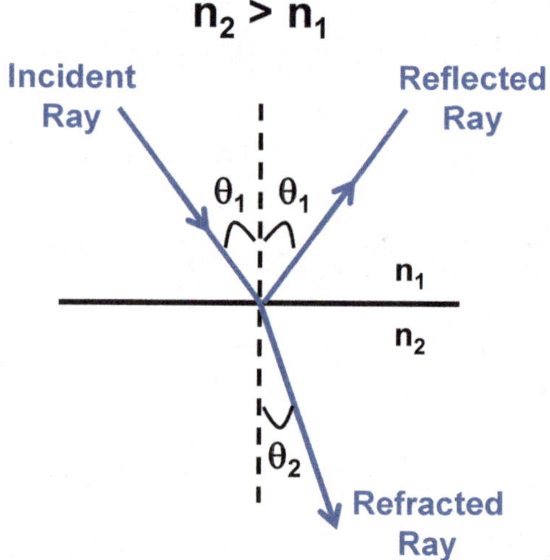

FIGURE 11.6: The angle of incidence is the same as the angle of reflection.

When we see an object, we perceive light reflecting off the surface of the object (obvious exceptions are objects that produce their own light, such as the sun or a light bulb). The light reflected from an object is reflected from all positions along the surface of the object and each of these reflections obeys the law of reflection. However, since most surfaces are not perfectly smooth (such as the flat boundary in Figure 11.6), incident light that is reflected off an object is typically reflected in many different directions. This is referred to as *diffuse reflection*.

Now let's consider the situation shown in the right panel of Figure 11.3, in which an electromagnetic wave is incident on a boundary with a medium of lower index of refraction. We can determine the angle of refraction using Equation 11-4.

$$\sin\theta_2 = \frac{n_1 \sin\theta_1}{n_2}$$

As the θ_1 increases, θ_2 must also increase. Once $\theta_2 = 90°$, however, there will no longer be any refracted ray. Under these conditions, all of the incident energy will be reflected from the boundary[5]. We refer to this phenomenon as *total internal reflection.* The corresponding angle of incidence, referred to as the critical angle (denoted by θ_c), is thus

$$\theta_2 = 90° \quad \rightarrow \quad \frac{n_1 \sin\theta_C}{n_2} = 1 \quad \rightarrow \quad \theta_C = \sin^{-1}\left(\frac{n_2}{n_1}\right)$$

Example 11-2

Problem: What is the critical angle for total internal reflection for light passing from glass ($n = 1.5$) to air ($n = 1$)?

Solution: The critical angle is

$$\theta_C = \sin^{-1}\left(\frac{1}{1.5}\right) \quad \rightarrow \quad \theta_C = 41.8°$$

If the light incident on a surface is unpolarized, the reflected light is partially polarized. The extent of this polarization depends upon the angle of incidence and the geometry of the surface. For example, sunlight reflected from a horizontal surface (such as pavement or water) will be partially linearly polarized parallel to the surface. Polarized sunglasses contained polarizers (Section 10-6) which specifically absorb this polarization of light and, consequently, specifically absorb this reflected light. This is how these sunglasses can reduce glare.

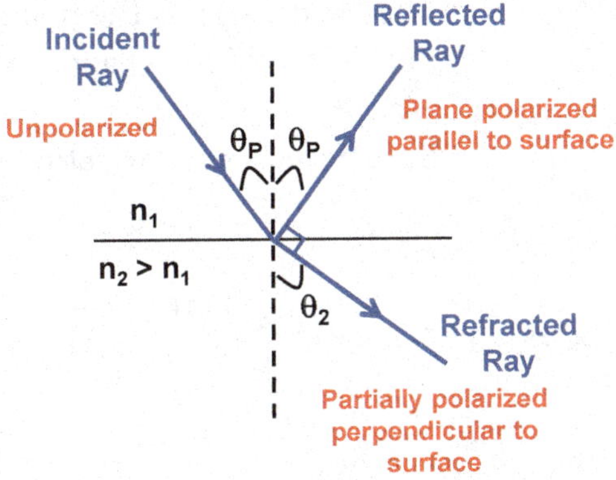

FIGURE 11.7: When unpolarized light is incident on a surface at the polarization angle, the reflected ray is plane polarized parallel to the surface.

The angle of incidence at which the reflected light is completely polarized is called the polarizing angle,

5 Things are actually a bit more complicated than this. Please take an advanced course in electricity and magnetism to learn more.

denoted by θ_P. For this angle of incidence, the angle between the refracted and reflected rays will be exactly 90° (Figure 11.7)[6].

The polarizing angle can be determined using Equation 11-4.

$$n_1 \sin\theta_P = n_2 \sin\theta_2 \quad \rightarrow \quad n_1 \sin\theta_P = n_2 \sin\left(180° - \theta_P - 90°\right)$$

$$n_1 \sin\theta_P = n_2 \sin\left(90° - \theta_P\right) \quad \rightarrow \quad n_1 \sin\theta_P = n_2 \cos\theta_P$$

$$\tan\theta_P = \frac{n_2}{n_1} \qquad \textbf{(11-5)}$$

Equation 11-5 is called Brewster's law, after David Brewster, and the polarizing angle θ_P is sometimes called Brewster's angle.

Example 11-3

Problem: What is the polarization angle for sunlight reflecting off a boundary between air (n ~ 1) and water (n ~ 1.3)?

Solution: The polarization angle is found using Equation 11-5.

$$\tan\theta_P = \frac{1.3}{1} \quad \rightarrow \quad \theta_P = \tan^{-1}\left(1.3\right) \quad \rightarrow \quad \theta_P = 52.4°$$

11-5 Thin Lenses

A ***lens*** is an optical device that *refracts* an electromagnetic wave; most lenses rely upon curved surfaces to impart further refraction of an incident electromagnetic wave. The ***optical axis*** of a lens is the axis passing through the center of the lens and perpendicular to the surface of a lens.

6 You should feel inspired to take more advanced physics classes to learn why this is true.

Lens: an optical device that refracts electromagnetic waves.

Optical axis: the axis passing through the center of a lens perpendicular to the surface of the lens.

Incident electromagnetic waves parallel to the optical axis of a ***converging lens*** are refracted toward a point. Incident electromagnetic waves parallel to the optical axis of a ***diverging lens*** are refracted away from a point. For both lenses, this point is referred to as the *focal point* of the lens.

Converging lens: a lens that refracts incident electromagnetic waves parallel to its optical axis toward its focal point.

Diverging lens: a lens that refracts incident electromagnetic waves parallel to its optical axis away from its focal point.

Ray diagrams of the interaction of electromagnetic rays with converging and diverging lenses are shown in Figure 11.8 and Figure 11.9. The rules for ray tracing for a *converging lens* are:

1. Any ray initially parallel to the optical axis on one side of a lens will refract through the focal point on the other side of the lens.
2. Any ray passing through the focal point on one side of the lens is refracted into a ray parallel to the optical axis on the other side of the lens.
3. Any ray directed at the center of the lens passes through in a straight line.

The rules for ray tracing for a *diverging* lens are:

1. Any ray initially parallel to the optical axis on one side of a lens will diverge along a line through the focal point on the same side of the lens.
2. Any ray directed along a line toward the focal point on the other side of the lens will emerge from the lens parallel to the optical axis.
3. Any ray directed at the center of the lens passes through in a straight line.

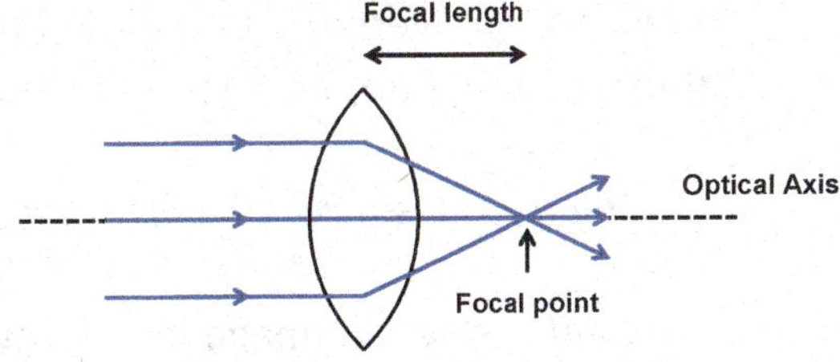

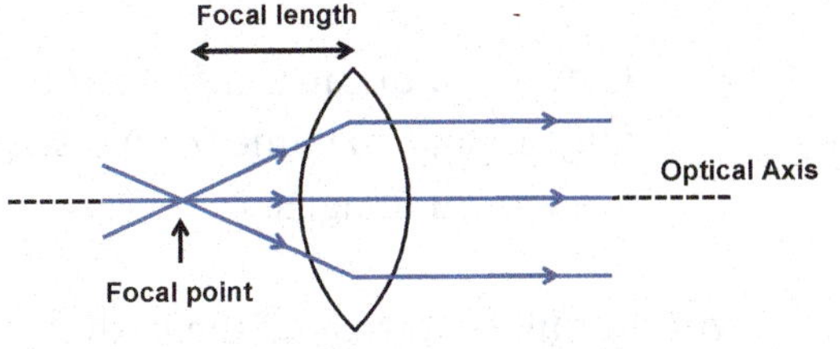

FIGURE 11.8: Ray tracing for a converging lens.

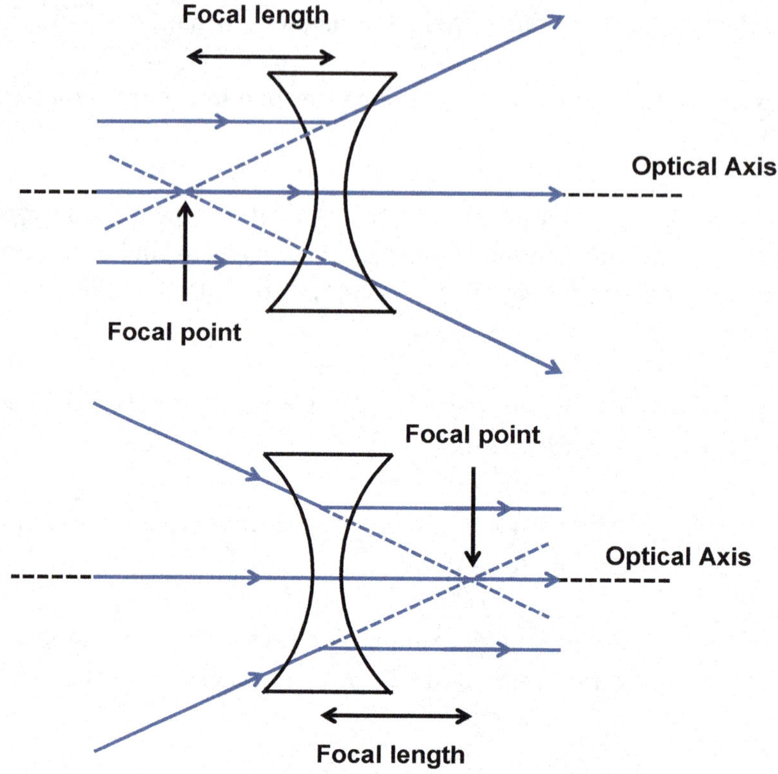

FIGURE 11.9: Ray tracing for a diverging lens.

An ***image*** is defined to be the source of a diverging group of rays. A ***real image*** is formed when rays converge at a point and then diverge from that point. A ***virtual image*** is created when rays appear to originate at a point (*i.e.*, appear to diverge from a single point).

> *Image:* the source of a diverging group of rays.
>
> *Real image:* an image formed where refracted rays intersect. These rays then diverge from this point.
>
> *Virtual image:* an image formed where refracted rays do not intersect. The rays only appear to come from a single point. All images formed by a diverging lens are virtual images.

An example of a real image is shown in Figure 11.10. The rays from the object converge at a single point on the far side of the lens.

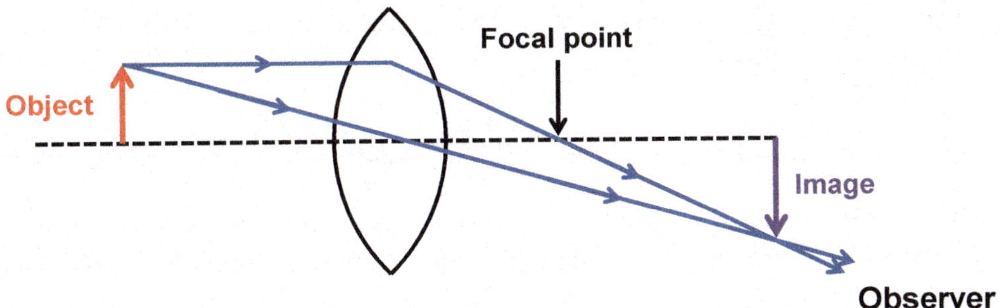

FIGURE 11.10: A real image is formed when the rays converge at a point.

Examples of virtual images are shown in Figure 11.11. The rays from the object in these situations do not converge at a single point, but appear to originate from a single point on the same side of the lens as the object.

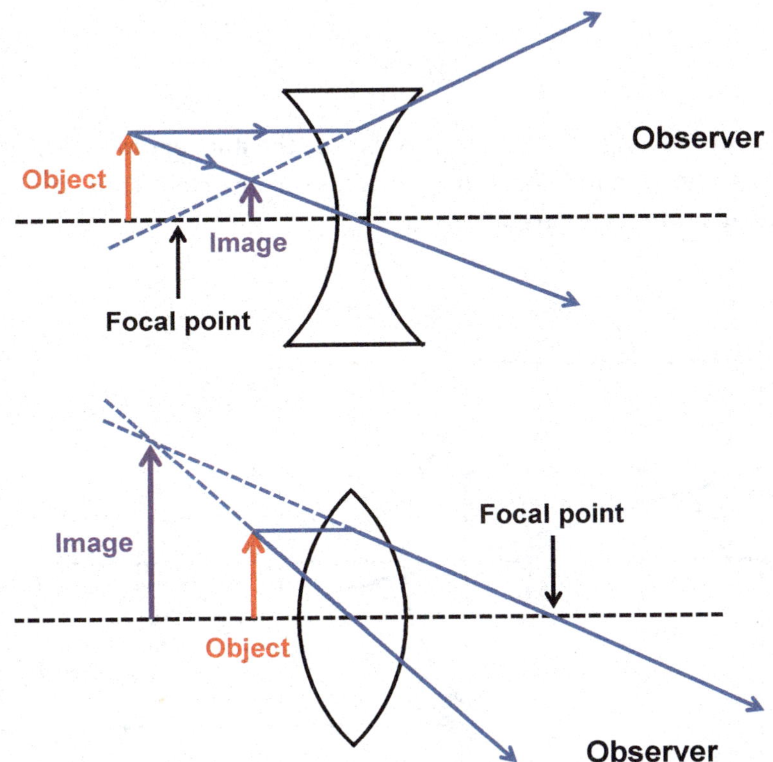

FIGURE 11.11: Virtual images are formed when the rays do not intersect. Rather, the rays appear to come from a point.

The *lens equation* (Equation 11-6) allows us to determine the position of an image from a lens.

$$\frac{1}{s}+\frac{1}{s'}=\frac{1}{f} \tag{11-6}$$

As shown in Figure 11.12, the variables s and s' denote the positions of the object and the image, respectively, relative to the lens. The object position is always a positive number. *The image position is positive for real images and negative for virtual images.* The variable f is the focal length of the lens, which is the distance from the lens to the focal point of the lens. The focal length of a converging lens is positive and the focal length of a diverging lens is negative.

The magnification of the object is defined to be the negative of the ratio of image size to object size, which can also be expressed in terms of the image and object positions using geometry. For the object and image in Figure 11.12, with heights h and h', respectively, the magnification is

$$M=-\frac{h'}{h}=-\frac{s'}{s} \tag{11-7}$$

A positive value for the magnification indicates that the image is upright, and a negative value for the magnification indicates that the image is inverted. Thus, all virtual images are upright and all real images are inverted.

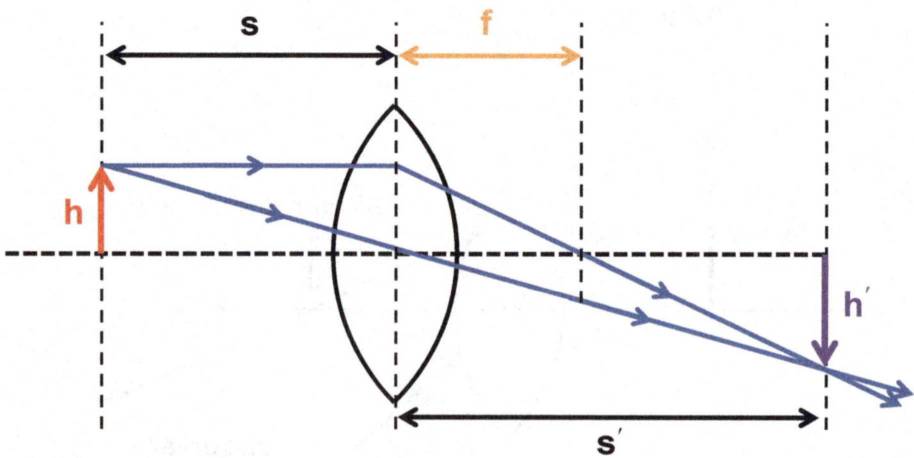

FIGURE 11.12: The variables in the lens equation.

Example 11-4

Problem: An object is located 12 cm away from a converging lens with a focal length of 4 cm. Where is the image located, is it real or virtual, what is its magnification, and is it upright or inverted?

Solution: Substitution into Equation 11-6 gives us

$$\frac{1}{12\,\text{cm}}+\frac{1}{s'}=\frac{1}{4\,\text{cm}} \quad\rightarrow\quad \frac{1}{s'}=\frac{1}{4\,\text{cm}}-\frac{1}{12\,\text{cm}} \quad\rightarrow\quad \frac{1}{s'}=\frac{2}{12\,\text{cm}}$$

$$s'=6\,\text{cm}$$

The image is located 6 cm from the lens on the other side of the lens from the object. Since the image distance is positive, the image is real. The magnification of the image is determined using Equation 11-7.

$$M=-\frac{6\,\text{cm}}{12\,\text{cm}} \quad\rightarrow\quad M=-\frac{1}{2}$$

The image is thus inverted and one-half the size of the object.

Example 11-5

Problem: An object is located 12 cm away from a diverging lens with a focal length of -4 cm. Where is the image located, is it real or virtual, what is its magnification, and is it upright or inverted?

Solution: Substitution into Equation 11-6 gives us

$$\frac{1}{12\,\text{cm}}+\frac{1}{s'}=-\frac{1}{4\,\text{cm}} \quad\rightarrow\quad \frac{1}{s'}=-\frac{1}{4\,\text{cm}}-\frac{1}{12\,\text{cm}} \quad\rightarrow\quad \frac{1}{s'}=-\frac{4}{12\,\text{cm}}$$

$$s'=-3\,\text{cm}$$

The image is located 3 cm from the lens on the same side of the lens as the object. Since the image distance is negative, the image is virtual. The magnification of the image is determined using Equation 11-7.

$$M = -\frac{-3\,\text{cm}}{12\,\text{cm}} \quad \rightarrow \quad M = \frac{1}{4}$$

The image is thus upright and one-fourth the size of the object.

As mentioned above, lenses form images through refraction of light. Refraction from a flat surface can also form a virtual image, as shown in Figure 11.13. This image will be closer to the surface or farther from the surface than the object depending upon the change in the index of refraction across the surface.

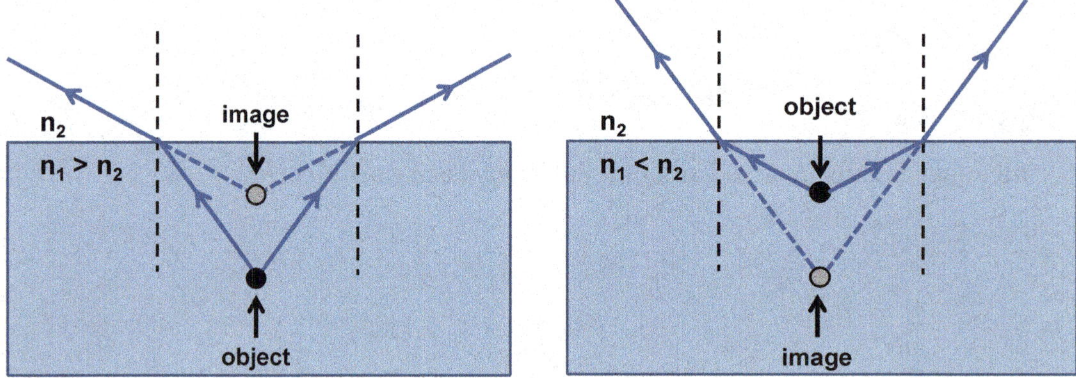

FIGURE 11.13: Refraction from a flat surface can form a virtual image.

Finally, since the index of refraction depends upon the frequency of the light, the focal length of a lens will also depend upon the frequency of the light. As shown in Figure 11.14, the focal length of red light is larger than the focal length of violet light; the other colors have focal lengths in-between these two. This inability of the lens to focus all colors of light to the same point is called ***chromatic aberration***.

Chromatic aberration: the failure of a lens to focus all colors of light to the same point.

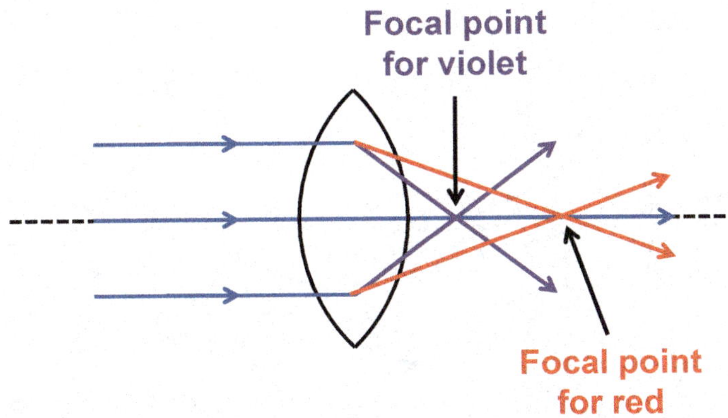

FIGURE 11.14: Chromatic aberration for a converging lens.

11-6 Combinations of Lenses

A simple magnifier can be made using only a single converging lens. Combining Equation 11-6 and Equation 11-7 gives us

$$\frac{1}{s'} = \frac{1}{f} - \frac{1}{s} \rightarrow s' = \frac{fs}{s-f} \rightarrow M = \frac{f}{f-s}$$

If the object distance is smaller than the focal length of the lens (*i.e.*, if $s < f$), a virtual, upright, and magnified image will be formed. More generally, however, magnifiers and other optical instruments are constructed with multiple lenses. We can extend our analysis of lenses in Section 11-5 to multiple lens systems by allowing the image of one lens to be the object for another lens. Rays from the object in Figure 11.15 are focused by the first converging lens (with focal length f_1) to form a real image. *This real image is then the object for the second converging lens* (with focal length f_2). As shown in Figure 11.15, the image created by the second converging lens is a virtual image that is inverted and magnified. Since someone looking through this collection of lenses from the right side of Figure 11.15 would see only this final magnified image, this orientation of these lenses forms an optical magnifier.

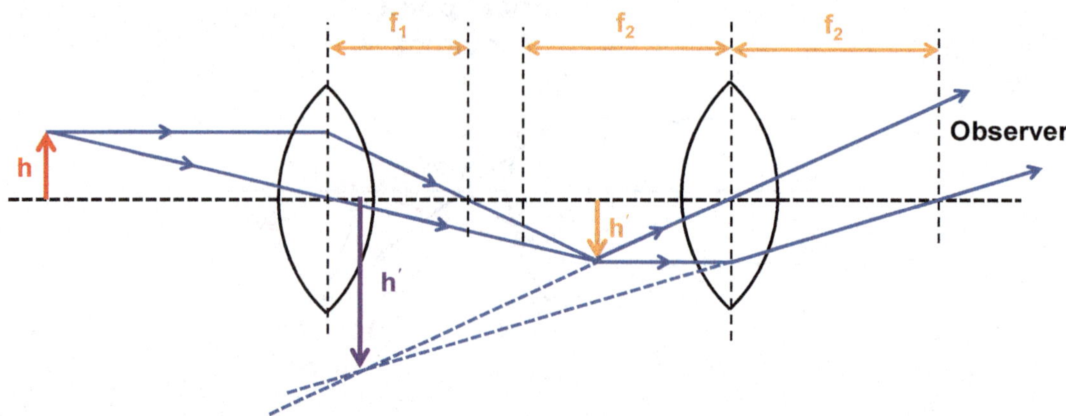

FIGURE 11.15: An optical magnifier with two converging lenses. The real image formed by the first lens is the object for the second lens. The final image is virtual and magnified.

Example 11-6

Problem: The focal lengths for the converging lenses in the optical system shown in Figure 11.15 are $f_1 = 13$ cm and $f_2 = 19$ cm. The two lenses are 31 cm apart from each other. What is the magnification for the final virtual image formed by this system if the object is 26 cm from the first converging lens?

Solution: We first solve for the location of the real image formed by the first converging lens.

$$\frac{1}{26\,\text{cm}} + \frac{1}{s'} = \frac{1}{13\,\text{cm}} \rightarrow \frac{1}{s'} = \frac{1}{13\,\text{cm}} - \frac{1}{26\,\text{cm}} \rightarrow \frac{1}{s'} = \frac{1}{26\,\text{cm}}$$

$$s' = 26\,\text{cm}$$

The magnification of the image is determined using Equation 11-7.

$$M = -\frac{26\,\text{cm}}{26\,\text{cm}} \rightarrow M = -1$$

This real image is then the object for the second converging lens. Since the two lenses are 31 cm apart from each other and the image is 26 cm from the first converging lens, the distance from the "object" to the second converging lens is 5 cm.

$$\frac{1}{5\,\text{cm}}+\frac{1}{s'}=\frac{1}{19\,\text{cm}} \quad\rightarrow\quad \frac{1}{s'}=\frac{1}{19\,\text{cm}}-\frac{1}{5\,\text{cm}} \quad\rightarrow\quad \frac{1}{s'}=-\frac{14}{95\,\text{cm}}$$

$$s'=-6.8\,\text{cm}$$

This is a virtual image, since the image distance is negative. The magnification of this image is

$$M=-\frac{-6.8\,\text{cm}}{5\,\text{cm}} \quad\rightarrow\quad M=1.4$$

The overall magnification of the original object is simply the product of the two magnifications in the system.

$$M=(-1)(1.4) \quad\rightarrow\quad M=-1.4$$

Therefore, when compared to the original object, the final virtual image is inverted and larger.

Example 11-7

Problem: An optical system consists of a diverging lens with focal length $f=-120$ mm and a converging lens with focal length $f=20$ mm. The two lenses are separated by 60 mm. Light from an object first passes through the diverging lens and then the converging lens. What are the location, size, and orientation of the image if the object is 360 mm away from the diverging lens?

Solution: We first solve for the location of the virtual image formed by the diverging lens.

$$\frac{1}{360\,\text{mm}}+\frac{1}{s'}=-\frac{1}{120\,\text{mm}} \rightarrow \frac{1}{s'}=-\frac{1}{120\,\text{mm}}-\frac{1}{360\,\text{mm}} \rightarrow \frac{1}{s'}=-\frac{4}{360\,\text{mm}}$$

$$s'=-90\,\text{mm}$$

The magnification of this image is

$$M=-\frac{-90\,\text{mm}}{360\,\text{mm}} \rightarrow M=\frac{1}{4}$$

This image is then the object for the converging lens. Since the two lenses are separated by 60 mm, the object distance for the converging lens is 100 mm. The image distance for the converging lens is therefore

$$\frac{1}{100\,\text{mm}}+\frac{1}{s'}=\frac{1}{20\,\text{mm}} \rightarrow \frac{1}{s'}=\frac{1}{20\,\text{mm}}-\frac{1}{100\,\text{mm}} \rightarrow \frac{1}{s'}=\frac{4}{100\,\text{mm}}$$

$$s'=25\,\text{mm}$$

The image formed by the converging lens is therefore a real image. The magnification of this image is

$$M=-\frac{25\,\text{mm}}{100\,\text{mm}} \rightarrow M=-\frac{1}{4}$$

The overall magnification of the original object is simply the product of the two magnifications in the system.

$$M=\left(\frac{1}{4}\right)\left(-\frac{1}{4}\right) \rightarrow M=-\frac{1}{16}$$

Therefore, when compared to the original object, the final image is inverted and one-sixteenth the size.

Combinations of lenses made from different material can also be used to mitigate chromatic aberration (Section 11-5) in an optical system.

11-7 Mirrors

The same ray-tracing approach used to determine the location and size of images formed by lenses can be used to determine the location and size of images formed by mirrors. Whereas lenses form images through refraction, mirrors form images through reflection. Incident electromagnetic waves parallel to the optical axis of a ***concave spherical mirror*** are reflected toward a point. Incident electromagnetic waves parallel to the optical axis of a ***convex spherical mirror*** are reflected away from a point. For both mirrors, this point is referred to as the *focal point* of the lens.

> *Concave spherical mirror:* a mirror that reflects incident electromagnetic waves parallel to its optical axis toward its focal point. Concave mirrors form real images.

> *Convex spherical mirror:* a mirror that reflects incident electromagnetic waves parallel to its optical axis away from its focal point. Convex mirrors form virtual images.

Pictorial representations of the interaction of electromagnetic rays with converging and diverging lenses are shown in Figure 11.16.

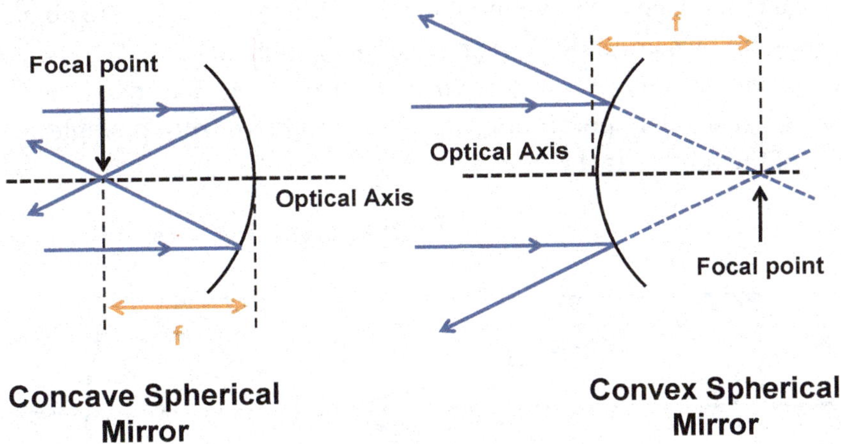

FIGURE 11.16: Concave and convex spherical mirrors. The focal length, *f*, of each mirror is also indicated.

The *optical axis* of a mirror is the axis passing through the center of the mirror and perpendicular to the surface of the mirror. The focal length of a mirror is the distance between surface of the mirror and the focal point as measured along the optical axis of the mirror.

We can qualitatively determine the location and size of the images formed by mirrors using ray tracing. The rules for ray tracing for a *concave mirror* are:

1. Any ray initially parallel to the optical axis on one side of the mirror will reflect back toward the focal point of the mirror.
2. Any ray initially passing through the focal point of the mirror will reflect back parallel to the optical axis of the mirror.
3. Any ray intersecting the mirror where the optical axis intersects the mirror will be reflected back at an equal angle on the opposite side of the optical axis.

The rules for ray tracing for a *convex mirror* are:

1. Any ray initially parallel to the optical axis on one side of the mirror will reflect back away the focal point of the mirror.
2. Any ray initially directed toward the focal point of the mirror will reflect back parallel to the optical axis of the mirror.
3. Any ray intersecting the mirror where the optical axis intersects the mirror will be reflected back at an equal angle on the opposite side of the optical axis.

An example of a real image is shown in Figure 11.17. The rays from the object converge at a single point on the same side of the mirror as the object.

Examples of virtual images are shown in Figure 11.18. The rays from the object do not converge at a single point, but appear to originate from a single point on the opposite side of the mirror as the object.

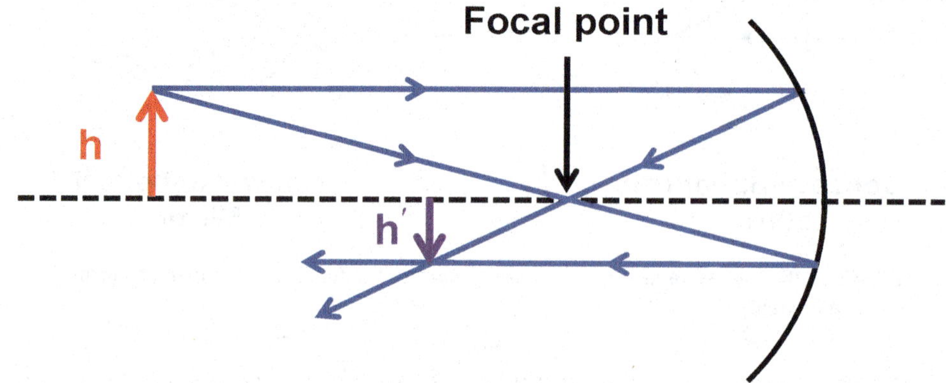

FIGURE 11.17: A real image is formed when the rays converge at a point.

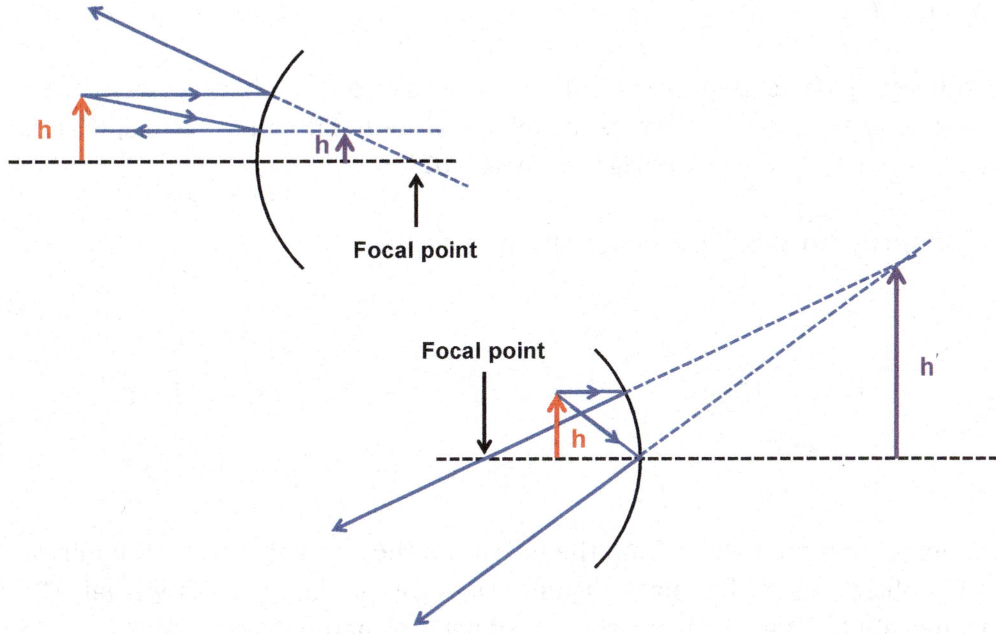

FIGURE 11.18: A virtual image is formed where the rays do not intersect. Rather, the rays appear to come from a point.

The mirror equation (Equation 11-8) relates the object and image locations to the focal length of the mirror.

$$\frac{1}{s}+\frac{1}{s'}=\frac{1}{f} \tag{11-8}$$

Concave mirrors have positive focal lengths, and convex mirrors have negative focal lengths. It is worth noting that the mirror equation (Equation 11-8) is identical to the lens equation (Equation 11-6)[7]. However, now our interpretation of the location and real and virtual images have switched. Namely, an object and real image are on opposite sides of a lens, but on the same side for a mirror (compare Figure 11.10 and Figure 11.17). Similarly, an object and virtual image are on the same side of a lens, but on opposite sides for a mirror (compare Figure 11.11 and Figure 11.18). Nevertheless, it is still true that a positive image distance denotes a real image and a negative image distance denotes a virtual image. Fortunately, the equation for the magnification of an image formed by a mirror is the same as Equation 11-7.

7 That's one fewer equation to remember.

Example 11-8

Problem: An object is located 4 cm away from a concave spherical mirror with a focal length of 12 cm. Where is the image located, is it real or virtual, what is its magnification, and is it upright or inverted?

Solution: Substitution into Equation 11-8 gives us

$$\frac{1}{4\,\text{cm}}+\frac{1}{s'}=\frac{1}{12\,\text{cm}} \rightarrow \frac{1}{s'}=\frac{1}{12\,\text{cm}}-\frac{1}{4\,\text{cm}} \rightarrow \frac{1}{s'}=-\frac{2}{12\,\text{cm}}$$

$$s'=-6\,\text{cm}$$

The image is located 4 cm from the mirror on the opposite side of the mirror as the object. Since the image distance is negative, the image is virtual. The magnification of the image is determined using Equation 11-7.

$$M=-\frac{-6\,\text{cm}}{4\,\text{cm}} \rightarrow M=\frac{3}{2}$$

The image is thus upright and 1.5 times the size of the object.

Example 11-9

Problem: An object is located 12 cm away from a convex spherical mirror with a focal length of −6 cm. Where is the image located, is it real or virtual, what is its magnification, and is it upright or inverted?

Solution: Substitution into Equation 11-8 gives us

$$\frac{1}{12\,\text{cm}}+\frac{1}{s'}=-\frac{1}{6\,\text{cm}} \rightarrow \frac{1}{s'}=-\frac{1}{6\,\text{cm}}-\frac{1}{12\,\text{cm}} \rightarrow \frac{1}{s'}=-\frac{3}{12\,\text{cm}}$$

$$s'=-4\,\text{cm}$$

The image is located 4 cm from the mirror on the opposite side of the mirror as the object. Since the image distance is negative, the image is virtual. The magnification of the image is determined using Equation 11-7.

$$M = -\frac{-4\,\text{cm}}{12\,\text{cm}} \quad \rightarrow \quad M = \frac{1}{3}$$

The image is thus upright and one-third the size of the object.

The focal length of a spherical mirror is approximately equal to one-half of the radius of curvature of the mirror. Therefore, a plane mirror, which has zero curvature, would have a focal length of infinity. It follows from Equation 11-8 that all images formed by a plane mirror must be virtual images.

$$f = \infty \quad \rightarrow \quad \frac{1}{s} + \frac{1}{s'} = 0 \quad \rightarrow \quad \frac{1}{s} = -\frac{1}{s'} \quad \rightarrow \quad s' = -s$$

Each incident ray on a plane mirror will reflect from the mirror at an equal angle on the opposite side of the normal axis for the mirror at that point (see Figure 11.19).

The mirrors on the passenger side of cars are convex mirrors, and consequently the images formed by these mirrors have fractional magnification. In other words, the image formed in the mirror is smaller than the object. As a result, *objects viewed through the passenger side mirror are smaller than they appear*. The driver side mirror is a flat plane mirror, and so the images formed by that mirror are equal in size to their corresponding objects (*i.e.*, the magnification for the mirror is 1). Convex mirrors were chosen for the passenger side since they are more effective at minimizing blind spots than plane mirrors.

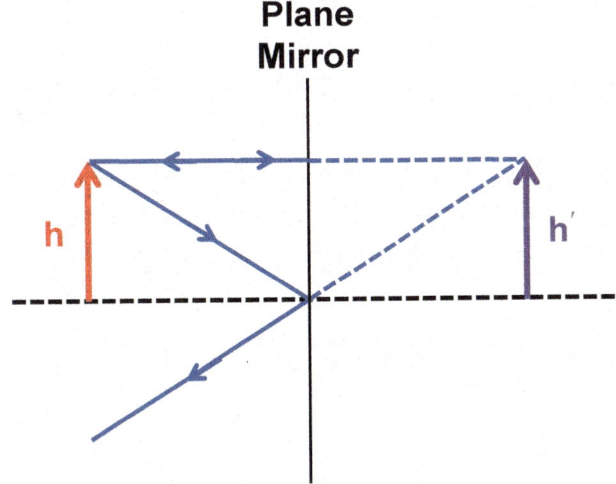

FIGURE 11.19: A plane mirror always forms virtual images.

Example 11-10

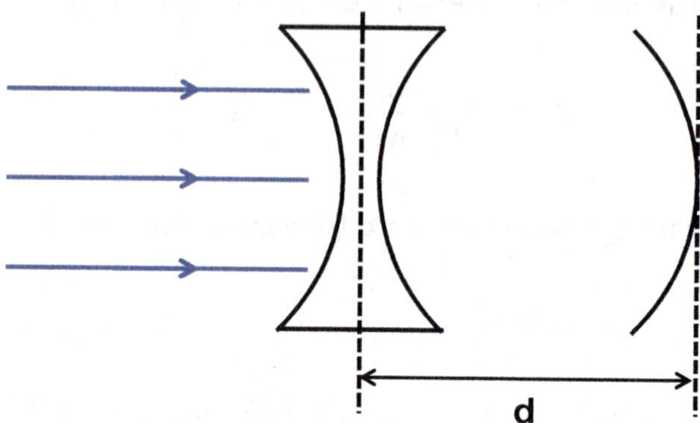

FIGURE 11.20: The optical system in Example 11-10.

Problem: The diverging lens and concave mirror in the optical system shown in Figure 11.20 have focal lengths −10 cm and 10 cm, respectively, are separated by a distance $d = 20$ cm, and have the same optical axis. Incident light is parallel to this optical axis. Where are the incident rays focused (if at all)?

Solution: The source of the parallel incident rays must be infinitely far from the diverging lens. Therefore, the position of the image formed by the diverging lens is

$$\frac{1}{\infty}+\frac{1}{s'}=-\frac{1}{10\,\text{cm}} \quad\rightarrow\quad s'=-10\,\text{cm}$$

This is a virtual image and also the object for the mirror. Since this virtual image is to the left of the diverging lens, in Figure 11.30 the object distance of the mirrors is 30 cm. The position of the image formed by the mirror is

$$\frac{1}{30\,\text{cm}}+\frac{1}{s'}=\frac{1}{10\,\text{cm}} \quad\rightarrow\quad \frac{1}{s'}=\frac{1}{10\,\text{cm}}-\frac{1}{30\,\text{cm}} \quad\rightarrow\quad \frac{1}{s'}=\frac{2}{30\,\text{cm}}$$

$$s'=15\,\text{cm}$$

The image is thus in-between the lens and the mirror at a distance of 5 cm from the lens (15 cm from the mirror).

Summary

- **Electromagnetic spectrum:** the range of all possible wavelengths/frequencies for electromagnetic waves
- **Ray model:** an idealized model of electromagnetic waves in which each wave is represented by a ray that points in the direction of energy flow for that wave.
- **Ray diagram:** a pictorial representation of the propagation of electromagnetic waves. The process of making a ray diagram is called *ray tracing*.
- **Refraction:** the change in the direction of an electromagnetic wave as it moves from a medium with one index of refraction to a medium with a different index of refraction.
- The index of refraction is the ratio of the speed of light in vacuum to the speed of light in another material; the index of refraction of vacuum is therefore one.

  $$n = \frac{c}{v} \quad \rightarrow \quad v = \frac{c}{n}$$

  The index of refraction depends upon the relative permittivity and relative permeability of the material and is also a function of the frequency (or wavelength) of the light.

  $$n = \left(\kappa_m \kappa_e\right)^{\frac{1}{2}}$$

  The index of refraction is always greater than or equal to one.
- Snell's law relates the angle of incidence to the angle of refraction.

  $$n_1 \sin\theta_1 = n_2 \sin\theta_2$$
- **Law of reflection:** the angle of reflection is equal to the angle of incidence.
- Reflected light is partially linear polarized parallel to the surface from which it reflects. When light is incident at the polarizing angle (or Brewster's angle) the

reflected light is completely polarized parallel to the surface. This angle is determined from the indices of refraction on either side of the surface.

$$\tan\theta_P = \frac{n_2}{n_1}$$

- The wavelength (λ_n) and wavenumber (k_n) of an electromagnetic wave in a medium with index of refraction n are

$$\lambda_n = \frac{\lambda}{n} \qquad k_n = \frac{k}{n}$$

  The variables λ and k in this equation are the wavelength and wavenumber, respectively, of the electromagnetic wave in vacuum.

- **Lens***:* an optical device that refracts electromagnetic waves.

- **Optical axis:** the axis passing through the center of a lens or mirror perpendicular to the surface of the lens or mirror.

- **Converging lens:** a lens that refracts incident electromagnetic waves parallel to its optical axis toward its focal point. The focal length of a converging lens is positive.

- **Diverging lens:** a lens that refracts incident electromagnetic waves parallel to its optical axis away from its focal point. The focal length for a diverging lens is negative.

- The rules for ray tracing for a *converging lens* are:

  1. Any ray initially parallel to the optical axis on one side of a lens will refract through the focal point on the other side of the lens.
  2. Any ray passing through the focal point on one side of the lens is refracted into a ray parallel to the optical axis on the other side of the lens.
  3. Any ray directed at the center of the lens passes through in a straight line.

- The rules for ray tracing for a *diverging* lens are:

  1. Any ray initially parallel to the optical axis on one side of a lens will diverge along a line through the focal point on the same side of the lens.
  2. Any ray directed along a line toward the focal point on the other side of the lens will emerge from the lens parallel to the optical axis.

3. Any ray directed at the center of the lens passes through in a straight line.

- **Image:** the source of a diverging group of rays.

- **Real image:** an image formed where refracted rays intersect. These rays then diverge from this point.

- **Virtual image:** an image formed where refracted rays do not intersect. The rays only appear to come from a single point. All images formed by a diverging lens are virtual images.

- **Chromatic aberration:** the failure of a lens to focus all colors of light to the same point.

- **Concave spherical mirror**: a mirror that reflects incident electromagnetic waves parallel to its optical axis toward its focal point. Concave mirrors have positive focal lengths.

- **Convex spherical mirror:** a mirror that reflects incident electromagnetic waves parallel to its optical axis away from its focal point. Convex mirrors have negative focal lengths.

- The rules for ray tracing for a *concave mirror* are:

  1. Any ray initially parallel to the optical axis on one side of the mirror will reflect back toward the focal point of the mirror.
  2. Any ray initially passing through the focal point of the mirror will reflect back parallel to the optical axis of the mirror.
  3. Any ray intersecting the mirror where the optical axis intersects the mirror will be reflected back at an equal angle on the opposite side of the optical axis.

- The rules for ray tracing for a *convex mirror* are:

  1. Any ray initially parallel to the optical axis on one side of the mirror will reflect back away the focal point of the mirror.
  2. Any ray initially directed toward the focal point of the mirror will reflect back parallel to the optical axis of the mirror.
  3. Any ray intersecting the mirror where the optical axis intersects the mirror will be reflected back at an equal angle on the opposite side of the optical axis.

- The location of an image is determined using the following equation:

$$\frac{1}{s}+\frac{1}{s'}=\frac{1}{f}$$

The variable s and s' denote the positions of the object and the image, respectively, relative to the lens or mirror. The object position is always a positive number. *The image position is positive for real images and negative for virtual images.* The variable f is the focal length of the lens or mirror.

- The magnification of the object is defined to be the negative of the ratio of image size to object size, which can also be expressed in terms of the image and object positions using geometry.

$$M=-\frac{h'}{h}=-\frac{s'}{s}$$

Problems

1. An object is located 12 cm away from a converging lens with a focal length of 3 cm. Where is the image located, is it real or virtual, what is its magnification, and is it upright or inverted?

2. An object is located 4 cm away from a diverging lens with a focal length of -12 cm. Where is the image located, is it real or virtual, what is its magnification, and is it upright or inverted?

3. An object is located 4 cm away from a converging lens with a focal length of 6 cm. Where is the image located, is it real or virtual, what is its magnification, and is it upright or inverted?

4. An object is located 100 cm away from a diverging lens with a focal length of -25 cm. Where is the image located, is it real or virtual, what is its magnification, and is it upright or inverted?

5. An object is located 3 cm away from a concave spherical mirror with a focal length of 12 cm. Where is the image located, is it real or virtual, what is its magnification, and is it upright or inverted?

6. An object is located 12 cm away from a concave spherical mirror with a focal length of 3 cm. Where is the image located, is it real or virtual, what is its magnification, and is it upright or inverted?

7. An object is located 6 cm away from a convex spherical mirror with a focal length of -12 cm. Where is the image located, is it real or virtual, what is its magnification, and is it upright or inverted?

8. An object is located 20 cm away from a convex spherical mirror with a focal length of −5 cm. Where is the image located, is it real or virtual, what is its magnification, and is it upright or inverted?

9. A 2 cm tall object is 20 cm to the left of a converging lens with a focal length of 10 cm. A second lens (a diverging lens) with a focal length of −5 cm is 30 cm to the right of the first lens. What is the position and height of the image viewed through the diverging lens?

10. What is the polarization angle for a boundary if the critical angle for total internal reflection at that boundary is 30°?

CHAPTER 12

Wave Optics

12-1 Introduction

In our discussion of geometrical optics in Chapter 11, we largely ignored the wave nature of the electromagnetic waves. In this chapter, we will discuss specific phenomena which are dependent upon these wave characteristics. However, we conclude this chapter with a discussion of the photoelectric effect, which cannot be explained by the wave properties of electromagnetic waves. Einstein's explanation for this phenomenon will lead us to a final refinement of the Bohr model for the hydrogen atom and, by extension, to the initial development of quantum mechanics.

12-2 Huygens's principle

The laws of reflection and refraction, presented without proof in Chapter 11, were derived through a geometric method by Christiaan Huygens. The main premise of his model for the propagation of light is referred to as ***Huygens's Principle***.

> *Huygens's principle*: each position on a wave front is the source of spherical waves (called wavelets) that spread out at the wave speed. At a later time, the shape of the wave front is the line tangent to all the wavelets.
>
> *Wave front:* a surface over which the phase[1] of a wave is constant.

An example of Huygens's principle is shown in Figure 12.1. In Figure 12.1, each position on the wave front of a plane wave is the source of spherical wavelets which then propagate away from those positions; for clarity, only five positions are indicated in Figure 12.1. For an infinite plane wave, there will be an infinite number of these positions and associated wavelets. The wave front at a later time is the tangent to all of these wavelets and will also correspond to a plane wave since the wavelets all move at the same speed. The corresponding light ray (Section 11-2) is perpendicular to the wave front and points in the direction of propagation of the wave.

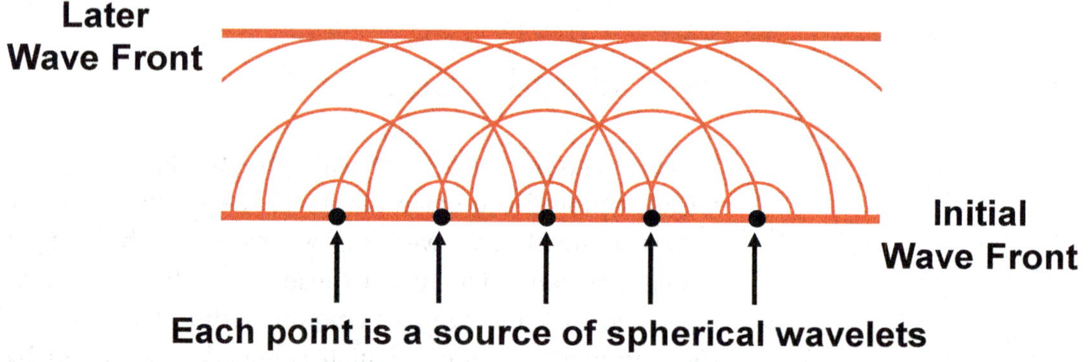

FIGURE 12.1: Huygens's principle.

Now let's consider the case of plane wave reflection shown in Figure 12.2. The different colors in Figure 12.2 and Figure 12.3 denote different time points at which the wave and wavelets are shown; specifically, the waves and wavelets shown in green occur later in time than those shown in blue, which occur later in time than those shown in red, which occur later in time than those shown in purple. Since the speed of the wavelets in the reflected wave is the same as the speed of the wavelets in the incident wave, the incident and reflected waves obey the law of reflection (Section 11-4).

1 See Section 10-4.

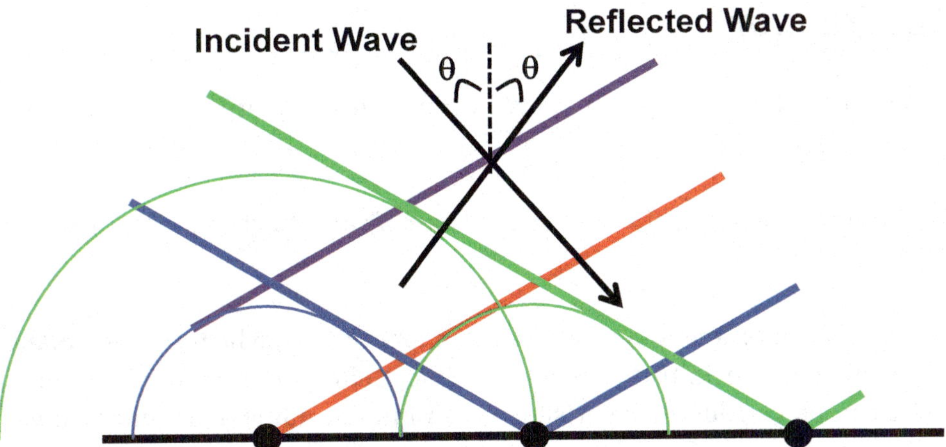

FIGURE 12.2: Reflection of a plane wave. The angle of incidence equals the angle of reflection. The direction of the incident and reflected light rays are shown as arrows.

In contrast, the speed of the wavelets in a refracted wave is different from the speed of the wavelets in the incident wave. Thus, as shown in Figure 12.3, although the refracted wave is also a plane wave, its propagation direction is different from the propagation direction of the incident wave.

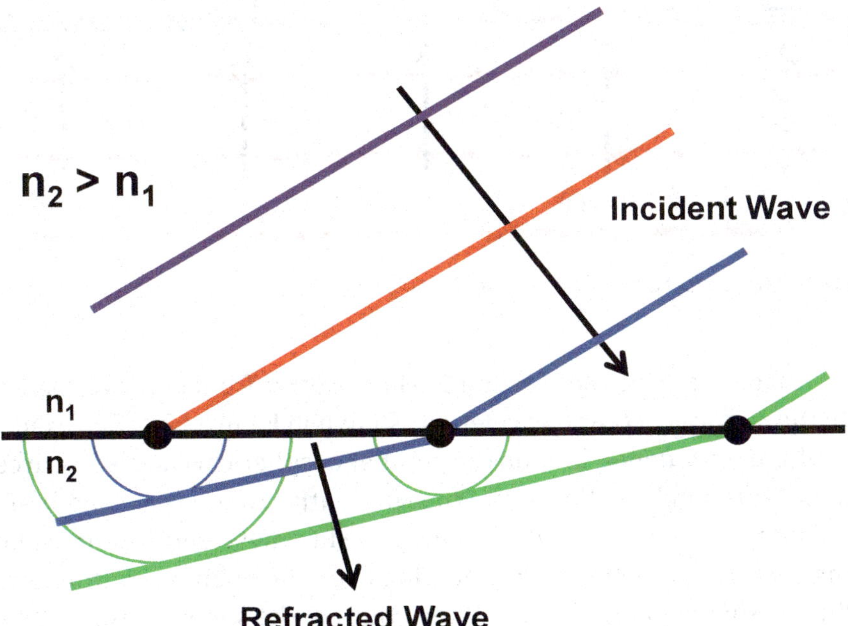

FIGURE 12.3: Refraction of a plane wave. The propagation direction of the refracted wave is different from the propagation direction of the incident wave. The directions of the incident and refracted light rays are shown as arrows.

12-3 Diffraction

When waves pass by an object or move through an aperture, they may experience ***diffraction***.

> *Diffraction*: the bending of waves as they pass by an object or move through an aperture.

A simple example of diffraction is illustrated in Figure 12.4, which shows a plane wave incident on an aperture, in this case a single linear slit. The wave on the other side of the barrier can be considered to arise from an infinite number of spherical wavelets emitted by an infinite number of sources within the slit.

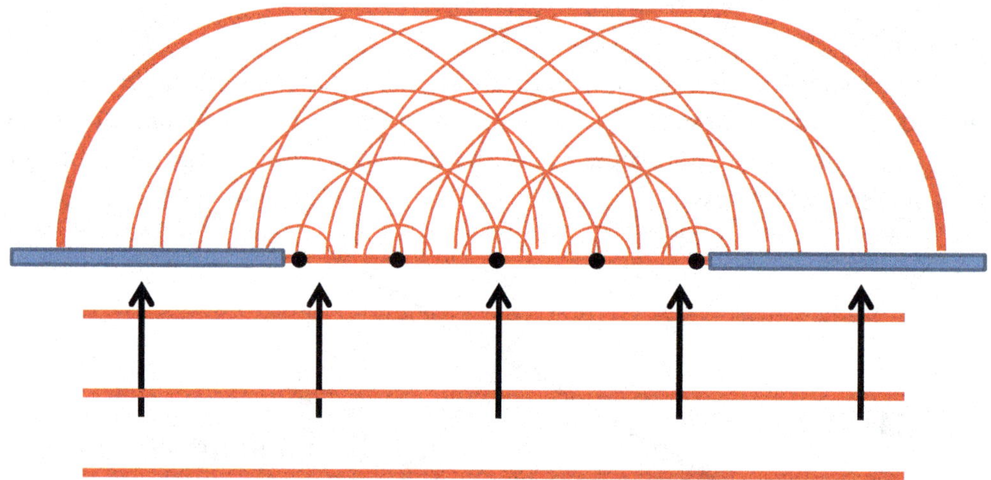

FIGURE 12.4: Diffraction of a plane wave through a single slit.

According to Huygens's principle (Section 12-1), the wave front of the transmitting wave is tangent to the surfaces of these wavelets, as shown in Figure 12.4. The exact structure of this transmitted wave depends upon the number of spherical wavelet "sources" within the slit and thus ultimately on the size of the slit relative to the wavelength of the wave. If a wave encounters a barrier in which there is a slit whose width is much larger than the wavelength of the wave, the wave emerging from the opening continues to move in a straight line, as shown in Figure 12.5. If the width of the slit is approximately equal to the wavelength of the wave, the wave emerging from the slit will spread out from the opening in all directions. If the width of the slit is much less than the wavelength of the light, the emerging waves will be spherical and centered on the opening.

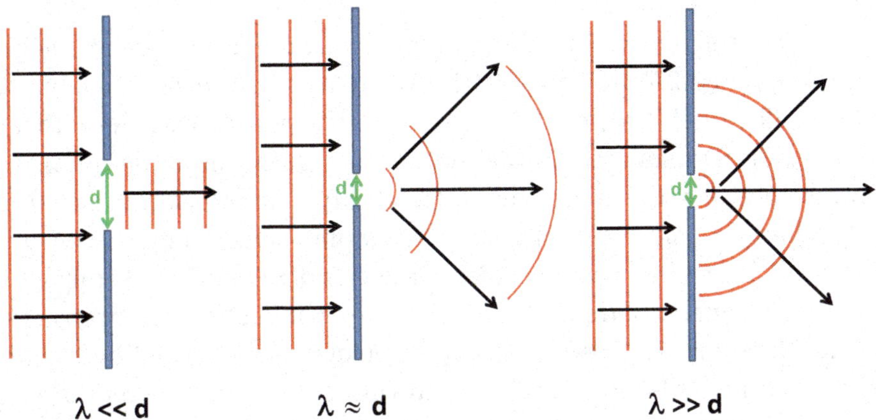

FIGURE 12.5: A plane wave incident upon a single slit. The diffraction depends upon the relative size of the slit and the wavelength of the wave.

In addition to causing a bending of the wave front, the individual wavelets produced by the diffraction of a wave through a slit will *interfere* with each other (Section 10-8). As shown in Figure 12.6, a pattern of repeating constructive and destructive interference will occur among the wavelets. This pattern of interference, typically referred to as a *diffraction pattern*, also depends upon the distance between the slit and the point of observation. We will limit our discussion here to diffraction patterns observed far from the slit; this region is referred to as the *far-field*. The diffraction pattern observed in the *near-field* (close to the slit) will not be discussed here[2].

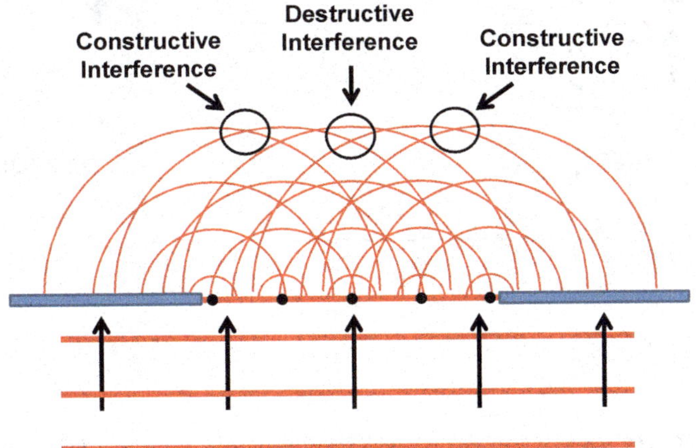

FIGURE 12.6: Interference of wavelets following diffraction through a single slit.

2 You should feel inspired to take an advanced course in optics to learn about what happens in the near-field.

We can determine an equation for the far-field diffraction pattern using the geometric model shown in Figure 12.7. In Figure 12.7, a plane wave diffracts through a single slit of width a. We will model three equally spaced sources within this slit with a distance of $a/2$ between adjacent sources; in reality the slit would contain an infinite number of these sources but only three are shown for clarity. The interference of these three waves will produce a *single-slit diffraction pattern* on a screen a distance L away from the slit. This diffraction pattern will consist of a series of bright lines (or fringes) where constructive interference occurs and dark lines (or fringes) where destructive interference occurs. In Figure 12.7, the position of a fringe is described by the distance y separating it from the mid-point of the screen (directly opposite the mid-point of the slit) or, equivalently, in terms of an angle θ between the line connecting the mid-point of the slit to the location of the fringe and a line connecting the mid-point of the slit with the mid-point of the screen.

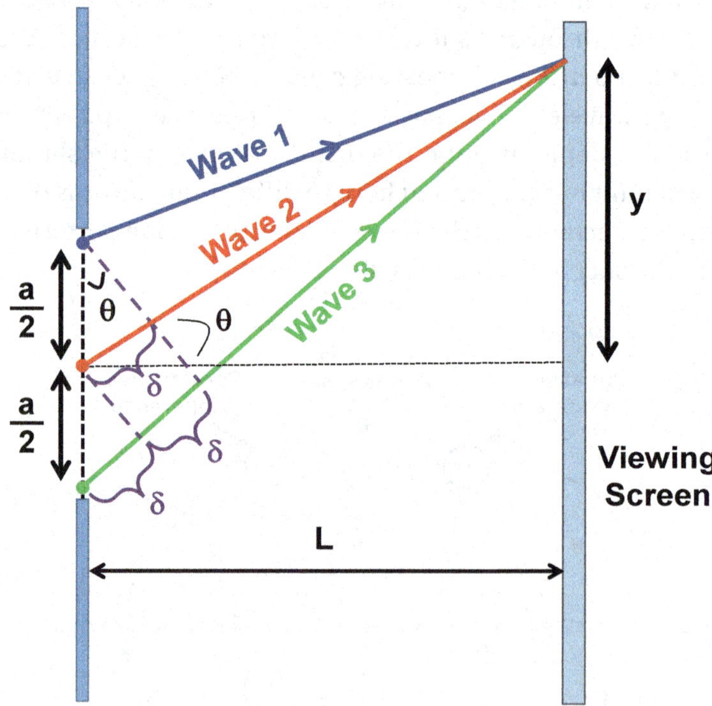

FIGURE 12.7: Geometric model for determining the single-slit diffraction pattern.

As shown in Figure 12.7, wavelets from the three difference sources will travel different distances (or path lengths) to reach the same point on the screen. If we assume that the distance between the slits and the observation screen is large, then the paths of all of the electromagnetic waves from the slits should be approximately parallel to each other, as shown in Figure 12.7[3]. Let's denote the difference in distance between any two adjacent sources as δ. The resultant wave at the position *y* in Figure 12.7 is simply the sum of the waves from each of the sources in the slit, which we can denote using the subscripts 1, 2, and 3. In the case of electromagnetic waves, the resultant electric field at the position *y* would be[4]

$$E = E_1 + E_2 + E_3 \quad \rightarrow \quad E = (E_0)_1 e^{i(kr-\omega t)} + (E_0)_2 e^{i(k(r+\delta)-\omega t)} + (E_0)_3 e^{i(k(r+2\delta)-\omega t)}$$

In this expression, the distance *r* is the distance from source 1 (the blue source in Figure 12.7) to position *y*. Since each of these sources originates from the same plane wave (incident on the slit from the left in Figure 12.7), the electric field magnitude of each source should be equal to each other and equal to one-third of the magnitude of the electric field of the incident wave.

$$(E_0)_1 = (E_0)_2 = (E_0)_3 = \frac{E_0}{3} \quad \rightarrow \quad E = \frac{E_0}{3}\left(e^{i(kr-\omega t)} + e^{i(k(r+\delta)-\omega t)} + e^{i(k(r+2\delta)-\omega t)}\right)$$

The distance δ can be related to the spacing between the slits and the angle θ in Figure 12.7 using trigonometry.

$$\delta = \frac{a}{2}\sin\theta \qquad \textbf{(12-1)}$$

Now let's imagine a more realistic situation in which there are *N* sources of wavelets within the slit. We will retain our previous notation and denote the width of the slit as *a* and the difference in path length between adjacent sources as δ. The resultant electric field at a position *y* on the screen for *N* sources is

$$E = E_1 + E_2 + \ldots + E_N$$

3 These paths cannot be exactly parallel, of course, because the electromagnetic waves would never converge (and interfere) if they were.

4 We will ignore the phase constant in our calculations for the sake of simplicity.

The intensity of the resultant electromagnetic wave is given by Equation 12-2[5].

$$I = I_{\max}\left(\frac{\sin\left(\frac{\pi a \sin\theta}{\lambda}\right)}{\frac{\pi a \sin\theta}{\lambda}}\right)^2 \quad \textbf{(12-2)}$$

Equation 12-2 is called the *Fraunhofer diffraction intensity equation* after Joseph Fraunhofer, who first derived an expression for it. The variable $I_{\max}$ is the maximum intensity in the interference pattern, which occurs when $\theta = 0°$. It follows from Equation 12-2 that dark lines (*i.e.*, complete destructive interference) occur whenever

$$\sin\left(\frac{\pi a \sin\theta}{\lambda}\right) = 0 \quad \rightarrow \quad \frac{\pi a \sin\theta}{\lambda} = m\pi \quad m = \pm 1, \pm 2, \ldots$$

$$\text{Complete destructive interference } \sin\theta = m\frac{\lambda}{a} \quad m = \pm 1, \pm 2, \ldots \quad \textbf{(12-3)}$$

A plot of the intensity of the light observed from diffraction through a single slit (Equation 12-2) is shown in Figure 12.8; this plot is also referred to as the single-slit diffraction pattern or the Fraunhofer diffraction pattern of a single-slit.

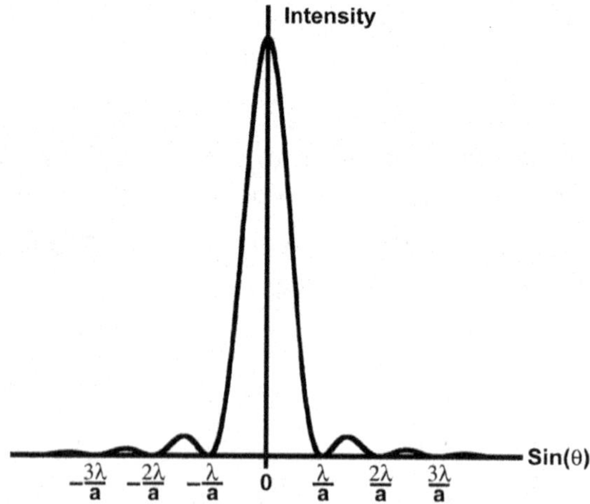

FIGURE 12.8: The diffraction pattern of a single-slit (Equation 12-2).

5 See Section C-18 for derivation of Equation 12-2.

Now let's consider the situation shown in Figure 12.9, in which plane waves from a single source are incident on a barrier containing two identical slits of equal width. This is referred to as double-slit diffraction.

In double-slit diffraction, we have two effective sources of travelling waves (one from each slit) and these travelling waves can interfere with each other to form an interference pattern.

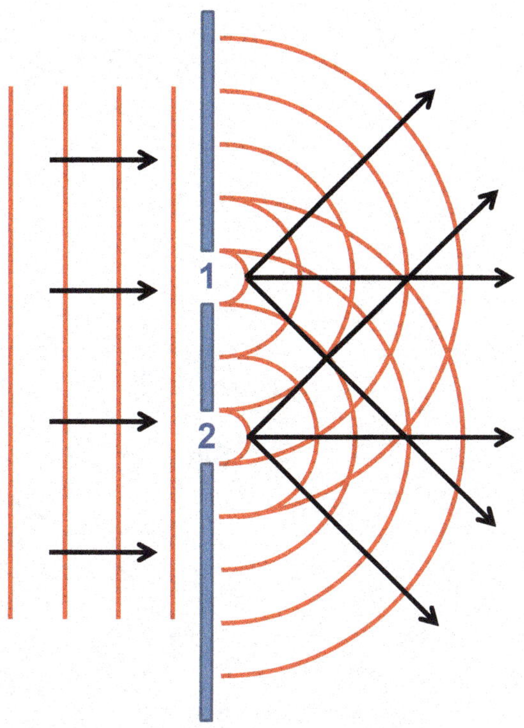

FIGURE 12.9: Double-slit diffraction.

Interestingly, if light bulbs were placed at the locations of the two slits, no interference pattern would be observed. The reason why is because there would be no constant phase relationship between the electromagnetic waves emitted by the two bulbs; there are continual changes in the phase of the emitted electromagnetic waves from a lightbulb (these changes occur on the nanosecond timescale), and these changes are not correlated between the two light bulbs. Therefore, the conditions for constructive and destructive interference are maintained at any location for only a short time interval (also on the nanosecond timescale) and since our eyes cannot follow such rapid changes, we do not see any interference pattern. In order for an interference pattern to be detected, the electromagnetic waves from both slits must maintain a constant phase with respect to each other. This condition is frequently stated as the two sources must be ***coherent***. The lightbulbs, in contrast, are *incoherent* sources.

> *Coherent sources*: two sources of electromagnetic waves are coherent if the waves they emit have a zero or constant phase difference.

The sources must also be *monochromatic* (*i.e.*, they should be of a single wavelength) in order for interference to be observed. If monochromatic plane electromagnetic waves that originated from a single source diffract through the two slits in Figure 12.9, the two slits will act like coherent sources and an interference pattern will be observed.

The spacing between the locations of constructive and destructive interference can be determined using the simple geometrical model shown in Figure 12.10, which is similar to the model in Figure 12.7. As before, if we assume that the difference between the slits and the screen is large, the difference in path length required to reach a particular position on the screen, denoted by δ, can be related to the distance between the slits, denoted by d, and the angle θ using Equation 12-4.

$$\delta = d\sin\theta \tag{12-4}$$

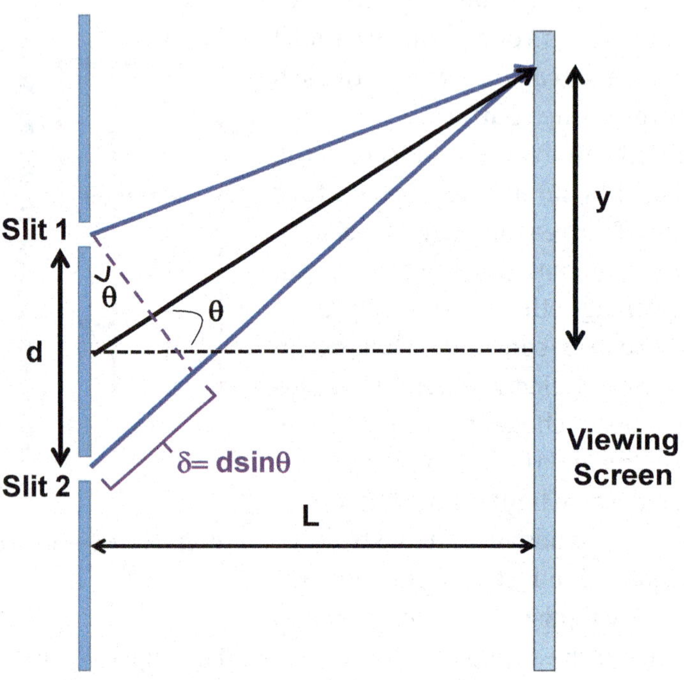

FIGURE 12.10: Geometrical model of double-slit diffraction.

As before, the magnitude of the electric field of the resultant electromagnetic wave at the position y is the sum of the electric fields of the electromagnetic waves from each of the two slits. As shown in Figure 12.9 and Figure 12.10, we will denote the two slits as 1 and 2.

$$E = E_1 + E_2 \quad \rightarrow \quad E = \left(E_0\right)_1 e^{i(kx-\omega t)} + \left(E_0\right)_2 e^{i(k(x+\delta)-\omega t)}$$

We will assume the magnitudes of these two electric fields are the same.

$$\left(E_0\right)_1 = \left(E_0\right)_2 = E_0 \quad \rightarrow \quad E = E_0\left(e^{i(kx-\omega t)} + e^{i(k(x+\delta)-\omega t)}\right)$$

Simplifying this expression[6] gives us

$$E = E_0 e^{i\left(kx + \frac{k\delta}{2} - \omega t\right)}\left(e^{-i\frac{k\delta}{2}} + e^{i\frac{k\delta}{2}}\right) \rightarrow E = E_0 e^{i\left(kx + \frac{k\delta}{2} - \omega t\right)}\left(2\cos\left(\frac{k\delta}{2}\right)\right)$$

The real part of this electric field is therefore

$$E = E_0 \cos\left(kx + \frac{k\delta}{2} - \omega t\right)\left(2\cos\left(\frac{k\delta}{2}\right)\right)$$

The average intensity of the resultant electromagnetic wave is thus[7]

$$I = I_{\max}\cos^2\left(\frac{k\delta}{2}\right) \rightarrow I = I_{\max}\cos^2\left(\frac{\left(\frac{2\pi}{\lambda}\right)(d\sin\theta)}{2}\right)$$

$$I = I_{\max}\cos^2\left(\frac{\pi d\sin\theta}{\lambda}\right) \qquad \textbf{(12-5)}$$

In the limit that the angle θ in Equation 12-5 is small, Equation 12-5 can be written as Equation 12-6.

$$\sin\theta \approx \frac{y}{L} \rightarrow I = I_{\max}\cos^2\left(\frac{\pi d}{\lambda L}y\right) \qquad \textbf{(12-6)}$$

Under these conditions we have

$$\text{Constructive interference: } y = m\left(\frac{L\lambda}{d}\right) \quad m = 0, \pm 1, \pm 2, \ldots \qquad \textbf{(12-7)}$$

6 Equation D-8.
7 Using Equation 12-4 and Equation 10-7.

$$\text{Destructive interference: } y = \left(m + \frac{1}{2}\right)\left(\frac{L\lambda}{d}\right) \quad m = 0, \pm 1, \pm 2, \ldots \qquad \textbf{(12-8)}$$

There is always a bright line (constructive interference) at the center of the screen directly opposite the two slits ($y = 0$). The other bright and dark lines are denoted by the corresponding integer m, referred to as the order of the diffraction.

Example 12-1

Problem: A viewing screen is separated from a barrier containing a double-slit by 6 m. The distance between the two slits is 0.04 mm. Monochromatic light incident on the double-slit forms a diffraction pattern on the screen. The first order dark line on the screen occurs 4.5 cm from the center of the screen. What is the wavelength of the light?

Solution: Since the distance between the slits and the screen is much larger than the distance between the dark line and the center of the screen, we can safely use the small angle approximation to determine positions in the interference pattern. For the destructive interference associated with the first dark line in the pattern, we have

$$y = \left(\frac{L}{d}\right)\left(0 + \frac{1}{2}\right)\lambda \;\rightarrow\; \lambda = \frac{2dy}{L} \;\rightarrow\; \lambda = \frac{2\left(4\times10^{-5}\,\text{m}\right)\left(4.5\times10^{-2}\,\text{m}\right)}{6\,\text{m}}$$

$$\lambda = 600\,\text{nm}$$

Example 12-2

Problem: A viewing screen is separated from a barrier containing a double-slit by 6 m. The distance between the two slits is 0.04 mm. Coherent light of two different wavelengths, $\lambda_1 = 450$ nm and $\lambda_2 = 550$ nm, is incident on the double-slit and forms a diffraction pattern on the screen. What is the separation distance between the second order bright lines of the two different colors of light on the screen?

Solution: Since the distance between the slits and the screen is much larger than the distance between the dark line and the center of the screen, we can safely use the small angle approximation to determine positions in the interference pattern. For the constructive interference associated with the second bright spot in the pattern we have

$$y = \left(\frac{L}{d}\right)(2)\lambda \rightarrow y = 2\frac{L\lambda}{d}$$

The difference between the first bright spots for the two colors of light is

$$\Delta y = 2\frac{L}{d}(\lambda_2 - \lambda_1) \rightarrow \Delta y = 2\left(\frac{6\,\text{m}}{4.5\times10^{-2}\,\text{m}}\right)(550\times10^{-9}\,\text{m} - 450\times10^{-9}\,\text{m})$$

$$\Delta y = 26.7\ \mu\text{m}$$

Of course, the intensity of the wave from each of the slits at position y in Figure 12.10 is also a function of the angle θ, due to the diffraction from each slit itself. Thus, the true expression for the intensity of the resultant wave is the product of Equation 12-2 and Equation 12-5, and is given in Equation 12-9.

$$I = I_{\text{max}} \cos^2\left(\frac{\pi d \sin\theta}{\lambda}\right)\left(\frac{\sin\left(\frac{\pi a \sin\theta}{\lambda}\right)}{\frac{\pi a \sin\theta}{\lambda}}\right)^2 \qquad \textbf{(12-9)}$$

In Equation 12-9, the variable d denotes the spacing between the two slits and the variable a denotes the width of each slit. A plot of the intensity of the light observed from double-slit diffraction (Equation 12-9) is shown in Figure 12.11; this plot is also referred to as the double-slit diffraction pattern or the Fraunhofer diffraction pattern of a double-slit.

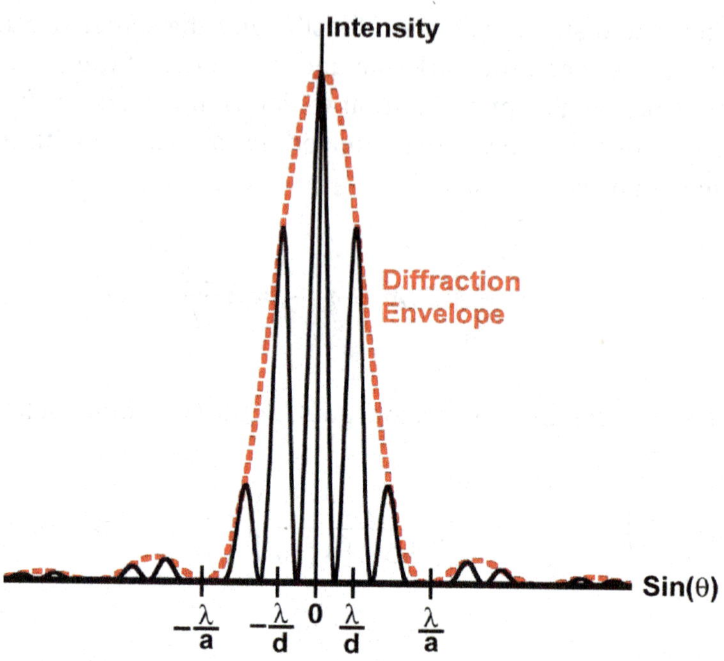

FIGURE 12.11: The diffraction pattern of a double-slit (Equation 12-9). The diffraction pattern of each slit forms an envelope around the interference pattern of the two slits.

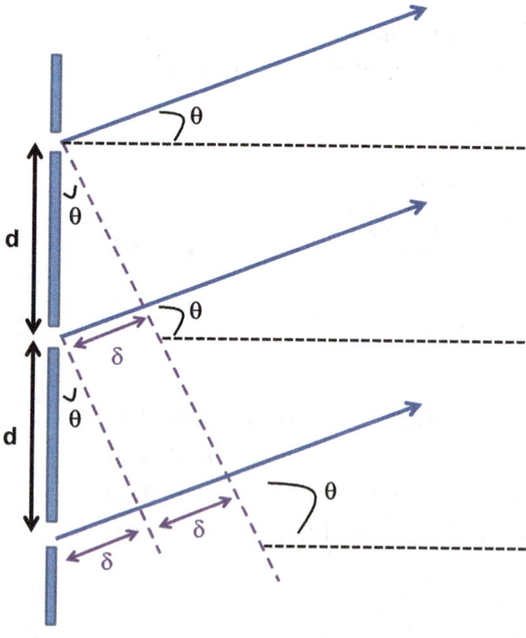

FIGURE 12.12: A diffraction grating.

Now let's increase the number of slits to N. Each of the N slits will act as a source for electromagnetic waves, and the interference pattern will be the superposition of the N electromagnetic waves at any position. If we assume that the distance between the slits and the observation screen is large, then the paths of all of the electromagnetic waves from the slits should be approximately parallel to each other, as shown in Figure 12.12.

Under these assumptions, the difference in path length for electromagnetic waves originating from any two adjacent slits will be equal and given by Equation 12-4. Furthermore, the conditions for constructive and destructive interference are given by Equation 12-7 and Equation 12-8. A system such as this, consisting of multiple slits, is referred to as a diffraction grating. Diffraction gratings are typically described in terms of the density of slits then contain, with the word "lines" denoting the number of slits. Thus, a diffraction grating with 1000 lines per mm would be a diffraction grating whose slits are separated by 1 μm.

Example 12-3

Problem: Monochromatic light of wavelength 500 nm illuminates a diffraction grating. The second order bright line above the center is at angle 30°. How many lines per mm does this grating have?

Solution: Combining Equation 12-6 and Equation 12.7 gives us

$$d\sin\theta = 2\lambda \quad \rightarrow \quad d = \frac{2\lambda}{\sin\theta} \quad \rightarrow \quad d = \frac{2\left(500\times10^{-9}\,\text{m}\right)}{\sin 30°}$$

$$d = 2\times10^{-6}\,\text{m} \quad \rightarrow \quad d = 2\times10^{-3}\,\text{mm} \quad \rightarrow \quad \frac{1}{d} = 500\,\frac{\text{lines}}{\text{mm}}$$

12-4 Other Sources of Interference Patterns

Interference patterns are also frequently observed in thin films, such as soap bubbles. The different colors of light observed when white light is incident upon such films results from the interference of the electromagnetic waves reflecting from the surfaces of the film. The change in phase associated with the different distances traveled by the electromagnetic waves contributes to the interference pattern, as we previously saw with double-slit interference patterns, but in addition there can be a change in phase associated with the process of reflection itself. *An electromagnetic wave undergoes a change in phase of π radians when reflected from a boundary leading to a material with higher index of refraction.* There is no change in phase when the electromagnetic wave reflects from a boundary leading to a material with lower index of refraction[8].

8 You should feel inspired to take an advanced course in electricity and magnetism to learn how to derive these rules.

Let's consider the system shown in Figure 12.13, in which incident light interacts with a thin film of thickness L. Incident light that initially reflected off the surface of the film (such as ray 1 in Figure 12.13) will undergo a change in phase of π radians since $n_2 > n_1$. Incident light that refracted into the film and then reflected off the other surface of the film (such as ray 2 in Figure 12.13) will experience a phase change associated with the path it transverses through the film, but no phase change associated with the reflection for this light since $n_1 < n_2$. For small incident angles, the path taken by ray 2 is approximately equal to twice the width of the film.

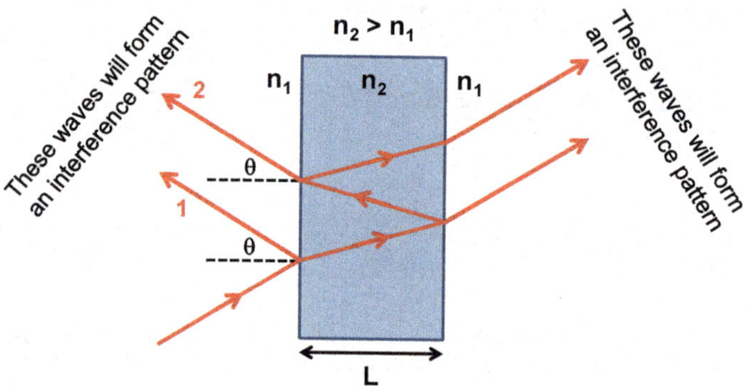

FIGURE 12.13: Interference from thin films.

The changes in phase for the different rays are therefore[9]

$$(\Delta\phi)_1 = \pi$$

$$(\Delta\phi)_2 = k_{n_2}(2L)$$

Ray 1 and ray 2 will constructively interfere with each other if these differences in phase are equal to an integer multiple of 2π.

$$(\Delta\phi)_2 - (\Delta\phi)_1 = 2\pi m \quad\rightarrow\quad k_{n_2}(2L) - \pi = 2\pi m \quad\rightarrow\quad \left(\frac{2\pi}{\lambda_{n_2}}\right)(2L) = \pi(2m+1)$$

For simplicity, let's assume that $n_1 = 1$. Then, substitution of Equation 11-3 gives us

9 The difference in phase depends upon the wavenumber of the wave in the material. Therefore, we need to use k_n (Equation 11-3), rather that the wavenumber for the wave in vacuum.

$$\text{Constructive interference } 2Ln_2 = \left(m + \frac{1}{2}\right)\lambda \quad m = 0, \pm 1, \pm 2, \ldots$$

$$\text{Destructive interference } 2Ln_2 = m\lambda \quad m = 0, \pm 1, \pm 2, \ldots$$

The wavelength λ in these equations is the wavelength of the light in medium 1 (with index of refraction n_1), which we have assumed to be vacuum. This calculation was performed for only two rays of light and is valid only when the angle of incidence is small. The total interference pattern for the film comes from all of the rays of light that reflect and refract, and is a function of the angle of incidence of the light. The associated calculation, however, is beyond the scope of this textbook. Nevertheless, the simplified explanation presented above demonstrates how interference patterns can be generated from a thin film.

Example 12-4

Problem: The reflected light and refracted light from a soap film constructively interfere for green light ($\lambda = 510$ nm). The index of refraction of the soap film at this wavelength is 1.3. What is the minimum thickness of the soap film?

Solution: Constructive interference occurs when

$$2Ln_2 = \left(m + \frac{1}{2}\right)\lambda \quad m = 0, \pm 1, \pm 2, \ldots$$

The minimum thickness corresponds to $m = 0$. Substitution gives us

$$2L(1.3) = \left(0 + \frac{1}{2}\right)(510\,\text{nm})$$

$$L = 98\,\text{nm}$$

The system shown in Figure 12.14 consists of a single source of coherent light and a plane mirror. Light from the source can reflect off the mirror and interfere with itself to form an interference pattern on the screen. We can use geometric optics to model the interference pattern as resulting from two coherent sources of light (one real and one virtual), as shown in Figure 12.14.

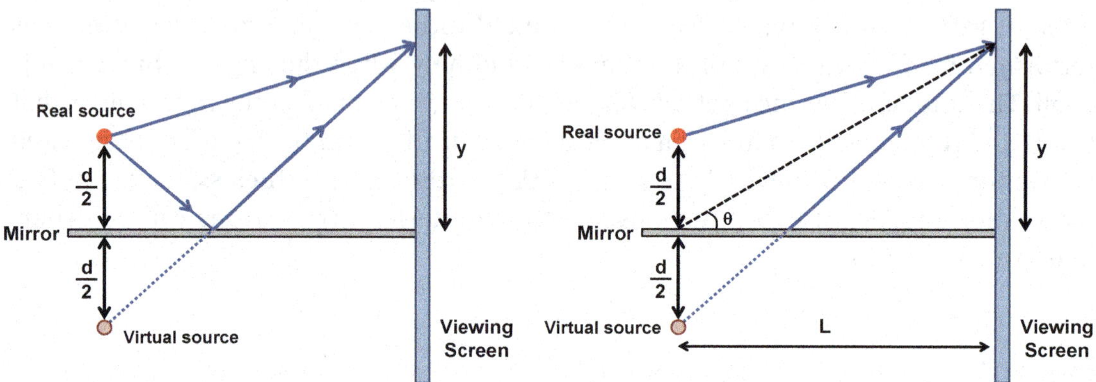

FIGURE 12.14: Forming an interference pattern through reflection from a mirror.

The magnitude of the electric field of the resultant electromagnetic wave at the position y in Figure 12.14 is the sum of the electric fields of the electromagnetic waves from each of the two sources. The magnitudes of these two electric fields are the same, so this summation gives us

$$E = E_0\left(e^{i(kx-\omega t)} + e^{i(k(x+\delta)-\omega t+\pi)}\right)$$

The additional phase change of π radians results from the reflection of that electromagnetic wave off the mirror. Following the previous derivation for the two-slit interference pattern gives us

$$E = E_0 e^{i\left(kx+\frac{k\delta}{2}-\omega t+\frac{\pi}{2}\right)}\left(e^{-i\left(\frac{k\delta}{2}+\frac{\pi}{2}\right)} + e^{i\left(\frac{k\delta}{2}+\frac{\pi}{2}\right)}\right)$$

$$E = E_0 e^{i\left(kx+\frac{k\delta}{2}-\omega t+\frac{\pi}{2}\right)}\left(2\cos\left(\frac{k\delta}{2}+\frac{\pi}{2}\right)\right)$$

The real part of this electric field is therefore

$$E = E_0 \cos\left(kx + \frac{k\delta}{2} - \omega t + \frac{\pi}{2}\right)\left(2\cos\left(\frac{k\delta}{2} + \frac{\pi}{2}\right)\right)$$

$$E = E_0 \sin\left(kx + \frac{k\delta}{2} - \omega t\right)\left(2\sin\left(\frac{k\delta}{2}\right)\right)$$

The average intensity of the resultant electromagnetic wave is thus[10]

$$I = I_{\max} \sin^2\left(\frac{k\delta}{2}\right) \quad \rightarrow \quad I = I_{\max} \sin^2\left(\frac{\left(\frac{2\pi}{\lambda}\right)(d\sin\theta)}{2}\right)$$

$$I = I_{\max} \sin^2\left(\frac{\pi d \sin\theta}{\lambda}\right)$$

In the limit that the angle θ in is small, this equation can be written as

$$I = I_{\max} \sin^2\left(\frac{\pi d}{\lambda L} y\right)$$

Under these conditions we have

$$\text{Constructive interference: } y = \left(m + \frac{1}{2}\right)\left(\frac{L\lambda}{d}\right) \qquad m = 0, \pm 1, \pm 2, \ldots$$

$$\text{Destructive interference: } y = m\left(\frac{L\lambda}{d}\right) \qquad m = 0, \pm 1, \pm 2, \ldots$$

The pattern of bright and dark fringes is reversed from what is observed in the double-slit experiment because of the change in phase associated with reflection from the mirror. There is also no "diffraction envelope" in the equation for the intensity of the interference pattern since there is no aperture.

10 Using Equation 12-4 and Equation 10-7.

12-5 Resolution

The ability of optical systems to distinguish between closely spaced sources of light is a direct result of the diffraction of the light from these sources. Consider the system shown in Figure 12.15, in which two independent sources of light pass through a slit and are then imaged on a screen. Since the two sources of light are independent they are incoherent, and thus no interference pattern will occur. However, the light from each source will diffract through the slit and produce a diffraction pattern on the screen. The two sources of light can thus be resolved (*i.e.*, distinguished as two separate sources) only if their corresponding diffraction patterns can be resolved/distinguished. This is known as ***Rayleigh's criterion***.

> *Rayleigh criterion*: when the central maximum of the diffraction pattern of one image falls on the first minimum of the diffraction pattern of another image, the images are said to be just resolved.

Using Rayleigh's criterion, we can determine the minimum angular separation between two sources of light for which the images of these sources are resolvable. According to Equation 12-3 this angle is

$$\sin\theta_{\min} = \frac{\lambda}{a}$$

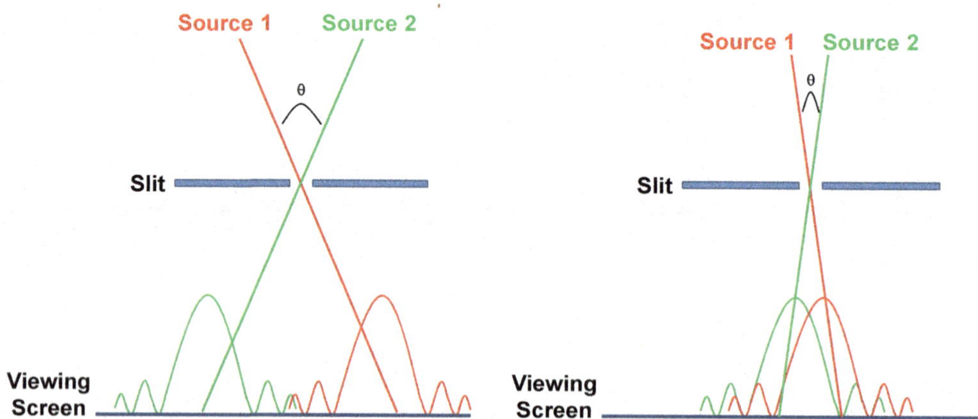

FIGURE 12.15: The light from two different sources makes two different diffraction patterns. If these diffraction patterns are separated from each other (left panel), the two sources of light can be resolved. If the diffraction patterns are not separated from each other (right panel), the two sources of light cannot be resolved.

In this equation, the variable a is the width of the slit. For most optical systems $a << \lambda$ and $\sin\theta_{min} << 1$. Under these conditions, the equation for resolution becomes

$$\theta_{min} = \frac{\lambda}{a} \tag{12-10}$$

For a circular "slit" of diameter D Equation 12-10 becomes

$$\theta_{min} = 1.22\frac{\lambda}{D} \tag{12-11}$$

Example 12-5

Problem: The ability of your eye to resolve different sources of light is limited by the diffraction of light through your pupil. Let's assume that the diameter of your pupil is 2 mm and you are looking at two sources of monochromatic light of wavelength 500 nm that are 50 cm away from you. What is the minimum separation distance of the two objects for your eye to resolve them as two separate objects?

Solution: Substitution into Equation 12-11 gives us

$$\theta_{min} = 1.22\left(\frac{500\times10^{-9}\,\text{m}}{2\times10^{-3}\,\text{m}}\right) \rightarrow \theta_{min} = 3.05\times10^{-4}\,\text{rad}$$

The corresponding minimum distance between the sources of light is the product of this angle, and the separation distance between the eye and the sources (*i.e.*, it is the corresponding arc length subtended by this angle). Let's denote this distance as d_{min}.

$$d_{min} = (50\,\text{cm})\theta_{min} \rightarrow d_{min} = (50\,\text{cm})(3.05\times10^{-4}\,\text{rad})$$

$$d_{min} = 152.5\,\mu\text{m}$$

Diffraction will also occur when light passes through a lens (an effect we ignored in Chapter 11). Thus, for any optical system, there will be a fundamental lower limit to the ability of the system to resolve different sources of light. This fundamental limit is a consequence of the diffraction of light which results from the wave nature of light. For this reason, different wavelengths of light are used to observe different structures. For example, x-rays are frequently used to resolve atomic structure since the wavelength of x-rays is on the same order as the diameter of an atom.

12-6 Photoelectric Effect and Wave/Particle Duality

The electric circuit in Figure 12.16 consists of a battery connected to two conducting plates separated from each other and enclosed in a vacuum chamber; the two conducting plates are effectively forming a capacitor with a large separation distance. Once all transient currents[11] have stopped, no additional current will flow in the circuit.

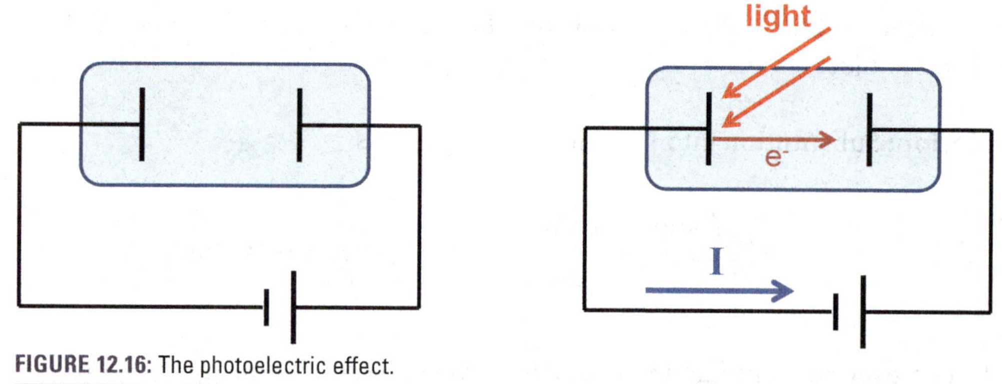

FIGURE 12.16: The photoelectric effect.

Interestingly, if the plates are exposed to light, then current will flow in the circuit. This occurs because the interaction of the light with the conducting plates ejects electrons from the plates, which can then flow as a current in the circuit. This process is known as the *photoelectric effect.*

At first, this observation was not considered especially interesting, since it was already known that electromagnetic waves could carry energy (Equation 10-16). Indeed, the current resulting from the photoelectric effect was directly proportional to the intensity and thus the average energy of the light (Equation 10-19). However,

11 Section 8-6.

additional characteristics of the photoelectric effect could not be so readily explained. For example, as shown in Figure 12.17, it was observed that the photoelectric effect would not occur unless the frequency of the light was above some minimum frequency, f_0. No matter how intense the light, there was no photoelectric effect (*i.e.*, no current in the circuit) unless the frequency of the light was above this minimum threshold frequency. Furthermore, this threshold frequency depended upon the metal used in the conducting plates, and thus is a characteristic of the metal itself.

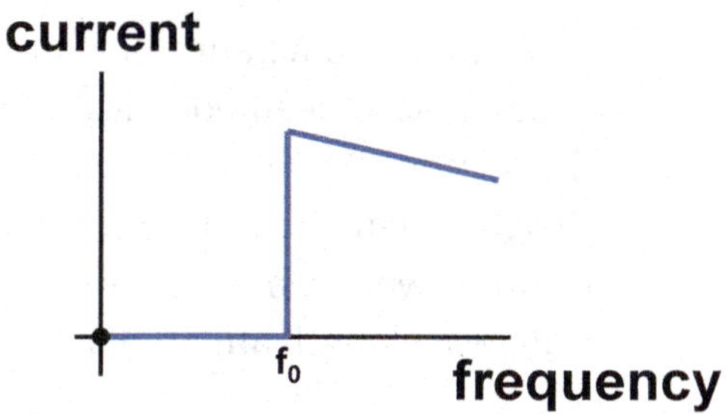

FIGURE 12.17: The photoelectric effect occurs only when the frequency of light is above a minimum threshold.

Albert Einstein's explanation of the photoelectric effect was that light was not a wave, but rather a collection of massless particles, called *photons*, which possess both kinetic energy and momentum. He postulated that the energy of a photon was quantized and proportional to the frequency of the light.

$$E = hf \quad \textbf{(12-12)}$$

The constant h in Equation 12-12 is Planck's constant, named for Max Planck.

$$h = 6.63 \times 10^{-34}\,\text{Js}$$

Einstein argued that photons are emitted or absorbed in an "all-or-nothing" process. In other words, the conducting plate in the photoelectric effect experiment must absorb an integer number of photons. Once absorbed by the conducting plate, all of the energy of each photon is transferred to a single electron.

If the energy of the photon is larger than the binding energy of the electron within the conducting plate, the absorption of energy of a single photon will be sufficient to "free" or "eject" the electron from the conducting plate and form a current. However, if this energy is less than the binding energy of the electron, the electron remains bound to the conducting plate. This is why there is a minimum frequency required for the photoelectric effect. Similarly, since the binding energy of electrons varies from metal to metal, this minimum frequency is characteristic of the material of the conducting plates.

A more intense light corresponds to more photons. Thus, the number of photons and ejected electrons should increase as the intensity of the light increases. It follows that the magnitude of the current should also be proportional to the intensity of light.

Although Einstein's particle model for light agreed with the photoelectric effect, it nevertheless seemed unsatisfactory since other phenomena involving light, such as diffraction (Section 12-3) were explained well treating light as a wave. Louis-Victor de Broglie reconciled these two competing models for light by postulating that everything (light as well as matter) was simultaneously both a particle *and* a wave. In his formulation, wavelength and linear momentum are related by

$$\lambda = \frac{h}{p} \tag{12-13}$$

The wavelength in Equation 12-13 is frequently referred to as the *de Broglie wavelength*. The variable h in Equation 12-13 is Planck's constant.

Example 12-6

Problem: What is the wavelength of a 0.145 kg baseball moving at 40 m/s (about 90 mph)?

Solution: Substitution into Equation 12-13 gives us

$$\lambda = \frac{6.63 \times 10^{-34}\,\mathrm{Js}}{(0.145\,\mathrm{kg})\left(40\,\frac{\mathrm{m}}{\mathrm{s}}\right)} \quad \rightarrow \quad \lambda = 1.14 \times 10^{-34}\,\mathrm{m}$$

If everything is both a particle and a wave, then why don't we notice the wave properties of everyday objects? The answer to this question requires us to reconsider the topic of resolution from Section 12-5. The ability to resolve different sources of light was found to be proportional to the wavelength of the light (Equation 12-10 and Equation 12-11), which was a direct result of the diffraction of the light. As shown in Figure 12.5, diffraction will occur only when the size of the slit used to diffract the wave is on the order of the wavelength of the light. Thus, in order to resolve the wave nature of the baseball in Example 12-6, we would need a diffraction slit on the order of 1×10^{-34} m, which is far too small for the baseball. Thus, although the baseball may actually be a wave, the wavelength of that wave is too small to detect. In contrast, high energy (and therefore high momentum) photons can have wavelengths on the order of 100 pm and can be used to resolve the structure of molecules.

If follows from Equation 12-13 that photons (*i.e.*, light) must also possess momentum[12]. The energy of a photon can also be expressed in terms of its momentum using Equation 12-14[13].

$$E = pc \tag{12-14}$$

When a photon is absorbed by an object (or a surface), the momentum and energy of the photon are transferred to that object (or surface). The change in momentum associated with photon absorption by a surface can further be described in terms of a pressure exerted by the photons on the surface using Newton's 2nd and 3rd laws[14]; this pressure is commonly referred to as *radiation pressure.*

$$F_{avg} = \frac{\Delta p}{\Delta t} \rightarrow F_{avg} = \frac{\Delta E}{c\Delta t}$$

The change in the energy of the object (or surface) is equal to the energy of the photon. This energy can be expressed in terms of the intensity (*i.e.*, the average Poynting vector[15]) and the area of the surface.

$$\Delta E = IA\Delta t \rightarrow F_{avg} = \frac{IA\Delta t}{c\Delta t} \rightarrow F_{avg} = \frac{IA}{c}$$

The radiation pressure on the surface is equal to the average force exerted on the surface divided by the area of the surface.

$$P = \frac{F_{avg}}{A} \rightarrow P = \frac{I}{c} \tag{12-15}$$

12 This is interesting since photons have no mass. You should feel inspired take additional physics classes to learn about this.

13 You should feel inspired to take additional physics classes to learn why Equation 12-14 is valid.

14 As you recall from thermodynamics, this is exactly how the pressure of a gas was determined from the collisions of the gas molecules with the container containing them.

15 See Equation 10-19. This is also similar to how we calculated the average power absorbed by an antenna in Equation 10-26.

If the photon reflects off the surface, rather than being absorbed by the surface, the radiation pressure would be

$$P = \frac{2I}{c} \tag{12-16}$$

Example 12-7

Problem: The intensity of electromagnetic radiation from the sun near the orbit of the Earth is approximately 1300 W/m². A 20,000 kg spacecraft will use a solar sail to accelerate away from the Earth. The sail is designed to perfectly reflect incident photons. What surface area must the sail have in order for the spacecraft to obtain an acceleration of 0.05 m/s²?

Solution: The pressure exerted by the electromagnetic radiation on the sail is determined using Equation 12-16.

$$P = \frac{2\left(1300\frac{\text{W}}{\text{m}^2}\right)}{3\times10^8\frac{\text{m}}{\text{s}}} \rightarrow P = 8.7\times10^{-6}\frac{\text{N}}{\text{m}^2}$$

The magnitude of the net force required for the spacecraft's acceleration is determined from Newton's 2nd law.

$$F_{net} = ma \rightarrow F_{net} = \left(2\times10^4\,\text{kg}\right)\left(0.05\frac{\text{m}}{\text{s}^2}\right) \rightarrow F_{net} = 1\times10^3\,\text{N}$$

The area required for the solar sail is therefore

$$A = \frac{F}{P} \rightarrow A = \frac{1\times10^3\,\text{N}}{8.7\times10^{-6}\frac{\text{N}}{\text{m}^2}} \rightarrow 1.2\times10^8\,\text{m}^2$$

If the solar sail is square each side would be 10.7 km long.

12-7 The Bohr Model for the Hydrogen Atom, Part V

According to Equation 12-13, the electron must also be both a wave and a particle. Indeed, in his model for the hydrogen atom, Bohr argued that the electron should be treated as a wave and that the only allowed "orbits" for the electron corresponded to standing waves of the electron. This requires that the circumference of an orbit must be an integer multiple of the wavelength of the electron.

$$2\pi R_{orbit} = m\lambda \quad m = 1, 2, \ldots$$

Substitution of Equation 12-13 then gives us

$$2\pi R_{orbit} = m\frac{h}{p} \quad \rightarrow \quad R_{orbit}\,p = m\frac{h}{2\pi}$$

We recognize the product of the radius of the orbit and the linear momentum of the electron as the angular momentum of the electron.

$$L = m\frac{h}{2\pi} \quad m = 1, 2, \ldots$$

The angular momentum of the electron is therefore quantized.

The accessible orbits of the electron can be calculated from the speed of the orbiting electron determined in Section 3-9.

$$v_{electron}^2 = \frac{e^2}{4\pi\varepsilon_0 R_{orbit} m_{electron}}$$

As derived previously,

$$R_{orbit}\,p = m\frac{h}{2\pi} \quad \rightarrow \quad R_{orbit} m_{electron} v_{electron} = m\frac{h}{2\pi} \quad \rightarrow \quad v_{electron} = m\left(\frac{h}{2\pi R_{orbit} m_{electron}}\right)$$

Combining these two equations for $v_{electron}$ gives us

$$m^2\left(\frac{h}{2\pi R_{orbit}m_{electron}}\right)^2 = \frac{e^2}{4\pi\varepsilon_0 R_{orbit}m_{electron}}$$

$$R_{orbit} = m^2\left(\frac{h^2\varepsilon_0}{\pi m_{electron}e^2}\right) \quad m = 1,2,\ldots$$

The corresponding accessible energies of the electron from Section 3-9 are thus

$$E = -\frac{1}{2}\left(\frac{e^2}{4\pi\varepsilon_0 R_{orbit}^2}\right) \rightarrow E = -\frac{1}{2}\left(\frac{e^2}{4\pi\varepsilon_0}\right)\left(\frac{\pi m_{electron}e^2}{h^2\varepsilon_0 m^2}\right)$$

$$E = -\left(\frac{e^4 m_{electron}}{8\varepsilon_0^2 h^2}\right)\left(\frac{1}{m^2}\right) \quad m = 1,2,\ldots$$

Substitution gives us

$$\frac{e^4 m_{electron}}{8\varepsilon_0^2 h^2} = \frac{\left(1.6\times10^{-19}\,\text{C}\right)^4\left(9.11\times10^{-31}\,\text{kg}\right)}{8\left(8.85\times10^{-12}\,\frac{\text{C}^2}{\text{J m}}\right)^2\left(6.63\times10^{-34}\,\text{Js}\right)^2} \rightarrow \frac{e^4 m_{electron}}{8\varepsilon_0^2 h^2} = 2.17\times10^{-18}\,\text{J}$$

Thus, the energies of the electron orbits are

$$E = -\left(\frac{2.17\times10^{-18}\,\text{J}}{m^2}\right) \quad m = 1,2,\ldots \tag{12-17}$$

The energies of the electron orbits decrease as the integer m increases. Based upon the 2nd law of thermodynamics, the largest population of atoms would have electrons in the $m = 1$ orbit. We can define this to be the *ground state* (or lowest energy) state for the hydrogen atom. The energy required to remove an electron from the ground state is the ionization energy (see Section 3-9) and from Equation 12-17 is equal to

$$E = 2.17\times10^{-18}\,\text{J}$$

This value for the ionization energy is exactly what is measured experimentally. Thus, by treating an electron as a wave, we are not only able to accurately predict the ionization energy of the hydrogen atom but are also able to avoid the problem of having the centripetal acceleration of the electron radiate away all the energy of atom (Section 9-7). Furthermore, this model of the hydrogen atom can also explain the light absorption properties of the atom[16].

Unfortunately, the Bohr model of the atom cannot be easily extended to more complicated atoms containing multiple protons and electrons. Nevertheless, the concepts of wave/particle duality, energy quantization, and momentum quantization did become the basis for the development of an entirely new branch of physics called quantum mechanics.

16 You should feel inspired to take additional physics courses to learn more about this.

Summary

- **Wave front:** a surface over which the phase of a wave is constant.
- **Huygens's principle:** each position on a wave front is the source of spherical waves (called wavelets) that spread out at the wave speed. At a later time, the shape of the wave front is the line tangent to all the wavelets.
- **Diffraction:** the bending of waves as they pass by an object or move through an aperture.
- **Coherent sources:** two sources of electromagnetic waves are coherent if the waves they emit have a zero or constant phase difference.
- The intensity of an electromagnetic wave with wavelength λ diffracting through a single slit with width a is

$$I = I_{max}\left(\frac{\sin\left(\frac{\pi a \sin\theta}{\lambda}\right)}{\frac{\pi a \sin\theta}{\lambda}}\right)^2$$

  This equation is also known as the Fraunhofer diffraction intensity equation.

- The intensity of the electromagnetic wave resulting from the diffraction of two coherent electromagnetic waves with wavelength λ through two slits with width a separated by a distance d is

$$I = I_{max}\cos^2\left(\frac{\pi d \sin\theta}{\lambda}\right)\left(\frac{\sin\left(\frac{\pi a \sin\theta}{\lambda}\right)}{\frac{\pi a \sin\theta}{\lambda}}\right)^2$$

- An electromagnetic wave undergoes a change in phase of π radians when reflected from a boundary leading to a material with higher index of refraction. There is no change in phase when the electromagnetic wave reflects from a boundary leading to a material with lower index of refraction

- **Rayleigh criterion:** when the central maximum of one image falls on the first minimum of another image, the images are said to be just resolved. The minimum angular separation between two sources of light is

$$\sin\theta_{\text{min}} = \frac{\lambda}{a}$$

In this equation the variable a is the width of the slit. For most optical systems, $a << \lambda$ and $\sin\theta_{\text{min}} << 1$. Under these conditions the equation for resolution becomes

$$\theta_{\text{min}} = \frac{\lambda}{a}$$

For a circular "slit" of diameter D Equation this equation is

$$\theta_{\text{min}} = 1.22\frac{\lambda}{D}$$

- **de Broglie wavelength:** The wavelength associated with the momentum of a particle.

$$\lambda = \frac{h}{p}$$

The variable h in this equation is Planck's constant.

- **Photon:** a massless particle of light carrying energy and momentum. The energy of a photon is

$$E = hf \quad E = pc$$

The variable h in this equation is Planck's constant.

- **Radiation pressure:** The pressure exerted on a surface by photons. If the photons are absorbed by the surface, the pressure is

$$P = \frac{I}{c}$$

If the photon reflects off the surface, the radiation pressure is

$$P = \frac{2I}{c}$$

The variable I in these equations is the intensity of the electromagnetic wave associated with the photons.

Problems

1. In a double-slit diffraction experiment, the separation between the slits is 4 times the width of each slit. How many bright interference fringes will be in the central peak of the diffraction envelope?

2. A diffraction grating has 2000 lines per mm. What is the second largest angle at which an intensity maximum appears on a distant viewing screen?

3. What are the requirements for constructive and destructive interference for light passing through a thin film with an index of refraction n? Assume that the index of refraction outside of the film is 1.

4. An astronaut is in orbit 450 km above the surface of the Earth. If the diameter of the astronaut's pupil is 5 mm, can she resolve the width (10 m) of the Great Wall of China? What is the minimum resolution of the astronaut's eye on the surface of the Earth? Assume that the wavelength of the light is 550 nm. Could a superhero using x-ray vision (0.1 nm wavelength light) resolve the Great Wall of China from this altitude?

5. What is the wavelength of an electron moving at 1×10^4 m/s?

6. A 2 kg object is at rest. How many photons of light with wavelength 500 nm must the object absorb to obtain a speed of 1 m/s? How many photons would be necessary if the wavelength of the light was 0.5 nm?

Appendix A

A-1 SI System of Units

All measurements in this book will use the International System of Units (known by its French acronym SI); this system is commonly referred to as the metric system. There are seven base quantities that constitute the SI units.

Quantity	Unit Name	Unit Symbol
Length	meter	m
Time	second	s
Mass	kilogram	kg
Temperature	kelvin	K
Amount of Substance	mole	mol
Electric Current	ampere	A
Luminous Intensity	candela	cd

There are several additional SI units that are derived from this basis set. A few of the more common are shown in the table below.

Quantity	Unit Name	Unit Symbol	Derivation
Energy	joule	J	$1\,\text{J} = 1\frac{\text{kg}\,\text{m}^2}{\text{s}^2}$
Power	watt	W	$1\,\text{W} = 1\frac{\text{kg}\,\text{m}^2}{\text{s}^3}$
Force	newton	N	$1\,\text{N} = 1\frac{\text{kg}\,\text{m}}{\text{s}^2}$
Frequency	hertz	Hz	$1\,\text{Hz} = 1\frac{1}{\text{s}}$
Electric Charge	coulomb	C	$1\,\text{C} = 1\,\text{As}$
Electric Potential	volt	V	$1\,\text{V} = 1\frac{\text{kg}\,\text{m}^2}{\text{As}^3} = 1\frac{\text{J}}{\text{C}}$
Capacitance	farad	F	$1\,\text{F} = 1\frac{\text{A}^2\text{s}^4}{\text{kg}\,\text{m}^2} = 1\frac{\text{C}}{\text{V}}$
Magnetic Field Intensity	tesla	T	$1\,\text{T} = 1\frac{\text{kg}}{\text{As}^2} = 1\frac{\text{Ns}}{\text{Cm}} = 1\frac{\text{Vs}}{\text{m}^2}$
	gauss[1]	G	$1\,\text{G} = 10^{-4}\,\text{T}$
Magnetic Flux	weber	Wb	$1\,\text{Wb} = 1\,\text{Tm}^2 = 1\,\text{Vs} = 1\frac{\text{J}}{\text{A}}$
Resistance	ohm	Ω	$1\,\Omega = 1\frac{\text{V}}{\text{A}} = 1\frac{\text{Js}}{\text{C}^2}$

1 Gauss is not an SI unit, but it is nevertheless commonly used.

Quantity	Unit Name	Unit Symbol	Derivation
Inductance	henry	H	$1\text{H} = 1\frac{\text{Vs}}{\text{A}}$
Pressure	pascal	Pa	$1\text{Pa} = 1\frac{\text{N}}{\text{m}^2}$

The following prefixes can be used when describing very large or very small quantities.

Prefix	Symbol	Factor
tera	T	10^{12}
giga	G	10^{9}
mega	M	10^{6}
kilo	K	10^{3}
hecto	H	10^{2}
deca	da	10^{1}
deci	D	10^{-1}
centi	C	10^{-2}
milli	M	10^{-3}
micro	μ	10^{-6}
nano	N	10^{-9}
pico	P	10^{-12}

Thus, $9.5\,\text{ns} = 9.5\times10^{-9}\,\text{s}$ and $2.6\,\text{MW} = 2.6\times10^{6}\,\text{W}$.

A-2 Physical Constants

Quantity	Variable	Accepted Value
Gravitational Constant	G	$6.67\times10^{-11}\,\frac{\text{Jm}}{\text{kg}^2}$
Fundamental Electric Charge	e	$1.6\times10^{-23}\text{C}$
Vacuum Permittivity	ε_0	$8.85\times10^{-12}\,\frac{\text{C}^2}{\text{Jm}}$

Quantity	Variable	Accepted Value
Vacuum Permeability	μ_0	$4\pi\times10^{-7}\,\frac{\text{kgm}}{\text{C}^2}$
Speed of Light in Vacuum	c	$3\times10^{8}\,\frac{\text{m}}{\text{s}}$
Avogadro Constant	N_A	$6.02\times10^{-23}\,\frac{1}{\text{mol}}$
Universal Gas Constant	R	$8.314\,\frac{\text{J}}{\text{mol K}}$
Boltzmann Constant	k_B	$1.38\times10^{-23}\,\frac{\text{J}}{\text{K}}$
Mass of the Earth	M_E	5.98×10^{24} kg
Radius of the Earth	R_E	6.37×10^{6} m
Mass of the Sun	M_S	1.99×10^{30} kg

A-3 Useful Mathematics

Solution for a quadratic equation

$$ax^2+bx+c=0 \quad \rightarrow \quad x=\frac{-b\pm\sqrt{b^2-4ac}}{2a}$$

Trigonometry

c, a, θ, b		
	$\sin\theta=\frac{a}{c}$	$\sin(-\theta)=-\sin\theta$
	$\cos\theta=\frac{b}{c}$	$\cos(-\theta)=\cos\theta$
	$\tan\theta=\frac{a}{b}$	$\tan(-\theta)=-\tan\theta$

$$\cos(\theta \pm \phi) = \cos(\theta)\cos(\phi) \mp \sin(\theta)\sin(\phi) \tag{A-1}$$

$$\sin(\theta \pm \phi) = \sin(\theta)\cos(\phi) \pm \cos(\theta)\sin(\phi) \tag{A-2}$$

$$\sin 2\theta = 2\sin\theta\cos\theta \tag{A-3}$$

$$\cos 2\theta = \cos^2\theta - \sin^2\theta \quad \rightarrow \quad \cos 2\theta = 2\cos^2\theta - 1 \tag{A-4}$$

$$\cos(\theta) + \cos(\phi) = 2\cos\left(\frac{\theta - \phi}{2}\right)\cos\left(\frac{\theta + \phi}{2}\right) \tag{A-5}$$

Calculus

$$\frac{d}{dx}x^n = nx^{n-1} \qquad \frac{d}{dx}e^{ax} = ae^{ax}$$

$$\frac{d}{dx}\sin(ax) = a\cos(ax) \qquad \frac{d}{dx}\cos(ax) = -a\sin(ax)$$

$$\frac{d}{dx}\tan(ax) = \frac{a}{\cos^2(ax)} \qquad \frac{dy}{dx} = \frac{dy}{du}\frac{du}{dx}$$

$$\frac{d}{dx}\ln(ax) = \frac{1}{x} \qquad \frac{d}{dx}a^x = a^x \ln a$$

$$\frac{d}{dx}\left(f(x)g(x)\right) = \left(\frac{df(x)}{dx}\right)g(x) + \left(\frac{dg(x)}{dx}\right)f(x)$$

$$\int \frac{dx}{\left(x^2 + y^2\right)^{\frac{1}{2}}} = \ln\left(x + \left(x^2 + y^2\right)^{\frac{1}{2}}\right)$$

$$\int \frac{dx}{\left(x^2 + y^2\right)^{\frac{3}{2}}} = \frac{x}{y^2\left(x^2 + y^2\right)^{\frac{1}{2}}}$$

$$\int \ln\left(\left(x^2 + y^2\right)^{\frac{1}{2}} - x\right)dx = \left(x^2 + y^2\right) + x\ln\left(\left(x^2 + y^2\right)^{\frac{1}{2}} - x\right)$$

Taylor series expansion

$$f(x) = f(a) + \frac{1}{1!}\left(\left.\frac{df}{dx}\right|_{x=a}\right)(x-a) + \frac{1}{2!}\left(\left.\frac{d^2 f}{dx^2}\right|_{x=a}\right)(x-a)^2 + \frac{1}{3!}\left(\left.\frac{d^3 f}{dx^3}\right|_{x=a}\right)(x-a)^3 + \ldots$$

$$\sin(\theta) = \theta - \frac{\theta^3}{3!} + \frac{\theta^5}{5!} - \frac{\theta^7}{7!} + \ldots$$

$$\cos(\theta) = 1 - \frac{\theta^2}{2!} + \frac{\theta^4}{4!} - \frac{\theta^6}{6!} + \ldots$$

$$e^x = 1 + x + \frac{x^2}{2!} + \frac{x^3}{3!} + \ldots$$

Binomial expansion

$$(1+x)^n = 1 + nx + \frac{n(n-1)}{2!}x^2 + \frac{n(n-1)(n-2)}{3!}x^3 + \ldots$$

$$(1+x)^n \approx 1 + nx \qquad x << 1$$

Geometric series expansion

$$a + ar + ar^2 + \ldots + ar^n = a\frac{1-r^{n+1}}{1-r}$$

A-4 Legendre Polynomials

The Legendre polynomials, $P_n(x)$, are solutions of the Legendre differential equation.

$$\frac{d}{dx}\left[(1-x)^2 \frac{d}{dx} P_n(x)\right] + n(n+1)P_n(x) = 0$$

Why this particular equation is useful or how the Legendre polynomials are determined is beyond the scope of this book. What is relevant to this book, however, is that the Legendre polynomials can be related to an expression that we encounter frequently.

$$\frac{1}{\left(1-2xy+x^2\right)^{\frac{1}{2}}}=\sum_{n=0}^{\infty}P_n\left(y\right)x^n \tag{A-6}$$

Consider the system shown in Figure A.1, in which the distance between a point particle and a position P where a property of that particle (its gravitational potential, *e.g.*) is measured is denoted by the variable *R*, the distance between the origin and the point particle is denoted by the variable *a*, and the distance between the origin and the position P is denoted by the variable *r*.

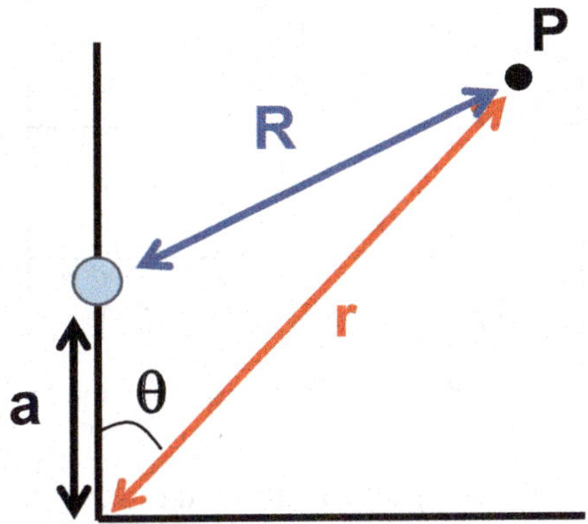

FIGURE A.1: Geometry for which Legendre polynomials are helpful.

Using trigonometry, we can relate *R* to *a*, *r*, and the angle θ as

$$R^2=r^2+a^2-2ra\cos\theta$$

$$\frac{1}{R}=\frac{1}{\left(r^2+a^2-2ra\cos\theta\right)^{\frac{1}{2}}} \rightarrow \frac{1}{R}=\frac{1}{r\left(1+\left(\frac{a}{r}\right)^2-2\frac{a}{r}\cos\theta\right)^{\frac{1}{2}}}$$

Upon comparison to Equation A-6, we see that

$$\frac{1}{R} = \frac{1}{r}\sum_{n=0}^{\infty}\left(\frac{a}{r}\right)^n P_n(\cos\theta) \rightarrow \frac{r}{R} = \sum_{n=0}^{\infty}\left(\frac{a}{r}\right)^n P_n(\cos\theta)$$

Legendre polynomials are thus the coefficients of a Taylor series expansion. A list of the first few Legendre polynomials is shown in Table A-1.

TABLE A-1. Legendre Polynomials

N	$P_n(x)$
0	1
1	x
2	$\frac{1}{2}(3x^2-1)$
3	$\frac{1}{2}(5x^3-3x)$
4	$\frac{1}{8}(35x^4-30x^2+3)$

A few useful properties of Legendre polynomials are

$$P_n(-y) = (-1)^n P_n(y) \tag{A-7}$$

$$(n+1)P_{n+1}(y) = (2n+1)yP_n(y) - nP_{n-1}(y) \tag{A-8}$$

Appendix B

B-1 1-Dimensional Vectors

When expressing a 1-dimensional vector, a positive (+) or negative (−) sign is commonly used to denote the direction of the vector with respect to the origin of the corresponding coordinate axis (Figure B.1).

It is worth noting that the positive (+) sign is frequently omitted when writing 1-dimensional vectors.

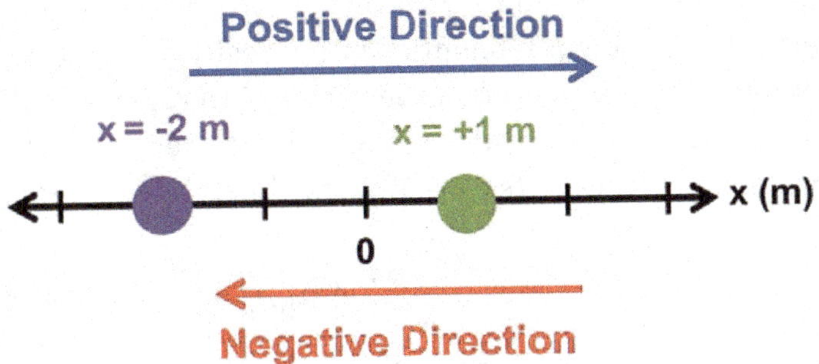

FIGURE B.1: Coordinate system for using describing the position of an object using 1-dimensional vectors.

B-2 General Descriptions of Vectors

Vector quantities are usually labeled with an arrow over their symbol, but this is sometimes neglected for 1-dimensional vectors. Using this notation, we can write the vector $\vec{A}$ with magnitude A and direction along $\hat{a}$ as

$$\vec{A} = A\hat{a} \quad \rightarrow \quad \hat{a} = \frac{\vec{A}}{A}$$

In this example, $\hat{a}$ denotes a *unit vector*, which points in the direction of $\vec{A}$ and has a magnitude of one. Unit vectors are denoted by the ^ symbol and are commonly used to express the components of a vector along the coordinate axis relevant to a particular problem. This is mathematically permissible since they have a magnitude of 1; in other words, they specify direction only.

B-3 Components and Decomposition

Working with vectors is made easier by expressing the vector in terms of its projections or components along a set of unit vectors (*i.e.*, its components along the axes of a coordinate system). We refer to this process as *decomposing* the vector. For example, in a standard Cartesian coordinate system, the vector $\vec{A}$ can also be written (or decomposed) as

$$\vec{A} = A_x\hat{x} + A_y\hat{y} + A_z\hat{z}$$

In this example, $\hat{x}$, $\hat{y}$, and $\hat{z}$ are the unit vectors pointing along the Cartesian x, y, and z axes, respectively, and A_x, A_y, and A_z are the components of the vector along those directions. These components can be positive, negative, or zero depending upon the orientation (*i.e.*, direction) of the vector with regards to the fixed Cartesian coordinates. When written this way, the magnitude of the vector $\vec{A}$, denoted by A or $\left\|\vec{A}\right\|$, can be expressed in terms of its components along the coordinate axes.

$$A = \left\|\vec{A}\right\| = \left(A_x^2 + A_y^2 + A_z^2\right)^{\frac{1}{2}} \qquad \textbf{(B-1)}$$

It is worth noting that although the components of a vector can be either positive or negative, but the magnitude of a vector is always positive (see Figure B.2).

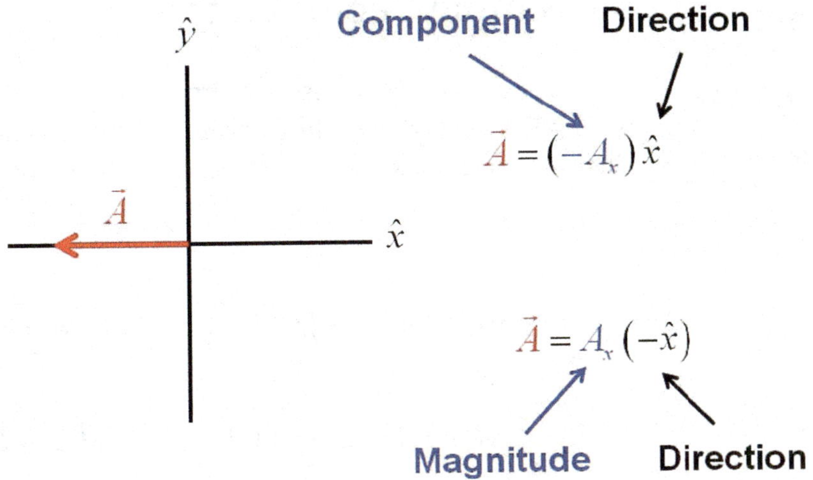

FIGURE B.2: The difference between describing a vector in terms of its components and magnitudes.

The components of a vector can also be used to determine the angles between the vector and the coordinate axes. Consider the simple case of the 2-dimensional vector shown in Figure B.3.

The following trigonometric identities apply to the vector shown in Figure B.3.

$$\sin\theta = \frac{A_y}{A} = \frac{A_y}{\|\vec{A}\|} = \frac{A_y}{\sqrt{A_x^2 + A_y^2}}$$

$$\cos\theta = \frac{A_x}{A} = \frac{A_x}{\|\vec{A}\|} = \frac{A_x}{\sqrt{A_x^2 + A_y^2}}$$

$$\tan\theta = \frac{A_y}{A_x}$$

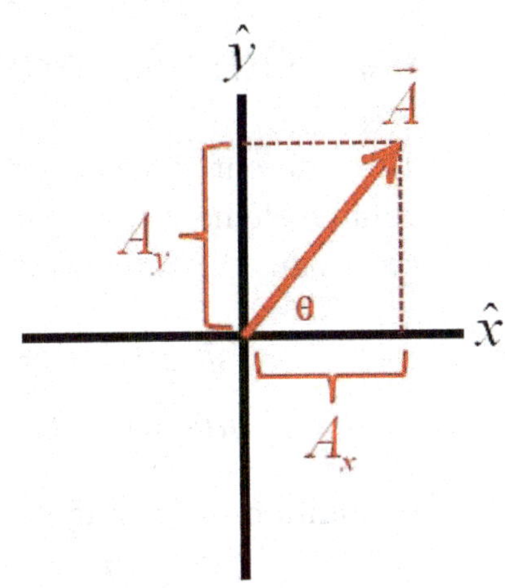

FIGURE B.3: A 2-dimensional vector with its components along both coordinate axes.

In addition to arrows, we will also use specific notation (Figure B.4) to pictorially represent vectors orientated perpendicular to the page.

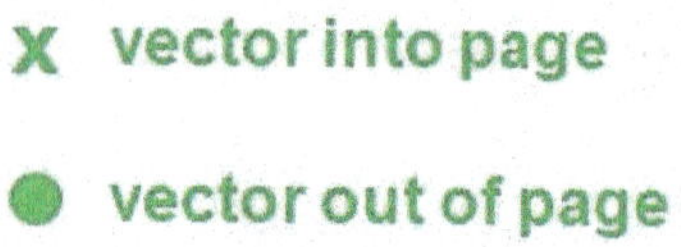

X vector into page

● vector out of page

FIGURE B.4: Pictorial representation of the direction of vectors orientated perpendicular to the page.

B-4 Simple Vector Mathematics

Expressing a vector in terms of its components (*i.e.*, decomposing the vector) simplifies vector arithmetic and calculus since all operations are carried out component by component.

Addition $$\vec{A}+\vec{B}=\left(A_x+B_x\right)\hat{x}+\left(A_y+B_y\right)\hat{y}+\left(A_z+B_z\right)\hat{z}$$

Subtraction $$\vec{A}-\vec{B}=\left(A_x-B_x\right)\hat{x}+\left(A_y-B_y\right)\hat{y}+\left(A_z-B_z\right)\hat{z}$$

Multiplication by a scalar $$c\vec{A}=\left(cA_x\right)\hat{x}+\left(cA_y\right)\hat{y}+\left(cA_z\right)\hat{z}$$

Differentiation $$\frac{d}{dt}\vec{A}=\left(\frac{d}{dt}A_x\right)\hat{x}+\left(\frac{d}{dt}A_y\right)\hat{y}+\left(\frac{d}{dt}A_z\right)\hat{z}$$

B-5 Other Coordinate Systems

The differential displacement vector and differential volume element in rectangular coordinates (*i.e.*, the Cartesian *x*, *y*, and *z* axes) are

$$d\vec{r}=\left(dx\right)\hat{x}+\left(dy\right)\hat{y}+\left(dz\right)\hat{z}$$
$$dV=\left(dx\right)\left(dy\right)\left(dz\right)$$

In addition to rectangular coordinates, we will also occasionally use cylindrical and spherical coordinate systems. In a cylindrical coordinate system, the location of a point is specified by the quantities ρ, φ, and *z*, as shown in Figure B.5.

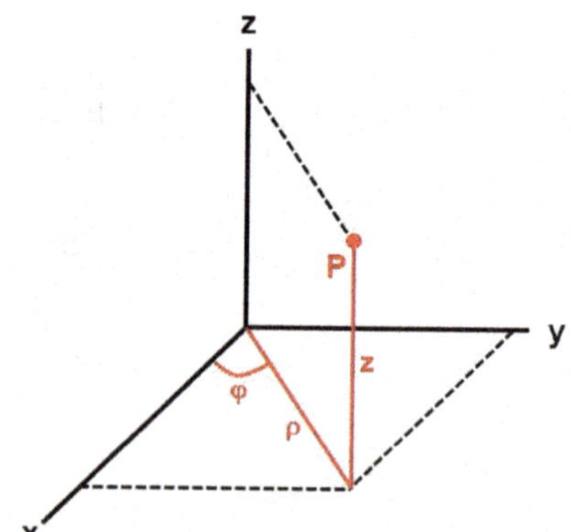

FIGURE B.5: Cylindrical coordinates.

The cylindrical coordinates are related to the rectangular coordinates through the following equations:

$$x=\rho\cos\varphi \qquad y=\rho\sin\varphi \qquad z=z$$

$$\rho=\left(x^2+y^2\right)^{\frac{1}{2}} \qquad \tan\varphi=\frac{y}{x}$$

The cylindrical coordinate unit vectors are related to the rectangular coordinate unit vectors by the following equations:

$$\hat{\rho} = (\cos\varphi)\hat{x} + (\sin\varphi)\hat{y} \qquad \hat{\varphi} = (-\sin\varphi)\hat{x} + (\cos\varphi)\hat{y}$$

$$\hat{x} = (\cos\varphi)\hat{\rho} - (\sin\varphi)\hat{\varphi} \qquad \hat{y} = (\sin\varphi)\hat{\rho} + (\cos\varphi)\hat{\varphi}$$

It follows that the position vector expressed in cylindrical coordinates is

$$\vec{r} = (\rho)\hat{\rho} + (z)\hat{z}$$

$$\hat{r} = \left(\frac{\rho}{\left(\rho^2 + z^2\right)^{\frac{1}{2}}}\right)\hat{\rho} + \left(\frac{z}{\left(\rho^2 + z^2\right)^{\frac{1}{2}}}\right)\hat{z}$$

The differential displacement vector and differential volume element in cylindrical coordinates are

$$d\vec{r} = (d\rho)\hat{\rho} + (\rho d\varphi)\hat{\varphi} + (dz)\hat{z}$$
$$dV = \rho(d\rho)(d\varphi)(dz)$$

The differential volume of a cylindrical shell of radius ρ and length z is thus

$$dV_{shell} = \int_0^{2\pi} \rho(d\rho)(d\varphi)(dz)$$

$$dV_{shell} = \rho(d\rho)(dz)\left(\int_0^{2\pi} d\varphi\right) \quad \rightarrow \quad dV_{shell} = 2\pi\rho(d\rho)(dz)$$

In a spherical coordinate system, the location of a point is specified by the quantities r, θ, and φ, as shown in Figure B.6. The spherical coordinates are related to the rectangular coordinates through the following equations:

$$x = r\sin\theta\cos\varphi \qquad y = r\sin\theta\sin\varphi \qquad z = r\cos\theta$$

$$r = \left(x^2 + y^2 + z^2\right)^{\frac{1}{2}} \qquad \tan\theta = \frac{\left(x^2 + y^2\right)^{\frac{1}{2}}}{z} \qquad \tan\varphi = \frac{y}{x}$$

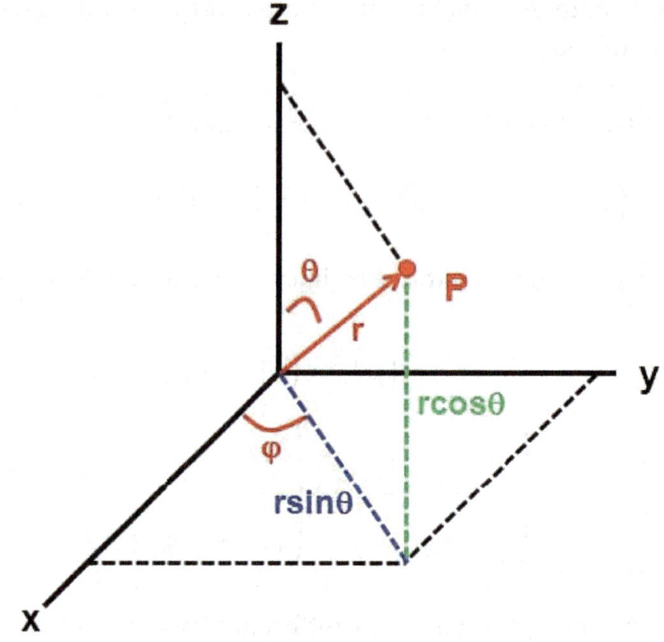

FIGURE B.6: Spherical coordinates.

The spherical coordinate unit vectors are related to the rectangular coordinate unit vectors by the following equations:

$$\hat{r} = (\sin\theta\cos\varphi)\hat{x} + (\sin\theta\sin\varphi)\hat{y} + (\cos\theta)\hat{z}$$
$$\hat{\theta} = (\cos\theta\cos\varphi)\hat{x} + (\cos\theta\sin\varphi)\hat{y} - (\sin\theta)\hat{z}$$
$$\hat{\varphi} = (-\sin\varphi)\hat{x} + (\cos\varphi)\hat{y}$$

$$\hat{x} = (\sin\theta\cos\varphi)\hat{r} + (\cos\theta\cos\varphi)\hat{\theta} - (\sin\varphi)\hat{\varphi}$$
$$\hat{y} = (\sin\theta\sin\varphi)\hat{r} + (\cos\theta\sin\varphi)\hat{\theta} + (\cos\varphi)\hat{\varphi}$$
$$\hat{z} = (\cos\theta)\hat{r} - (\sin\theta)\hat{\theta}$$

The differential displacement vector and differential volume element in spherical coordinates are

$$d\vec{r} = (dr)\hat{r} + (rd\theta)\hat{\theta} + (r\sin\theta d\varphi)\hat{\varphi}$$
$$dV = r^2\sin\theta(dr)(d\theta)(d\varphi)$$

The differential volume of a spherical shell of radius r is thus

$$dV_{shell} = \int_0^{2\pi}\int_0^{\pi} r^2 \sin\theta (dr)(d\theta)(d\varphi) \quad \rightarrow \quad dV_{shell} = (r^2 dr)\left(\int_0^{2\pi} d\varphi\right)\left(\int_0^{\pi} \sin\theta\, d\theta\right)$$

$$dV_{shell} = (r^2 dr)(2\pi)\left(-\cos\theta\Big|_0^{\pi}\right)$$

$$dV_{shell} = (2\pi r^2 dr)(-(-1-1)) \quad \rightarrow \quad dV_{shell} = 4\pi r^2 dr$$

B-6 The Dot Product

The dot product (also called scalar product or the inner product) of two vectors is the related to the magnitude of the projection of one vector onto the other; the dot product is thus a scalar quantity. The expression for the dot product of the two vectors $\vec{A}$ and $\vec{B}$ is

$$\vec{A} \bullet \vec{B} = AB\cos\theta \qquad \textbf{(B-2)}$$

The angle θ in Equation B-2 is the angle between the two vectors. The dot product is a commutative operation so $\vec{A} \bullet \vec{B} = \vec{B} \bullet \vec{A}$. Because the unit vectors of the standard rectangular coordinate system are orthonormal, we have the following identities:

$$\hat{x} \bullet \hat{x} = 1 \quad \hat{y} \bullet \hat{y} = 1 \quad \hat{z} \bullet \hat{z} = 1$$

$$\hat{x} \bullet \hat{y} = 0 \quad \hat{x} \bullet \hat{z} = 0 \quad \hat{y} \bullet \hat{z} = 0$$

The dot product of a vector with itself is the square of the magnitude of that vector.

$$\vec{A} \bullet \vec{A} = (A_x\hat{x} + A_y\hat{y} + A_z\hat{z}) \bullet (A_x\hat{x} + A_y\hat{y} + A_z\hat{z})$$

$$\vec{A} \bullet \vec{A} = A_x^2(\hat{x} \bullet \hat{x}) + A_y^2(\hat{y} \bullet \hat{y}) + A_z^2(\hat{z} \bullet \hat{z}) + A_xA_y(\hat{x} \bullet \hat{y}) + A_xA_z(\hat{x} \bullet \hat{z}) +$$
$$A_yA_x(\hat{y} \bullet \hat{x}) + A_yA_z(\hat{y} \bullet \hat{z}) + A_zA_x(\hat{z} \bullet \hat{x}) + A_zA_y(\hat{z} \bullet \hat{y})$$

$$\vec{A} \bullet \vec{A} = A_x^2 + A_y^2 + A_z^2 = A^2$$

Similarly,

$$\vec{A} \bullet \vec{B} = A_x B_x + A_y B_y + A_z B_z \quad \textbf{(B-3)}$$

For the unit vectors in cylindrical and spherical coordinates we have

$$\hat{\rho} \bullet \hat{\rho} = 1 \quad \hat{\varphi} \bullet \hat{\varphi} = 1 \quad \hat{z} \bullet \hat{z} = 1$$

$$\hat{\rho} \bullet \hat{\varphi} = 0 \quad \hat{\rho} \bullet \hat{z} = 0 \quad \hat{\varphi} \bullet \hat{z} = 0$$

$$\hat{r} \bullet \hat{r} = 1 \quad \hat{\theta} \bullet \hat{\theta} = 1 \quad \hat{\varphi} \bullet \hat{\varphi} = 1$$

$$\hat{r} \bullet \hat{\theta} = 0 \quad \hat{r} \bullet \hat{\varphi} = 0 \quad \hat{\theta} \bullet \hat{\varphi} = 0$$

B-7 The Cross Product

The cross product (or vector product) of two vectors $\vec{A}$ and $\vec{B}$ results in a new vector whose magnitude is the area of a parallelogram with $\vec{A}$ and $\vec{B}$ as its sides. The equation for the magnitude of this vector is

$$\left\| \vec{A} \times \vec{B} \right\| = AB \sin\theta \quad \textbf{(B-4)}$$

The angle θ in Equation B-4 is the angle between the two vectors $\vec{A}$ and $\vec{B}$. By convention the angle θ is always drawn *counterclockwise* from $\vec{A}$ and $\vec{B}$, as shown in Figure B.7.

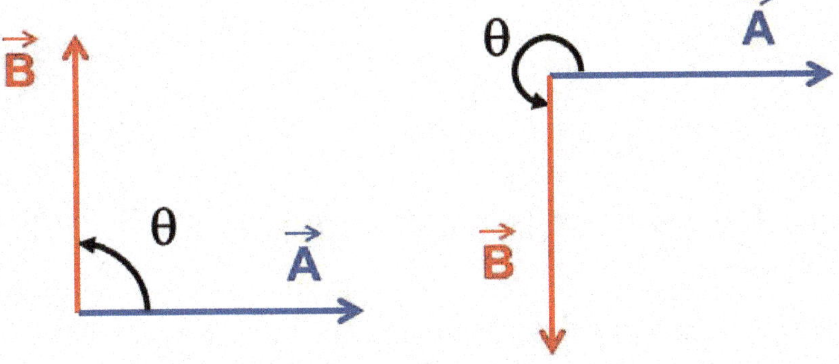

FIGURE B.7: The definition of the angle θ between the vectors $\vec{A}$ and $\vec{B}$ in the vector cross product $\vec{A} \times \vec{B}$.

The direction of the vector resulting from the cross product of $\vec{A}$ and $\vec{B}$ is determined using the right-hand rule. Point the figures of your right hand in the direction of the first vector in the cross product ($\vec{A}$), bend them to point in the direction of the second vector in the cross product ($\vec{B}$), and your thumb will then point in the direction of the resultant vector ($\vec{A} \times \vec{B}$). Two examples of vector cross product are shown in Figure B.8.

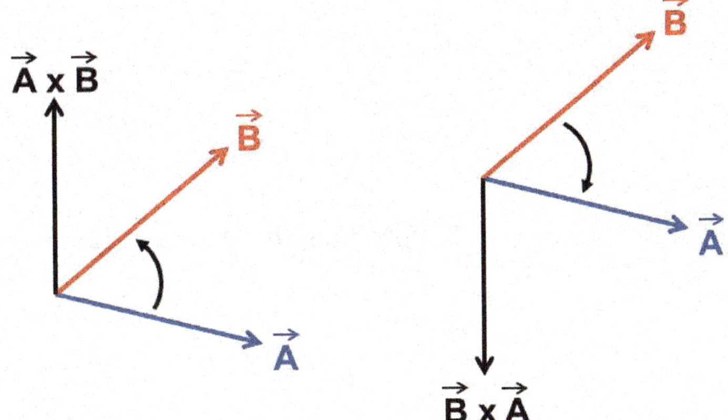

FIGURE B.8: The direction of the vector cross product determined using the right hand rule.

Alternatively, for 3-dimensional vectors, the cross product of $\vec{A}$ and $\vec{B}$ can be calculated using a simple determinant.

$$\vec{A}\times\vec{B}=\begin{vmatrix}\hat{x} & \hat{y} & \hat{z}\\ A_x & A_y & A_z\\ B_x & B_y & B_z\end{vmatrix}$$

$$\vec{A}\times\vec{B}=\left(A_yB_z-A_zB_y\right)\hat{x}+\left(A_zB_x-A_xB_z\right)\hat{y}+\left(A_xB_y-A_yB_x\right)\hat{z}$$

Incidentally, we define a *right-handed coordinate system* in terms of the cross product of the associated unit vectors. For a right-handed rectangular system, we have

$$\hat{x}\times\hat{y}=\hat{z}\quad \hat{y}\times\hat{z}=\hat{x}\quad \hat{z}\times\hat{x}=\hat{y}$$

$$\hat{x}\times\hat{x}=0\quad \hat{y}\times\hat{y}=0\quad \hat{z}\times\hat{z}=0$$

In cylindrical and spherical coordinates, we have

$$\hat{\rho}\times\hat{\varphi}=\hat{z}\quad \hat{\varphi}\times\hat{z}=\hat{\rho}\quad \hat{z}\times\hat{\rho}=\hat{\varphi}$$

$$\hat{\rho}\times\hat{\rho}=0\quad \hat{\varphi}\times\hat{\varphi}=0\quad \hat{z}\times\hat{z}=0$$

$$\hat{r}\times\hat{\theta}=\hat{\varphi}\quad \hat{\theta}\times\hat{\varphi}=\hat{r}\quad \hat{\varphi}\times\hat{r}=\hat{\theta}$$

$$\hat{r}\times\hat{r}=0\quad \hat{\theta}\times\hat{\theta}=0\quad \hat{\varphi}\times\hat{\varphi}=0$$

It's important to note that unlike the dot product, the cross product is not commutative (see Figure B.8).

$$\vec{A}\times\vec{B}=-\vec{B}\times\vec{A}$$

B-8 The Gradient of a Scalar Field

A *scalar field* is defined to be an assignment of a scalar to each point in a region of space. Suppose we have a scalar field f that is a function of position so that we can write

$$f=f\left(x,y,z\right)$$

The function f might be the temperature at a particular point in a room, for example. The infinitesimal change in f associated with an infinitesimal change in position is thus

$$df = \left(\frac{\partial f}{\partial x}\right)dx + \left(\frac{\partial f}{\partial y}\right)dy + \left(\frac{\partial f}{\partial z}\right)dz$$

We can write this expression as

$$df = \left[\left(\frac{\partial f}{\partial x}\right)\hat{x} + \left(\frac{\partial f}{\partial y}\right)\hat{y} + \left(\frac{\partial f}{\partial z}\right)\hat{z}\right] \cdot \left[(dx)\hat{x} + (dy)\hat{y} + (dz)\hat{z}\right]$$

$$df = \nabla f \bullet d\vec{r}$$

In this equation, we have introduced the *gradient* of f as the quantity

$$\nabla f = \left(\frac{\partial f}{\partial x}\right)\hat{x} + \left(\frac{\partial f}{\partial y}\right)\hat{y} + \left(\frac{\partial f}{\partial z}\right)\hat{z} \qquad \textbf{(B-5)}$$

The infinitesimal change in f is then simply equal to the dot product of the gradient of f and the infinitesimal change in position $(d\vec{r})$[1]. The direction of the gradient is the direction in which the scalar has its maximum rate of change and the magnitude of the gradient is that maximum rate of change.

In cylindrical coordinates the gradient is

$$\nabla f = \left(\frac{\partial f}{\partial \rho}\right)\hat{\rho} + \left(\frac{1}{\rho}\frac{\partial f}{\partial \varphi}\right)\hat{\varphi} + \left(\frac{\partial f}{\partial z}\right)\hat{z} \qquad \textbf{(B-6)}$$

In spherical coordinates the gradient is

$$\nabla f = \left(\frac{\partial f}{\partial r}\right)\hat{r} + \left(\frac{1}{r}\frac{\partial f}{\partial \theta}\right)\hat{\theta} + \left(\frac{1}{r\sin\theta}\frac{\partial f}{\partial \varphi}\right)\hat{\varphi} \qquad \textbf{(B-7)}$$

1 See Section B-5.

B-9 The Divergence and Curl

A *vector field* is defined to be an assignment of a vector to each point in a region of space. For example, the wind speed in different towns and cities in a state form a vector field of wind velocity for that state. The components of the vector in a vector field are often functions of the position of the measurement of the vector in the region of space. An example of such a two-dimensional vector field is

$$\vec{A} = (2x)\hat{x} + (3xy^2)\hat{y}$$

The x-axis and y-axis components of this vector field are functions of the position of the measurement along the x-axis and y-axis. The *divergence* of a vector field is equal to

$$\nabla \bullet \vec{A} = \frac{\partial A_x}{\partial x} + \frac{\partial A_y}{\partial y} + \frac{\partial A_z}{\partial z} \tag{B-8}$$

The divergence of a vector field is a measure of rate of change of a vector field. If the magnitude of the vector field is constant along a line in the same direction as the direction of the vector field, then the divergence of the vector field is zero. On the other hand, if the magnitude of the vector field varies along a line in the same direction as the direction of the vector field, then the divergence of the vector field is non-zero.

The *curl* of a 2-dimensional vector field is equal to

$$\nabla \times \vec{A} = \begin{vmatrix} \hat{x} & \hat{y} & \hat{z} \\ \frac{\partial}{\partial x} & \frac{\partial}{\partial y} & \frac{\partial}{\partial z} \\ A_x & A_y & A_z \end{vmatrix}$$

$$\nabla \times \vec{A} = \left(\frac{\partial A_z}{\partial y} - \frac{\partial A_y}{\partial z} \right)\hat{x} + \left(\frac{\partial A_x}{\partial z} - \frac{\partial A_z}{\partial x} \right)\hat{y} + \left(\frac{\partial A_y}{\partial x} - \frac{\partial A_x}{\partial y} \right)\hat{z} \tag{B-9}$$

The curl of a vector field is a measure of the "rotation" of that vector field. If the magnitude of the vector field is constant along a line perpendicular to the direction of the vector field, then the curl of the vector field is zero. If the magnitude of the vector field varies along a line perpendicular to the direction of the vector field, then the curl of the vector field is non-zero. These properties of the divergence and the curl are demonstrated pictorially in Figure B.9.

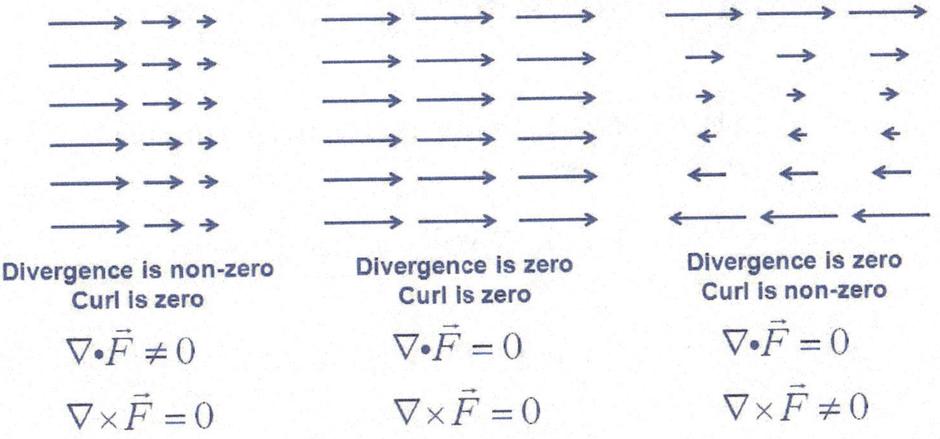

FIGURE B.9: Pictorial representations of the divergence and the curl.

In cylindrical and spherical coordinates, the divergence and curl are

$$\nabla \bullet \vec{A} = \frac{1}{\rho}\frac{\partial}{\partial \rho}\left(\rho A_\rho\right) + \frac{1}{\rho}\frac{\partial A_\varphi}{\partial \varphi} + \frac{\partial A_z}{\partial z} \tag{B-10}$$

$$\nabla \times \vec{A} = \left(\frac{1}{\rho}\frac{\partial A_z}{\partial \varphi} - \frac{\partial A_\varphi}{\partial z}\right)\hat{\rho} + \left(\frac{\partial A_\rho}{\partial z} - \frac{\partial A_z}{\partial \rho}\right)\hat{\varphi} + \left(\frac{1}{\rho}\frac{\partial}{\partial \rho}\left(\rho A_\varphi\right) - \frac{1}{\rho}\frac{\partial A_\rho}{\partial \varphi}\right)\hat{z} \tag{B-11}$$

$$\nabla \bullet \vec{A} = \frac{1}{r^2}\frac{\partial}{\partial r}\left(r^2 A_r\right) + \frac{1}{r\sin\theta}\frac{\partial}{\partial \theta}\left(\sin\theta\, A_\theta\right) + \frac{1}{r\sin\theta}\frac{\partial A_\varphi}{\partial \varphi} \tag{B-12}$$

$$\nabla \times \vec{A} = \left(\frac{\partial}{\partial \theta}\left(\sin\theta A_\varphi\right) - \frac{\partial A_\theta}{\partial \varphi}\right)\frac{\hat{r}}{r\sin\theta} + \left(\frac{1}{\sin\theta}\frac{\partial A_r}{\partial \varphi} - \frac{\partial}{\partial r}\left(rA_\varphi\right)\right)\frac{\hat{\theta}}{r} + \left(\frac{\partial}{\partial r}\left(rA_\theta\right) - \frac{\partial A_r}{\partial \theta}\right)\frac{\hat{\varphi}}{r} \tag{B-13}$$

B-10 Additional Useful Vector Relationships

$$\oint_{surface} \vec{B} \bullet d\vec{A} = \int_{volume} \left(\nabla \bullet \vec{B}\right) dV \text{ (Divergence theorem)} \quad \textbf{(B-14)}$$

$$\oint_{line} \vec{B} \bullet d\vec{s} = \int_{surface} \left(\nabla \times \vec{B}\right) \bullet d\vec{A} \text{ (Stoke's theorem)} \quad \textbf{(B-15)}$$

$$\nabla \bullet \left(\vec{A} + \vec{B}\right) = \nabla \bullet \vec{A} + \nabla \bullet \vec{B} \quad \textbf{(B-16)}$$

$$\nabla \times \left(\vec{A} + \vec{B}\right) = \nabla \times \vec{A} + \nabla \times \vec{B} \quad \textbf{(B-17)}$$

$$\nabla \bullet \left(\vec{A} \times \vec{B}\right) = \vec{B} \bullet \left(\nabla \times \vec{A}\right) - \vec{A} \bullet \left(\nabla \times \vec{B}\right) \quad \textbf{(B-18)}$$

$$\nabla \times \left(\nabla \times \vec{A}\right) = \nabla\left(\nabla \bullet \vec{A}\right) - \nabla^2 \vec{A} \quad \textbf{(B-19)}$$

$$\vec{A} \times \vec{B} \times \vec{C} = \vec{B}\left(\vec{A} \bullet \vec{C}\right) - \vec{C}\left(\vec{A} \bullet \vec{B}\right) \quad \textbf{(B-20)}$$

$$\nabla \times \left(c\vec{A}\right) = \left(\nabla c\right) \times \vec{A} + c\left(\nabla \times \vec{A}\right) \quad \textbf{(B-21)}$$

Appendix C

C-1 Example 2-4

We begin with the following expression:

$$V_g = -\frac{GM}{L}\ln\left(\frac{\left(L^2+4y^2\right)^{\frac{1}{2}}+L}{\left(L^2+4y^2\right)^{\frac{1}{2}}-L}\right)$$

Then multiply the numerator and denominator by the denominator.

$$\ln\left(\frac{\left(L^2+4y^2\right)^{\frac{1}{2}}+L}{\left(L^2+4y^2\right)^{\frac{1}{2}}-L}\right) = \ln\left(\left(\frac{\left(L^2+4y^2\right)^{\frac{1}{2}}+L}{\left(L^2+4y^2\right)^{\frac{1}{2}}-L}\right)\left(\frac{\left(L^2+4y^2\right)^{\frac{1}{2}}-L}{\left(L^2+4y^2\right)^{\frac{1}{2}}-L}\right)\right)$$

$$\ln\left(\frac{\left(L^2+4y^2\right)^{\frac{1}{2}}+L}{\left(L^2+4y^2\right)^{\frac{1}{2}}-L}\right) = \ln\left(\frac{L^2+4y^2-L\left(L^2+4y^2\right)^{\frac{1}{2}}+L\left(L^2+4y^2\right)^{\frac{1}{2}}-L^2}{\left(\left(L^2+4y^2\right)^{\frac{1}{2}}-L\right)^2}\right)$$

$$\ln\left(\frac{\left(L^2+4y^2\right)^{\frac{1}{2}}+L}{\left(L^2+4y^2\right)^{\frac{1}{2}}-L}\right)=\ln\left(\frac{4y^2}{\left(\left(L^2+4y^2\right)^{\frac{1}{2}}-L\right)^2}\right)$$

Both the numerator and the denominator are now perfect squares.

$$\ln\left(\frac{\left(L^2+4y^2\right)^{\frac{1}{2}}+L}{\left(L^2+4y^2\right)^{\frac{1}{2}}-L}\right)=\ln\left(\frac{2y}{\left(L^2+4y^2\right)^{\frac{1}{2}}-L}\right)^2$$

$$\ln\left(\frac{\left(L^2+4y^2\right)^{\frac{1}{2}}+L}{\left(L^2+4y^2\right)^{\frac{1}{2}}-L}\right)=2\ln\left(\frac{2y}{\left(L^2+4y^2\right)^{\frac{1}{2}}-L}\right)$$

Thus,

$$V_g=-\frac{2GM}{L}\ln\left(\frac{2y}{\left(L^2+4y^2\right)^{\frac{1}{2}}-L}\right)$$

This equation can also be written as

$$V_g=-\frac{2GM}{L}\ln\left(\frac{2y\left(\left(L^2+4y^2\right)^{\frac{1}{2}}+L\right)}{4y^2}\right)$$

In this limit that $L >> y$ this equation becomes

$$\lim_{L>>y} V_g=-\frac{2GM}{L}\ln\left(\frac{L}{y}\right)$$

C-2 Example 2-5

What is the gravitational potential energy of a system shown in Figure C.1 consisting of a 2 kg point particle and a uniformly dense 4 kg rod? The 2 kg point particle is located on the bisector of the rod.

The gravitational potential energy of this system can be determined by integrating, with respect to the length of the rod, the product of the gravitational potential of the point particle and each infinitesimal bit of mass that constitutes the rod.

$$dU_g = (dm)V_g \quad \rightarrow \quad U_g = \int V_g \, dm$$

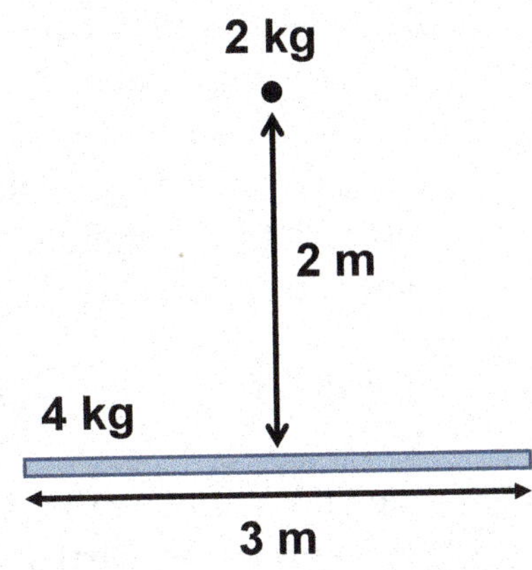

FIGURE C.1: The system in Example 2-5.

Substitution of the gravitational potential of the point particle (Equation 2-2) gives us

$$U_g = \int \left(-G \frac{m_{particle}}{r} \right) dm \quad \rightarrow \quad U_g = \left(-Gm_{particle} \right) \int \frac{dm}{r}$$

Finally, as shown in Example 2-4, the integral with respect to the mass of the rod in this expression can be changed into an integral with respect to the length of the rod using the linear mass density of the rod (λ).

$$U_g = \left(-Gm_{particle} \right) \int \frac{\lambda dx}{r} \quad \rightarrow \quad U_g = -G\lambda m_{particle} \int_{-\frac{L_{rod}}{2}}^{\frac{L_{rod}}{2}} \frac{dx}{\left(x^2 + y^2 \right)^{\frac{1}{2}}}$$

$$U_g = -\frac{2GM_{rod}m_{particle}}{L_{rod}} \ln \left(\frac{2y}{\left(L_{rod}^2 + 4y^2 \right)^{\frac{1}{2}} - L_{rod}} \right)$$

Substitution gives us

$$U_g = -\frac{2\left(6.67 \times 10^{-11} \frac{\text{J m}}{\text{kg}^2} \right)(4\text{kg})(2\text{kg})}{3\text{m}} \ln \left(\frac{2(2\text{m})}{\left((3\text{m})^2 + 4(2\text{m})^2 \right)^{\frac{1}{2}} - 3\text{m}} \right)$$

$$U_g = -2.46 \times 10^{-10} \text{ J}$$

This is the same result as obtained in Example 2-5.

C-3 Derivation of Equation 2-11

Consider the point particle moving along the trajectory shown in Figure C.2.

The work done by the gravitational field on the particle can be calculating from the gravitational force that the gravitational field exerts on the particle (Equation 2-10).

$$W = \int \vec{F} \cdot d\vec{s} \quad \rightarrow \quad W = \int m\vec{g} \cdot d\vec{s} \quad \rightarrow \quad W = m\int \vec{g} \cdot d\vec{s}$$

The differential vector $d\vec{s}$ in this equation is a small segment of the particle's trajectory, as shown in Figure C.2. Since the gravitational force is a conservative force we can relate the work done by the gravitational force on the particle to the change in the particle's gravitational potential energy.

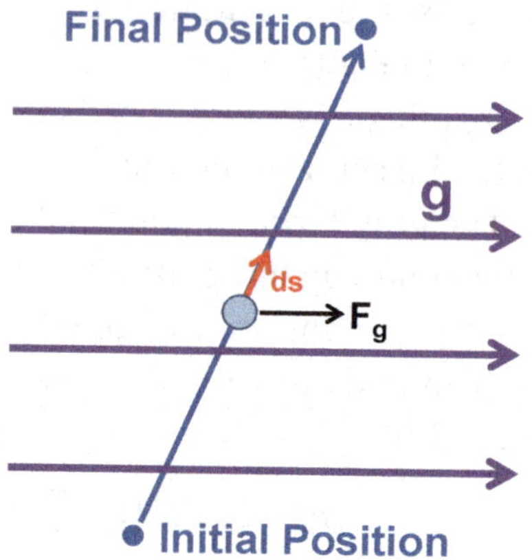

FIGURE C.2: A point particle moves through a gravitational field.

$$W = -\Delta U_g \quad \rightarrow \quad -\Delta U_g = m\int \vec{g} \cdot d\vec{s} \quad \rightarrow \quad \frac{\Delta U_g}{m} = -\int \vec{g} \cdot d\vec{s}$$

Substitution of Equation 2-5 then gives us

$$\Delta V_g = -\int \vec{g} \cdot d\vec{s}$$

C-4 Multipole Expansion of the Magnetic Field of a Current-Carrying Loop

The expression for the magnetic field of the current carrying loop in Figure 4.18 can be written as

$$\vec{B} = \left[\frac{\mu_0 I R^2}{2\left(R^2 + z^2\right)^{\frac{3}{2}}}\right]\hat{z} \quad \rightarrow \quad \vec{B} = \left[\frac{\mu_0 I R^2}{2z^3\left(1 + \frac{R^2}{z^2}\right)^{\frac{3}{2}}}\right]\hat{z} \quad \rightarrow \quad \vec{B} = \left[\frac{\mu_0 I R^2}{2z^3}\left(1 + \frac{R^2}{z^2}\right)^{-\frac{3}{2}}\right]\hat{z}$$

The term in parenthesis in this expression can be rewritten using the binomial series expansion (Section A-3)

$$\left(1+\frac{R^2}{z^2}\right)^{-\frac{3}{2}}=1-\frac{3}{2}\frac{R^2}{z^2}+\frac{\left(-\frac{3}{2}\right)\left(-\frac{3}{2}-1\right)}{2!}\left(\frac{R^2}{z^2}\right)^2+\ldots$$

$$\left(1+\frac{R^2}{z^2}\right)^{-\frac{3}{2}}=1-\frac{3}{2}\frac{R^2}{z^2}+\frac{15}{8}\left(\frac{R^2}{z^2}\right)^2+\ldots$$

Substitution of this expansion then gives us

$$\vec{B}=\left[\frac{\mu_0 IR^2}{2z^3}\left(1-\frac{3}{2}\frac{R^2}{z^2}+\frac{15}{8}\left(\frac{R^2}{z^2}\right)^2+\ldots\right)\right]\hat{z}$$

$$\vec{B}=\left[\frac{\mu_0 IR^2}{2z^3}-\frac{3\mu_0 IR^4}{4z^5}+\frac{15\mu_0 IR^6}{16z^7}+\ldots\right]\hat{z}$$

This is an equation for the multipole expansion of the magnetic field of the current carrying loop. The first term in this expansion is the dipole term, the second term is the quadrupole term, and the remaining terms in the series are higher order multipoles. Notice that there is no monopole term since magnetic monopoles do not exist[1].

C-5 Derivation of Equation 4-12

We begin with the identity

$$\nabla\left(\frac{1}{r}\right)=-\frac{\hat{r}}{r^2}\quad\rightarrow\quad\nabla\left(\frac{1}{r}\right)=-\frac{\vec{r}}{r^3}$$

Therefore

$$\frac{d\vec{s}\times\vec{r}}{r^3}=d\vec{s}\times\left(\frac{\vec{r}}{r^3}\right)\quad\rightarrow\quad\frac{d\vec{s}\times\vec{r}}{r^3}=-d\vec{s}\times\nabla\left(\frac{1}{r}\right)\quad\rightarrow\quad\frac{d\vec{s}\times\vec{r}}{r^3}=\nabla\left(\frac{1}{r}\right)\times d\vec{s}$$

1 You should feel inspired to take advanced physics courses to learn a more elegant way of using the magnetic multipole expansion to show that magnetic monopoles do not exist.

This expression can be simplified using Equation B-21.

$$\nabla\left(\frac{1}{r}\right)\times d\vec{s} = \nabla\times\left(\frac{d\vec{s}}{r}\right) - \frac{1}{r}\left(\nabla\times d\vec{s}\right)$$

The last term in this expression is zero since $d\vec{s}$ is a constant with respect to the curl operation.

$$\nabla\times d\vec{s} = 0 \quad\rightarrow\quad \nabla\left(\frac{1}{r}\right)\times d\vec{s} = \nabla\times\left(\frac{d\vec{s}}{r}\right) \quad\rightarrow\quad \frac{d\vec{s}\times\vec{r}}{r^3} = \nabla\times\left(\frac{d\vec{s}}{r}\right)$$

Substitution into Equation 4-5 then gives us

$$d\vec{B} = \left(\frac{\mu_0 I}{4\pi}\right)\nabla\times\left(\frac{d\vec{s}}{r}\right) \quad\rightarrow\quad d\vec{B} = \nabla\times\left(\frac{\mu_0 I}{4\pi}\frac{d\vec{s}}{r}\right) \quad\rightarrow\quad \vec{B} = \nabla\times\left(\frac{\mu_0}{4\pi}\int\frac{Id\vec{s}}{r}\right)$$

Comparison to Equation 4-11 then gives us

$$\vec{V}_m = \left(\frac{\mu_0}{4\pi}\right)\int\frac{Id\vec{s}}{r}$$

C-6 Further Discussion of Equation 5-8

The substitution of Equation 5-7 into Equation 5-8 gives us

$$C = -\frac{q}{\Delta V_e} \quad\rightarrow\quad \Delta U_e = \frac{q^2}{2\left(-\frac{q}{\Delta V_e}\right)} \quad\rightarrow\quad \Delta U_e = -\frac{1}{2}\left(q\Delta V_e\right)$$

This result is different from what we would expect from Equation 3-5, however. This difference results from the fact that Equation 5-7 represents the internal energy associated with constructing the charge separation shown in Figure 5.7, whereas Equation 3-5[2] represents the energy associated with the interaction of a charged point particle with an external electric field.

2 More specifically, Equation 3-3.

Let's consider two charged point particles with net electric charge q_1 and q_2. The total electric potential energy of this charge distribution is found using Equation 3-3 or Equation 3-4.

$$(U_e)_{12} = \frac{q_1 q_2}{4\pi\varepsilon_0 r_{12}}$$

The variable r_{12} in this equation is the distance between the two charged point particles. For a set of N charged point particles each pair of particles has a similar electric potential energy. We can therefore express the total electric potential energy of the system of particles as

$$U_e = \frac{1}{2}\sum_{i \neq j} \frac{q_i q_j}{4\pi\varepsilon_0 r_{ij}} \quad \rightarrow \quad U_e = \frac{1}{2}\sum_{i \neq j} q_i (V_e)_j$$

The summation in this equation is over all pairs of particles, excluding the interaction of a particle with itself (*i.e.*, the values of *i* and *j* in the summation cannot be the same) and the factor of one-half accounts for the appearance of each pair twice in the summation. This is the same reason why the factor of one-half appears in Equation 5-8 (in addition to the obvious calculus reason).

For a continuous distribution of net electric charge, the summation in the above expression can be written as an integral.

$$U_e = \frac{1}{2}\int \rho(\vec{r}) V_e(\vec{r})\, dV \qquad \textbf{(C-1)}$$

C-7 Further Discussion of Validity of Equation 6-1 and Equation 6-2

Let's consider the system shown in Figure C.3 that consists of two positively charged point particles moving with the same velocity; the net electric charges of the two particles are q_1 and q_2. The electric and magnetic fields of each particle are shown in Figure C.3

The electric and magnetic fields of the point particles in reference frame 1 (that is stationary with respect to the motion of the particles) are found using Equation 3-9 and Equation 4-1.

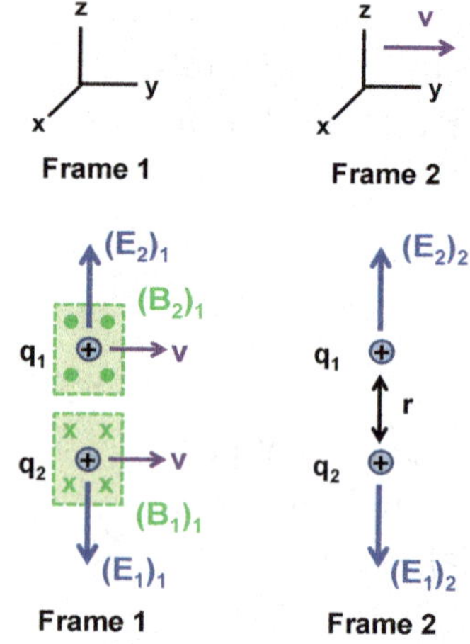

FIGURE C.3: Transformation of the electric and magnetic fields of two positively charged point particles moving with the same velocity.

$$\left(\vec{E}_1\right)_1 = \left(\frac{q_1}{4\pi\varepsilon_0 r^2}\right)(-\hat{z}) \qquad \left(\vec{E}_2\right)_1 = \left(\frac{q_2}{4\pi\varepsilon_0 r^2}\right)\hat{z}$$

$$\left(\vec{B}_1\right)_1 = \left(\frac{\mu_0 q_1 v}{4\pi r^2}\right)(-\hat{x}) \qquad \left(\vec{B}_2\right)_1 = \left(\frac{\mu_0 q_2 v}{4\pi r^2}\right)\hat{x}$$

The final subscripts in these equations denote that these fields are measured in reference frame 1; the initial subscript denotes the particle that is the source of the field. The corresponding magnetic fields in reference frame 2 (that is moving at the same velocity as the particles) are determined using Equation 6-2. For the particle with net electric charge q_1, we have

$$\left(\vec{B}_1\right)_2 = \left(\vec{B}_1\right)_1 - (\mu_0\varepsilon_0)\vec{v}\times\left(\vec{E}_1\right)_1$$

$$\vec{v} = (v)\hat{y} \rightarrow \left(\vec{B}_1\right)_2 = \left(\frac{\mu_0 q_1 v}{4\pi r^2}\right)(-\hat{x}) - (\mu_0\varepsilon_0)(v)\hat{y}\times\left(\frac{q_1}{4\pi\varepsilon_0 r^2}\right)(-\hat{z})$$

$$\left(\vec{B}_1\right)_2 = \left(\frac{q_1\mu_0 v}{4\pi r^2} - \frac{\mu_0 q_1 v}{4\pi r^2}\right)(\hat{x}) \rightarrow \left(\vec{B}_1\right)_2 = 0$$

As expected, the magnetic field is zero in reference frame 2, since the particle is stationary in this reference frame. The electric fields in reference frame 2 are found using Equation 6-1. For the particle with net electric charge q_1, we have

$$\left(\vec{E}_1\right)_2 = \left(\vec{E}_1\right)_1 + \vec{v}\times\left(\vec{B}_1\right)_1$$

$$\vec{v} = (v)\hat{y} \rightarrow \left(\vec{E}_1\right)_2 = \left(\frac{q_1}{4\pi\varepsilon_0 r^2}\right)(-\hat{z}) + (v)\hat{y}\times\left(\frac{\mu_0 q_1 v}{4\pi r^2}\right)(-\hat{x})$$

$$\vec{v} = (v)\hat{y} \rightarrow \left(\vec{E}_1\right)_2 = \left(\frac{q_1}{4\pi\varepsilon_0 r^2}\right)(-\hat{z}) + (v)\hat{y}\times\left(\frac{\mu_0 q_1 v}{4\pi r^2}\right)(-\hat{x})$$

$$\left(\vec{E}_1\right)_2 = \left(\frac{\mu_0 q_1 v^2}{4\pi r^2} - \frac{q_1}{4\pi\varepsilon_0 r^2}\right)(\hat{z}) \rightarrow \left(\vec{E}_1\right)_2 = \left(\frac{q_1}{4\pi\varepsilon_0 r^2}\right)(1-\mu_0\varepsilon_0 v^2)(-\hat{z})$$

Based upon Equation 3-9, we would have anticipated the field to be

$$\left(\vec{E}_1\right)_2 = \left(\frac{q_1}{4\pi\varepsilon_0 r^2}\right)(-\hat{z})$$

So the transformation (Equation 6-1) has given us the wrong result. The Galilean transformations used to derive Equation 6-1 and Equation 6-2 are "correct" only when the additional term $\mu_0\varepsilon_0 v^2$ is much less than 1. We can express this constraint mathematically as

$$v^2 << \frac{1}{\mu_0\varepsilon_0} \rightarrow v << \frac{1}{\left(\left(4\pi\times10^{-7}\frac{\text{kgm}}{\text{C}^2}\right)\left(8.85\times10^{-12}\frac{\text{C}^2}{\text{Jm}}\right)\right)^{\frac{1}{2}}}$$

$$v << 3\times10^8\,\frac{\text{m}}{\text{s}}$$

Hence, Equation 6-1 and Equation 6-2 are correct only when the speed of the particles is much less than 3×10^8 m/s. You should feel encouraged to take additional physics courses to learn the correct general transformation for electric and magnetic fields.

C-8 Derivation of Equation 6-14

Consider the system shown in Figure C.4 in which two loops of current are brought together by externally applied forces. Each loop contains an EMF that maintains a current in the loop and these currents are in opposite directions.

Since the currents in the loops are in opposite directions, the magnetic force acting between the loops pushes the loops apart (Figure 4.23). Thus, work must have been done on this system to bring the two loops together, and this work resulted in an increase in the magnetic potential energy of the system.

Rather than calculating the magnetic potential energy of the system from the work required to push the loops together, let's instead calculate it by the determining the

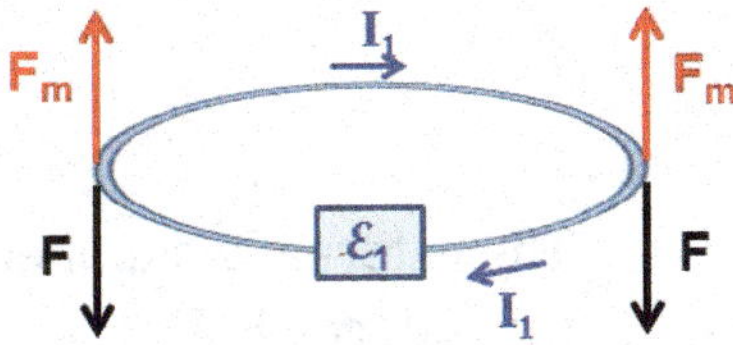

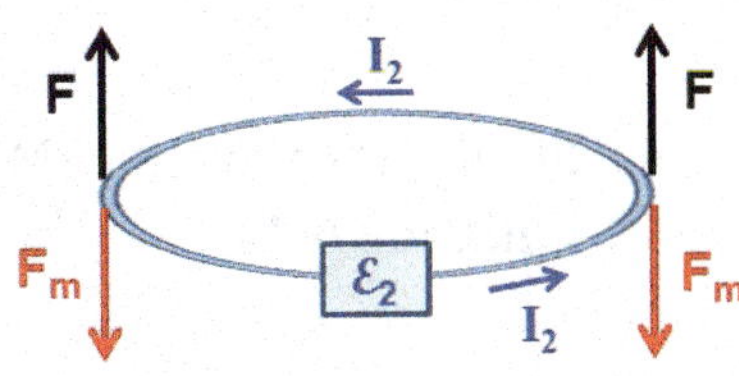

FIGURE C.4: Two parallel loops of current held together by an external applied force.

work required to increase the currents in the loops while the loops are held in place a fixed distance away from each other[3]. We begin by assuming that the two loops are initially close to each other, but carry no current. As the current in each loop starts to increase, the associated magnetic field will change the magnetic flux through the other loop. This changing magnetic flux will induce a new EMF and associated current in each loop.

In order to maintain the same current in the loops, each original EMF must do additional work to overcome the induced EMF. To accomplish this, the EMF must increase in magnitude by gaining energy from a source external to the system. Let's denote this increase in the EMF as $\mathcal{E}_{ext}$, where the subscript reminds us that this source of energy is external to our original system. For an infinitesimal change in magnetic flux (*i.e.*, an infinitesimal induced EMF), we can express the associated infinitesimal work done by $\mathcal{E}_{ext}$ using Equation 6-8.

$$đW = \mathcal{E}_{ext} dq$$

In this expression, dq denotes a small amount of electric charge that travels around the loop. We can express this small amount of charge in terms of the current in the loop.

$$i = \frac{dq}{dt} \quad \rightarrow \quad đW = \mathcal{E}_{ext} i dt$$

We are temporarily using the variable i to denote the current in a loop at some instant while the current is being increased to its final value of I. As stated previously, this work is equal to the change in the magnetic potential energy of the system.

$$dU_m = \mathcal{E}_{ext} i dt$$

The externally supplied EMF must be equal in magnitude, but opposite in direction to the induced EMF.

$$\mathcal{E}_{ext} = -\mathcal{E}_{induced} \quad \rightarrow \quad dU_m = -\mathcal{E}_{induced} i dt$$

Finally, the induced EMF can be expressed in terms of the change in magnetic flux using Equation 6-11.

3 Since potential energy is a state function the path by which the system came to have the potential energy is irrelevant.

$$\mathcal{E}_{induced}dt = -d\Phi_B \quad \rightarrow \quad dU_m = i(d\Phi_B)$$

For the system shown in Figure C.4 the magnetic flux can be written in terms of the mutual inductances and self-inductances of the circuits using Equation 6-5 and Equation 6-6.

$$dU_m = i_1(d\Phi_B)_1 + i_2(d\Phi_B)_2$$

$$dU_m = i_1(M_{12}di_2 + L_1di_1) + i_2(M_{21}di_1 + L_2di_2)$$

The total magnetic potential energy of this system is then found by summing the infinitesimal changes in magnetic potential energy associated with increasing the currents in the loops from zero to their final values. There are many ways that this process can happen, but the *final result must be independent of the process by which it occurred*[4]. For simplicity, let's assume that at any instant of time during this process each current is the same fraction, *f*, of its final value.

$$i_1(t) = f(t)I_1 \qquad i_2(t) = f(t)I_2$$

Substitution then gives us

$$dU_m = fI_1(M_{12}I_2df + L_1I_1df) + fI_2(M_{21}I_1df + L_2I_2df)$$

$$dU_m = (M_{12}I_2I_1 + L_1I_1I_1 + M_{21}I_1I_2 + L_2I_2I_2)f\,df$$

The total magnetic potential energy is then found by integrating this expression. In this integration we will assume that the initial magnetic potential energy of the system is zero.

$$U_m = \int dU_m \quad \rightarrow \quad U_m = \int(M_{12}I_2I_1 + L_1I_1I_1 + M_{21}I_1I_2 + L_2I_2I_2)f\,df$$

$$U_m = (M_{12}I_2I_1 + L_1I_1I_1 + M_{21}I_1I_2 + L_2I_2I_2)\int f\,df$$

Since the variable f in this expression represents fraction of the final value of the current, the limits of the integration are from 0 to 1.

4 Magnetic potential energy is a state function so the path is irrelevant.

$$U_m = \left(M_{12}I_2I_1 + L_1I_1I_1 + M_{21}I_1I_2 + L_2I_2I_2\right)\int_0^1 f\,df$$

$$U_m = \left(M_{12}I_2I_1 + L_1I_1I_1 + M_{21}I_1I_2 + L_2I_2I_2\right)\left(\frac{f^2}{2}\bigg|_0^1\right)$$

$$U_m = \frac{1}{2}\left(M_{12}I_2I_1 + L_1I_1I_1 + M_{21}I_1I_2 + L_2I_2I_2\right)$$

We can then use Equation 6-6 again to rewrite this expression in terms of the magnetic fluxes through the loops; this includes contributions to the fluxes from both currents.

$$U_m = \frac{1}{2}\left(\left(\Phi_B\right)_{12}I_1 + \left(\Phi_B\right)_{11}I_1 + \left(\Phi_B\right)_{21}I_2 + \left(\Phi_B\right)_{22}I_2\right)$$

$$U_m = \frac{1}{2}\left[I_1\left(\left(\Phi_B\right)_{12} + \left(\Phi_B\right)_{11}\right) + I_2\left(\left(\Phi_B\right)_{21} + \left(\Phi_B\right)_{22}\right)\right]$$

We can generalize this result and write the equation for the magnetic potential energy of a system as

$$U_m = \frac{1}{2}\sum_i I_i\left(\Phi_B\right)_i$$

The variable $\left(\Phi_B\right)_i$ in this expression is the total flux through loop *i* from all sources of magnetic field, including itself.

C-9 Further Discussion of Equation 6-14.

Substitution of Equation 5-13 into Equation 6-14 gives us

$$U_m = \frac{1}{2}\sum_i \oint_{loop\,i} \vec{V}_m\left(\vec{r}_i\right)\cdot I_i d\vec{s}_i \qquad \textbf{(C-2)}$$

In this expression, the variable $\vec{V}_m(\vec{r}_i)$ is the total magnetic potential at the location $\vec{r}_i$ of the current element $I_i d\vec{s}_i$ of loop *i*. The integral in this expression gives the contributions to U_m from all current elements in loop *i* and the summation over *i* produces the sum of the contributions to U_m from all current elements in the whole system. The factor of one-half accounts for any double-counting in the summation (See Section C-6). In this way, Equation 6-14 can be derived directly from Equation 4-14 (or Equation 4-13) just as Equation 5-8 can be derived directly from Equation 3-3.

Indeed, we can use Equation C-2 to derive Equation 6-15. The magnetic potential of a solenoid was determined in Example 5-13. The value of the magnetic potential at the location of the current in the solenoid (*i.e.*, at the radius of the solenoid) is

$$\vec{V}_m = \left(\frac{\mu_0 nIR}{2}\right)\hat{\varphi}$$

Let's choose a path of integration in Equation C-2 around a single loop of wire in the solenoid. Since the current in each loop of the solenoid is constant we have

$$d\vec{s}_n = (Rd\varphi)\hat{\varphi} \quad\rightarrow\quad U_m = \frac{1}{2}\sum_i \oint_{loop\,i} \left[\left(\frac{\mu_0 nIR}{2}\right)\hat{\varphi}\right]\bullet\left[I(Rd\varphi)\hat{\varphi}\right]$$

$$U_m = \frac{1}{2}\sum_i \frac{\mu_0 nI^2R^2}{2}\oint_{loop\,i} d\varphi$$

The integral in this expression is simply the total angle in a single loop of wire in the solenoid.

$$\oint_{loop\,i} d\varphi = 2\pi \quad\rightarrow\quad U_m = \frac{1}{2}\sum_i \frac{\mu_0 nI^2R^2}{2}(2\pi) \quad\rightarrow\quad U_m = \frac{1}{2}\sum_i \mu_0 nI^2\pi R^2$$

The summation is simply the number of loops of wire in the solenoid. Thus,

$$U_m = \frac{1}{2}\mu_0 nI^2\pi R^2 N \quad\rightarrow\quad U_m = \frac{1}{2}\mu_0 nI^2\pi R^2(nl) \quad\rightarrow\quad U_m = \frac{1}{2}\mu_0 n^2\left(\pi R^2 l\right)I^2$$

The term in parenthesis is the volume of the solenoid (Section 5-6).

$$U_m = \frac{1}{2}\mu_0 n^2 VI^2 \quad\rightarrow\quad U_m = \frac{1}{2}LI^2$$

C-10 Derivation of Equation 7-17

We begin with the general equation for the drift velocity.

$$\vec{v}_d = \mu\vec{E} + \mu\left(\vec{v}_d \times \vec{B}\right)$$

Taking the cross product of both sides of this equation with $\vec{B}$ gives us

$$\vec{v}_d \times \vec{B} = \left[\mu\vec{E} + \mu\left(\vec{v}_d \times \vec{B}\right)\right] \times \vec{B} \quad \rightarrow \quad \vec{v}_d \times \vec{B} = \mu\vec{E} \times \vec{B} + \mu\left(\vec{v}_d \times \vec{B}\right) \times \vec{B}$$

We can simplify this equation using Equation B-20.

$$\mu\left(\vec{v}_d \times \vec{B}\right) \times \vec{B} = -\mu\left(\vec{v}_d\left(\vec{B} \bullet \vec{B}\right) - \vec{B}\left(\vec{v}_d \bullet \vec{B}\right)\right)$$
$$\mu\left(\vec{v}_d \times \vec{B}\right) \times \vec{B} = -\mu\vec{v}_d B^2 + \mu\vec{B}\left(\vec{v}_d \bullet \vec{B}\right)$$

Thus,

$$\vec{v}_d \times \vec{B} = \mu\vec{E} \times \vec{B} - \mu\vec{v}_d B^2 + \mu\vec{B}\left(\vec{v}_d \bullet \vec{B}\right)$$

Now let's substitute the equation for the drift velocity into the final term of this expression.

$$\vec{v}_d \times \vec{B} = \mu\vec{E} \times \vec{B} - \mu\vec{v}_d B^2 + \mu\vec{B}\left(\left(\mu\left(\vec{E} + \vec{v}_d \times \vec{B}\right)\right) \bullet \vec{B}\right)$$

Simplifying gives us

$$\vec{v}_d \times \vec{B} + \mu\vec{v}_d B^2 = \mu\vec{E} \times \vec{B} + \mu^2\vec{B}\left(\vec{E} \bullet \vec{B} + \left(\vec{v}_d \times \vec{B}\right) \bullet \vec{B}\right)$$

$$\left(\vec{v}_d \times \vec{B}\right) \bullet \vec{B} = 0 \quad \rightarrow \quad \vec{v}_d \times \vec{B} + \mu\vec{v}_d B^2 = \mu\vec{E} \times \vec{B} + \mu^2\vec{B}\left(\vec{E} \bullet \vec{B}\right)$$

$$\frac{\vec{v}_d}{\mu} - \vec{E} = \vec{v}_d \times \vec{B} \quad \rightarrow \quad \frac{\vec{v}_d}{\mu} - \vec{E} + \mu\vec{v}_d B^2 = \mu\vec{E} \times \vec{B} + \mu^2\vec{B}\left(\vec{E} \bullet \vec{B}\right)$$

$$\vec{v}_d - \mu\vec{E} + \mu^2\vec{v}_d B^2 = \mu^2\vec{E} \times \vec{B} + \mu^3\vec{B}\left(\vec{E} \bullet \vec{B}\right)$$

$$\vec{v}_d\left(1 + \mu^2 B^2\right) = \mu\vec{E} + \mu^2\vec{E} \times \vec{B} + \mu^3\vec{B}\left(\vec{E} \bullet \vec{B}\right)$$

$$\vec{v}_d = \frac{\mu}{1 + \left(\mu B\right)^2}\left(\vec{E} + \mu\vec{E} \times \vec{B} + \mu^2\left(\vec{E} \bullet \vec{B}\right)\vec{B}\right)$$

C-11 Derivation of Equation 9-12.

We can determine the average power over a time interval T using the following equation:

$$P_{avg} = \frac{\int_0^T P\,dt}{\int_0^T dt} \rightarrow P_{avg} = \frac{\int_0^T P\,dt}{T}$$

The instantaneous power supplied by the EMF is

$$P = I_0\mathcal{E}_0\left(\cos^2(\omega t)\cos(\phi) - \sin(\omega t)\cos(\omega t)\sin(\phi)\right)$$

We therefore have two integrals to consider.

$$\int_0^T \cos^2(\omega t)\,dt = \left(\frac{t}{2} + \frac{\sin(2\omega t)}{4\omega}\Bigg|_0^T\right) \rightarrow \int_0^T \cos^2(\omega t)\,dt = \frac{T}{2} + \frac{\sin(2\omega T)}{4\omega}$$

$$\int_0^T \sin(\omega t)\cos(\omega t)\,dt = \left(-\frac{\cos^2(\omega t)}{2\omega}\Bigg|_0^T\right) \rightarrow \int_0^T \cos^2(\omega t)\,dt = \frac{1}{2\omega} - \frac{\cos^2(\omega T)}{2\omega}$$

The average power over the time interval T is thus

$$P = \frac{I_0\mathcal{E}_0}{T}\left(\left(\frac{T}{2} + \frac{\sin(2\omega T)}{4\omega}\right)\cos(\phi) - \left(\frac{1}{2\omega} - \frac{\cos^2(\omega T)}{2\omega}\right)\sin(\phi)\right)$$

$$P_{avg} = I_0\mathcal{E}_0\left(\left(\frac{1}{2} + \frac{\sin(2\omega T)}{4\omega T}\right)\cos(\phi) - \left(\frac{1}{2\omega T} - \frac{\cos^2(\omega T)}{2\omega T}\right)\sin(\phi)\right)$$

The true average power for the EMF is the limit of this expression for an infinite time interval. In other words, when we talk about average power we mean the average power over the entire time the current is flowing in the circuit.

$$\lim_{T\to\infty} P_{avg} = I_0\mathcal{E}_0\left(\frac{1}{2}\cos(\phi)\right) \rightarrow \lim_{T\to\infty} P_{avg} = \frac{1}{2}I_0\mathcal{E}_0\cos(\phi)$$

C-12 Derivation of the Self-Inductance of a Circular Loop of Current.

Let's use Equation 9-18 to determine the self-inductance of the circular loop of current shown in Figure C.5.

Let's assume that this loop of wire has no resistance so that the only impedance results from the self-inductance. The magnetic potential for this current loop is found by integrating Equation 9-18 over the circumference of the loop.

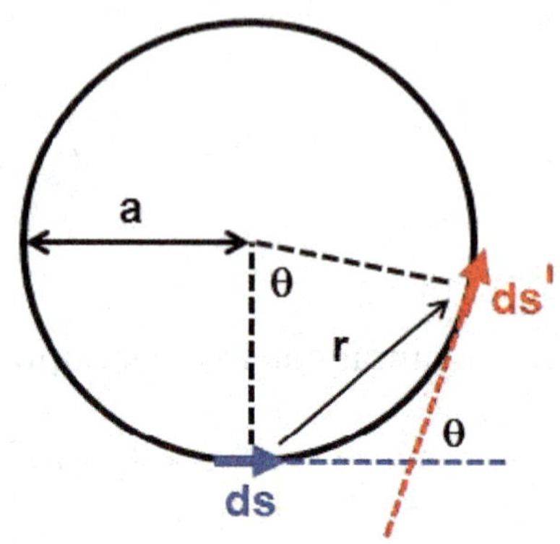

FIGURE C.5: Determining the self-inductance of a circular loop of current.

$$\vec{V}_m = \left(\frac{\mu_0}{4\pi}\right)\oint \frac{I_0 e^{i\omega\left(t-\frac{r}{c}\right)} d\vec{s}'}{r}$$

This integration is carried out over wire segments $d\vec{s}'$ that are a distance r away from an "origin" segment $d\vec{s}$. The magnetic flux through the loop can be determined using Equation 5-13 and Equation 9-18 by integrating over all segments $d\vec{s}$ or the wire.

$$\Phi_B = \oint_{line}\left(\left(\frac{\mu_0}{4\pi}\right)\oint \frac{I_0 e^{i\omega\left(t-\frac{r}{c}\right)} d\vec{s}'}{r}\right) \bullet d\vec{s}$$

The self-inductance of the loop is calculated from this magnetic flux using Equation 6-7.

$$L = \frac{\oint_{line}\left(\left(\frac{\mu_0}{4\pi}\right)\oint \frac{I_0 e^{i\omega\left(t-\frac{r}{c}\right)} d\vec{s}'}{r}\right) \bullet d\vec{s}}{I_0 e^{i\omega t}} \rightarrow L = \oint_{line}\left(\left(\frac{\mu_0}{4\pi}\right)\oint \frac{e^{-i\frac{\omega r}{c}} d\vec{s}'}{r}\right) \bullet d\vec{s}$$

$$L = \left(\frac{\mu_0}{4\pi}\right)\oint_{line}\oint \frac{e^{-i\frac{\omega r}{c}}}{r} d\vec{s}' \bullet d\vec{s}$$

We can express the dot-product in this expression in terms of the angle θ between $d\vec{s}'$ and $d\vec{s}$ (Figure C.5)

$$L=\left(\frac{\mu_0}{4\pi}\right)\oint_{line}\oint\frac{e^{-i\frac{\omega r}{c}}}{r}\cos\theta\, ds'\, ds$$

Substitution using Equation D-5 yields

$$L=\left(\frac{\mu_0}{4\pi}\right)\oint_{line}\oint\frac{\cos\left(\frac{\omega r}{c}\right)-i\sin\left(\frac{\omega r}{c}\right)}{r}\cos\theta\, ds'\, ds$$

$$L=\left[\left(\frac{\mu_0}{4\pi}\right)\oint_{line}\oint\frac{\cos\left(\frac{\omega r}{c}\right)}{r}\cos\theta\, ds'\, ds\right]-i\left[\left(\frac{\mu_0}{4\pi}\right)\oint_{line}\oint\frac{\sin\left(\frac{\omega r}{c}\right)}{r}\cos\theta\, ds'\, ds\right]$$

Let's rewrite this equation as

$$L=L'-iL''$$

In this expression, the quantities L' and L'' are real and imaginary parts of the self-inductance.

$$L'=\left(\frac{\mu_0}{4\pi}\right)\oint_{line}\oint\frac{\cos\left(\frac{\omega r}{c}\right)}{r}\cos\theta\, ds'\, ds \qquad L''=\left(\frac{\mu_0}{4\pi}\right)\oint_{line}\oint\frac{\sin\left(\frac{\omega r}{c}\right)}{r}\cos\theta\, ds'\, ds$$

The impedance associated with this self-inductance is

$$Z=i\omega L \quad\rightarrow\quad Z=i\omega\left(L'-iL''\right) \quad\rightarrow\quad Z=i\omega L'+\omega L''$$

This real part of this impedance will result in energy dissipation from the system as though it were a resistor[5]. Thus, even though there is no resistance in the loop of wire energy will nevertheless be dissipated. Indeed, the average power dissipated by the self-inductance is

5 Recall from Section 9-4 that energy is dissipated because of the real part of the impedance, usually resulting from resistance in the circuit. The imaginary part of the impedance, associated with inductance and capacitance, causes energy storage.

$$P_{avg} = I_{rms}^2 R \quad \rightarrow \quad P_{avg} = I_{rms}^2 \omega L'' \quad \rightarrow \quad P_{avg} = I_{rms}^2 \operatorname{Re}(Z)$$

We can estimate the magnitude of the real part of the impedance (*i.e.*, the effective resistance of the loop) using a Taylor series expansion[6].

$$\operatorname{Re}(Z) = \left(\frac{\mu_0 \omega}{4\pi}\right) \oint_{line} \oint \frac{\left(\frac{\omega r}{c} - \frac{\omega^3 r^3}{3! c^3} + \ldots\right)}{r} \cos\theta \, ds' \, ds$$

$$\operatorname{Re}(Z) = \left(\frac{\mu_0 \omega}{4\pi}\right) \oint_{line} \oint \left(\frac{\omega}{c} - \frac{\omega^3 r^2}{3! c^3} + \ldots\right) \cos\theta \, ds' \, ds$$

Using a little trigonometry, we see from Figure C.5 that

$$r = 2a \sin\left(\frac{\theta}{2}\right) \qquad ds' = a d\theta$$

$$\operatorname{Re}(Z) = \left(\frac{\mu_0 \omega}{4\pi}\right) \oint_{line} \int_0^{2\pi} \left(\frac{\omega}{c} - \frac{\omega^3}{6c^3}\left(4a^2 \sin^2\left(\frac{\theta}{2}\right)\right) + \ldots\right) \cos\theta a d\theta \, ds$$

Now let's assume that the speed c is very large so that higher order terms of the expansion can be neglected.

$$\operatorname{Re}(Z) \approx \left(\frac{\mu_0 \omega}{4\pi}\right) \oint_{line} \int_0^{2\pi} \left(\frac{\omega}{c} - \frac{\omega^3}{6c^3}\left(4a^2 \sin^2\left(\frac{\theta}{2}\right)\right)\right) \cos\theta a d\theta \, ds$$

We now have two integrals to perform

$$\frac{\omega a}{c} \int_0^{2\pi} \cos\theta \, d\theta = 0$$

$$\frac{4a^3 \omega^3}{6c^3} \int_0^{2\pi} \sin^2\left(\frac{\theta}{2}\right) \cos\theta \, d\theta = \frac{4a^3 \omega^3}{6c^3}\left(-\frac{\pi}{2}\right)$$

Hence,

6 See Section A-3.

$$\mathrm{Re}(Z) \approx \left(\frac{\mu_0 \omega}{4\pi}\right) \oint_{line} -\left(\frac{4a^3\omega^3}{6c^3}\left(-\frac{\pi}{2}\right)\right) ds \rightarrow \mathrm{Re}(Z) \approx \left(\frac{\mu_0 a^3 \omega^4}{12c^3}\right) \oint_{line} ds$$

The integral in this expression is simply the circumference of the loop.

$$\mathrm{Re}(Z) \approx \left(\frac{\mu_0 a^3 \omega^4}{12c^3}\right)(2\pi a) \rightarrow \mathrm{Re}(Z) \approx \frac{\mu_0 \pi a^4 \omega^4}{6c^3}$$

Finally, we can write this expression in terms of the area of the loop, denoted by A.

$$\mathrm{Re}(Z) \approx \frac{\mu_0 A^2 \omega^4}{6\pi c^3}$$

C-13 Derivation of Differential Form of Faraday's Law of Induction

$$\oint_{line} \vec{E} \bullet d\vec{s} = -\frac{d\Phi_B}{dt} \rightarrow \oint_{line} \vec{E} \bullet d\vec{s} = -\frac{d}{dt}\left(\int_{surface} \vec{B} \bullet d\vec{A}\right)$$

Applying Stoke's theorem (Equation B-15) gives us

$$\oint_{line} \vec{E} \bullet d\vec{s} = \int_{surface} \left(\nabla \times \vec{E}\right) \bullet d\vec{A} \rightarrow \int_{surface} \left(\nabla \times \vec{E}\right) \bullet d\vec{A} = -\frac{d}{dt}\left(\int_{surface} \vec{B} \bullet d\vec{A}\right)$$

We can then combine the two integrals into one integral.

$$\int_{surface} \left(\nabla \times \vec{E} + \frac{\partial \vec{B}}{\partial t}\right) \bullet d\vec{A} = 0$$

In order for this integral to always be zero, the term in the parenthesis must be zero.

$$\left(\nabla \times \vec{E} + \frac{\partial \vec{B}}{\partial t}\right) = 0 \rightarrow \nabla \times \vec{E} = -\frac{\partial \vec{B}}{\partial t} \qquad \text{(C-3)}$$

Substitution of Equation 4-11 then gives us

$$\nabla\times\vec{E}=-\frac{\partial}{\partial t}\left(\nabla\times\vec{V}_m\right)\ \rightarrow\ \nabla\times\vec{E}=\nabla\times\left(-\frac{\partial\vec{V}_m}{\partial t}\right)\ \rightarrow\ \vec{E}=-\frac{\partial\vec{V}_m}{\partial t}$$

As derived previously (Equation 6-12). Of course, from Equation 3-8 we know that an electric field can also result from a spatially varying electric potential. Combining these two sources allows us to write a general expression for the electric field.

$$\vec{E}=-\nabla V_e-\frac{\partial\vec{V}_m}{\partial t} \qquad \text{(C-4)}$$

C-14 Derivation of Differential Form of Ampère's Law

$$\oint_{line}\vec{B}\bullet d\vec{s}=\mu_0\varepsilon_0\frac{d\Phi_E}{dt}\ \rightarrow\ \oint_{line}\vec{B}\bullet d\vec{s}=\mu_0\varepsilon_0\frac{d}{dt}\left(\int_{surface}\vec{E}\bullet d\vec{A}\right)$$

Applying Stoke's theorem (Equation B-15) gives us

$$\oint_{line}\vec{B}\bullet d\vec{s}=\int_{surface}\left(\nabla\times\vec{B}\right)\bullet d\vec{A}\ \rightarrow\ \int_{surface}\left(\nabla\times\vec{B}\right)\bullet d\vec{A}=\mu_0\varepsilon_0\frac{d}{dt}\left(\int_{surface}\vec{E}\bullet d\vec{A}\right)$$

We can then combine the two integrals into one integral.

$$\int_{surface}\left(\nabla\times\vec{B}-\mu_0\varepsilon_0\frac{\partial\vec{E}}{\partial t}\right)\bullet d\vec{A}=0$$

In order for this integral to always be zero, the term in the parenthesis must be zero.

$$\nabla\times\vec{B}-\mu_0\varepsilon_0\frac{\partial\vec{E}}{\partial t}=0\ \rightarrow\ \nabla\times\vec{B}=\mu_0\varepsilon_0\frac{\partial\vec{E}}{\partial t} \qquad \text{(C-5)}$$

Finally, if we include the presence of free currents we have

$$\nabla\times\vec{B}=\mu_0\vec{J}_f+\mu_0\varepsilon_0\frac{\partial\vec{E}}{\partial t} \qquad \text{(C-6)}$$

C-15 Derivation of Equation 10-10

Let's assume that the solution to Equation 10-9 can be written as the product of two functions, A and B. One of these functions will be a function of position (x) only and the other function will be a function of time (t) only.

$$y = A(x)B(t)$$

Substitution of this expression into Equation 10-9 gives us

$$B(t)\frac{d^2A(x)}{dx^2} = \frac{1}{v^2}A(x)\frac{d^2B(t)}{dt^2}$$

Divide through both sides by $y = A(x)B(t)$.

$$\frac{1}{A(x)}\frac{d^2A(x)}{dx^2} = \frac{1}{v^2}\frac{1}{B(t)}\frac{d^2B(t)}{dt^2}$$

The left-hand side of this equation is a function of position only and the right-hand side of this equation is a function of time only. We could therefore argue that one way for this equation to be valid (*i.e.*, for the two sides of the equation to be equal) is for each side of the equation to be equal to a constant. Let's denote this constant as $-k^2$.

$$\frac{1}{A(x)}\frac{d^2A(x)}{dx^2} = \frac{1}{v^2}\frac{1}{B(t)}\frac{d^2B(t)}{dt^2} = -k^2$$

Hence,

$$\frac{d^2A(x)}{dx^2} = -k^2A(x) \rightarrow A(x) = A_1e^{ikx} + A_2e^{-ikx}$$

$$\frac{d^2B(t)}{dt^2} = -v^2k^2B(t) \rightarrow B(t) = B_1e^{ikvt} + B_2e^{-ikvt}$$

In these equations A_1, A_2, B_1, and B_2 are constants determined by the initial conditions of the wave (*i.e.*, the initial magnitude and phase). Putting everything together gives us

$$y = \left(A_1e^{ikx} + A_2e^{-ikx}\right)\left(B_1e^{ikvt} + B_2e^{-ikvt}\right)$$

$$y = A_1B_1e^{ik(x+vt)} + A_1B_2e^{ik(x-vt)} + A_2B_1e^{-ik(x-vt)} + A_2B_2e^{-ik(x+vt)}$$

Each of these exponential terms represents a sinusoidal oscillating wave. Therefore, the solution to the one-dimensional wave equation is a superposition of sinusoidal waves. These waves are moving in different directions (in the positive x-axis direction or the negative x-axis direction).

Positive x-axis direction: $y = A_1B_2e^{ik(x-vt)} + A_2B_1e^{-ik(x-vt)}$

Negative x-axis direction: $y = A_1B_1e^{ik(x+vt)} + A_2B_2e^{-ik(x+vt)}$

We could also determine the direction of the motion from the (±) sign of the wave number (k). If k is negative, the directions of the waves listed above switch. If we thus ignore the complex conjugate terms and use the wavenumber to determine the direction, we can write the solution to the one-dimensional wave equation as

$$y = y_0e^{ik(x-vt)}$$

The constant y_0 in this equation is determined from the initial conditions of the wave. We recognize the complex exponential in this solution as representing an oscillating function (Equation D-5), as expected for a wave.

C-16 Derivation of Equation 10-17

If an external electric potential difference is supplied to a conductor, a current will flow in the conductor. If left alone, current will eventually cease to exist and the conductor will attain a state of electrostatic equilibrium (Section 7-2); in other words, the electric charges carried by the current will eventually build up a distribution of electric charge and associated electric potential difference that will cancel the external electric potential difference. Therefore, in order to maintain a constant current in the conductor a constant electric potential, difference must be maintained. This requires a continuous supply of energy to the conductor.

But in the case of a constant current, there must also not be an accumulation of electric potential energy in the system. As we discussed in Section 7-5, the excess electric potential energy supplied to the system is dissipated by the resistance of the conductor in the form of heat. The electric energy supplied can be calculated from the work done on the electric charges constituting the current by the electric potential through which they move. The instantaneous infinitesimal work done by the electric field to move an infinitesimal amount of charge through an electric potential difference is found using Equation 3-5[7]

$$đW = -\Delta U_e \quad \rightarrow \quad đW = -(dq)\Delta V_e$$

Therefore, the rate at which this work is done is

$$\frac{đW}{dt} = -\left(\frac{dq}{dt}\right)\Delta V_e \quad \rightarrow \quad \frac{đW}{dt} = -I\Delta V_e$$

Let's assume that the conductor has a cross-sectional area A and length l. The rate of work per volume of the conductor, defined by the variable w, would then be

$$\frac{\frac{đW}{dt}}{V} = \frac{-I\Delta V_e}{V} \quad \rightarrow \quad w = \frac{-I\Delta V_e}{Al} \quad \rightarrow \quad w = \left(\frac{I}{A}\right)\left(\frac{-\Delta V_e}{l}\right) \quad \rightarrow \quad w = JE$$

The variable J in this equation is the current density (Equation 7-13). Since the current is parallel to the electric field, the current density is parallel to the electric field. Thus, we can rewrite our expression for w as

$$w = \vec{J} \bullet \vec{E}$$

The direction for the current density vector is the same as the direction for the current and the electric field. The work done by the electric field must be equal to the energy dissipated by the conductor as heat. The total rate of energy dissipation (*i.e.*, the total power dissipated) by the conductor is therefore

$$P = \int_{volume} \left(\vec{J} \bullet \vec{E}\right) dV$$

In this expression, the integral can be taken with respect to the entire volume of the conductor to determine the total power dissipated, or with respect to a smaller volume to determine the power dissipated from that smaller volume alone.

Rearranging the differential form of Ampère's Law (Equation C-6) gives us

7 Recall from classical mechanics that the work done by a conservative force is equal to the negative of the change in the potential energy associated with that force.

$$\vec{J}_f = \nabla \times \left(\frac{\vec{B}}{\mu_0}\right) - \varepsilon_0 \frac{d\vec{E}}{dt} \rightarrow P = \int_{volume} \left(\left(\nabla \times \left(\frac{\vec{B}}{\mu_0}\right) - \varepsilon_0 \frac{d\vec{E}}{dt}\right) \bullet \vec{E}\right) dV$$

$$P = \int_{volume} \left(\left(\nabla \times \left(\frac{\vec{B}}{\mu_0}\right)\right) \bullet \vec{E}\right) dV + \int_{volume} \left(\left(-\varepsilon_0 \frac{d\vec{E}}{dt}\right) \bullet \vec{E}\right) dV$$

Applying Equation B-18 then yields

$$P = \int_{volume} \left(\nabla \bullet \left(\vec{E} \times \frac{\vec{B}}{\mu_0}\right) + \frac{\vec{B}}{\mu_0} \bullet (\nabla \times \vec{E})\right) dV - \int_{volume} \left(\left(-\varepsilon_0 \frac{d\vec{E}}{dt}\right) \bullet \vec{E}\right) dV$$

Substitution of the differential form of Faraday's Law (Equation C-3) then gives us

$$P = \int_{volume} \left(\nabla \bullet \left(\vec{E} \times \frac{\vec{B}}{\mu_0}\right) + \frac{\vec{B}}{\mu_0} \bullet \left(-\frac{d\vec{B}}{dt}\right)\right) dV + \int_{volume} \left(\left(-\varepsilon_0 \frac{d\vec{E}}{dt}\right) \bullet \vec{E}\right) dV$$

Rearranging this expression then yields

$$P = \int_{volume} \left(\nabla \bullet \left(\vec{E} \times \frac{\vec{B}}{\mu_0}\right)\right) dV + \int_{volume} \left(\left(-\varepsilon_0 \frac{d\vec{E}}{dt}\right) \bullet \vec{E} + \left(-\frac{1}{\mu_0} \frac{d\vec{B}}{dt}\right) \bullet \vec{B}\right) dV$$

$$P = \int_{volume} \left(\nabla \bullet \left(\vec{E} \times \frac{\vec{B}}{\mu_0}\right)\right) dV - \frac{d}{dt}\left[\int_{volume} \left(\frac{\varepsilon_0 \vec{E} \bullet \vec{E}}{2} + \frac{\vec{B} \bullet \vec{B}}{2\mu_0}\right) dV\right]$$

Now since[8]

$$\frac{\varepsilon_0 \vec{E} \bullet \vec{E}}{2} + \frac{\vec{B} \bullet \vec{B}}{2\mu_0} = \frac{\varepsilon_0 E^2}{2} + \frac{B^2}{2\mu_0} \rightarrow \frac{\varepsilon_0 \vec{E} \bullet \vec{E}}{2} + \frac{\vec{B} \bullet \vec{B}}{2\mu_0} = u_e + u_m$$

$$\frac{\varepsilon_0 \vec{E} \bullet \vec{E}}{2} + \frac{\vec{B} \bullet \vec{B}}{2\mu_0} = u_{total} \rightarrow P = \int_{volume} \left(\nabla \bullet \left(\vec{E} \times \frac{\vec{B}}{\mu_0}\right)\right) dV - \frac{d}{dt}\left[\int_{volume} u_{total}\, dV\right]$$

The second integral in this expression is the rate of change of the total energy of the system.

8 Equation 10-14 and Equation 10-15.

$$P = \int_{volume} \left(\nabla \bullet \left(\vec{E} \times \frac{\vec{B}}{\mu_0} \right) \right) dV - \frac{dU_{total}}{dt}$$

Finally, applying the divergence theorem (Equation B-14) to the remaining integral gives us

$$P = \oint_{surface} \left(\vec{E} \times \frac{\vec{B}}{\mu_0} \right) \bullet dA - \frac{dU_{total}}{dt}$$

The power, P, in this expression is the rate of energy dissipation in the form of heat. From energy conservation we know that any energy that is not dissipated as heat must pass out of the volume V through the surface of this volume. Since the remaining integral is an integral with respect to that surface we can interpret that

$$\oint_{surface} \left(\vec{E} \times \frac{\vec{B}}{\mu_0} \right) \bullet dA = \left(\frac{dU_{total}}{dt} \right)_{through\ surface}$$

We now define the Poynting vector to be

$$\vec{S} = \vec{E} \times \frac{\vec{B}}{\mu_0} \quad \rightarrow \quad \oint_{surface} \vec{S} \bullet dA = \left(\frac{dU_{total}}{dt} \right)_{through\ surface}$$

The Poynting vector is the rate of flow of electromagnetic energy per unit area. In other words, the Poynting vector is the energy current density or power flux. This is the valid expression for energy flow for any distribution of electric and magnetic fields. The direction of the Poynting vector is thus also the direction of propagation of an electromagnetic wave.

C-17 Derivation of Electric and Magnetic Fields for a Linear Dipole Antenna

Let's define the axis of the linear dipole antenna (Figure 10.15 and Figure 10.16) to be the z-axis. The magnetic potential of the antenna can be determined from the retarded alternating current using Equation 4-12.

$$I = I_0 e^{i(\omega t - kr)} \quad \rightarrow \quad \vec{V}_m = \left[\left(\frac{\mu_0}{4\pi} \right) \int \frac{I_0 e^{i(\omega t - kr)} dz}{r} \right] \hat{z} \quad \rightarrow \quad \vec{V}_m = \left[\left(\frac{\mu_0}{4\pi} \right) \frac{I_0 e^{i(\omega t - kr)}}{r} \int dz \right] \hat{z}$$

The integral in this expression is simply the length of the antenna, *l*.

$$\vec{V}_m = \left[\frac{\mu_0 I_0 l e^{i(\omega t - kr)}}{4\pi r}\right]\hat{z}$$

In spherical coordinates we have

$$\vec{V}_m = \left[\frac{\mu_0 I_0 l e^{i(\omega t - kr)}}{4\pi r}\right]\left[(\cos\theta)\hat{r} - (\sin\theta)\hat{\theta}\right] \tag{C-7}$$

The electric potential of the antenna shown in Figure 10.16 is

$$V_e = \frac{I_0}{i\omega 4\pi\varepsilon_0}\left(\frac{e^{i(\omega t - kr_A)}}{r_A} - \frac{e^{i(\omega t - kr_B)}}{r_B}\right)$$

The variables r_A and r_B can be determined using trigonometry.

$$r_A = r\left(1 + \left(\frac{l}{2r}\right)^2 - 2\left(\frac{l}{2r}\right)\cos\theta\right)^{\frac{1}{2}} \qquad r_B = r\left(1 + \left(\frac{l}{2r}\right)^2 + 2\left(\frac{l}{2r}\right)\cos\theta\right)^{\frac{1}{2}}$$

Expanding these expressions in a series gives us

$$r_A = r\left(1 - \left(\frac{l}{2r}\right)\cos\theta + \frac{1}{2}\left(\frac{l}{2r}\right)^2 + \ldots\right) \qquad r_B = r\left(1 + \left(\frac{l}{2r}\right)\cos\theta + \frac{1}{2}\left(\frac{l}{2r}\right)^2 + \ldots\right)$$

Under conditions that $r >> l$, we can approximate the values of r_A and r_B as

$$r_A = r - \frac{l}{2}\cos\theta \qquad r_B = r + \frac{l}{2}\cos\theta$$

Substitution into the expression for the electric potential gives us

$$V_e = \frac{I_0}{i\omega 4\pi\varepsilon_0}\left(\frac{e^{i\left(\omega t - k\left(r - \frac{l}{2}\cos\theta\right)\right)}}{r - \frac{l}{2}\cos\theta} - \frac{e^{i\left(\omega t - k\left(r + \frac{l}{2}\cos\theta\right)\right)}}{r + \frac{l}{2}\cos\theta}\right)$$

Simplifying this expression yields

$$V_e = \frac{I_0 e^{i(\omega t - kr)}}{i\omega 4\pi\varepsilon_0}\left(\frac{\left(r + \frac{l}{2}\cos\theta\right)e^{i\left(\frac{kl}{2}\cos\theta\right)} - \left(r - \frac{l}{2}\cos\theta\right)e^{-i\left(\frac{kl}{2}\cos\theta\right)}}{r^2}\right)$$

Now assume $\lambda >> l$

$$\lambda >> l \;\rightarrow\; \frac{2\pi}{k} >> l \;\rightarrow\; 2\pi >> lk \;\rightarrow\; lk << 1$$

Then from the small angle approximation we have

$$lk << 1 \;\rightarrow\; \cos\left(\frac{kl}{2}\cos\theta\right) \approx 1$$

$$lk << 1 \;\rightarrow\; \sin\left(\frac{kl}{2}\cos\theta\right) \approx \frac{kl}{2}\cos\theta$$

Hence,

$$\left(r + \frac{l}{2}\cos\theta\right)e^{i\left(\frac{kl}{2}\cos\theta\right)} = \left(r + \frac{l}{2}\cos\theta\right)\left(1 + i\frac{kl}{2}\cos\theta\right)$$

$$\left(r - \frac{l}{2}\cos\theta\right)e^{-i\left(\frac{kl}{2}\cos\theta\right)} = \left(r - \frac{l}{2}\cos\theta\right)\left(1 - i\frac{kl}{2}\cos\theta\right)$$

Putting everything back together again gives us

$$V_e = \frac{I_0 e^{i(\omega t - kr)}}{i\omega 4\pi\varepsilon_0}\left(\frac{\left(r + \frac{l}{2}\cos\theta\right)\left(1 + i\frac{kl}{2}\cos\theta\right) - \left(r - \frac{l}{2}\cos\theta\right)\left(1 - i\frac{kl}{2}\cos\theta\right)}{r^2}\right)$$

$$V_e = \frac{I_0 l e^{i(\omega t - kr)}\cos\theta}{i\omega 4\pi\varepsilon_0}\left(\frac{ikr + 1}{r^2}\right) \;\rightarrow\; V_e = \frac{I_0 l e^{i(\omega t - kr)}\cos\theta}{4\pi\varepsilon_0}\left(\frac{1}{cr} + \frac{1}{i\omega r^2}\right)$$

$$V_e = \frac{I_0 l e^{i(\omega t - kr)}\cos\theta}{4\pi\varepsilon_0 c}\left(\frac{1}{r} + \frac{c}{i\omega r^2}\right) \qquad \textbf{(C-8)}$$

Substitution of Equation C-7 and Equation C-8 into Equation C-4[9] then gives us

$$\vec{E} = -\left[\left(\frac{\partial V_e}{\partial r}\right)\hat{r} + \left(\frac{1}{r}\frac{\partial V_e}{\partial \theta}\right)\hat{\theta} + \left(\frac{1}{r\sin\theta}\frac{\partial V_e}{\partial \varphi}\right)\hat{\varphi}\right] - \frac{d\vec{V}_m}{dt}$$

$$\frac{\partial V_e}{\partial r} = \frac{(-ik)I_0 le^{i(\omega t - kr)}\cos\theta}{4\pi\varepsilon_0 c}\left(\frac{1}{r} + \frac{c}{i\omega r^2}\right) + \frac{I_0 le^{i(\omega t - kr)}\cos\theta}{4\pi\varepsilon_0 c}\left(-\frac{1}{r^2} - \frac{2c}{i\omega r^3}\right)$$

$$\frac{\partial V_e}{\partial \theta} = -\frac{I_0 le^{i(\omega t - kr)}\sin\theta}{4\pi\varepsilon_0 c}\left(\frac{1}{r} + \frac{c}{i\omega r^2}\right)$$

$$\frac{\partial V_e}{\partial \varphi} = 0$$

$$\frac{d\vec{V}_m}{dt} = \left[\frac{i\omega\mu_0 I_0 le^{i(\omega t - kr)}}{4\pi r}\right]\left[(\cos\theta)\hat{r} - (\sin\theta)\hat{\theta}\right]$$

Putting it all together gives us

$$\vec{E} = \frac{I_0 le^{i(\omega t - kr)}}{2\pi\varepsilon_0}\left\{\cos\theta\left[\frac{1}{cr^2} + \frac{1}{i\omega r^3}\right]\hat{r} + \frac{\sin\theta}{2\pi}\left[\frac{i\omega}{c^2 r} + \frac{1}{cr^2} + \frac{1}{i\omega r^3}\right]\hat{\theta}\right\}$$

The magnetic field is found by substituting Equation C-7 into Equation 4-11 (using Equation B-13).

$$\vec{V}_m = \left[(V_m)_r\hat{r} + (V_m)_\theta\hat{\theta}\right]$$

$$\nabla\times\vec{V}_m = \left(-\frac{\partial (V_m)_\theta}{\partial\varphi}\right)\frac{\hat{r}}{r\sin\theta} + \left(\frac{1}{\sin\theta}\frac{\partial (V_m)_r}{\partial\varphi}\right)\frac{\hat{\theta}}{r} + \left(\frac{\partial}{\partial r}\left(r(V_m)_\theta\right) - \frac{\partial (V_m)_r}{\partial\theta}\right)\frac{\hat{\varphi}}{r}$$

$$\frac{\partial (V_m)_\theta}{\partial\varphi} = 0 \qquad \frac{\partial (V_m)_r}{\partial\varphi} = 0$$

$$\nabla\times\vec{V}_m = \left(\frac{\partial}{\partial r}\left(r(V_m)_\theta\right) - \frac{\partial (V_m)_r}{\partial\theta}\right)\frac{\hat{\varphi}}{r}$$

9 Since we are in spherical coordinates, we will use Equation B-7 for the gradient.

$$\frac{\partial}{\partial r}\left(r\left(V_m\right)_\theta\right)=\frac{ik\mu_0 I_0 le^{i(\omega t-kr)}\sin\theta}{4\pi}$$

$$\frac{\partial\left(V_m\right)_r}{\partial\theta}=-\frac{\mu_0 I_0 le^{i(\omega t-kr)}\sin\theta}{4\pi r}$$

Putting it all together gives us

$$\vec{B}=\frac{\mu_0 I_0 le^{i(\omega t-kr)}\sin\theta}{4\pi}\left(\frac{i\omega}{cr}+\frac{1}{r^2}\right)\hat{\varphi}$$

C-18 Derivation of Equation 12-2

The resultant electric field at a position y on the screen for N sources is

$$E=E_1+E_2+\ldots+E_N$$

The magnitude of the electric field of each source is equal to 1/N of the magnitude of the electric field of the incident wave.

$$E=\frac{E_0}{N}e^{i(kx-\omega t)}+\frac{E_0}{N}e^{i(k(x+\delta)-\omega t)}+\ldots+\frac{E_0}{N}e^{i(k(x+(N-1)\delta)-\omega t)}$$

If we factor out a common exponential term in this summation, we have

$$E=\frac{E_0}{N}e^{i(kx-\omega t)}\left(1+e^{ik\delta}+e^{i2k\delta}+\ldots+e^{ik(N-1)\delta}\right)$$

The term in the parenthesis is a geometric series[10]. The equivalent expression of this series is[11]

$$E=\frac{E_0}{N}e^{i(kx-\omega t)}\left(\frac{1-e^{iNk\delta}}{1-e^{ik\delta}}\right)$$

10 Another example of why series expansions are useful and cool.

11 See Section A-3.

Simplifying this expression gives us

$$E = \frac{E_0}{N} e^{i(kx-\omega t)} \left(\frac{e^{-i\frac{Nk\delta}{2}} - e^{i\frac{Nk\delta}{2}}}{e^{-i\frac{k\delta}{2}} - e^{i\frac{k\delta}{2}}} \right) \left(\frac{e^{i\frac{Nk\delta}{2}}}{e^{i\frac{k\delta}{2}}} \right)$$

This difference in complex exponentials in the parenthesis can be written in terms of the sine function (Equation D-7).

$$E = \frac{E_0}{N} e^{i(kx-\omega t)} \left(\frac{2i\sin\left(\frac{Nk\delta}{2}\right)}{2i\sin\left(\frac{k\delta}{2}\right)} \right) e^{i\left(\frac{Nk\delta}{2} - \frac{k\delta}{2}\right)}$$

$$E = \frac{E_0}{N} e^{i\left(kx-\omega t+\frac{k\delta}{2}(N-1)\right)} \left(\frac{\sin\left(\frac{Nk\delta}{2}\right)}{\sin\left(\frac{k\delta}{2}\right)} \right)$$

The real part of the electric field is therefore

$$E = \frac{E_0}{N} \cos\left(kx - \omega t + \frac{k\delta}{2}(N-1) \right) \left(\frac{\sin\left(\frac{Nk\delta}{2}\right)}{\sin\left(\frac{k\delta}{2}\right)} \right)$$

The intensity of the resultant electromagnetic wave[12] is then found using Equation 10-18.

$$I = \frac{1}{2} c\varepsilon_0 \left(\frac{E_0}{N} \right)^2 \cos^2\left(kx - \omega t + \frac{k\delta}{2}(N-1) \right) \left(\frac{\sin\left(\frac{Nk\delta}{2}\right)}{\sin\left(\frac{k\delta}{2}\right)} \right)^2$$

The average intensity is thus

12 We will assume that the wave is propagating through vacuum.

$$I_{avg} = \frac{1}{4}c\varepsilon_0\left(\frac{E_0}{N}\right)^2\left(\frac{\sin\left(\frac{Nk\delta}{2}\right)}{\sin\left(\frac{k\delta}{2}\right)}\right)^2$$

The variable δ in this equation is

$$\delta = \frac{a}{N-1}\sin\theta \quad \to \quad I_{avg} = \frac{1}{4}c\varepsilon_0\left(\frac{E_0}{N}\right)^2\left(\frac{\sin\left(\frac{Nka\sin\theta}{2(N-1)}\right)}{\sin\left(\frac{ka\sin\theta}{2(N-1)}\right)}\right)^2$$

Substitution of Equation 10-7 into our expression for the average intensity of the resultant wave gives us

$$I_{avg} = \frac{1}{4}c\varepsilon_0\left(\frac{E_0}{N}\right)^2\left(\frac{\sin\left(\frac{N\pi a\sin\theta}{\lambda(N-1)}\right)}{\sin\left(\frac{\pi a\sin\theta}{\lambda(N-1)}\right)}\right)^2$$

In the limit that N becomes large

$$N >> 1 \quad \to \quad \frac{\pi a\sin\theta}{\lambda(N-1)} << 1 \quad \to \quad \sin\left(\frac{\pi a\sin\theta}{\lambda(N-1)}\right) \approx \frac{\pi a\sin\theta}{\lambda(N-1)}$$

$$N >> 1 \quad \to \quad N \approx (N-1) \quad \to \quad \sin\left(\frac{N\pi a\sin\theta}{\lambda(N-1)}\right) \approx \sin\left(\frac{\pi a\sin\theta}{\lambda}\right)$$

$$N >> 1 \quad \to \quad I_{avg} = \frac{1}{4}c\varepsilon_0\left(\frac{E_0}{N}\right)^2\left(\frac{\sin\left(\frac{\pi a\sin\theta}{\lambda}\right)}{\frac{\pi a\sin\theta}{\lambda(N-1)}}\right)^2$$

Simplifying this expression further[13] gives us

$$I_{avg} = \frac{1}{4}c\varepsilon_0\left(\frac{E_0}{N}\right)^2(N-1)^2\left(\frac{\sin\left(\frac{\pi a\sin\theta}{\lambda}\right)}{\frac{\pi a\sin\theta}{\lambda}}\right)^2 \rightarrow I_{avg} = \frac{1}{4}c\varepsilon_0(E_0)^2\left(\frac{\sin\left(\frac{\pi a\sin\theta}{\lambda}\right)}{\frac{\pi a\sin\theta}{\lambda}}\right)^2$$

$$I = I_{max}\left(\frac{\sin\left(\frac{\pi a\sin\theta}{\lambda}\right)}{\frac{\pi a\sin\theta}{\lambda}}\right)^2$$

13 Still assuming that $N >> 1$ so $N \approx (N-1)$

Appendix D

D-1 Definition of a Complex Number

A complex number is a number that can be written as

$$r = x + iy \tag{D-1}$$

The variables x and y in this equation are real numbers and

$$i = \sqrt{-1}$$

The variables x and y are frequently referred to as the real and imaginary parts, respectively, of the complex number r.

$$x = \mathrm{Re}(r) \qquad y = \mathrm{Im}(r)$$

A complex number can be represented as a vector in a two-dimensional reference frame, as shown in Figure D.1 The horizontal axis in Figure D.1 is the real axis, and the component of the vector along the real axis is the real part of the complex number (the variable x in Equation D-1). Similarly, the vertical axis in Figure D.1 is the imaginary axis, and the component of the vector along the imaginary axis is the imaginary part of the complex number (the variable y in Equation D-1).

The components of the vector in Figure D.1 (*i.e.*, the real and imaginary parts of the complex number) can also be written in terms of the angle between the vector and the real axis (the angle θ in Figure D.1).

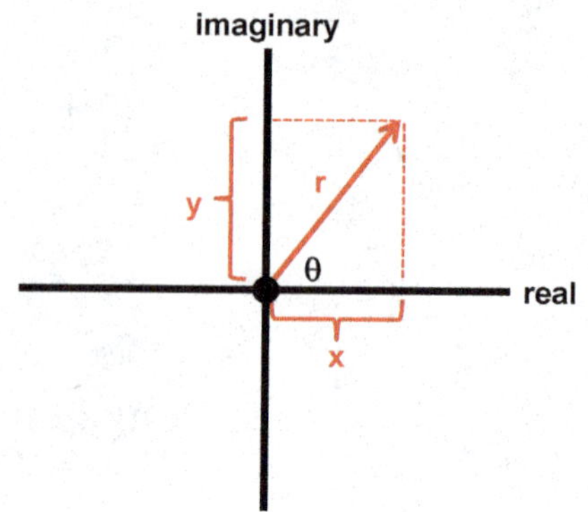

FIGURE D.1: Trigonometric representation of a complex number.

$$r = \left(\left(x^2 + y^2\right)^{\frac{1}{2}} \cos(\theta)\right) + i\left(\left(x^2 + y^2\right)^{\frac{1}{2}} \sin(\theta)\right)$$

$$r = \left(x^2 + y^2\right)^{\frac{1}{2}} \left[\cos(\theta) + i\sin(\theta)\right] \tag{D-2}$$

The angle θ in Equation D-2 is given by the equation

$$\tan\theta = \frac{y}{x} = \frac{\mathrm{Im}(r)}{\mathrm{Re}(r)} \tag{D-3}$$

Similarly,

$$\sin\theta = \frac{\mathrm{Im}(r)}{\left[\mathrm{Re}(r)\right]^2 + \left[\mathrm{Im}(r)\right]^2} \qquad \cos\theta = \frac{\mathrm{Re}(r)}{\left[\mathrm{Re}(r)\right]^2 + \left[\mathrm{Im}(r)\right]^2} \tag{D-4}$$

D-2 Exponential Notation for Complex Numbers

Euler's formula for complex numbers is

$$e^{i\theta} = \cos(\theta) + i\sin(\theta) \tag{D-5}$$

Thus, Equation D-2 can be rewritten as

$$r = \left(x^2 + y^2\right)^{\frac{1}{2}} e^{i\theta} \quad \rightarrow \quad r = \left[\left(\mathrm{Re}(r)\right)^2 + \left(\mathrm{Im}(r)\right)^2\right]^{\frac{1}{2}} e^{i\theta} \tag{D-6}$$

It follows from Equation D-4 that the sine and cosine functions can be written in terms of complex exponentials.

$$\sin(\theta)=\frac{e^{i\theta}-e^{-i\theta}}{2i} \tag{D-7}$$

$$\cos(\theta)=\frac{e^{i\theta}+e^{-i\theta}}{2} \tag{D-8}$$

Finally, we see that

$$e^{in\pi}=(-1)^n\ ;\ n=0,1,2,\ldots \tag{D-9}$$

Equation D-9 is just fun.

D-3 Use of Complex Exponentials for Solving Differential Equations

Complex exponentials are especially useful in solving differential equations. Consider the simple electric circuit shown in Figure D.2.

Applying Kirchhoff's loop rule to the loop ABCDA in this electric circuit yields

$$\mathcal{E}-L\frac{dI}{dt}=0$$

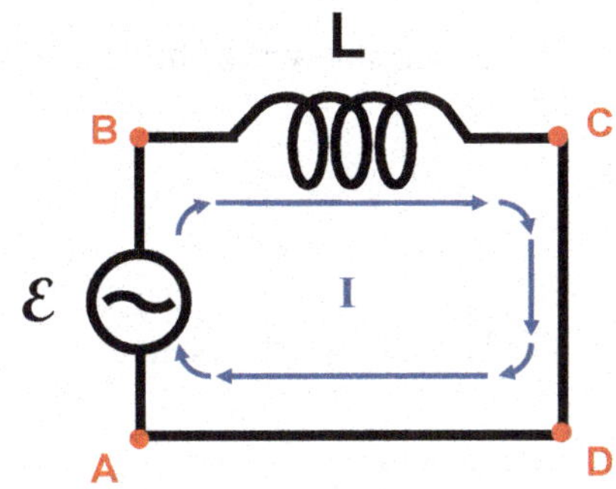

FIGURE D.2: An AC circuit with a single inductor.

Let's assume that the EMF is oscillating at an angular frequency ω according to the equation

$$\mathcal{E}=\mathcal{E}_0\cos(\omega t)$$

Substitution then gives us

$$\mathcal{E}_0\cos(\omega t)-L\frac{dI}{dt}=0\ \rightarrow\ \frac{dI}{dt}=\frac{\mathcal{E}_0}{L}\cos(\omega t)\ \rightarrow\ dI=\frac{\mathcal{E}_0}{L}\cos(\omega t)\,dt$$

Integrating both sides then yields

$$\int_0^I dI = \int_0^t \frac{\mathcal{E}_0}{L}\cos(\omega t)\,dt \;\rightarrow\; I = \frac{\mathcal{E}_0}{L\omega}\left(\sin(\omega t)\Big|_0^t\right) \;\rightarrow\; I = \frac{\mathcal{E}_0}{L\omega}\sin(\omega t)$$

We see that although the EMF is oscillating as $\cos(\omega t)$, the current is oscillating as $\sin(\omega t)$. In other words, the oscillation of the current is out of phase with the oscillation of the EMF; the difference in the phase is $\pi/2$ radians.

Now let's use a complex exponential to describe the oscillations of the EMF.

$$\mathcal{E} = \mathcal{E}_0 e^{i\omega t}$$

Substitution then gives us

$$\mathcal{E}_0 e^{i\omega t} - L\frac{dI}{dt} = 0 \;\rightarrow\; \frac{dI}{dt} = \frac{\mathcal{E}_0}{L}e^{i\omega t} \;\rightarrow\; dI = \frac{\mathcal{E}_0}{L}e^{i\omega t}dt$$

Integrating both sides then yields

$$\int_0^I dI = \int_0^t \frac{\mathcal{E}_0}{L}e^{i\omega t}\,dt \;\rightarrow\; I = \frac{\mathcal{E}_0}{iL\omega}\left(e^{i\omega t}\Big|_0^t\right) \;\rightarrow\; I = \frac{\mathcal{E}_0}{iL\omega}\left(e^{i\omega t} - 1\right)$$

We can rewrite this expression using Equation D-5.

$$I = \frac{\mathcal{E}_0}{iL\omega}\left(\cos(\omega t) + i\sin(\omega t) - 1\right) \;\rightarrow\; I = \frac{\mathcal{E}_0}{L\omega}\left(\sin(\omega t) + i\left(1 - \cos(\omega t)\right)\right)$$

The real part of the current is therefore

$$\mathrm{Re}(I) = \frac{\mathcal{E}_0}{L\omega}\sin(\omega t)$$

This solution is identical to that derived previously. It might seem at first glance, however, that it is more difficult to use complex exponentials to solve this problem, rather than using sine and cosine. While that may be true for this particular problem, using complex exponentials is easier than using sine and cosine for more complicated electric circuits. For example, let's determine the current in the electric circuit shown in Figure D.3.

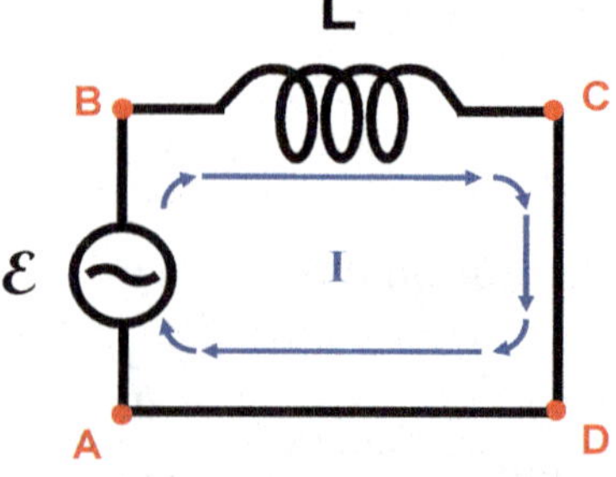

FIGURE D.3: An AC circuit with resistance and inductance.

Applying Kirchhoff's loop rule to the loop ABCDA of this electric circuit gives us

$$\mathcal{E} - L\frac{dI}{dt} - RI = 0$$

As before, let's first assume that the EMF is oscillating at an angular frequency ω according to the equation

$$\mathcal{E} = \mathcal{E}_0 \cos(\omega t)$$

Substitution then gives us

$$\mathcal{E}_0 \cos(\omega t) - L\frac{dI}{dt} - RI = 0$$

It is clear that an equation of the form $I = I_0 \sin(\omega t)$ is not a solution to this equation. This suggests that the difference in the phase between the oscillation of the EMF and the oscillation of the current is not $\pi/2$, as for the electric circuit in Figure D.2. Let's try a solution with a current separated from the EMF by a phase difference ϕ.

$$I = I_0 \cos(\omega t + \phi)$$

Substitution then gives us

$$\mathcal{E}_0 \cos(\omega t) + LI_0\omega \sin(\omega t + \phi) - RI_0 \cos(\omega t + \phi) = 0$$

Our next step is to simply this expression using trigonometric identities (Equation A-1 and Equation A-2). After substitution of these expressions and a bit of simplification, we have

$$\left[\mathcal{E}_0 + LI_0\omega \sin(\phi) - RI_0 \cos(\phi)\right]\cos(\omega t) + \left[RI_0 \sin(\phi) + LI_0\omega \cos(\phi)\right]\sin(\omega t) = 0$$

In order for this expression to always be true, each bracketed term must be separately equal to zero.

$$RI_0 \sin(\phi) + LI_0\omega \cos(\phi) = 0 \quad \rightarrow \quad RI_0 \sin(\phi) = -LI_0\omega \cos(\phi) \quad \rightarrow \quad \tan(\phi) = -\frac{\omega L}{R}$$

$$\phi = \tan^{-1}\left(-\frac{\omega L}{R}\right) \quad \rightarrow \quad \phi = -\tan^{-1}\left(\frac{\omega L}{R}\right)$$

It follows from trigonometry that

$$\tan(\phi) = -\frac{\omega L}{R} \quad \rightarrow \quad \sin(\phi) = -\frac{\omega L}{\left(R^2 + L^2\omega^2\right)^{\frac{1}{2}}} \quad \rightarrow \quad \cos(\phi) = \frac{R}{\left(R^2 + L^2\omega^2\right)^{\frac{1}{2}}}$$

Thus,

$$\mathcal{E}_0 + LI_0\omega\sin(\phi) - RI_0\cos(\phi) = 0 \;\rightarrow\; I_0\left[-L\omega\sin(\phi) + R\cos(\phi)\right] = \mathcal{E}_0$$

$$I_0\left[-L\omega\left(-\frac{\omega L}{\left(R^2+L^2\omega^2\right)^{\frac{1}{2}}}\right) + R\left(\frac{R}{\left(R^2+L^2\omega^2\right)^{\frac{1}{2}}}\right)\right] = \mathcal{E}_0 \;\rightarrow\; I_0\left(\frac{L^2\omega^2+R^2}{\left(R^2+L^2\omega^2\right)^{\frac{1}{2}}}\right) = \mathcal{E}_0$$

$$I_0\left(R^2+L^2\omega^2\right)^{\frac{1}{2}} = \mathcal{E}_0 \;\rightarrow\; I_0 = \frac{\mathcal{E}_0}{\left(R^2+L^2\omega^2\right)^{\frac{1}{2}}}$$

Our final answer is therefore

$$I = \frac{\mathcal{E}_0}{\left(R^2+L^2\omega^2\right)^{\frac{1}{2}}}\cos\left(\omega t - \tan^{-1}\left(\frac{L\omega}{R}\right)\right)$$

Now let's approach the problem using complex exponentials. Let's assume that the EMF and the current are oscillating as

$$\mathcal{E} = \mathcal{E}_0 e^{i\omega t} \qquad I = I_0 e^{i\omega t}$$

Substitution gives us

$$\mathcal{E}_0 e^{i\omega t} - LI_0 e^{i\omega t} i\omega - RI_0 e^{i\omega t} = 0 \;\rightarrow\; \left(\mathcal{E}_0 - i\omega LI_0 - RI_0\right)e^{i\omega t} = 0$$

In order for this expression to be zero the term in the parenthesis must be zero.

$$\mathcal{E}_0 - i\omega LI_0 - RI_0 = 0 \;\rightarrow\; I_0\left(R + i\omega L\right) = \mathcal{E}_0 \;\rightarrow\; I_0 = \frac{\mathcal{E}_0}{R + i\omega L}$$

We can rewrite this expression using Equation D-6.

$$I_0 = \frac{\mathcal{E}_0}{\left(R^2+L^2\omega^2\right)^{\frac{1}{2}} e^{i\tan^{-1}\left(\frac{\omega L}{R}\right)}} \;\rightarrow\; I_0 = \frac{\mathcal{E}_0}{\left(R^2+L^2\omega^2\right)^{\frac{1}{2}}} e^{-i\tan^{-1}\left(\frac{\omega L}{R}\right)}$$

Substitution then gives us

$$I = I_0 e^{i\omega t} \quad \rightarrow \quad I = \frac{\mathcal{E}_0}{\left(R^2 + L^2\omega^2\right)^{\frac{1}{2}}} e^{-i\tan^{-1}\left(\frac{\omega L}{R}\right)} e^{i\omega t} \quad \rightarrow \quad I = \frac{\mathcal{E}_0}{\left(R^2 + L^2\omega^2\right)^{\frac{1}{2}}} e^{i\left(\omega t - \tan^{-1}\left(\frac{\omega L}{R}\right)\right)}$$

The real part of the current is therefore

$$I = \frac{\mathcal{E}_0}{\left(R^2 + L^2\omega^2\right)^{\frac{1}{2}}} \cos\left(\omega t - \tan^{-1}\left(\frac{\omega L}{R}\right)\right)$$

This is the same answer that we derived previously, but it was obtained more easily. The benefit of using complex exponentials becomes even more pronounced as the complexity of the electric circuits increases (through the addition of more components, *etc.*), as demonstrated in Chapter 9.

CPSIA information can be obtained
at www.ICGtesting.com
Printed in the USA
FSOW04n0432110117
29493FS

9 781631 898563